Nanomedicine and Nanotoxicology

Series Editor:

V. Zucolotto

For further volumes:
http://www.springer.com/series/10620

Nelson Durán • Silvia S. Guterres •
Oswaldo L. Alves
Editors

Nanotoxicology

Materials, Methodologies, and Assessments

Foreword by Valtencir Zucolotto

Springer

Editors
Nelson Durán
Biological Chemistry Laboratory
Institute of Chemistry
Universidade Estadual de Campinas – UNICAMP
13083-970, Campinas, São Paulo, Brazil
and
Center of Natural and Human Sciences
Universidade Federal do ABC
Santo André, São Paulo, Brazil

Silvia S. Guterres
Faculdade de Farmácia
Universidade Federal do Rio Grande do Sul
Avenida Ipiranga 2752
90610-00, Porto Alegre,
RS, Brazil

Oswaldo L. Alves
Solid State Chemistry Laboratory
Institute of Chemistry
Universidade Estadual de Campinas – UNICAMP
13083-970, Campinas, São Paulo, Brazil

ISSN 2194-0452 ISSN 2194-0460 (electronic)
ISBN 978-1-4614-8992-4 ISBN 978-1-4614-8993-1 (eBook)
DOI 10.1007/978-1-4614-8993-1
Springer New York Heidelberg Dordrecht London

Library of Congress Control Number: 2013950381

Printed on acid-free paper

Springer is part of Springer Science+Business Media (www.springer.com)

Foreword

In the last decades, nanotechnology has emerged as an important new field that has found application in a number of scientific and technological areas, including materials science, electronics, biotechnology, and energy, to name a few. The interactions between nanomaterials and biological systems—an area that has been referred to as Nanotoxicology—is one of the promising fields that evolved from the convergence between nano- and biotechnologies. Today, a major challenge related to nanotoxicology is to ensure the safety of engineered nanomaterials to human health and the environment.

It is a great pleasure to recommend this wonderful book, in which a myriad of important topics on the theme of Nanotoxicology is presented and discussed based upon the state-of-the-art literature. The overall organization of the book allows readers to find specific topics of interest moving from the basic concepts on the synthesis and characterization of nanomaterials to the most recent methods and techniques employed for toxicity evaluation. Relevant data regarding cyto- and genotoxicity are also presented and discussed for each class of nanomaterials, including polymeric, lipid, and metallic nanoparticles, as well as carbon nanotubes and graphenes. A discussion on the pharmacokinetics/phamacodynamics, and the chemical stability of nano-structured systems of interest in clinical areas, complements the topics covered by the authors.

I believe this captivating book is an excellent reference for all researchers involved in the field of nano-biotechnology, who will certainly benefit from the up-to-date materials presented here.

São Paulo, Brazil　　　　Valtencir Zucolotto

Preface

Nanotoxicology appears to have emerged as a distinct research field, and it is possible to see both an accumulation of data and an enhanced understanding of relevant toxicity factors through the physicochemical characterization of new nanomaterials. Many new initiatives are now focusing on defining appropriate cytotoxicity and genotoxicity testing protocols.

The first part of this book introduces fundamental concepts of nanostructured materials, their interactions with cellular systems, and nanostabilities. Part 2 critically examines relevant methodologies and approaches to assessing the cytotoxicity and genotoxicity of nanomaterials, such as pharmacokinetics and pharmacodynamics, induced genotoxic effects, new approaches to the development and toxicological profiling of targeted nanomedicines, cellular mechanisms in nanomaterial internalization, and intracellular trafficking and toxicity.

Cyto- and genotoxicity in specific nanoparticles are examined in Part 3 of the book which presents detailed in vitro and in vivo case studies for specific nanomaterials including solid lipid nanoparticles, biogenically synthesized silver nanoparticles, iron oxide nanoparticles, poloxamers, polymeric nanoparticles, and copper nanoparticles. Other aspects of cytotoxicity such as accessing the erythrocyte toxicity of nanomaterials and toxicity of nanomaterials to microorganisms are also discussed.

Our major concern in this book was to review and validate current assessments of nanomaterials safety as well as the standard assays and protocols applied to nanotoxicology. Didactic aspects were emphasized in all chapters, making the book suitable for a broad audience ranging from advanced undergraduate and graduate students to materials scientists and toxicologists in academia and industry.

Campinas, São Paulo, Brazil — Nelson Durán
Porto Alegre, Rio Grande do Sul, Brazil — Silvia S. Guterres
Campinas, São Paulo, Brazil — Oswaldo L. Alves

Contents

List of Contributors

Alessandra Cristina Santos Akkari Center of Natural and Human Sciences, Universidade Federal do ABC – UFABC, Santo André, São Paulo, Brazil

Oswaldo L. Alves Solid State Chemistry Laboratory, Institute of Chemistry, Universidade Estadual de Campinas—UNICAMP, Campinas, São Paulo, Brazil

Daniele R. de Araújo Centro de Ciências Naturais e Humanas, Universidade Federal do ABC – UFABC, Santo André, São Paulo, Brazil

Kyl Assaf Department of Structural and Functional Biology, Institute of Biology, University of Campinas – UNICAMP, Campinas, São Paulo, Brazil

Guilherme B. Bubols Laboratório de Toxicologia (LATOX), Departamento de Análises, Faculdade de Farmácia, Universidade Federal do Rio Grande do Sul, Porto Alegre, Rio Grande do Sul, Brazil

Rachel P. Bulcão Laboratório de Toxicologia (LATOX), Departamento de Análises, Faculdade de Farmácia, Universidade Federal do Rio Grande do Sul, Porto Alegre, Rio Grande do Sul, Brazil

Guillermo R. Castro Laboratory of Nanobiomaterials, Institute of Applied Biotechnology (CINDEFI), Universidad Nacional de La Plata—CONICET, La Plata, Argentina

Ana Cauerhff Laboratory of Nanobiomaterials, Institute of Applied Biotechnology (CINDEFI), Universidad Nacional de La Plata—CONICET, La Plata, Argentina

Helder J. Ceragioli Faculty of Electric Engineering and Computation (FEEC), University of Campinas – UNICAMP, Campinas, São Paulo, Brazil

Mariele F. Charão Laboratório de Toxicologia (LATOX), Departamento de Análises, Faculdade de Farmácia, Universidade Federal do Rio Grande do Sul, Porto Alegre, Rio Grande do Sul, Brazil

Nelson Durán Biological Chemistry Laboratory, Institute of Chemistry, Universidade Estadual de Campinas, Campinas, São Paulo, Brazil

Center of Natural and Human Sciences, Universidade Federal do ABC – UFABC, Santo André, São Paulo, Brazil

Andreia Fonseca de Faria Solid State Chemistry Laboratory, Institute of Chemistry, Universidade Estadual de Campinas-UNICAMP, Campinas, Sao Paulo, Brazil

Odair P. Ferreira Advanced Functional Materials Laboratory, Department of Physics, Universidade Federal do Ceará—UFC, Fortaleza, Ceará, Brazil

Antonio G. Souza Filho Departamento de Física, Universidade Federal do Ceará—UFC, Fortaleza, Ceará, Brazil

Leandro Carneiro Fonseca Solid State Chemistry Laboratory, Institute of Chemistry, Universidade Estadual de Campinas—UNICAMP, Campinas, São Paulo, Brazil

Leonardo Fernandes Fraceto Department of Environmental Engineering, São Paulo State University - UNESP, Sorocaba, São Paulo, Brazil

Leonardo P. Franchi Department of Genetics, Faculty of Medicine of Ribeirão Preto, Universidade de São Paulo—USP, Ribeirão Preto, São Paulo, Brazil

Camila M. Freria Department of Structural and Functional Biology, Institute of Biology, Universidade Estadual de Campinas—UNICAMP, Campinas, São Paulo, Brazil

Aniket Gade Department of Biotechnology, Sant Gadge Baba Amravati University, Amravati, Maharashtra, India

Swapnil Gaikwad Department of Biotechnology, Sant Gadge Baba Amravati University, Amravati, Maharashtra, India

Solange C. Garcia Laboratório de Toxicologia (LATOX), Departamento de Análises, Faculdade de Farmácia, Universidade Federal do Rio Grande do Sul, Porto Alegre, Rio Grande do Sul, Brazil

Indarchand Gupta Department of Biotechnology, Sant Gadge Baba Amravati University, Amravati, Maharashtra, India

Department of Biotechnology, Institute of Science, Aurangabad, Maharashtra, India

Silvia S. Guterres Faculdade de Farmácia, Universidade Federal do Rio Grande do Sul, Porto Alegre, Rio Grande do Sul, Brazil

Paula S. Haddad Exact and Earth Sciences Department, Universidade Federal de São Paulo—UNIFESP, Diadema, São Paulo, Brazil

Raphael Dias Holtz Solid State Chemistry Laboratory, Institute of Chemistry, Universidade Estadual de Campinas—UNICAMP, Campinas, São Paulo, Brazil

Avinash Ingle Department of Biotechnology, Sant Gadge Baba Amravati University, Amravati, Maharashtra, India

German A. Islan Laboratory of Nanobiomaterials, Institute of Applied Biotechnology (CINDEFI), Universidad Nacional de La Plata—CONICET, La Plata, Argentina

Marcelo Bispo de Jesus Department of Biochemistry, Institute of Biology, State University of Campinas—UNICAMP, Campinas, São Paulo, Brazil

Giselle Z. Justo Departamento de Bioquímica (Campus São Paulo) and Departamento de Ciências Biológicas (Campus Diadema), Universidade Federal de São Paulo, São Paulo, São Paulo, Brazil

Yvonne L. Kapila Department of Periodontics and Oral Medicine, School of Dentistry, University of Michigan, Ann Arbor, MI, USA

Marcelo A. Lima Departamento de Bioquímica (Campus São Paulo), Universidade Federal de São Paulo, São Paulo, São Paulo, Brazil

Department of Structural and Chemical Biology, University of Liverpool, Liverpool, UK

Renata de Lima Department of Biotechnology, University of Sorocaba, Sorocaba, São Paulo, Brazil

Universidade Federal de São Carlos, Sorocaba, São Paulo, Brazil

Luis A.Visani de Luna Solid State Chemistry Laboratory, Institute of Chemistry, University of Campinas—UNICAMP, Campinas, São Paulo, Brazil

Priscyla D. Marcato Faculty of Pharmaceutical Sciences of Riberão Preto, Universidade de São Paulo, Ribeirão Preto, São Paulo, Brazil

Diego Stéfani T. Martinez Solid State Chemistry Laboratory, Institute of Chemistry, Universidade Estadual de Campinas—UNICAMP, Campinas, São Paulo, Brazil

Yanina N. Martinez Laboratory of Nanobiomaterials, Institute of Applied Biotechnology (CINDEFI), Universidad Nacional de La Plata—CONICET, La Plata, Argentina

Joyce Cristine de Mello Center of Natural and Human Sciences, Universidade Federal do ABC – UFABC, Santo André, São Paulo, Brazil

Carina Melo Departamento de Bioquímica (Campus São Paulo), Universidade Federal de São Paulo, São Paulo, São Paulo, Brazil

Patricia S. Melo METROCAMP, Campinas, São Paulo, Brazil

Ana Carolina Mazarin de Moraes Solid State Chemistry Laboratory, Institute of Chemistry, Universidade Estadual de Campinas—UNICAMP, Campinas, São Paulo, Brazil

Helena B. Nader Departamento de Bioquímica (Campus São Paulo), Universidade Federal de São Paulo, São Paulo, São Paulo, Brazil

Rafaella Oliveira do Nascimento Solid State Chemistry Laboratory, Institute of Chemistry, Universidade Estadual de Campinas—UNICAMP, Campinas, São Paulo, Brazil

Alexandre L.R. de Oliveira Department of Structural and Functional Biology, Institute of Biology, University of Campinas – UNICAMP, Campinas, São Paulo, Brazil

Alisson Oshiro Center of Natural and Human Sciences, Universidade Federal do ABC – UFABC, Santo André, São Paulo, Brazil

Maria Aparecida S. Pinhal Departamento de Bioquímica (Campus São Paulo), Universidade Federal de São Paulo, São Paulo, São Paulo, Brazil

Departamento de Morfologia e Fisiologia, Faculdade de Medicina do ABC, Universidade Federal de São Paulo, São Paulo, São Paulo, Brazil

Adriana R. Pohlmann Departamento de Química Orgânica, Instituto de Química, Universidade Federal do Rio Grande do Sul, Porto Alegre, Rio Grande do Sul, Brazil

Mahendra Rai Department of Biotechnology, Sant Gadge Baba Amravati University, Amravati, Maharashtra, India

Biological Chemistry Laboratory, Universidade Estadual de Campinas, Campinas, São Paulo, Brazil

Tiago Rodrigues Center of Natural and Human Sciences, Universidade Federal do ABC – UFABC, Santo André, São Paulo, Brazil

Olga Rubilar Department of Chemical Engineering, University of La Frontera, Temuco, Chile

Scientifical & Technological Bioresource Nucleus, University of La Frontera, Temuco, Chile

Amedea B. Seabra Exact and Earth Sciences Department, Universidade Federal de São Paulo—UNIFESP, Diadema, São Paulo, Brazil

Deyse Cardoso da Silva Center of Natural and Human Sciences, Universidade Federal do ABC – UFABC, Santo André, São Paulo, Brazil

Mateus Batista Simões Solid State Chemistry Laboratory, Institute of Chemistry, Universidade Estadual de Campinas—UNICAMP, Campinas, São Paulo, Brazil

Eloah R. Suarez Departamento de Bioquímica (Campus São Paulo), Universidade Federal de São Paulo, São Paulo, São Paulo, Brazil

Catarina S. Takahashi Department of Genetics, Faculty of Medicine of Ribeirão Preto, Universidade de São Paulo—USP, Ribeirão Preto, São Paulo, Brazil

Chapter 1
Nanomaterials

Oswaldo L. Alves, Ana Carolina Mazarin de Moraes, Mateus Batista Simões, Leandro Carneiro Fonseca, Rafaella Oliveira do Nascimento, Raphael Dias Holtz, and Andreia Fonseca de Faria

Abstract Much of the extensive development of nanotechnology is related to nanomaterials. In this chapter nanomaterials are presented non-exhaustively, aiming to show that their definitions are still under construction. We present some families of nanomaterials with emphasis on quantum dots, carbon nanotubes, graphene, fullerene, inorganic nanotubes, and metallic nanoparticles due to their importance in both basic and applied sciences. The main applications are introduced in various industrial sectors and also in biology and medicine in order to show that nanomaterials are part of several industrial products. The central question of the characterization of properties by using several physicochemical techniques, many of them ISO (International Organization for Standardization) recommended, and the implications for the emerging field of nanotoxicology are shown. Finally, comments are made about the regulation and we indicate an extensive bibliography based on the main documents that explore this critical issue for the marketing of products containing nanomaterials.

1.1 Nanomaterials: Concepts and Definitions

Nanotechnology can be defined as the study, manipulation, and/or building of materials, devices, and objects that are normally at the nanoscale (1 nm = 10^{-9} m) and that have properties strongly size dependent on this scale.

According to Ratner and Ratner (2003), it is important to keep in mind that "nanoscale implies not only a question of to be small, it is, yes, a special kind of small." Such consideration is important, as it leads us to the understanding that there are fundamental chemical, physical, mechanical, and biological properties of

O.L. Alves (✉)
Solid State Chemistry Laboratory, Institute of Chemistry, Universidade Estadual de Campinas-UNICAMP, Cidade Universitária "Zeferino Vaz"—Barão Geraldo, Caixa Postal 6154, Campinas, Sao Paulo 13083-970, Brazil
e-mail: oalves@iqm.unicamp.br

N. Durán et al. (eds.), *Nanotoxicology*, Nanomedicine and Nanotoxicology,
DOI 10.1007/978-1-4614-8993-1_1,

materials which depend on the size or, in other words, which maintain “complicity” with it, complicity that constitutes the key to the nanoscience.

Regarding the materials and substances, these entities are called nanomaterial or nanoengineered systems. It is important to mention that nanomaterials can be produced deliberately by certain chemical or physical processes (*bottom-up* production), creating materials with properties that do not appear in their macroscale (*bulk*). Nanomaterials may also be produced by manufacturing processes, such as milling or grinding (*top-down* production), generating nano-sized particles which may or may not have different properties from the bulk materials which they are originated. Generally, it is accepted as nanomaterials those materials that were produced by synthetic processes or manufacturing, i.e., were “intentionally produced.”

According to US Environmental Protection Agency (2005), “the definition does not include unintentionally produced nanomaterials such as particles of nanometric size or materials that occur naturally in the environment, such as viruses or volcanic ash, byproducts and nanoparticles, from human activity, such as particulates from diesel engines or other byproducts from the combustion or automotive friction.”

The definitions of nanomaterials have gone through several revisions since their introduction in the mid-1990s. These revisions have been made in the sense not only to accompany the development of nanotechnology itself but also to enable advances in its regulation issues.

The European Union (2011) has adopted the following definition for nanomaterials: “Nanomaterial” as “a natural, incidental or manufactured materials containing particles, in an unbound state or to an aggregate or an agglomerate and where, for 50 % or more of the particles in the number size distribution, one or more external dimensions is in the size range 1–100 nm. In specific cases and where warranted by concerns for the environment, health, safety or competitiveness the number size distribution threshold of 50 % may be replaced by a threshold between 1 and 50 %.” This definition was revised in 2012 (EU 2012) in order to make it more embracing and more adjusted with the provisions of REACH (Registration, Evaluation, Authorization and Restriction of Chemicals Regulation).

In a report published by the Scientific Committee on Emerging and Newly Identified Health Risks (SCENIHR) entitled “*Scientific Basis for the Definition of the Term ‘Nanomaterial’*” (2010), the following conclusion regarding definitions for nanomaterials is presented—“the definitions/descriptions of nanomaterials formulated so far have: (i) given a general size frame for nanomaterials in both external and internal dimensions; and (ii) some have referred to the unique physicochemical characteristics of the specific material under discussion. Whilst such broad definitions can be scientifically justified, they are not easy to apply within the context of a regulatory framework.”

In our opinion the debate concerning definitions of nanotechnology is open, mainly due to the lack of consensus among different stakeholders: researchers, innovators, industrial producers, and regulatory and environmental protection agencies. Authors like Desmoulin-Canselier (2010) estimate that this situation has led to great difficulties in legal concerns, e.g., the elaboration of a text to a situation where several uncertainties still linger.

1.2 Main Types of Nanomaterials

Nanomaterials are at the origin of an important scientific and technological activity. In fact some of them are already in our day-to-day life because they are already embedded in several commercial products ranging from cosmetics, medicines, electronics, lighting, and so on. Therefore, we can expect that these materials could greatly impact on human and the environment.

Some families of frequently used nanomaterials, both in fundamental research and applications, have been briefly described here.

1.2.1 Quantum Dots

The first quantum dots obtained were studied in the beginning of 1980s and were based on semiconductors of groups II–IV (CdS, CdSe, CdTe, and others), III–V (InP, GaAs, GaP, and others), IV–VI (PbSe, PbS, SnS, and others), and IV (Si and Ge). These materials were first synthetized by Ekimov et al. (1985) by using microcrystal growth technique on glassy matrix aiming to study the size quantum effects observed for these microcrystals. Only in 2002 the quantum dots started to be used in a practical application for fluorescent imaging labels of some biomolecules (Bourzac 2013), and the later studies revealed the potential application on optoelectronic and other biomedical devices (Smith and Nie 2010). Basically, quantum dots are semiconductors in nanometric scale that exhibit pronounced differences on their optoelectronic properties considering the material on the bulk state.

These different properties observed on a bulk material and the same material on nanoscale can be assigned to two main reasons:

- Surface effects caused by the majority presence of atoms at the surface of the material. These atoms present fewer neighbors than atoms in the interior of the material, generating a lower coordination number and dangling bonds. As consequence, the superficial atoms are less stable in comparison with those inside the solid, altering the melting point, phases transitions, and thermodynamic principles (e.g., negative heat capacity) of the solid (Roduner 2006).
- Quantum effects caused by quantum confinement of delocalized electrons, in which the energy and density of states will have direct relation in particle size.

Bulk semiconductors present a minimum energy to excite one electron from valence shell to conduct shell-denominated band-gap energy (E_g), which is characteristic from each semiconductor. When one electron crosses this barrier, a “hole” arises at valence shell, creating a hole–electron pair, called exciton. This exciton has a finite size defined by Bohr radius (a_B), also dependent on the material. If the particle size is smaller than Bohr radius, this will entail in a spatial confinement of the charge carriers, raising the energy required to excite the exciton. In this way, the optoelectronic properties will present strict relationship with the particle size (Smith and Nie 2010).

Moreover, some quantum mechanical aspects should be taken into account. Each atom has its own atomic orbital, and when two atoms are combined with each other, the formation of two molecular orbitals will be observed: one with lower energy that stabilizes the bonding (bonding orbital) and another with higher energy that destabilizes the bonding (antibonding orbital). When N atoms are combined, there will be the formation of N molecular orbital, and if N is big enough, there will be a great quantity of states of energy-forming bands. These bands are centered close to atomic energy levels. When the particle size is very tiny, generally smaller than 25 nm (Nozik et al. 2010), discontinuities begin to be observed in the bands, generating delocalized electrons, modifying the density of energy states. This question can be treated such as "a particle in a box" from quantum mechanics, considering the dimensions in which occurs the spatial confinement: one for quantum wells, two for quantum wires, and three for quantum dots. The electronic structure of the material is modified, and discrete levels of energy for quantum dots may be observed. Further, the gap of these materials will be between the bulk solid and discrete molecules, being possible to modulate the electronic energy state, controlling some properties of the material (Smith and Nie 2010). For example, CdS present a band gap of 1.7 eV, when in particles of 20 nm, against a band gap of 2.4 eV, when in particles of 2 nm (Alivisatos 1996).

A correlation among particle size, quantum confinement, exciton energy, and wavelength absorption is illustrated on Fig. 1.1 (Zrazhevskiy and Gao 2009). Smaller the particle size, wider the quantum confinement, larger the exciton energy, and, consequently, lower the wavelength of the absorption and emission.

In the 1990s, the synthesis of quantum dots advanced considerably, requiring milder conditions and with better control of the obtained material (Nozik et al. 2010). Nowadays, quantum dots are obtained from top-down (bulk material to nanoparticles) and bottom-up methods (atoms, molecules, or clusters to nanoparticles). The main top-down methods are based on molecular beam epitaxy, ion implantation, and lithography, while the main bottom-up approaches are divided in two categories: wet-chemical methods (microemulsion, sol–gel, hot-solution decomposition, electrochemistry, microwaves, and hydrothermal methods) or vapor-phase methods (self-assembly of nanostructures in a matrix by molecular beam epitaxy, sputtering liquid metal ion sources, and aggregation of gaseous monomers) (Valizadeh et al. 2012). Further, quantum dots of carbon and carbon nanotubes (CNTs) (Xu et al. 2004), graphene (Yan et al. 2010), silicon (Holmes et al. 2011), and other heavy metal-free composition were obtained in 2000s, ordering mainly bioapplications, opening a new search field for chemists, toxicologists, pharmacists, biochemists, engineers, and so many other professionals.

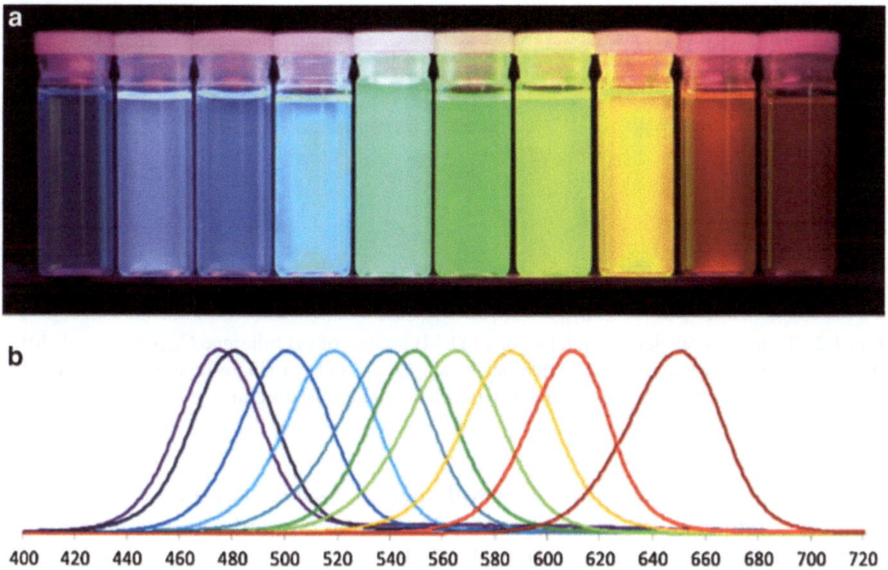

Fig. 1.1 (**a**) Photograph showing the tunable emission of quantum dots in function of the particle size. (**b**) Emission spectra of quantum dots of different sizes (from Zrazhevskiy and Gao 2009)

1.2.2 Fullerenes

Fullerenes are carbon molecular forms spherically arranged in hexagonal and pentagonal fused rings with high symmetry (Fig. 1.2a, Aich et al. 2012). In 1985, Kroto et al. performed simple spray of a graphite disk surface by irradiating a pulsed laser of high power under a flow of helium gas. The experiment indicated the formation of stable clusters containing 60 atoms of carbon. Regarding the structure adopted, scientists proposed that only a closed arrangement would have such stability and symmetry displayed, then this could be a spherical icosahedron. Under these circumstances, the discovery of C_{60} or buckminsterfullerene opened a new field in materials science. Experimentally it is found that the most abundant fullerene is C_{60}, which can be obtained in larger quantities by electric arc discharge (Krätschmer et al. 1990). The obtained soot is composed by a mix of higher fullerenes, and eight forms of stable molecules were isolated hitherto, e.g., C_{76}, C_{84}, and C_{90}. Fullerenes are commonly separated and purified by organic solvent extraction and chromatographic methods (high-performance liquid chromatography) (Dos Santos et al. 2010).

As well as in the graphite network, the carbon atoms in fullerenes are bonded through single and double bonds. The pentagonal structure introduces a curvature in the molecule; therefore, the bonds between the adjacent carbons deviate from planarity. It generates much tension in the molecule, making it more reactive (Prato 1997).

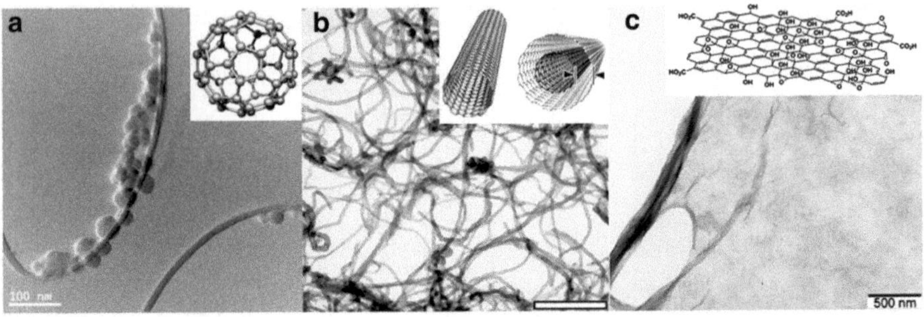

Fig. 1.2 Transmission electron microscopy (TEM) images of (**a**) fullerene C_{60} (Aich et al. 2012), (**b**) multiwalled carbon nanotubes (Stéfani et al. 2011), and (**c**) single-layered graphene oxide (Faria et al. 2012). *Insets* represent the proposed nanomaterials structures (*inset* (**b**) from Cheung et al. 2010)

Fullerene has low solubility in a wide range of solvents, especially polar. It is partially soluble in alkanes and the solubility increases significantly in the benzenic and naphthalenic solvents (Ruoff et al. 1993).

The chemical functionalization of fullerenes is able to offer relevant solutions to improve the solubility. The C_{60} is a nonaromatic molecule by far. It can behave as electron-deficient olefins and practically all the chemistry of fullerenes is based on this important feature. Thus, the fullerene reacts easily with nucleophiles in addition reactions, but does not suffer electrophilic addition reactions, typical of simple alkenes. The C_{60} molecule can also undergo several further cycloaddition reactions (Lopez et al. 2011).

Prato (1997), Maggini et al. (1993), Da Ros and Prato (1999), and Lopez et al. (2011) has been studying extensively fullerene cycloadditions, aiming applications in materials science and medicinal chemistry. For instance, C_{60} can be incorporated into polymers for photoconductive devices, thin films, and electro-optical devices. Fullerene derivatives can also be promising in the field of photodynamic therapy, as inhibitor of HIV protease, as neuroprotectors in neurodegenerative diseases, and in the inhibition of apoptosis (Da Ros and Prato 1999).

A wide range of molecules can also be added to the C_{60} through nucleophilic addition reactions (cyclopropanation), e.g., acetylenes, peptides, DNA fragments, porphyrins, and dendrimers, making it biologically active, amphiphilic, and water soluble (Hirsch and Brettreich 2005).

Felder et al. (2000) reported the synthesis and addition of a water-soluble amphiphilic fullerene prepared by cyclopropanation in a glass matrix via sol–gel process and their properties as optical limiter.

Kumar and Menon (2009) reported the synthesis of s-triazine fullerene derivatives via Prato's cycloaddition (Maggini et al. 1993). These compounds were tested against strains of Gram-positive and Gram-negative bacteria by the agar diffusion method, obtaining inhibition halos of microbial growth. All compounds showed antibacterial activity at low concentrations (12.5 μg/mL), comparable to ciprofloxacin.

Non-covalent functionalization can also occur through π–π interactions, hydrogen bonds, and coordination with metals. Moreover, atoms, molecules, or ions can be trapped within the fullerene "cage," leading to endohedral fullerenes (Lopez et al. 2011).

1.2.3 Carbon Nanotubes

CNTs are one of the most important carbon allotropic forms and were firstly reported by Iijima (1991). CNTs can be understood as graphene sheets rolled into tubular nanostructure, as seen in Fig. 1.2b (Stéfani et al. 2011). High instability of a graphene sheet can be explained by the presence of dandling bonds; and the tubular structure has minimized energy (Dresselhaus et al. 1996). The graphene sheet wrapping is determined by unitary vectors combination (n, m) onto xy graphene plane (Dresselhaus et al. 1995). These vectors determine the chirality and conductive properties (conductor or semiconductor) of CNTs (Jishi et al. 1994). Moreover, the combinations of vectors (n, m) result in different chiralities. Thus, when $n = m \neq 0$, the armchair CNTs are formed. Although when $n \neq 0$ and $m = 0$, the zigzag CNTs are obtained. Other n, m combinations promote chiral CNT generation (Samsonidze et al. 2003).

CNTs can be classified according to their number of walls. In this regard, CNTs formed by only a graphene tubular nanostructure are called *single-walled carbon nanotubes* (SWCNT). Furthermore, CNTs possessing two or more graphite concentric cylinders are called *multiwalled carbon nanotubes* (MWCNT) (Scarselli et al. 2012).

The CNTs are obtained by distinct synthesis methodologies. The arc discharge method was used to obtain the first related CNTs. This method consists into the generation of arc discharge between two graphite electrodes into vacuum chamber and high temperature, 4,000 °C (Ebbesen and Ajayan 1992). The synthesis by chemical vapor deposition (CVD) is based on the decomposition of a carbon source, e.g., hydrocarbons, and its reaction with a reductant gas (hydrogen) and a catalyst like metal oxide (Kumar and Ando 2010). This synthesis method is considered one of the most economic when compared with arc discharge due to lower process temperatures (around 950 °C). The laser ablation is another route to obtain CNTs. This method consists of a laser discharge into ultrapure graphite sample under inert atmosphere. The carbonaceous vapor generated is deposited under a collector surface (reaction chamber). In this method, MWCNT are generated by the use of graphite samples, and SWCNT are obtained by doped graphite source (Jost et al. 2001; Nikolaev et al. 2010). Additionally, there is the occurrence of residues in all CNT synthesis methods. These residues can be amorphous carbon, catalyst, and complex structures, among others.

The CNT growth characteristic from CVD with supported catalyst occurs by two mechanisms: the first is named extrusion and CNTs grow upwards from metal particles that remain attached on substrate; the second mechanism is called "tip growth" in which the catalyst particles detach of support and move into the CNTs' growth direction (Sinnott et al. 1999).

A novel CNT synthesis route uses graphene layers and ferrocene aldehyde (Quintana et al. 2012). The combination among reaction compounds and ultrasound leads to the conversion of nanoscroll formed by graphene/ferrocene aldehyde interaction to MWCNTs. The residues obtained from this method can be graphene sheets as well as ferrocene aldehydes, inside or outside MWCNTs.

The van der Waals interactions in CNTs promote their aggregation, disturbing the dispersion in polar solvents, e.g., water. Therefore, CNT modifications/interactions are important for applications of these nanostructures into polar solvents and biological medium, for example. The non-covalent interactions can occur by adsorptions and dopings (Souza Filho and Terrones 2008). The first is based on π–π interactions, whereas the second is based on electronic transfer between CNTs and the molecules (Souza Filho et al. 2006). Surfactants are used to disperse CNTs in aqueous medium; their hydrophobic portion interacts with nanotubes and their polar chain interacts with water (O'Connell et al. 2002; Vaisman et al. 2006). Additionally, the ultrasound activity improves the CNT/surfactant interactions. The CNT dopings are divided into endohedral, exohedral or intercalation, and in-plane doping (Souza Filho and Terrones 2008). The endohedral doping occurs when the molecules or atoms are inside CNTs. The exohedral consists of electronic transfer between CNTs and external molecules. The in-plane doping occurs when one atom interacts with CNTs in the same plane of carbon atoms.

The covalent functionalization of CNTs generates functional groups onto CNTs surface, improving the dispersion of these nanostructures in water. Besides, functional groups are able to bond with several molecules. The most common functionalization is the chemical oxidation of CNTs in the presence of acids (Ramanathan et al. 2005). These oxidations promote the debris generation (Fogden et al. 2008), which are carbon fragments having molecular structure similar to humic and fulvic acids (Stéfani et al. 2011). Other functionalization onto CNTs promotes the insert of amide (Awasthi et al. 2009), amines, alcohols, and biomolecules (Zhang et al. 2010), among other functional groups.

1.2.4 Graphene and Graphene Oxide

Graphene consists of a single sp2-hybridized carbon atom layer arranged in 2D hexagonal lattice, in which atoms are covalently and highly bonded in a uniform surface, slightly flat, with undulations corresponding to single-atom thick (Geim and Novoselov 2007). Its isolation was absolutely unexpected since 2D crystals are thermodynamically unstable. The π-electron conjugation along the 2D lattice led to unique and superlative electronic, optical, thermal, and mechanical properties, wherein electrons have remarkable high charge-carrier mobility. Besides, it exhibits large superficial area, great Young's modulus, fracture resistance, thermal conductivity, chemical stability, optical transmittance, and anomalous quantum Hall effect (Guo and Dong 2011). This wide range of singular properties open up interesting possibilities of graphene application in electronics manufacturing, storage and gas sensing, catalysis, batteries, fuel cells, sensors, high-performance ultracapacitors, nanocomposites, and so forth (Mas-Balleste et al. 2011). Moreover, graphene is a promising next generation of electronics (Johns and Hersam 2013).

Novoselov et al. (2004) firstly isolated graphene through a *top-down* approach. The "Scotch-tape" method involves the micromechanical exfoliation of a 3D bulk graphite into 2D graphene monolayers with good quality. The monosheets can also be generated through epitaxial growth from crystalline SiC (Berger et al. 2004) and CVD from hydrocarbons gases (Zhu et al. 2010). On the other hand, those methodologies are low yield and not scalable (Allen et al. 2010). An alternative to overcome those drawbacks includes the preparation of graphene sheets through chemical exfoliation of graphite. This approach gave the rise of the so-called graphene oxide, a water-soluble chemically derived graphene monolayer that can undergo a reduction process to generate graphene with good yield and high throughput (Fig. 1.2c, Faria et al. 2012). However, the complete graphene exfoliation in suspension is strongly influenced by remaining defects in its structure, becoming more interesting to applications that do not require crystalline homogeneity of the sheets, e.g., nanocomposites, sensors, molecular vehicles, and nanomedicine platforms (Faria et al. 2012).

The surface of graphene oxide contains a wide range of reactive functional groups, e.g., hydroxyl, carboxyl, carbonyl, and epoxy, which provide high reactivity and good water stability (hydrophilic character), increasing the field of its applications through chemical functionalization. Moreover, GO exhibits solubility in a wide range of polar and nonpolar organic solvents. The most important solution-based route to generate GO single sheets is the modified Hummers method (Hummers and Offeman 1958; Kovtyukhova et al. 1999), which involves strongly oxidative treatment of graphite with potassium permanganate ($KMnO_4$) in the presence of concentrated sulfuric acid (H_2SO_4).

Recently, Rourke et al. (2011) reported that the oxidative process to generate GO, as well as in the case of CNTs (Verdejo et al. 2007), is also able to introduce carbonaceous by-products on its surface, named oxidation debris—highly oxygenated polycyclic aromatic sheets. These oxidized fragments can adsorb on to GO surface through π–π stacking and act as a surfactant stabilizing aqueous GO suspensions. Furthermore, the chemical composition of the carbon nanomaterial may vary significantly, implicating on the chemical functionalization of this material.

In this way, Faria et al. (2012) have highlighted that the chemical nature of oxidation debris is different from GO. It was exploited through the silver nanoparticle (AgNP) GO decoration as a conceptual model. They have found that the presence of debris on GO surface influenced the nucleation and stabilization of AgNP, resulting in smaller and less crystalline nanoparticles. Therefore, a portion of the carboxylic reactive sites on GO available to chemical functionalization is, indeed, reactive sites on oxidized debris fragments.

Despite the importance of oxidation debris on chemical functionalization of GO sheets, this issue has not been related in previous works; however, the chemically modified graphene and its surface modification have been extensively investigated.

In this perspective, Sun et al. (2008) have obtained pegylated graphene oxide sheets through PEG (polyethylene glycol) molecules grafting onto carboxylic groups of GO, resulting in photoluminescent and serum/buffers soluble GO sheets. Stankovich et al. (2006) reported a conductive polystyrene–graphene nanocomposite. Xu et al. (2010) synthesized graphene functionalized with β-cyclodextrin which indicated high bio-recognition of hemoglobin. Das et al. (2011) demonstrated the antimicrobial properties of silver nanoparticle in graphene oxide against Gram-negative bacteria.

1.2.5 Inorganic Nanotubes

One year after the discovery of CNTs, Tenne et al. (1992) reported the synthesis of new type of nanotubes: WS_2 nanotubes. This development can be considered a milestone to the field of nanostructures based on inorganic systems. Inorganic nanotubes are generally formed by rolling and folding of metal oxide lamellas, similarly to graphite/graphene to produce CNTs. The weak van der Waals forces between interlayers and the strong intralayers covalent bonds are conditions necessary to form tubelike structures (Tenne 2002).

Several layered inorganic compounds have similar structures of graphite. One important example is the metal dichalcogenides, MX_2 (M = W, Nb, Mo, Hf; X = Se, S), that contain a metal layer between two layers of chalcogen, with metal in a coordination number of octahedral or trigonal pyramidal (Rao and Nath 2003). Many other types of inorganic nanotubes have been developed in the last decades, including metal oxides. These nanostructures have become a symbol of the advancement in the field of synthesis in nanotechnology. Some of important inorganic nanostructures tubelike (NTs), divided by family, are presented in Table 1.1.

The most relevant methods used for growing inorganic nanotubes are sulfurization, decomposition of precursor crystals, precursor-assisted pyrolysis, template growth, misfit rolling, and direct synthesis from the vapor phase and hydro- and solvo-thermal treatment (Rao and Nath 2003; Remskar 2004). The latter has experimented an exceptional reemergence, being considered the most important method for producing inorganic nanotubes. In this procedure the intermediate (or reagents) in bulk form is treated by heating and autogenic pressure in aqueous or solvent medium. This treatment offers the energy necessary to roll the layers of bulk material and produce nanostructures tubelike. Occasionally, the use of driving-structure agent is necessary to form this morphology.

There are other important inorganic nanostructures similar to nanotubes but non-hollow counterparts: nanorods and nanowires. Oxide nanorods and nanowires are currently one of the most important topics of nanotechnology research in

Table 1.1 Examples of inorganic nanotubes synthesized until date divided by chemical family (Xiong et al. 2005; Remskar 2004; Rao and Nath 2003; Patzke et al. 2002)

Chemical family	Structure
Transition metal chalcogenide	MoS_2, $MoSe_2$, WS_2, WSe_2, NbS_2, $NbSe_2$, TaS_2, ZrS_2, HfS_2, TiS_2, ZnS, NiS, CdSe, CdS
Transition metal halogeneous	$NiCl_2$
Transition metal oxides	TiO_2, ZnO, GaO/ZnO, VO_x, $W_{18}O_{49}$, V_2O_5, Al_2O_3, In_2O_3, Ga_2O_3, $BaTiO_3$, $PbTiO_3$, MoO_3, RuO_2, $BiVO_3$
Silicon oxide	SiO_2
Rare-earth oxides	(Er, Tm, Yb, Lu) oxide
Mixed phase and metal doped	$PbNb_nS_{2n+1}$, $Mo_{1-x}WS_2$, $W_xMo_yC_zS_z$, Nb–WS_2, WS_2–carbon NTs, NbS_2–carbon NTs, Au–MoS_2, Ag–WS_2, Ag–MoS_2, $Cu_{5,5}FeS_{6,5}$
Boron and silicon based	BN, BCN, Si
Metal	Au, Co, Fe, Cu, Ni, Te, Bi

materials science. Among several oxide nanorods, nanotubes, and nanowires, there is a special highlight to vanadium oxide. This compound has formed interesting nanostructures, such as nanotubes and nanowires, mainly when it is doped/intercalated with other elements.

Vanadium oxides have gained special attention, owing to their structural flexibility and interesting electrochemical and catalytic properties and more recently due its antibacterial activity (Holtz et al. 2010, 2012). Holtz et al. (2010) developed a novel type of antibacterial agent based on silver nanoparticles and silver vanadate nanowires. This nanostructured material was obtained by a hydrothermal treatment of silver vanadate and presented high antibacterial activity against strains of Gram-positive and Gram-negative bacteria. The same authors synthesized a similar nanostructure based on silver and vanadium oxides by a simple precipitation reaction between ammonium vanadate and silver nitrate, dismissing hydrothermal treatment and driving-structure agents.

This hybrid nanostructure was presented as a promising antibacterial additive to water-based paints (Holtz et al. 2012). This research opened the way to use inorganic systems with antibacterial property as additive to paints and varnishes, which in turn can be used to paint bathrooms, kitchens, or hospital environments and contribute to decrease the cases of bacterial infections.

1.2.6 Metallic Nanoparticles

The Lycurgus Cup, a famous glass vessel of Roman Era, is a remarkable example of how the properties of the metal in nanoscale are different from bulk solids. This vessel changes its color considering if the light is transmitted or reflected, and the main reason for this fact is the presence of gold and silver nanoparticles that have been found on its composition (Freestone et al. 2007). The explanation of color changes is the resonant interaction of free electrons of the metal nanoparticles with the electromagnetic light field (Ellenbogen et al. 2012). In other words, this phenomenon is the surface plasmon resonance, collective free electron transitions on the metallic surface (Kawamura et al. 2013).

The major plasma resonance bands can be found on the UV radiation region. Therefore, Pb, In, Hg, Sn, and Cd do not show considerably color effects. On the other hand, Cu, Ag, and Au present plasmon resonance bands displaced to the visible region of the spectrum due to d–d transitions. As well as those metallic nanoparticles are more stable, they are often used to perform plasmon experiments. Regardless of the element, differences on the plasmon resonance wavelength can be observed by changing particle size, shape, and morphology, besides the nature of surrounding medium (Liz-Marzán 2006).

It is worth to stress that those parameters can change depending on the synthetic methodology, e.g., silver nanoparticles obtained by using ascorbic acid as reducing agent of ions Ag^+ (provided by $AgNO_3$) and Daxad 19 as stabilizer led to stable dispersions of spherical nanoparticles, yielding particles with mean size of 26 nm (Fig. 1.3a, Sondi et al. 2003). On the other hand, the same reducing and stabilizing agents in the presence of perchloric acid ($HClO_4$) resulted in polydisperse silver nanoparticles, as seen in Fig. 1.3b (Sondi et al. 2003). The $HClO_4$ affects the nucleation and growth stages in the precipitation of silver (Sondi et al. 2003). Silver nanoparticles were spontaneously formed in the presence of poly(*N*-vinyl-2-pyrrolidone) (PVP) dissolved in pyridine (Deivaraj et al. 2005). In this case, PVP acts as the reducing and stabilizing particle agent leading to triangular and quadrilateral particles (Fig. 1.3c, Deivaraj et al. 2005). Gold nanospheres with 15 nm of diameter were formed by mixing sodium citrate and tetrachloroauric (III) acid ($HAuCl_4$) at 100 °C (Fig. 1.3d, Cho et al. 2011). Rod-shaped nanoparticles are usually obtained in the presence of surfactants, and gold nanorods can be obtained through the mixing of a hexadecyltrimethylammonium bromide (CTAB) solution and $HAuCl_4$ by using sodium borohydride ($NaBH_4$) as reducing agent (Fig. 1.3e, Cho et al. 2011).

Metallic nanoparticles, e.g., copper, gold, silver, palladium, titanium oxide, and magnetite, have large surface area and specific physical, chemical, electronic, and optical properties. Therefore, they feature a wide range of applications, such as surface-enhanced Raman scattering (SERS), display devices, catalysis, microelectronics, light-emitting diodes, solar cells, and biological purposes (Das et al. 2011; Slouf et al. 2012).

Silver nanoparticles (AgNPs) are well known due to its remarkable antimicrobial properties. Industries have been incorporating silver nanoparticles in textiles, food packaging, filters for water disinfection, and so forth (Chaudhry et al. 2008). Besides, silver nanoparticles present anti-inflammatory effect (Nadworny et al. 2008) and accelerate the healing of wounds (Wright et al. 2002), showing the potential use in burn treatment (Maneerung et al. 2008). Gulrajani et al. (2008) have produced AgNPs through hydrazine and glucose reducing of silver ions. The colloidal AgNPs were impregnated into silk fabrics, resulting in 100 % of

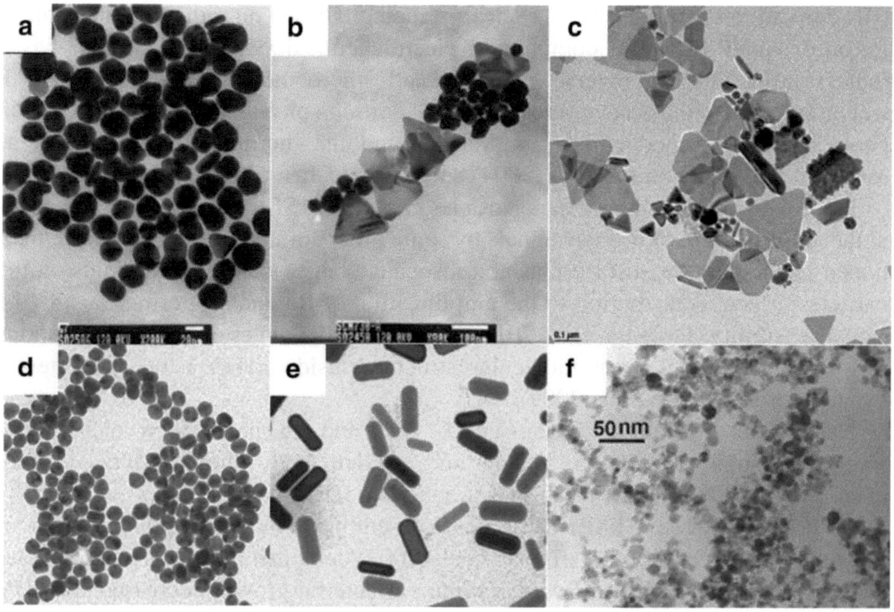

Fig. 1.3 Transmission electron microscopy (TEM) images of (**a**) spherical silver nanoparticles (Sondi et al. 2003), (**b**) silver nanoparticles of different shapes (Sondi et al. 2003), (**c**) triangular and quadrilateral silver nanoparticles (Deivaraj et al. 2005), (**d**) spherical gold nanoparticles (Cho et al. 2011), (**e**) gold nanorods (Cho et al. 2011), and (**f**) spherical TiO_2 nanoparticles (Tian et al. 2009)

antibacterial activity against Gram-positive *Staphylococcus aureus*. Alt et al. (2004) have reported AgNPs' high antibacterial activity against methicillin-resistant *S. aureus* (MRSA). Elechiguerra et al. (2005) have prepared silver nanoparticles through sodium borohydride reduction and demonstrated that the nanoparticles can interact with HIV by size-dependent interaction; solely nanoparticles ranging from 1 to 10 nm are able to attach to the virus. Beyond the chemical methods to preparing silver nanoparticles, some biological synthesis approaches have been exploited, as will be reported in the Chap. 12.

Recently, gold nanoparticles (AuNPs) have been demonstrating good potential for delivery of small drug molecules and biomolecules like proteins and DNA into their targets. The AuNP core is primarily inert and nontoxic (Ghosh et al. 2008). Gold nanoparticles have raised the development of antibacterial strategies due to nontoxicity, versatility in surface modification, polyvalent effects, and photothermal effects. Cui et al. (2012) obtained AuNPs through the borohydride reduction of gold ions. They found that AuNPs show high antibacterial properties against multidrug-resistant Gram-negative bacteria by transcriptomic and proteomic approaches.

A large number of different approaches for gold nanoparticle synthesis with controlled size and shape were described, highlighting the citrate reduction of gold

(III) derivatives; modified Brust–Schiffrin method, based on reduction and stabilization of AuNP by sulfur compounds; electrochemical synthesis; template-based routes; utilization of inverse micelles and microemulsions; seeding growth techniques and solid-state routes; and in addition, physical methods, based on photochemistry, sonochemistry, radiolysis, and the thermolysis approaches, as reviewed by Daniel and Astruc (2004) and Lu and Chen (2012).

Moreover, silver and gold nanoparticles can exhibit SERS effect, which consists in the amplification of Raman signals by some orders of magnitude. The aforementioned phenomena of surface plasmon resonance beget local electric fields on the particle surface, contributing to the amplification of Raman scattering, raising the signal sensibility (Rycenga et al. 2010). The rising challenges of SERS applications rely on information about molecular structure inside living cells and internal dynamic processes (Willets 2009).

Nanoparticles such as TiO_2, Fe_3O_4, Pd, and Pt have been also extensively investigated. TiO_2 is stable and present high photocatalytic activity (Tian et al. 2009). Figure 1.3f shows spherical TiO_2 nanoparticles prepared by flame spray pyrolysis (FSP) technique, presenting a broad size distribution of 5–25 nm (Tian et al. 2009). Armelao et al. (2007) demonstrated TiO_2 nanoparticles bactericidal activity toward *Bacillus subtilis*. Bacterial growth decreases upon UV light exposure onto TiO_2 nanoparticles. Fe_3O_4 nanoparticles possess magnetic properties and present compatible size with biological entities, enabling magnetic detection of biomolecules and target drug delivery, presenting promising potential on biomedicine (Hao et al. 2010).

Slouf et al. (2012) have prepared palladium nanoparticles (PdNPs) with well-defined cube shapes and average size of 15 nm. The palladium nanocubes may be a potential material for antibody conjugation for multiple immunolabeling. Platinum nanoparticles (PtNPs) possess antioxidant and antitumor activities. In this way, Rehman et al. (2012) have showed the anti-inflammatory effects of PtNPs toward lipopolysaccharide-stimulated macrophages cells.

Nevertheless, metallic nanoparticles properties can undergo superficial functionalization in order to gain novel applications, besides stabilization (Lu et al. 2007). Moreover, those particles can be incorporated into other material matrix, e g., polymers, CNTs, and graphene, resulting in synergistic hybrid materials with unique characteristics.

1.3 Applications of Nanomaterials

1.3.1 Cosmetics

Nanomaterials are widely used in cosmetics because, in principle, they can improve their properties, such as transparency, solubility, color, and chemical reactivity. In a recent review article, Raj et al. (2012) discussed different types of nanomaterials used in cosmetics which are listed below:

- Liposomes
- Nanoemulsions
- Nanocapsules
- Solid lipid nanoparticles
- Nanocrystals
- Nanosilver and nanogold
- Dendrimers
- Cubosomes
- Hydrogels
- Fullerenes
- Titanium and zinc oxide nanoparticles
- CNTs
- Quantum dots

Nanomaterials cited have been used by leading global companies in their cosmetic formulations. It is observed also that many products containing nanomaterials are already a significant part of patent portfolios of companies.

It is important to note that there is an ongoing discussion about the adverse health effects related to the use of nanomaterials in cosmetic formulations. To address this situation, the European Commission launched in 2012 the document "Guidance on the Safety Assessment of Nanomaterials in Cosmetics" (2012), which aims to assist in building a scientific basis for risk assessment and safe use of nanomaterials.

An important source of information on cosmetic products with embedded nanotechnology market is available in the online inventory of nanotechnology-based consumer products of *The Project on Emerging Nanotechnologies* (PEN). Of the 1,317 products listed, 143 and 33, respectively, are related to the items cosmetics and sunscreens (PEN 2013).

1.3.2 Textiles

In the textile industry nanomaterials are used to modify the surface of fabrics. These modifications allow reaching new and unique properties such as water repellence, soil resistance, wrinkle resistance, antibacteria, antistatic and UV protection, flame retardation, and improvement of dyeability.

Nanoparticles have a high aspect ratio (area-to-volume) and a high surface energy which determine, in most cases, a strong interaction with the fabric fibers, resulting in functionalities more permanent and little affected by the actions of wearing, cleaning, or washing.

Table 1.2 presents nanomaterials used to achieve some different properties.

Table 1.2 Functionality versus nanomaterial in textile industries (Wong et al. 2006; Patel and Chattopadhyay 2007)

Functionality/property	Nanomaterials/nanoparticles
Water repellence	Thin nanoparticulate plasma film, nano-whiskers of hydrocarbons, nanocellulose (whiskers)
UV protection	Semiconductor oxides, nano-TiO_2 and nano-ZnO, nano-SiO_2, nano-Al_2O_3
Antibacteria	Silver nanoparticles (AgNP), nano-TiO_2, nano-ZnO
Antistatic	Nano-TiO_2, nano-ZnO (whiskers), nano-antimony-doped tin oxide (ATO), silane nanosol, electrically conductive nanoparticles
Wrinkle resistance	Nano-TiO_2, nano-SiO_2, nanocellulose (whiskers)
Magnetic properties and remote heating	Fe nanoparticles
Resistance to abrasion, chemical resistance, electrical conductivity, and coloration of some textiles	Carbon black nanoparticles
High electrical, heat and chemical resistance	Clay nanoparticles
Stain resistance	Nanocellulose (whiskers)

Among nanomaterials shown in Table 1.2, the nanosol solutions that are nanometer-sized metal oxide particles in aqueous or organic solvents can be highlighted. With these materials, several of the features shown in Table 1.2 or others of interest (oleophobic, shielding of X/α-rays, luminescence, controlled release, enzyme immobilization, etc.) via sol–gel coatings are obtained (Mahtig and Textor 2008).

Other nanomaterials are being incorporated into textiles such as CNTs (Panhuis et al. 2007; Schrijver and Govaert 2011) and graphene oxide (Krishnamoorthy et al. 2011). These developments allow the manufacture of conductive fabrics that opens up possibilities toward the use of embedded electronics in ordinary clothing.

1.3.3 Electronics

Electronics is one of the major areas of application of nanomaterials with great economic impact. Many applications have had an almost instant commercial success and paved the route for the large-scale use. The approach to application use is based on composites, i.e., nanomaterials combined with other materials, or on nanostructured films deposited under many different substrates (glass, polymers, and metals).

Table 1.3 Common applications of nanomaterials

Nanomaterial	Applications	References
Quantum dots	Optoelectronic devices, photovoltaic devices, light-emitting diodes (LEDs), solar cells, displays, sensors	Tilley (2008), Coe et al. (2002), Maier-Flaig et al. (2013), Bourzac (2013), and Bera et al. (2010)
Carbon nanotubes	Field emission applications, implantable electrodes, field effect transistors, nanoelectronic devices, mirror actuators for space laser communications, flexible transparent conductive films, optoelectronic devices, radiation detectors, RF applications, sensors, superconductivity	Marulanda (2011), Bandaru (2007), Voutilainen et al. (2012), and Park et al. (2013)
Graphene	Touch screen, E-paper, foldable OLED, high-frequency transistor, logic transistor, electrodes, capacitors, field effect transistors, diodes, secondary batteries, sensors, electric switches, electroluminescent devices	Novoselov et al. (2012), Frazier et al. (2012), and Avouris and Xia (2012)

It is important to note that many applications in this area are due to unique properties of nanomaterials that generate new science absorbed by industry. An outstanding example is the semiconductor nanocrystal also called quantum dot (introduced in Sect. 1.2.1) where quantum confinement effects were observed.

Quantum dots, CNTs and graphene, and other nanomaterials have been the subject of intensive research that will lead to a large number of applications in electronics. Table 1.3 shows a comprehensive range of possible applications in development and many others already coming to our day to day. However, there is a long way for all these possibilities to achieve the condition of consumer and industrial products. Matters of economic nature related to the high price, availability, and also the toxicity are the major difficulties.

1.3.4 Others

According to their unique size-dependent properties, nanomaterials may have application in several other areas of basic and industrial research. In a review, Salata (2004) reported a series of applications related to medicine and biology which are cited below:

- Fluorescent biological labels
- Drug and gene delivery
- Biodetection of pathogens

- Detection of proteins
- Probing of DNA structure
- Tissue engineering
- Tumor destruction via heating (hyperthermia)
- Separation and purification of biological molecules and cells
- MRI contrast enhancement
- Phagokinetic studies

Other areas of application of nanomaterials can be cited. Among them we highlight catalysis (Machado et al. 2013), ceramics (Jayadevan et al. 2004), and polymers (Lazzari et al. 2006). Taking into consideration the different aspects presented in this section, there is no doubt about the crucial role of nanomaterials in the years to come. Indeed, nanomaterials will be present directly or indirectly in many products and future technologies.

1.4 Potential Risks of Nanomaterials

1.4.1 The Importance of the Physicochemical Properties on Toxicological Experiments

In nanobiotechnology, synthesized versatile materials for biological focus, such as silica nanoparticles (Lai et al. 2003), CNTs (Fletcher and Hamilton 2005), and liposomes (Torchilin 2007), among other materials, must undergo toxicological experiments as a way to ensure the lowest risk to human health, and ecotoxicological aspects in order to extinguish the exposure risks of the cited nano-objects to the external environment (Powers et al. 2006; Farré et al. 2009).

Toxicological experiment is important to avoid potential possible adverse effects by nano-objects on humans and the environment, because it evaluates the characteristics of health hazard, in vitro and in vivo dose–response process, the exposure time of the organism to the colloidal particle, and the hazard identification. This is a set of additional tests focused on safe consumption of nanomaterials owing to medicinal applications for human beings (National Research Council 1983; Kuempel et al. 2012).

The first step in biological experiments is started on hazard identification (Oberdorster 2009). Taking into evidence that materials can cause beneficial or deleterious effects to the organism, it is important to know its behavior in biological environment that guides its mechanism of action. In this context, nano-objects are tested in vitro, such as bacteria or their introduction into living tissue, in order to study the mechanisms of interaction of nanoparticles with biological environment (e.g., interaction with plasma membrane receptors). After obtaining satisfactory results, in vivo analysis are then performed. The second step of this test is the range of the dose–response test, whose purpose is to measure an adverse effect caused by nanomaterial in a certain population or environment in a specific time and amount of sample. The third step in this process consists of the exposure assessment, which covers the different ways in which nanomaterials may contact the body, either orally or dermally and by inhalation. For each type of contact, concentration of nano-object and exposure time which causes adverse effects to health is analyzed. Concluding the range of biological analysis, the fourth and final step in this process is the risk characterization, performed by calculating the probability of health risk from nano-object in the organism/environment, based on the above three steps (Oberdorster 2009).

Regarding the demand for toxicological tests aforementioned, another flagship factor is the physicochemical characterization of materials, considering them as complex systems which need to be investigated according to their size, shape, surface area, composition, purity, dispersion, solubility, and surface charges, among other features, which are directly or indirectly related to their mechanism of action in the living body, extolling the importance with regard to the investigation of their physicochemical properties and the need to implement techniques that provide such studies. It is important to mention the facility in characterizing a molecule such as a drug and its applications in biological systems and risks, since it is a relatively simple system. In the case of nanomaterials, their complexity must be studied, since each detail (desirable or undesirable) can be related to their action in the biological system.

Among the physicochemical properties that govern the behavior of nanoparticles in living organism and environment, we can highlight some of the most important particles whose subject is discussed below. The agglomeration of materials is a first feature to be analyzed because of their interspecific interactions (electrostatic or van der Waals interactions) (Grasso et al. 2002). Due to the formation of these aggregates (also called secondary particles), the average size of nano-objects increases, which can exacerbate their exposure and uptake by cells, a phenomenon linked to the toxicity of the organism (Limbach et al. 2005). Taking into account this assumption, the analysis of the average size of a supramolecular system to be developed for biological applications highlighted here is of substantial importance. The possible shapes being assumed by particles have been investigated in the scientific community. In this context, the toxicological effects depending on the shape of the nano-objects have not been fully studied; however, it is known that certain biological systems interact with nanomaterials arresting specific forms: sphere-like and fiber-like, among other kinds of shape (Tang et al. 2012). One of the most important physicochemical properties is the surface area inherent in porous materials widely studied for biological applications. The relevance of this feature is supported by the fact that the surface area is directly related to the chemical reactivity of these materials, resulting in higher specific interactions of these systems to biological fluids and structures (cells, blood proteins, tissues, etc.) (Lison et al. 1997; Tran et al. 2000).

The complete chemical composition of supramolecular system is crucial as it regards the need to elucidate the chemical elements that compose it, as these are toxicological indicators (ISO/TR 13014). The surface of nanoparticles fits the key item to acquire new physicochemical properties as surface-reactive groups of the nanoparticles can be involved in condensation reactions to organic groups of interest, providing new characteristics to the nanostructured system (Soller-Illia et al. 2002; Huh et al. 2003). From the point of view of the surface of the nano-objects, surface charge falls as an interesting physicochemical property regarding colloidal stabilization. In this context, the charged particles tend to repel each other by avoiding their aggregation. It is a phenomenon which explains the importance of this latter characteristic (Grasso et al. 2002). The solubility of nanoparticles is an important physicochemical parameter because they must have significant solubility in biological fluids to ensure their action. Its dispersion must also be ensured in order to avoid agglomeration, since this latter phenomenon may prevent the penetration of nano-objects inside the biological systems without generating toxicity (Powers et al. 2007).

It is important to mold a supramolecular system according to physicochemical parameters of interest and the using of techniques focused on confirmation of the characteristics of nano-objects as a complementary way is even more relevant. Accordingly, Table 1.4 presents the main techniques involved for each commented physicochemical property.

1.4.2 Regulation

Materials on nanoscale have different properties compared with the same material on *bulk* state. The explanation of this effect is related to the increased surface area and the domain of quantum size effects (Dhawan et al. 2011). The spectrum of possible applications of nanomaterials reached a magnitude and versatility which inaugurate a real technological revolution. Currently, there are 1,317 products with nanotechnology in the global market and is estimated that the annual production of materials with nanostructures passed 1,000–5,000 tons in the last 6 years, with a view to reaching 100,000 tons in the next decade (Paschoalino et al. 2010). Despite the numerous advantages and applications of nanomaterials, it is known that these structures can cause deleterious effects to humans and the environment as well. Thus, there is a current need of guidelines for the safe use of these nanomaterials in a manner to minimize exposure of researchers and minimize damage to the environment. However, it is observed that investment in studies assessing the toxicity of nanomaterials is still very small. This can be seen in the number of scientific papers published on the subject. There are approximately 250,000 indexed articles published that present the topics nanotechnology, nanomaterials, or nanoparticles, less than 4.5 % of them addressed the issue of toxicity and ecotoxicity (10,267 articles) (Web of Knowledge, March 2013). However, since 2003 there is a concern about the risks of nanomaterials in various regions and countries: European Community (EC), USA, Canada, and Australia.

There is a huge concern on the risks of nanotechnology on humans and the environment. In addition, numerous reviews on this topic show the significance of

Table 1.4 The main techniques involved for each physicochemical property

Physicochemical parameters	Characterization methods	References
Particle size	Dynamic light scattering	ISO 22412:2008, Persson et al. (1994)
	Small-angle X-ray scattering	ISO/TS 13762:2001, Pedersen (1997)
	Analysis of SEM or TEM	Lami and Lang (1998) and Dokoutchaev et al. (1999)
	Laser-induced incandescence	Schulz et al. (2006) and Mewes and Seitzman (1997)
Aggregation/agglomeration state	Static light scattering, X-ray diffraction, laser diffraction, and small-angle neutron scattering	ISO/TR 13097, ISO 13322-1:2004, ISO/DIS 12025
Surface area	Methods based on gas or liquid adsorption isotherms	Morais et al. (2012), ISO 15901-2:2005
	Liquid porosimetry	ISO 15901-2:2005
Shape	Analysis of SEM or TEM	Paula et al. (2012) and Wepasnick et al. (2010)
Composition	X-ray fluorescence—chemical purity	ISO 22489:2006
	X-ray photoelectron spectroscopy—chemical purity	ISO 22489:2006
	X-ray diffraction—crystallinity, purity	ISO Microbeam Analysis
	Scanning electron microscopy	ISO Microbeam Analysis
Surface chemistry	X-ray photoelectron spectroscopy	ISO/TR 19319:2003
	3D atom-probe tomography	Thompson et al. (2005)
	Raman and other molecular spectroscopies	Moskovits (2005)
	Secondary ion mass spectrometry	ISO 17560:2002, ISO 18114:2003
Surface charge	Isoelectric point, electrophoresis, electrosmosis, and zeta potential	ISO 20998-1:2006
Solubility	There are no specific methods for this kind of assessment	
Dispersibility	Based on particle size measurements	

this area today. A preoccupation in both aspects of environmental and human has been reported (Wani et al. 2011; Takhar and Mahant 2011; Clift et al. 2011; Jiang et al. 2011; Pumera 2011; Lee et al. 2011; Kumar et al. 2011).

According to the Centre for NanoBioSafety and Sustainability (CNBSS 2013), 2012 was an important year for issue of regulation of nanotechnology. The most important documents produced on regulation and recommendation, guidance reports, registry of nanomaterials, and standards are compiled in the site of the center.

1.5 Final Remarks

Nanomaterials are the soul of nanotechnologies and nanosciences because their unique physicochemical and natural interface with chemistry, physics, biology, and engineering have enabled a wide range of possibilities for various areas of science and technology. This privileged situation has led to the development of new products containing new features that, despite of legislation currently under construction, already form part of our daily lives. Nevertheless, many important challenges still need to be overcome, such as the industrial-scale processes, understanding the interactions with biosystems (impact against humans and environment), and the regulation issues. Over the past 10 years, there have been significant advances in the development of nanotechnology, but it is still little, considering the enormous scientific and technological challenges that nanoscale world even has to offer.

References

Aich N, Flora JRV, Saleh NB (2012) Preparation and characterization of stable aqueous higher-order fullerenes. Nanotechnology 23:1–9

Alivisatos AP (1996) Semiconductor clusters, nanocrystals and quantum dots. Science 271:933–937

Allen MJ, Tung VC, Kaner RB (2010) Honeycomb carbon: a review of graphene. Chem Rev 110:132–145

Alt V, Bechert T, Steinrücke P et al (2004) An *in vitro* assessment of the antibacterial properties and cytotoxicity of nanoparticulate silver bone cement. Biomaterials 25:4383–4391

Armelao L, Barreca D, Bottaro G et al (2007) Photocatalytic and antibacterial activity of TiO_2 and Au/TiO_2 nanosystems. Nanotechnology 18:1–7

Avouris P, Xia F (2012) Graphene applications in electronics and photonics. MRS Bull 37:1225–1234

Awasthi K, Singh DP, Singh SK et al (2009) Attachment of biomolecules (protein and DNA) to amino-functionalized carbon nanotubes. New Carbon Mater 24:301–306

Bandaru PR (2007) Electrical properties and applications of carbon nanotube structures. J Nanosci Nanotechnol 7:1239–1267

Bera D, Qian L, Tseng T-K et al (2010) Quantum dots and their multimodal applications: a review. Materials 3:2260–2345

Berger C, Song ZM, Li TB et al (2004) Ultrathin epitaxial graphite: 2D electron gas properties and a route toward graphene-based nanoelectronics. J Phys Chem B 108:19912–19916

Bourzac K (2013) Quantum dots go on display. Nature 493:283

Centre for NanoBioSafety and Sustainability—CNBSS (2013) Safety and regulation of Nanotechnology and nanomaterials in 2012: what's next? http://www.cnbss.eu/index.php/editorial/item/68-safety-and-regulation-of-nanotechnology-and-nanomaterials-in-2012-what%E2%80%99s-next. Accessed 28 Feb 2013

Chaudhry Q, Scotter M, Blackburn J et al (2008) Applications and implications of nanotechnologies for the food sector. Food Addit Contam 25:241–258

Cheung W, Pontoriero F, Taratula O et al (2010) DNA and carbon nanotubes as medicine. Adv Drug Deliv Rev 62:633–649

Cho EC, Zhang Q, Xia Y (2011) The effect of sedimentation and diffusion on cellular uptake of gold nanoparticles. Nat Nanotechnol 6:385–391
Clift MJD, Gehr P, Rothen-Rutishauser B (2011) Nanotoxicology: a perspective and discussion of whether or not in vitro testing is a valid alternative. Arch Toxicol 85:723–731
Coe S, Woo WK, Bawendi M et al (2002) Electroluminescence from single monolayers of nanocrystals in molecular organic devices. Nature 420:800–803
Cui Y, Zhao Y, Tian Y et al (2012) The molecular mechanism of action of bactericidal gold nanoparticles on *Escherichia coli*. Biomaterials 33:2327–2333
Da Ros T, Prato M (1999) Medicinal chemistry with fullerenes and fullerene derivatives. Chem Commun 8:663–669
Daniel MC, Astruc D (2004) Gold nanoparticles: assembly, supramolecular chemistry, quantum-size-related properties, and applications toward biology, catalysis and nanotechnology. Chem Rev 104:293–346
Das MR, Sarma RK, Saikia R et al (2011) Synthesis of silver nanoparticles in an aqueous suspension of graphene oxide sheets and its antimicrobial activity. Colloids Surf B Biointerfaces 1:16–22
Deivaraj TC, Lala NL, Lee JY (2005) Solvent-induced shape evolution of PVP protected spherical silver nanoparticles into triangular nanoplates and nanorods. J Colloid Interface Sci 289:402–409
Desmoulin-Canselier S (2010) La définition des notions fondamentales: mise em perspective juridique. In: Lacour S (ed) La Régulations des Nanotecnologies. Larcier, Brussels
Dhawan A, Shanker R, Das M et al (2011) Guidance for safe handling of nanomaterials. J Biomed Nanotechnol 7:218–224
Dokoutchaev A, James TJ, Koene SC et al (1999) Colloidal metal deposition onto functionalized polystyrene microspheres. Chem Mater 11:2389–2399
Dos Santos LJ, Rocha GP, Alves RB et al (2010) Fulereno[C60]: Química e aplicações. Quim Nova 33:680–693
Dresselhaus MS, Dresselhaus G, Saito R (1995) Physics of carbon nanotubes. Carbon 33:883–891
Dresselhaus MS, Dresselhaus G, Eklund PC (1996) Science of fullerenes and carbon nanotubes. Academic, New York
Ebbesen TW, Ajayan PM (1992) Large-scale synthesis of carbon nanotubes. Nature 358:220–221
Ekimov AI, Efros AL, Onushchenko AA (1985) Quantum size effect in semiconductor microcrystals. Solid State Commun 56:921–924
Elechiguerra JL, Burt JL, Morones JR et al (2005) Interaction of silver nanoparticles with HIV-1. J Nanobiotechnol 3:1–10
Ellenbogen T, Seo K, Crozier K (2012) Chromatic plasmonic polarizers for active visible color filtering and polarimetry. Nano Lett 12:1026–1031
European Commission (2012) Guidance on the safety assessment of nanomaterials in cosmetics. http://ec.europa.eu/health/scientific_committees/consumer_safety/docs/sccs_s_005.pdf. Accessed 22 Feb 2013
European Union (2011) Commission recommendation of 18 October 2011 on the definition of nanomaterial. http://eurlex.europa.eu/LexUriServ/LexUriServ.do?uri=OJ:L:2011:275:0038:0040:EN:PDF. Accessed 22 Feb 2013
European Union (2012) Communication from the commission to the European parliament, the council and the European economic and social committee. http://eurlex.europa.eu/LexUriServ/LexUriServ.do?uri=COM:2012:0572:FIN:en:PDF. Accessed 22 Feb 2013
Faria AF, Martinez DST, Moraes ACM et al (2012) Unveiling the role of oxidation debris on the surface chemistry of graphene through the anchoring of Ag nanoparticles. Chem Mater 24:4080–4087
Farré M, Schrantz KG, Kantiani L et al (2009) Ecotoxicity and analysis of nanomaterials in the aquatic environment. Anal Bioanal Chem 393:81–95

Felder D, Guillon D, Levy R et al (2000) A water soluble methanofullerene derivative: synthesis, micellar aggregation in aqueous solutions, and incorporation in sol-gel glasses for optical limiting applications. J Mater Chem 10:887–892
Fletcher S, Hamilton AD (2005) Protein surface recognition and proteomimetics: mimics of protein surface structure and function. Curr Opin Chem Biol 9:632–638
Fogden S, Verdejo R, Cottam B et al (2008) Purification of single walled carbon nanotubes: the problem with oxidation debris. Chem Phys Lett 460:162–167
Frazier RM, Hough WL, Chopra N et al (2012) Advances in graphene-related technologies: synthesis, devices and outlook. Recent Pat Nanotechnol 6:79–98
Freestone I, Meeks N, Sax M et al (2007) The Lycurgus cup—a Roman nanotechnology. Gold Bull 40:270–277
Geim AK, Novoselov KS (2007) The rise of graphene. Nat Mater 6:183–191
Ghosh P, Han G, De M et al (2008) Gold nanoparticles in delivery applications. Adv Drug Deliv Rev 60:1307–1315
Grasso D, Subramaniam K, Butkus M et al (2002) A review of non-DLVO interactions in environmental colloidal systems. Rev Environ Sci Biotechnol 1:17–38
Gulrajani ML, Gupta D, Periyasamy S et al (2008) Preparation and application of silver nanoparticles on silk for imparting antimicrobial properties. J Appl Polym Sci 108:614–623
Guo SJ, Dong SJ (2011) Graphene nanosheet: synthesis, molecular engineering, thin film, hybrids, and energy and analytical applications. Chem Soc Rev 40:2644–2672
Hao R, Xing R, Xu Z et al (2010) Synthesis, functionalization, and biomedical applications of multifunctional magnetic nanoparticles. Adv Mater 22:2729–2742
Hirsch A, Brettreich M (2005) Fullerenes: chemistry and reactions. Wiley-VCH, Weinheim
Holmes JD, Ziegler KJ, Doty RC et al (2011) Highly luminescent silicon nanocrystals with discrete optical transitions. J Am Chem Soc 123:3743–3748
Holtz RD, Souza Filho AG, Brocchi M et al (2010) Development of nanostructured silver vanadates decorated with silver nanoparticles as a novel antibacterial agent. Nanotechnology 21:1–8
Holtz RD, Lima BA, Souza Filho AG et al (2012) Nanostructured silver vanadate as a promising antibacterial additive to water-based paints. Nanomed Nanotechnol 8:935–940
Huh S, Wiench JW, Yoo JC et al (2003) Organic functionalization and morphology control of mesoporous silica via a co-condensation synthesis method. Chem Mater 15:4247–4256
Hummers WS, Offeman RE (1958) Preparation of graphitic oxide. J Am Chem Soc 80:1339
Iijima S (1991) Helical microtubules of graphitic carbon. Nature 354:56–58
ISO 13322-1:2004, Particle size analysis—image analysis method—Part 1: static image analysis methods
ISO 15901-2:2005, Pore size distribution and porosity of solid materials by mercury porosimetry and gas adsorption—Part 2: analysis of mesopores and macropores by gas adsorption
ISO 17560:2002, Surface chemistry analysis—secondary ion mass spectrometry—method for depth profiling of boron in silicon
ISO 18114:2003, Surface chemistry analysis—secondary ion mass spectrometry—determination of relative sensitivity factors from ion implanted reference materials
ISO 20998-1:2006, Measurement and characterization of particles by acoustic methods—Part 1: concepts and procedures in ultrasonic attenuation spectroscopy
ISO 22412:2008, Particle size analysis—dynamic light scattering (DLS)
ISO 22489:2006, Microbeam analysis—electron probe microanalysis—quantitative point analysis for bulk specimens using wavelength-dispersive X-ray spectroscopy
ISO Microbeam Analysis—guidelines for orientation measurement using electron backscatter diffraction
ISO/DIS 12025 Nanotechnologies, developed a general framework for determining the nano-object release from powders
ISO/TC 24 New work item on dispersion, ISO/TR 13097, Guidelines for the characterization of dispersion stability

ISO/TR 13014, Nanotechnologies—guidance on physico-chemical characterization of engineered nanoscale materials for toxicologic assessment
ISO/TR 19319:2003, Surface chemical analysis—Auger electron spectroscopy and X-ray photoelectron spectroscopy—determination of lateral resolution, analysis area and sample area viewed by the analyser
ISO/TS 13762:2001, Particle Size analysis–small angle X-ray scattering
Jayadevan KP, Liu CY, Tseng TY (2004) Dielectric characteristics of nanocrystalline Ag-$Ba_{0.5}Sr_{0.5}TiO_3$ composite thin films. Appl Phys Lett 85:1211–1213
Jiang GX, Shen ZY, Niu JF et al (2011) Nanotoxicity of engineered nanomaterials in the environment. Prog Chem 23:1769–1781
Jishi RA, Inomata D, Nakao K et al (1994) Electronic and lattice properties of carbon nanotubes. J Phys Soc Jpn 63:2252–2260
Johns JE, Hersam MC (2013) Atomic covalent functionalization of graphene. Acc Chem Res 46:77–86
Jost O, Gorbunov AA, Möller J et al (2001) Impact of catalyst coarsening on the formation of single-wall carbon nanotubes. Chem Phys Lett 339:297–304
Kawamura G, Nogami M, Matsuda A (2013) Shape-controlled metal nanoparticles and their assemblies with optical functionalities. J Nanomater 2013:1–17
Kovtyukhova NI, Ollivier PJ, Martin BR et al (1999) Layer-by-layer assembly of ultrathin composite films from micron-sized graphite oxide sheets and polycations. Chem Mater 11:771–778
Krätschmer W, Lamb LD, Fostiropoulos K et al (1990) Solid C-60—a new form of carbon. Nature 347:354–358
Krishnamoorthy K, Navaneethaiyer U, Mohan R et al (2011) Graphene oxide nanostructures modified multifunctional cotton fabrics. Appl Nanosci 2:119–126
Kroto HW, Heath JR, Obrien SC et al (1985) C-60—Buckminsterfullerene. Nature 318:162–163
Kuempel ED, Castranova V, Geraci CL et al (2012) Development of risk-based nanomaterial groups for occupational exposure control. J Nanopart Res 14:1–15
Kumar M, Ando Y (2010) Chemical vapor deposition of carbon nanotubes: a review on growth mechanism and mass production. J Nanosci Nanotechnol 10:3739–3758
Kumar A, Menon SK (2009) Fullerene derivatized s-triazine analogues as antimicrobial agents. Eur J Med Chem 44:2178–2183
Kumar A, Shanker R, Dhawan A (2011) The need for novel approaches in ecotoxicity of engineered nanomaterials. J Biomed Nanotechnol 7:79–80
Lai CY, Trewyn BG, Jeftinija DM et al (2003) A mesoporous silica nanosphere-based carrier system with chemically removable CdS nanoparticle caps for stimuli-responsive controlled release of neurotransmitters and drug molecules. J Am Chem Soc 125:4451–4459
Lami EB, Lang J (1998) Encapsulation of inorganic particles by dispersion polymerization in polar media. J Colloid Interface Sci 197:293–308
Lazzari M, Rodriguez-Abreu C, Rivas J et al (2006) Self-assembly: a minimal route to the fabrication of nanomaterials. J Nanosci Nanotechnol 6:892–905
Lee K, Lee H, Lee V et al (2011) Optical imaging of intracellular reactive oxygen species for the assessment of the cytotoxicity of nanoparticles. Biomaterials 32:2556–2565
Limbach LK, Li Y, Grass RN et al (2005) Oxide nanoparticle uptake in human lung fibroblasts: effects of particle size, agglomeration, and diffusion at low concentrations. Environ Sci Technol 39:9370–9376
Lison D, Lardot C, Huax F et al (1997) Influence of particle surface area on the toxicity of insoluble manganese dioxide dusts. Arch Toxicol 6:725–729
Liz-Marzán LM (2006) Tailoring surface plasmons through the morphology and assembly of metal nanoparticles. Langmuir 22:32–41
Lopez AM, Mateo-Alonso A, Prato M (2011) Materials chemistry of fullerene C60 derivatives. J Mater Chem 21:1305–1318

Lu Y, Chen W (2012) Sub-nanometre sized metal clusters: from synthesis to the unique property discoveries. Chem Soc Rev 41:3594–3623
Lu AH, Salabas EL, Schuth F (2007) Magnetic nanoparticles: synthesis, protection, functionalization and application. Angew Chem 46:1222–1244
Machado BF, Marchionni A, Bacsa RR et al (2013) Synergistic effect between few layer graphene and carbon nanotube supports for palladium catalysing electrochemical oxidation of alcohols. J Energy Chem 22:296–304
Maggini M, Scorrano G, Prato M (1993) Addition of azomethine ylides to C-60—synthesis, characterization, and functionalization of fullerene pyrrolidines. J Am Chem Soc 115:9798–9799
Mahtig B, Textor T (2008) Nanosols and textiles. World Scientific Publishing, Singapore
Maier-Flaig F, Rinck J, Stephan M et al (2013) Multicolor silicon light-emitting diodes (SiLEDs). Nano Lett 13:475–480
Maneerung T, Tokura S, Rujiravanit R (2008) Impregnation of silver nanoparticles into bacterial cellulose for antimicrobial wound dressing. Carbohydr Polym 72:43–51
Marulanda JM (2011) Carbon nanotubes applications on electron devices. InTech, New York
Mas-Balleste R, Gomez-Navarro C, Gomez-Herrero J et al (2011) 2D materials: to graphene and beyond. Nanoscale 3:20–30
Mewes B, Seitzman JM (1997) Soot volume fraction and particle size measurements with laser-induced incandescence. Appl Opt 36:709–717
Morais EC, Correa GG, Brambilla R et al (2012) Silica imprinted materials containing pharmaceuticals as a template: textural aspects. J Sol Gel Sci Technol 64:324–334
Moskovits M (2005) Surface-enhanced Raman spectroscopy: a brief retrospective. J Raman Spectrosc 36:485–496
Nadworny PL, Wang JF, Tredget EE et al (2008) Anti-inflammatory activity of nanocrystalline silver in a porcine contact dermatitis model. Nanomed Nanotechnol 4:241–251
National Research Council (1983) Risk assessment in the Federal Government: managing the process. National Academy Press, Washington, DC
Nikolaev P, Holmes W, Sosa E et al (2010) Effect of vaporization temperature on the diameter and chiral angle distributions of single-walled carbon nanotubes. J Nanosci Nanotechnol 10:3780–3789
Novoselov KS, Geim AK, Morozov SV et al (2004) Electric field effect in atomically thin carbon films. Science 306:666–669
Novoselov KS, Fal'ko VI, Colombo L et al (2012) A roadmap for graphene. Nature 490:192–200
Nozik AJ, Beard MC, Luther JM et al (2010) Semiconductor quantum dots and quantum dot arrays and applications of multiple exciton generation to third-generation photovoltaic solar cells. Chem Rev 110:6873–6890
O'Connell MJ, Bachilo SM, Huffman CB et al (2002) Band gap fluorescence from individual single-walled carbon nanotubes. Science 297:593–596
Oberdorster G (2009) Safety assessment for nanotechnology and nanomedicine: concepts of nanotoxicology. J Int Med 267:89–105
Panhuis M, Wu J, Ashraf SA et al (2007) Conducting textiles from single-walled carbon nanotubes. Synth Met 157:358–362
Park S, Vosguerichian M, Bao Z (2013) A review of fabrication and applications of carbon nanotube film-based flexible electronics. Nanoscale 5:1727–1752
Paschoalino MP, Marcone GPS, Jardim WF (2010) Nanomaterials and the environment. Quim Nova 33:421–430
Patel BH, Chattopadhyay DP (2007) Nano-particles & their uses in textiles. http://www.indiantextilejournal.com/articles/FAdetails.asp?id=693. Accessed 25 Feb 2013
Patzke GR, Krumeich F, Nesper R (2002) Oxidic nanotubes and nanorods—anisotropic modules for a future nanotechnology. Angew Chem Int Edit 41:2446–2461

Paula AJ, Montoro LA, Souza Filho AG et al (2012) Towards long-term colloidal stability of silica-based nanocarriers for hydrophobic molecules: beyond Stober method. Chem Commun 48:591–593

Pedersen JS (1997) Analysis of small-angle scattering data from colloids and polymer solutions: modeling and least-squares fitting. Adv Colloid Interface Sci 70:171–210

Persson AE, Schoeman BJ, Sterte J et al (1994) The synthesis of discrete colloidal particles of TPA-silicalite-1. Zeolites 14:557–567

Powers KW, Brown SC, Krishna VB et al (2006) Research strategies for safety evaluation of nanomaterials. Part VI. Characterization of nanoscale particles for toxicological evaluation. Toxicol Sci 90:296–303

Powers KW, Palazuelos M, Moudgil BM et al (2007) Characterization of the size shape and state of dispersion of nanoparticles for toxicological studies. Nanotoxicology 1:42–51

Prato M (1997) Fullerene chemistry for materials science applications. J Mater Chem 7:1097–1109

Pumera M (2011) Nanotoxicology: the molecular science point of view. Chem Asian J 6:340–348

Quintana M, Grzelczak M, Spyrou K et al (2012) A simple road for the transformation of few-layer graphene into MWNTs. J Am Chem Soc 134:13310–13315

Raj S, Jose S, Sumod U et al (2012) Nanotechnology in cosmetics: opportunities and challenges. J Pharm Bioallied Sci 4:186–193

Ramanathan T, Fisher FT, Ruoff RS et al (2005) Amino-functionalized carbon nanotubes for binding to polymers and biological systems. Chem Mater 17:1290–1295

Rao CNR, Nath M (2003) Inorganic nanotubes. Dalton Trans 1:1–24

Ratner M, Ratner D (2003) Nanotechnology. Prentice-Hall, Upper Saddle River, NJ

Rehman MU, Yoshihisa Y, Miyamoto Y, Shimizu T (2012) The anti-inflammatory effects of platinum nanoparticles on the lipopolysaccharide-induced inflammatory responde in RAW 264.7 macrophages. Inflammation Res 61:1177–1185

Remskar M (2004) Inorganic nanotubes. Adv Mater 16:1497–1504

Roduner E (2006) Size matters: why nanomaterials are different. Chem Soc Rev 35:583–592

Rourke JP, Pandey PA, Moore JJ et al (2011) The real graphene oxide revealed: stripping the oxidative debris from the graphene-like sheets. Angew Chem 50:3173–3177

Ruoff RS, Tse DS, Malhotra R et al (1993) Solubility of C-60 in a variety of solvents. J Phys Chem 97:3379–3383

Rycenga M, Camargo PHC, Li W et al (2010) Understanding the SERS effects of single silver nanoparticles and their dimers, one at a time. J Phys Chem Lett 1:696–703

Salata OV (2004) Applications of nanoparticles in biology and medicine. J Nanobiotechnol 2:3

Samsonidze GG, Saito R, Jorio A et al (2003) The concept of cutting lines in carbon nanotube science. J Nanosci Nanotechnol 3:431–458

Scarselli M, Castrucci P, De Crescenzi M (2012) Electronic and optoelectronic nano-devices based on carbon nanotubes. J Phys Condens Matter 24:1–36

Schrijver I, Govaert F (2011) Carbon nanotubes in composites and coatings for smart textile applications. http://www.euronanoforum2011.eu/wp-content/uploads/2011/09/enf2011_nanomaterials_govaert_fin.pdf. Accessed 25 Feb 2013

Schulz C, Kock BF, Hofmann M et al (2006) Laser-induced incandescence: recent trends and current questions. Appl Phys B 83:333–354

Scientific Committee on Emerging and Newly Identified Health Risks (SCENIHR) (2010) Scientific basis for the definition of the term "Nanomaterial". http://ec.europa.eu/health/scientific_committees/emerging/docs/scenihr_o_030.pdf. Accessed 22 Feb 2013

Sinnott SB, Andrews R, Qian D et al (1999) Model of carbon nanotube growth through chemical vapor deposition. Chem Phys Lett 315:25–30

Slouf M, Hruby M, Bakaeva Z et al (2012) Preparation of stable Pd nanocubes and their use in biological labeling. Colloids Surf B Interfaces 100:205–208

Smith AM, Nie SM (2010) Semiconductor nanocrystals: structure, properties, and band gap engineering. Acc Chem Res 43:190–200

Soller-Illia GJAA, Sanchez C, Lebeau B et al (2002) Chemical strategies to design textured materials: from microporous and mesoporous oxides to nanonetworks and hierarchical structures. Chem Rev 102:4093–4138
Sondi I, Goia DV, Matijevic E (2003) Preparation of highly concentrated stable dispersions of uniform silver nanoparticles. J Colloid Interface Sci 260:75–81
Souza Filho AG, Terrones M (2008) Properties and applications of doped carbon nanotubes. In: Jorio A, Dresselhaus M, Dresselhaus G (eds) Carbon nanotubes: advanced topics in synthesis, structure, properties and applications. Springer, Berlin
Souza Filho AG, Endo M, Muramatsu H et al (2006) Resonance Raman scattering studies in Br_2-adsorbed double-wall carbon nanotubes. Phys Rev B 73:235413/1–235413/12
Stankovich S, Dikin DA, Dommett JHB et al (2006) Graphene-based composite materials. Nature 442:282–286
Stéfani D, Paula AJ, Vaz BG et al (2011) Structural and proactive safety aspects of oxidation debris from multi-walled carbon nanotubes. J Hazard Mater 189:391–396
Sun X, Liu Z, Welsher K et al (2008) Nano-graphene oxide for cellular imaging and drug delivery. Nano Res 1:203–212
Takhar P, Mahant S (2011) *In vitro* methods for nanotoxicity assessment: advantages and applications. Arch Appl Sci Res 3:389–403
Tang F, Li L, Chen D (2012) Mesoporous silica nanoparticles: synthesis, biocompatibility and drug delivery. Adv Mater 24:1504–1534
Tenne R (2002) Fullerene-like materials and nanotubes from inorganic compounds with a layered (2-D) structure. Colloids Surf A 208:83–92
Tenne R, Margulis L, Genut M et al (1992) Polyhedral and cylindrical structures of tungsten disulfide. Nature 360:444–446
The Project on Emerging Nanotechnologies—PEN (2013) Consumer Products Inventory. http://www.nanotechproject.org/inventories/consumer/analysis_draft/. Accessed 22 Feb 2013
Thompson K, Booske J, Kelly DJLT (2005) Three-dimensional atom mapping of dopants in Si nanostructures. Appl Phys Lett 87:1–3
Tian B, Li C, Gu F et al (2009) Flame sprayed V-doped TiO_2 nanoparticles with enhanced photocatalytic activity under visible light irradiation. Chem Eng J 151:220–227
Tilley RD (2008) Synthesis and applications of nanoparticles and quantum dots. Chem N Z 72:146–150
Torchilin VP (2007) Micellar nanocarriers: pharmaceutical perspectives. Pharm Res 24:1–16
Tran CL, Buchanan D, Cullen RT et al (2000) Inhalation of poorly soluble particles. II. Influence of particle surface area on inflammation and clearance. Inhal Toxicol 12:1113–1126
U.S. Environmental Protection Agency—EPA (2005) Nanotechnology white paper. http://www.epa.gov/osa/pdfs/EPA_nanotechnology_white_paper_external_review_draft_12-02-2005.pdf. Accessed 22 Feb 2013
Vaisman L, Wagner HD, Marom G (2006) The role of surfactants in dispersion of carbon nanotubes. Adv Colloid Interface Sci 128:37–46
Valizadeh A, Mikaeili H, Samiei M et al (2012) Quantum dots: synthesis, bioapplications and toxicity. Nanoscale Res Lett 7:480
Verdejo R, Lamorinierea S, Cottam B et al (2007) Removal of oxidation debris from multi-walled carbon nanotubes. Chem Commun 5:513–515
Voutilainen MC, Seppala ET, Pasanen P et al (2012) Graphene and carbon nanotube applications in mobile devices. IEEE Trans Electron Dev 59:2876–2887
Wani MY, Hashim MA, Nabi F et al (2011) Nanotoxicity: dimensional and morphological concerns. Adv Phys Org Chem ID 450912:1–15
Wepasnick KA, Smith BA, Bitter JL et al (2010) Chemical and structural characterization of carbon nanotube surfaces. Anal Bioanal Chem 396:1003–1014
Willets KA (2009) Surface-enhanced Raman scattering (SERS) for probing internal cellular structure and dynamics. Anal Bioanal Chem 394:85–94

Wong YWH, Yuen CWM, Leung MYS et al (2006) Selected applications of nanotechnology in textiles. AUTEX Res J 6:1–8

Wright JB, Lam K, Buret AG et al (2002) Early healing events in a porcine model of contaminated wounds: effects of nanocrystalline silver on matrix metalloproteinases, cell apoptosis, and healing. Wound Repair Regen 10:141–151

Xiong YJ, Mayers BT, Xia YN (2005) Some recent developments in the chemical synthesis of inorganic nanotubes. Chem Commun 40:5013–5022

Xu X, Ray R, Gu Y et al (2004) Electrophoretic analysis and purification of fluorescent single-walled carbon nanotubes fragments. J Am Chem Soc 126:12736–12737

Xu C, Wang X, Wang J et al (2010) Synthesis and photoelectrical properties of b-cyclodextrin functionalized graphene materials with high bio-recognition capability. Chem Phys Lett 498:162–167

Yan X, Cui Y, Li LS (2010) Synthesis of large, stable colloidal graphene quantum dots with tunable size. J Am Chem Soc 132:5944–5945

Zhang B, Chen Q, Tang H et al (2010) Characterization of and biomolecule immobilization on the biocompatible multi-walled carbon nanotubes generated by functionalization with polyamidoamine dendrimers. Colloids Surf B Biointerfaces 80:18–25

Zhu YW, Murali S, Cai WW et al (2010) Graphene and graphene oxide: synthesis, properties, and applications. Adv Mater 22:3906–3924

Zrazhevskiy P, Gao X (2009) Multifunctional quantum dots for personalized medicine. Nano Today 4:414–428

Chapter 2
Concepts and Methodology of Interaction of Carbon Nanostructures with Cellular Systems

Alexandre L.R. de Oliveira, Helder J. Ceragioli, and Kyl Assaf

Abstract Carbon nanostructures have been shown to have a wide range of applications in several fields. However, characterization seems to be a key factor in order to make data comparable. In this sense, the present chapter addresses important techniques and approaches to provide fundamental information regarding purification and structure of carbon nanoparticles including SEM, FESEM, TEM, Raman, FTIR, and AFM, as tools for spectroscopy and morphology of the samples. Additionally to that, functionalization of nanotubes and graphene with some chemical groups is addressed providing basis for enhancing the biological and functional properties of the nanostructures. Finally, we address different applications of carbon nanoparticles in medical and biological areas, including tumor research, cell culture peripheral nerve regeneration, stem cell therapy, and applications as biosensors.

2.1 Functionalization

2.1.1 Attachment of Oxidic Groups

One of the most common functionalization techniques is the oxidation of nanotubes by liquid phase or gas phase, introducing carboxylic groups and some other oxygen-bearing functionalities such as hydroxyl, carbonyl, ester, and nitro groups (nitro is

---

A.L.R. de Oliveira (✉) • K. Assaf
Department of Structural and Functional Biology, Institute of Biology,
University of Campinas – UNICAMP, Campinas, São Paulo, Brazil
e-mail: alroliv@unicamp.br

H.J. Ceragioli
Faculty of Electric Engineering and Computation (FEEC),
University of Campinas – UNICAMP, Campinas, São Paulo, Brazil

N. Durán et al. (eds.), *Nanotoxicology*, Nanomedicine and Nanotoxicology,
DOI 10.1007/978-1-4614-8993-1_2,

not considered an oxidic group as hydroxyl or carbonyl ones, please see this concept) into the tubes (Hirsch and Vostrowsky 2005).

The development of an oxidation process for SWCNTs (single-walled carbon nanotubes) involves extensive ultrasonic treatment in a mixture of concentrated nitric and sulfuric acid. Such drastic conditions lead to the opening of the tube caps as well as the formation of holes in the sidewalls, followed by an oxidative etching along the walls with the concomitant release of carbon dioxide. The final products are nanotube fragments with lengths in the range of 100–300 nm, whose ends and sidewalls are decorated with a high density of various oxygen-containing groups (mainly carboxyl groups). Under less vigorous conditions, such as refluxing in nitric acid, the shortening of the tubes can be minimized. The chemical modification is then limited mostly to the opening of the tube caps and the formation of functional groups at defect sites along the sidewalls. Nanotubes functionalized in this manner basically retain their pristine electronic and mechanical properties (Zhang et al. 2003; Balasubramanian and Burghard 2005).

The oxidative introduced carboxyl groups represent useful sites for further modifications as they enable the covalent coupling of molecules through the creation of amide and ester bonds. By this method the nanotubes can be provided with a wide range of functional moieties for which purpose bifunctional molecules (e.g., diamines) are often utilized as linkers. Illustrative examples are nanotubes equipped with dendrimers, nucleic acids, enzymes, metal complexes, or semiconductor and metal nanoparticles (Zhang et al. 2003; Katz and Willner 2004; Balasubramanian and Burghard 2005). Another interesting application of the carboxyl groups is the formation of an anhydride at the tube ends through which rings of nanotubes are accessible. The presence of (modified) carboxyl groups leads to a reduction of van der Waals interactions between the CNTs, which strongly facilitates the separation of nanotube bundles into individual tubes. Additionally, the attachment of suitable groups renders the tubes soluble in aqueous or organic solvents, opening the possibility of further modifications through subsequent solution-based chemistry. A high water solubility of a few tenths of a gram per milliliter has recently been achieved on the basis of the carboxyl-based coupling of hydrophilic polymers such as poly(ethylene glycol) (PEG) (Zhang et al. 2003; Fernando et al. 2004; Katz and Willner 2004; Balasubramanian and Burghard 2005). SWCNTs with a good solubility in organic solvents can be obtained by covalent (Hamon et al. 1999; Zhang et al. 2003; Fernando et al. 2004; Katz and Willner 2004; Balasubramanian and Burghard 2005) or ionic (Hamon et al. 1999; Chen et al. 2001; Zhang et al. 2003; Fernando et al. 2004; Katz and Willner 2004; Balasubramanian and Burghard 2005) attachment of long-chain aliphatic amines onto the carboxyl groups.

2.1.2 Amidation

The introduction of carboxylic groups to nanotubes by oxidative procedures gives access to a large number of functional exploitations by transformation of the carboxylic functions and provides anchor groups for further modification. From carboxylic acids, carboxamides can be formed via carboxylic acid chlorides and allow for the decoration of oxidized tubes with aliphatic amines, aryl amines, amino acid derivatives, peptides, amino-group-substituted dendrimers, etc. as nucleophiles. The carboxylic groups can be activated by conversion into acyl chloride groups with thionyl chloride (Chan et al. 2006; Chen et al. 2001), and the acyl chlorides formed can be transformed to carboxamides by amidation. Similarly, carboxamide nanotubes have been prepared using dicyclohexylcarbodiimide (DCC) as dehydrating agent and allowing the direct coupling of amines and carboxylic functions under mild, neutral conditions (Hirsch and Vostrowsky 2005).

Hamon et al. (2002) were the first to report the functionalization of oxidatively treated SWCNTs with alkyl amines and less nucleophilic aniline derivatives (Chen et al. 1988, 2001; Liu et al. 1998; Hirsch and Vostrowsky 2005; Tasis et al. 2006). The conversion of the acid functionality to the amide of octadecylamine led to the first shortened, soluble SWCNTs. The analysis of (end group and defect site) octadecylamido (ODA)-functionalized SWCNTs by solution phase mid-IR spectroscopy showed that the weight percentage of the acylamido functionality was about 50 % (Func CNT) (Chen et al. 1988, 2001; Liu et al. 1998; Marcoux et al. 2004; Hirsch and Vostrowsky 2005; Tasis et al. 2006). In the seminal work of Liu et al. (1998), it was demonstrated that the groups generated by the acid-cut nanotubes were carboxylates, which could be derivatized chemically by thiolalkylamines through amidation reaction. The resulting material could be visualized by AFM imaging after tethering gold nanoparticles to the thiol moieties. It was demonstrated that nanotube tips can be created by coupling basic or biomolecular probes to the carboxylic groups that are present at the open ends (Woolley et al. 2000). These modified nanotubes were used as AFM tips to titrate acids and bases, to image patterned samples based on molecular interactions, and to measure binding forces between single protein-ligand pairs 139c SWCNT tips. Images recorded on patterned SAMs enabled true molecular-resolution imaging (Tasis et al. 2006).

Chen et al. (2001) treated oxidized nanotubes with long-chain alkylamines via acylation and made for the first time the functionalized material soluble in organic solvents (Fig. 2.1). Further studies showed that 4-alkylanilines could also give soluble material, 140 whereas the presence of the long alkyl chain played a critical role in the solubilization process.

Esterification reactions resulted also in soluble functionalized nanotubes (Hamon et al. 2002). The photochemical behavior of soluble alkyl ester-modified nanotubes gave rise to measurable photocurrents after illuminating solutions of these tubes. By using time-resolved spectroscopy (laser flash photolysis), the

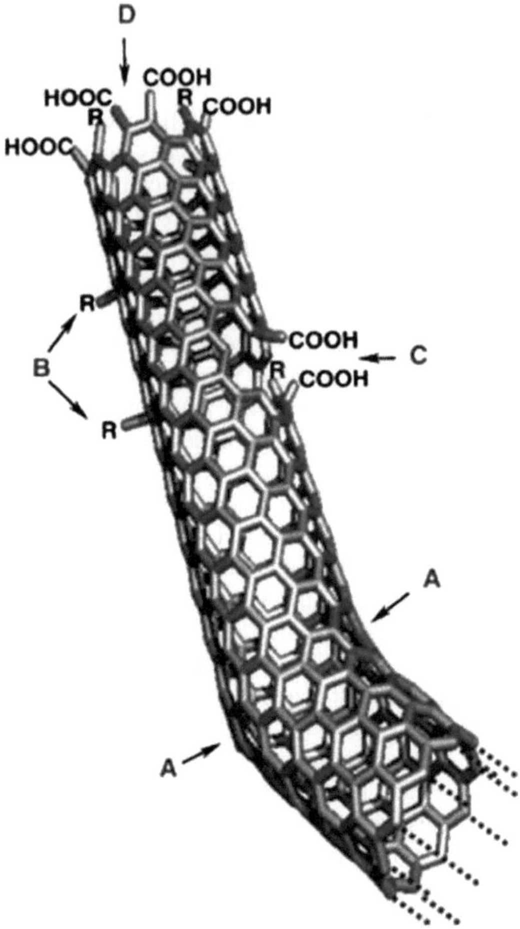

Fig. 2.1 Typical defects in a SWNT. (**a**) Five- or seven-membered rings in the carbon framework, instead of the normal six-membered ring, leads to a bend in the tube. (**b**) sp^3-hybridized defects (R=H and OH). (**c**) Carbon framework damaged by oxidative conditions, which leaves a hole lined with –COOH groups. (**d**) Open end of the SWNT, terminated with COOH groups. Besides carboxyl termini, the existence of which has been unambiguously demonstrated, other terminal groups such as $-NO_2$, –OH, –H, and = O are possible (Hirsch 2002) (www.entechopen.com)

transient spectrum of the charge separated state could be detected. Size and shape are very important issues in CNT chemistry. The change in shape from straight form to circles can have interesting implications in electronics (Liu et al. 1998; Martel et al. 1999; Woolley et al. 2000; Chen et al. 2001; Hamon et al. 2002; Marcoux et al. 2004; Tasis et al. 2006). The amidation or esterification of oxidized nanotubes has become one of the most popular ways of producing soluble materials either in organic solvents or in water.

2.1.3 *Covalent Bonding*

Chemical functionalization is based on the covalent bond of functional groups onto carbon form of CNTs. It can be performed at the end caps of nanotubes or at their sidewalls which have many defects. Direct covalent sidewall functionalization is associated with a change of hybridization from sp^2 to sp^3 and a simultaneous loss of *p*-conjugation system on graphene layer (Jeon et al. 2011). This process can be made by reaction with some molecules of a high chemical reactivity. In the first approach, fluorination of CNTs has become popular for initial investigation of the covalent functionalization because the CNTs sidewalls are expected to be inert (Prato et al. 2008; Jeon et al. 2011). The fluorinated CNTs have C–F bonds that are weaker than those in alkyl fluorides (Kelly et al. 1999; Prato et al. 2008; Jeon et al. 2011) and thus providing substitution sites for additional functionalization (Kelly et al. 1999; Inahara et al. 2002; Prato et al. 2008; Jeon et al. 2011). Successful replacements of the fluorine atoms by amino, alkyl, and hydroxyl groups have been achieved. Other methods, including cycloaddition, such as Diels–Alder reaction, carbene and nitrene addition, chlorination, bromination (Kelly et al. 1999; Inahara et al. 2002; Prato et al. 2008; Jeon et al. 2011), hydrogenation, and azomethine ylides (Kelly et al. 1999; Inahara et al. 2002; Unger et al. 2002; Tagmatarchis and Prato 2004; Prato et al. 2008; Jeon et al. 2011) have also been successfully employed.

Another method is defect functionalization of CNTs. These intrinsic defects are supplemented by oxidative damage to the nanotube framework by strong acids which leave holes functionalized with oxygenated functional groups (Chen et al. 1988; Kelly et al. 1999; Inahara et al. 2002; Unger et al. 2002; Tagmatarchis and Prato 2004; Prato et al. 2008; Jeon et al. 2011). In particular, treatment of CNTs with strong acid such as HNO_3, H_2SO_4, or a mixture of them (Chen et al. 1988; Esumi et al. 1996; Kelly et al. 1999; Inahara et al. 2002; Unger et al. 2002; Tagmatarchis and Prato 2004; Prato et al. 2008; Jeon et al. 2011) or with strong oxidants such as $KMnO_4$, ozone, and reactive plasma tend to open these tubes and to subsequently generate oxygenated functional groups such as carboxylic acid, ketone, alcohol, and ester groups that serve to tether many different types of chemical moieties onto the ends and defect sites of these tubes (Jeon et al. 2011). These functional groups have rich chemistry and the CNTs can be used as precursors for further chemical reactions, such as silanation, polymer grafting, esterification, thiolation, and even some biomolecules (Jeon et al. 2011). The CNTs functionalized by the covalent methods have the major advantage of exhibiting solubility in various organic solvents because the CNTs possess many functional groups such as polar or nonpolar groups (Jeon et al. 2011). For example (Fig. 2.1), the end caps of nanotubes tend to be composed of highly curved fullerene-like hemispheres, which are therefore highly reactive as compared with the side walls (Chen et al. 1988; Esumi et al. 1996; Kelly et al. 1999; Hirsch 2002; Inahara et al. 2002; Unger et al. 2002; Tagmatarchis and Prato 2004; Prato et al. 2008; Jeon et al. 2011). The sidewalls themselves contain defect sites such as pentagon–heptagon pairs called Stone–Wales defects, *sp3*-hybridized defects, and vacancies in the nanotube lattice (Chen et al. 1988; Esumi et al. 1996; Kelly et al. 1999; Hirsch 2002; Inahara et al. 2002; Unger et al. 2002; Tagmatarchis and Prato 2004; Prato et al. 2008; Jeon et al. 2011).

For graphene or graphene oxide (GO), several chemical procedures have been developed to afford dispersible graphene; in principle they are based on the exfoliation of graphite, chemical or thermal reduction of GO, intercalative expansion of graphite, chemical vapor deposition, and epitaxial growth (Georgakilas et al. 2012). Pristine graphene has been isolated by sonication of graphite in organic solvents. The dispersion of graphene in organic solvents helps in the functionalization of graphene by several functional groups (Georgakilas et al. 2012).

The main purpose is the dispersibility of graphene in common organic solvents that is usually obtained after attachment of certain organic groups. The dispersion of graphene sheets in organic solvents is a crucial move toward the formation of nanocomposite materials with graphene. In addition, organic functional groups such as chromophores offer new properties that could be combined with the properties of graphene such as conductivity. In most cases when organic molecules are covalently attached on the graphene surface, its extended aromatic character is perturbed, enabling the control of its electronic properties. The development of a band gap through chemical doping is a powerful method for the use of graphene in nanoelectronic devices (Niyogi et al. 2010; Georgakilas et al. 2012). The organic covalent functionalization reactions of graphene include two general routes: (a) the formation of covalent bonds between free radicals or dienophiles and C=C bonds of pristine graphene and (b) the formation of covalent bonds between organic functional groups and the oxygen groups of GO. Based on the previous experimental and theoretical experience with fullerene and carbon nanotubes, the most attractive organic species for the reaction with sp^2 carbons of graphene are organic free radicals and dienophiles. Usually both are intermediate reactive components that are produced under certain conditions in the presence of graphene (Georgakilas et al. 2012).

2.2 Characterization Techniques

2.2.1 SEM

Research development of electron microscopes began in the 1920s. Under the guidance of Max Knoll, Ernst Ruska began work on the development of electron lenses at the Technical University of Berlin, Germany, in 1928. His work was fundamental for the subsequent creation of an electron microscope in that these lenses were needed to channel the electrons of the beam. The first functional transmission electron microscope was developed in the early 1930s by Ruska who constructed a two stage electron microscope with three magnetic lenses, condenser, objective, and projector (Stadtländer 2007; Haguenau et al. 2003). Early electron microscopic studies were primarily focused on the study of the optical behavior of electron beams under various conditions. Thus, no biological applications were initially envisioned. However, it soon became clear that the superior magnifying power of an electron microscope could be applied to the study of the structure of various specimens, including those from plants, animals, microorganisms, and viruses (Stadtländer 2007; Haguenau et al. 2003). EM is considered today by many scientists as one of the most significant and useful developments for the ultrastructural investigation of specimens in the life sciences as well as in physics and material science (Haguenau et al. 2003). Because of the broad applications of the electron microscope to the fields of biology, medicine, and material science, Ruska received the prestigious Nobel Prize in Physics in 1986 for his fundamental work in electron optics and for the design of the first electron microscope (Stadtländer 2007).

The two basic types of electron microscopes are the scanning electron microscope and the transmission electron microscope.

Scanning electron microscopy (SEM) is a powerful method for the investigation of surface structures. This technique provides a large depth of field, which means, the area of the sample that can be viewed in focus at the same time is actually quite large. SEM has also the advantage that the range of magnification is relatively wide allowing the investigator to easily focus in on an area of interest on a specimen that was initially scanned at a lower magnification. Furthermore, the three-dimensional appearing images may be more appealing to the human eye than the two-dimensional images obtained with a transmission electron microscope. Therefore, an investigator may find it easier to interpret SEM images. Finally, the number of steps involved for preparing specimens for SEM investigation is lower and thus the entire process is less time consuming than the preparation of samples for investigation with a transmission electron microscope (Stadtländer 2007).

The first and basic part of the microscopes is the source of electrons. It is usually a V-shaped filament made of LaB6 or W (tungsten) that is wreathed with Wehnelt electrode (Wehnelt Cap). Due to negative potential of the electrode, the electrons are emitted from a small area of the filament (point source). A point source is important because it emits monochromatic electrons (with similar energy). The two usual types of electron guns are the conventional electron guns and the field emission guns (FEG) (Voutou and Stefanaki 2008).

When an electron beam interacts with the atoms in a sample, individual incident electron undergo two types of scattering—elastic and inelastic. In the former, only the trajectory changes and the kinetic energy and velocity remain constant. In the case of inelastic scattering, some incident electrons will actually collide with and displace electrons from their orbits (shells) around nuclei of atoms comprising the sample. This interaction places the atom in an excited (unstable) state. Specimen interaction is what makes electron microscopy possible. The interactions (inelastic) noted on the top side of the diagram are utilized when examining thick or bulk specimens (SEM) while on the bottom side are those examined in thin or foil specimens (transmission electron microscopy, TEM) (Voutou and Stefanaki 2008).

When a sample is bombarded with electrons, the strongest region of the electron energy spectrum is due to secondary electrons (Fig. 2.2). The secondary electron yield depends on many factors and is generally higher for high atomic number targets and at higher angles of incidence. Secondary electrons are produced when

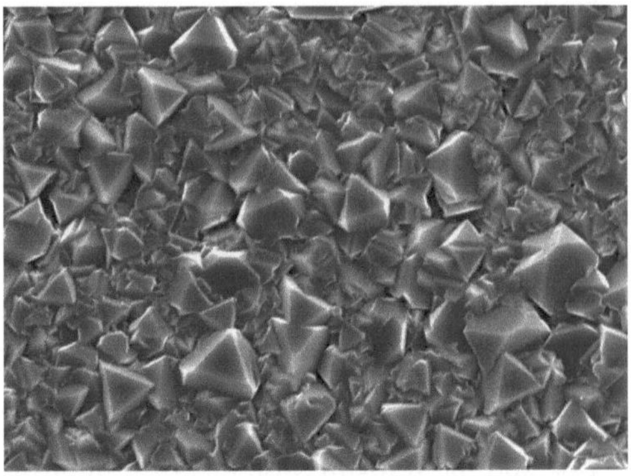

Fig. 2.2 Images of secondary electron (picture from Prof. Dr. Helder Ceragioli personal archive)

an incident electron excites an electron in the sample and loses most of its energy in the process. The excited electron moves toward the surface of the sample undergoing elastic and inelastic collisions until it reaches the surface, where it can escape if it still has sufficient energy (Voutou and Stefanaki 2008).

Backscattered electrons consist of high-energy electrons originating in the electron beam that are reflected or backscattered out of the specimen interaction volume. The production of backscattered electrons varies directly with the specimen's atomic number. This differing production rates cause higher atomic number elements to appear brighter than lower atomic number elements. This interaction is utilized to differentiate parts of the specimen that have different average atomic number (Voutou and Stefanaki 2008).

Auger electrons are electrons ejected by radiationless excitation of a target atom by the incident electron beam. When an electron from the L-shell drops to fill a vacancy formed by K-shell ionization, the resulting X-ray photon with energy EK–EL may not be emitted from the atom. If this photon strikes a lower energy electron (e.g., an M-shell electron), this outer electron may be ejected as a low energy Auger electron. Auger electrons are characteristic of the fine structure of the atom and have energies between 280 eV (carbon) and 2.1 keV (sulfur). By discriminating between Auger electrons of various energies, a chemical analysis of the specimen surface can be made. Auger electrons are exploited in Auger electron spectroscopy tools. The volume inside the specimen in which interactions occur while being struck with an electron beam is called specimen interaction volume.

2.2.2 FESEM

FESEM is the abbreviation of field emission scanning electron microscope. A FESEM is microscope that works with electrons (particles with a negative charge) instead of light. These electrons are liberated by a field emission source. The object is scanned by electrons according to a zigzag pattern. A FESEM is used to visualize very small topographic details on the surface or entire or fractioned objects.

Researchers in biology, chemistry, and physics apply this technique to observe structures that may be as small as 1 nm (= billion of a millimeter). The FESEM may be employed, for example, to study organelles and DNA material in cells, synthetical polymers, and coatings on microchips and nanomaterials (Yao and Kimura 2007).

Nanotechnology has strongly driven in the development of recent electron microscopy, with demands not only for increasing resolution but also for more information from the sample. Electron microscopes use a beam of highly energetic electrons to probe objects on a very fine scale (Pawley 1997; Wang 2000). In standard electron microscopes, electrons are mostly generated by "heating" a tungsten filament (electron gun). They are also produced by a crystal of LaB6 (define this). The use of LaB6 results in a higher electron density in the beam and a better resolution than that with the conventional device. In a field emission (FE) electron microscope, on the other hand, no heating but a so-called "cold" source is employed. Field emission is the emission of electrons from the surface of a conductor caused by a strong electric field. An extremely thin and sharp tungsten needle (tip diameter 10–100 nm) works as a cathode. The FE source reasonably combines with scanning electron microscopes (SEMs) whose development has been supported by advances in secondary electron detector technology. The acceleration voltage between cathode and anode is commonly in the order of magnitude of 0.5–30 kV, and the apparatus requires an extreme vacuum (~10^{-6} Pa) in the column of the microscope. Because the electron beam produced by the FE source is about 1,000 times smaller than that in a standard microscope with a thermal electron gun, the image quality will be markedly improved; for example, resolution is on the order of ~2 nm at 1 keV and ~1 nm at 15 keV. Therefore, the FESEM is a very useful tool for high-resolution surface imaging in the fields of nanomaterials science (Yao and Kimura 2007).

The electron beam is focused by the electromagnetic lenses (condenser lens, scan coils, stigmator coils, and objective lens) and the apertures in the column to a tiny sharp spot.

The image is formed when the primary probe bombards the object and secondary electrons are emitted from the object surface with a certain velocity that is determined by the levels and angles at the surface of the object. The secondary electrons, which are attracted by the Corona, strike the scintillator (fluorescing mirror) that produces photons. The location and intensity of illumination of the mirror vary depending on the properties of the secondary electrons. The signal produced by the scintillator is amplified and transduced to a video signal that is fed to a cathode ray tube in synchrony with the scan movement of the electron beam. The contrast in the "real time" image that appears on the screen reflects the structure on the surface of the object. Parallel to the analog image, a digital image is generated which can be further processed (Fig. 2.3).

Fig. 2.3 FESEM image of carbon nanoparticle (picture from Prof. Dr. Helder Ceragioli personal archive)

2.2.3 TEM

Transmission electron microscopy (TEM) is a technique where an electron beam interacts and passes through a specimen. The electrons are emitted by a source and are focused and magnified by a system of magnetic lenses. The electron beam is confined by the two condenser lenses which also control the brightness of the beam, passes the condenser aperture, and "hits" the sample surface. The electrons that are elastically scattered consist the transmitted beams, which pass through the objective lens. The objective lens forms the image display and the following apertures, the objective and selected area aperture are used to choose of the elastically scattered electrons that will form the image of the microscope. Finally, the beam goes to the magnifying system that is consisted of three lenses, the first and second intermediate lenses which control the magnification of the image and the projector lens. The formed image is shown either on a fluorescent screen or in monitor or both and is printed on a photographic film (Voutou and Stefanaki 2008).

The operation of TEM requires an ultrahigh vacuum and a high voltage. The first step is to find the electron beam, so the lights of the room must be turned off. Through a sequence of buttons and adjustments of focus and brightness of the beam, we can adjust the settings of the microscope so that by shifting the sample holder find the thin area of the sample. Then, tilting of the sample begins by rotating the holder (Voutou and Stefanaki 2008). This is a way to observe as much areas as we can, so we can obtain as much information as well. Different types of images are obtained in TEM using the apertures properly and the different types of electrons. As a result, diffraction patterns are shown because of the scattered electrons. If the unscattered beam is selected, we obtain the bright field image. Dark field images are attained if diffracted beams are selected by the objective aperture. Also in TEM, analysis is done with EDX (energy dispersive X-ray), EELS (electron energy loss spectrum), and EFTEM (energy-filtered transmission electron microscopy).

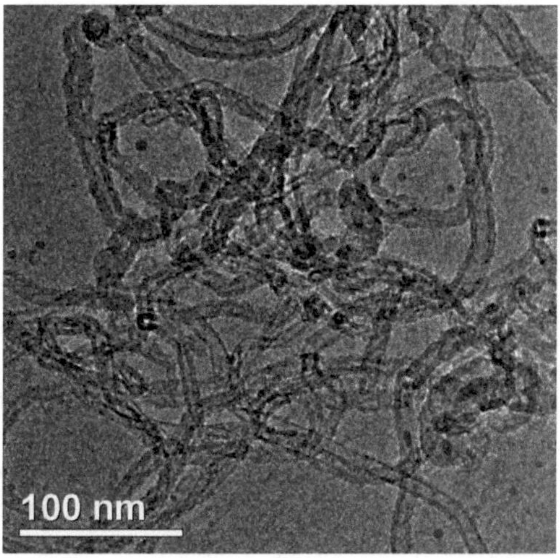

Fig. 2.4 Transmission electron microscopy image (picture from Prof. Dr. Helder Ceragioli personal archive)

In transmission microscopy, we can actually see the specimen's structure and its atomic columns; thus, compositional and crystallographic information is attained. However, expertise is needed and the sample preparation phase is too difficult so that very thin samples are achieved. The first step is to decide whether the sample is useful to be observed and in which view, plan or cross section. Due to the strong interaction between electrons and matter, the specimens have to be rather thin, less than 100 nm. This is achieved with several methods, depending on the material. The sample preparation is a precise and a severe procedure, which may affects the results of the microscopic analysis and study.

Transmission microscopy provides several types of images (Fig. 2.4) as reported above. The diffraction patterns show dots, regions, or circles originating from the sample area illuminated by the electron beam that depends on the material's structure. Monocrystals show distinguished dots in diffraction patterns, polycrystalline materials' common centered circles, and amorphous materials' diffused circles. Distortions and defects are visible in bright and dark field images, but expertise is needed in order to interpret whether they are defects or artifacts. Electron or ion beam damages must be considered in TEM analysis because of the sensibility of the sample and its really low thickness.

Additionally, there is always the problem of calibration and alignment of the instrument. Both of them require experience and skills, so the resulting images and data that emerge are reliable and free of objective astigmatism. These works have to be done in order to keep the instrument in excellent working condition. TEM provides accurate measurements and studies in different types of materials, given that observations are in atomic scale in HRTEM (define this). Transmission electron microscopes offer resolutions up to 0.1 nm at 300 kV and probe diameters up to 0.34 nm (Voutou and Stefanaki 2008).

2.2.4 Raman

Chandrasekhara Venkata Raman discovered the Raman effect; he began his scientific research into the scattering of light, the work that led to his discovery of the Raman effect in 1928, for which he was awarded the Nobel Prize in record time, just 2 years later.

The technique consists in an inelastic process that occurs when a sample is illuminated with a monochromatic light source, such as a laser beam. In this process, energy from the incident photons is transferred to the sample molecules, exciting them to high vibrational modes. Scattered photons have a lower frequency than the incident ones due to the energy transfer (Guimarães et al. 2006; Moreira et al. 2008).

The irradiation of a molecule with a monochromatic light always results in two types of light scattering, elastic and inelastic. In elastic scattering, there occurs no change in photon frequency or without any change in its wavelength and energy. Conversely, the other is inelastic scattering which is accompanied by the shift in photon frequency due to excitation or deactivation of molecular vibrations in which either the photon may lose some amount of energy or gains energy (Das and Agrawal 2011). Thus, three types of phenomena can occur (Das and Agrawal 2011; Fig. 2.5).

First, when light is incident on a molecule, it can interact with the molecule but the net exchange of energy (E) is zero, so the frequency of the scattered light is the same as that of the incident light ($E = E_o$). This process is known as Rayleigh scattering (Das and Agrawal 2011).

Second, the light can interact with the molecule and the net exchange of energy is the energy of one molecular vibration. If the interaction causes the light photon to gain vibrational energy from the molecule, then the frequency of the scattered light will be higher than that of the incident light, known as anti-Stokes Raman scattering (Das and Agrawal 2011).

Third, if the interaction causes the molecule to gain energy from the photon, then the frequency of the scattered light will be lower than that of the incident light ($E = E_o - E_v$); this process is known as Stokes Raman scattering. A Raman spectrometer is composed of light source, monochromator, sample holder and detector. The factors which affect the analysis on Raman spectra may include high signal-to-noise ratio, instrument stability, and sufficient resolution. The development of effective FT Raman spectrometers using NIR or red excitation lasers solved the problem of avoiding fluorescence that affects the Raman signals. On the other hand, the development of highly sensitive detectors in conjunction with coupling of optical fibers and microscopes enhanced the capacity of analysis (Parker 1994; Das and Agrawal 2011). Two major technologies are used to collect the Raman spectra, dispersive Raman spectroscopy and Fourier transform Raman spectroscopy, with difference in their laser sources and the way by which Raman scattering is detected and analyzed. Both these techniques have unique advantages

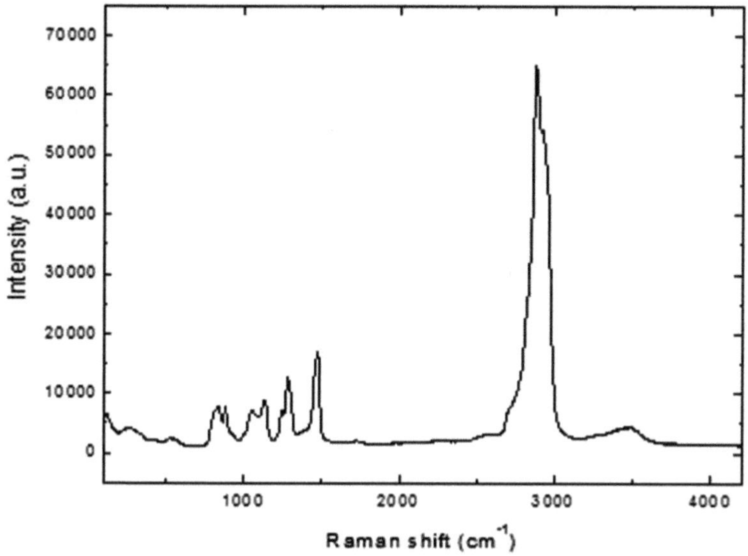

Fig. 2.5 Raman spectroscopy of poly(ethylene glycol) (picture from Prof. Dr. Helder Ceragioli personal archive)

and the method that best suit the sample should be preferred (Wang and McCreery 1989; Parker 1994; Das and Agrawal 2011).

Several type of lasers can be used as the excitation source, like argon ion (488.0 and 514.5 nm), krypton ion (530.9 and 647.1 nm), He:Ne (632.8 nm), Nd:YAG (1,064 and 532 nm), and diode laser (630 and 780 nm). Use of 1,064 nm near-IR (NIR) excitation laser causes lower fluorescent effect than visible wavelength lasers (Hanlon et al. 2000).

The Raman-scattered light can be collected by a spectrometer and displayed as a spectrum, in which its intensity is displayed as a function of its frequency change. Since each molecular species has its own unique set of molecular vibrations, the Raman spectrum of a particular species will consist of a series of peaks or bands, each shifted by one of the characteristic vibrational frequencies of that molecule. The spectrum is a plot of the scattered light intensity versus its change in frequency relative to that of the incident light (Hanlon et al. 2000). Raman frequency shifts are conventionally measured in wave numbers (cm^{-1}), a convenient unit for relating the change in vibrational energy of the scattering molecule to the change in frequency of the scattered light. 1 cm^{-1} equals 30,000 MHz; this is typically 10,000 times smaller than the frequency of the light itself. Each band of scattered light in Raman spectrum is characteristic of molecular vibrational motions, which, taken all together, are unique for PEG (Fig. 2.6). In other words, the molecular structure and composition of a material under study is encoded as a set of frequency shifts in the Raman-scattered light. Thus, the Raman spectrum can provide a fingerprint of a substance from which the molecular composition can be determined (Hanlon et al. 2000).

Although Raman spectroscopy had been widely used for chemical and molecular analysis for many years, its application to biomedical problems is relatively recent. Raman studies of biological tissue have been facilitated over the past 10 years by advancing technology, particularly in the areas of lasers and detectors. However, the real success of these studies at providing insight into the molecular basis of disease and showing the potential for medical diagnostic applications has relied on advanced methods of analyzing tissue Raman spectra. Raman spectroscopy is a nondestructive technique with a wide range of possible applications in the field of biomedical research (Hanlon et al. 2000).

2.2.5 FTIR

FTIR spectrometers have almost entirely replaced dispersive instruments because of their improved performance in nearly all respects. The application of this technique has improved the acquisition of IR spectra dramatically. The heart of the optical hardware in such FT spectrometers is the interferometer. The classic two-beam Michelson interferometer is shown schematically in and consists of two mutually perpendicular plane mirrors: a fixed mirror and a movable one. A semi-reflecting mirror, the *beam splitter*, bisects the planes of these two mirrors. A beam emitted by a source is split in two by the beam splitter (Bunaciu et al. 2010). The reflected part of the beam travels to the fixed mirror, is reflected there, and hits the beam splitter again. The same happens to the transmitted radiation. Because the two split beams are spatially coherent, they interfere on recombination. The beam, modulated by the movement of the mirror, leaves the interferometer and is finally focused on the detector. The signal actually registered by the detector, the interferogram, is, thus, the radiation intensity of the combined beams as a function of the position of the movable mirror. The mathematics of the conversion of an interferogram into a spectrum is the Fourier transformation. Based on fully developed software, a computer performs this transformation. The essential steps for obtaining an FTIR spectrum are to produce an interferogram with and without a sample in the beam and then to transform these interferograms into spectra of the source with sample absorption and the source without sample absorption (Bunaciu et al. 2010). The ratio of the former to the latter is the IR transmission spectrum of the sample. In the case of FTIR spectroscopy, the sample is usually placed between the interferometer and the detector. In transmission measurement, the source illuminates the sample, and the detector is placed behind the sample to acquire the fraction of light transmitted through the sample. Transmission analysis requires the sample to be partly transparent. In most cases, in the MIR range, samples must be diluted in nonabsorbing (or nonabsorbent?) matrix; otherwise, no light might be transmitted to the detector. Liquid can be prepared as a dilute solution in a cell. Solid samples are dispersed usually in a potassium bromide (KBr) disk or mull (Bunaciu

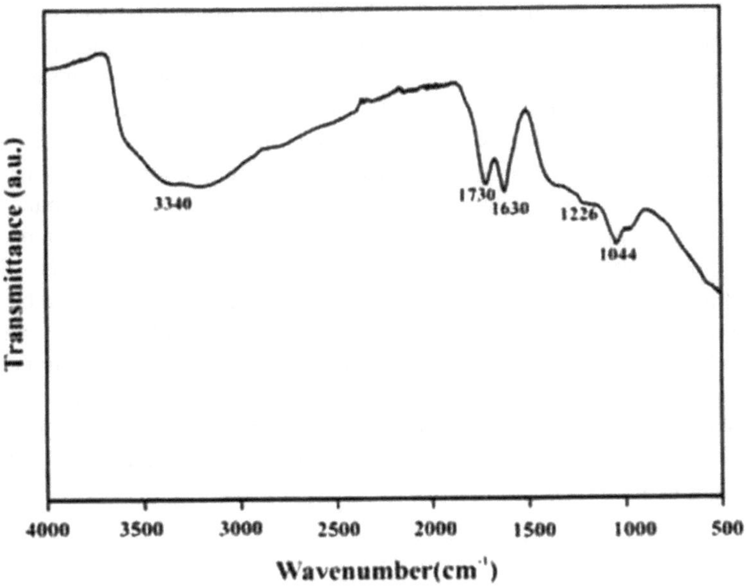

Fig. 2.6 FTIR spectrum of graphene oxide nanosheets (picture from Prof. Dr. Helder Ceragioli personal archive)

et al. 2010). Moreover, the powder particle size must be smaller than the radiation wavelength to avoid the Christiansen scattering effect, which appears as band distortion in the spectra (Bowie et al. 2002). Transmission has been extensively used to analyze thin samples such as films (Chan et al. 2006) or tissues (Krafft et al. 2006). It is not possible with thick samples such as tablets (Bunaciu et al. 2010). FTIR spectroscopy is a vibrational spectroscopic technique that can be used to optically probe the molecular changes associated with diseased tissues. The method is employed to find more conservative ways of analysis to measure characteristics within tumor tissue and cells that would allow accurate and precise assignment of the functional groups, bonding types, and molecular conformations. Spectral bands in vibrational spectra are molecule specific and provide direct information about the biochemical composition. FTIR peaks are relatively narrow and in many cases can be associated with the vibration of a particular chemical bond (or a single functional group) in the molecule (Bunaciu et al. 2010).

The characteristic FTIR spectrum of GO nanosheets (Rattanaa et al. 2012) is depicted in Fig. 2.6. It is seen with oxygen-containing groups in which the main absorption band at 3,340 cm^{-1} is assigned to the O–H group stretching vibrations. The absorption peak at 1,730 and 1,630 cm^{-1} can be assigned to C=O stretching of carboxylic and/or carbonyl moiety functional groups. The two absorption peaks at about 1,226 cm^{-1} and 1,044 cm^{-1} are assigned to the C–O stretching vibrations.

Although Raman spectroscopy and FTIR are relevant techniques, with their respective spectra complementary to one another, there are some differences between these two techniques. Probably the most important difference is the type of samples that can be investigated by each of these methods. FTIR mainly deals with nonaqueous samples, while Raman is as effective with aqueous samples as it is with nonaqueous ones (Movasaghi et al. 2008). This is because of the problem mostly taken place with FTIR spectroscopy; the problem is due to strong absorption bands of water. In Raman, however, fluorescence and the strong effect of glass (mostly as containers) are the most significant problems during analysis. Raman requires minimal sample preparation and can perform confocal imaging, whereas FTIR requires comparatively more sample preparation and does not have the ability of confocal imaging. Furthermore, the physical effect of infrared is created by absorption and mainly influences the dipole and ionic bands such as O–H, N–H, and C=O. Raman effect originates from scattering (emission of scattered light) and changing of the polarizability of covalent bands like C=C, C–S, S–S, and aromatics. In other words, FTIR spectroscopy is due to changes in dipole moment during molecular vibration, whereas Raman spectroscopy involves a change in polarizability (Movasaghi et al. 2008).

2.2.6 AFM

The atomic force microscope (AFM) was discovered by Binnig, Quate, and Gerber (Binnig et al. 1986) and has become an important tool for imaging surfaces. The principle of AFM is based on laser tracking the deflection of microscopic-sized probe (i.e., cantilever) through its nanometer tip interacting with sample surface. The system consists of the AFM probe with a sharp nanometer-size tip mounted on each soft cantilever with specific spring constant, the piezo-control (feedback loop) system that allows for monitoring the interaction forces between the tip and sample surface and controls the piezo scanner movements through digital/analog converter, the piezoelectric scanner that moves the tip on the surface of the sample in the x–y–z planes, segmented photodiode detector for measuring deflections of cantilever through laser, and analog/digital converter system for recording data. AFM probes consist of a microscopic-sized rectangular or/and "V"-shaped cantilevers, typically constructed from silicon or silicon nitride, with a sharp pyramidal tip (nanometer) or glass/polystyrene beads (micrometer). The property of the soft cantilever determines the sensitivity. The dimension of the tip determines the spatial resolution of the instrument, smaller in tip size and better resolution. The tips are typically made from Si_3N_4, or Si (Wu et al. 2012).

The AFM imaging can be obtained through contact and tapping (intermittent) mode. The cantilever in AFM probe is oscillated at its resonant frequency under an external electrical excitation, lightly "taps" on the sample surface, and contacting the surface at the bottom (z-axis) of each swing at each given x–y point. By maintaining constant oscillation amplitude, a constant tip–sample interaction is maintained and an image of the surface is obtained (Pereira 2001). The advantages of tapping mode are that it allows high-resolution scanning of samples that are prone to be easily damaged and could be used for freshly isolated cells that loosely held to a dish bottom. The disadvantages of tapping mode are that it does not offer a good image in liquids and slower scan speeds are needed for tapping mode operation (Wu et al. 2012).

2.3 Applications in Medical and Biological Areas

2.3.1 Tumor

The use of nanobiotechnology to reduce growth and to deliver substances directly to tumor cells is one of the most promising approaches to achieve highly efficient treatments against cancer. In this sense, nanoparticle drug delivery has been actively investigated in the recent years. The interaction between CNT with proteins and nucleic acids indicates that they may become adsorbed to the nanotube surface. Depending on the size of the molecules, they may be loaded to the interior cavity of open nanotubes as well.

Since the intracellular delivery of drugs is very complicated, CNTs may be used as carrier due to their capacity of penetrating the plasma membrane. As shown by Pantarotto et al. (2004a), functionalized carbon nanotubes accumulate in the cytoplasm and reach the nucleus in a nontoxic way up to a concentration of 10 μm. This was achieved by combining modified SWNTs with FITC (Fluorescein isothiocyanate). The process of cell penetration seems to be fast, reaching saturation points at 60 min.

The same authors have also provided data regarding plasmid DNA gene delivery (Pantarotto et al. 2004b) by the use of ammonium-functionalized CNTs. In this sense, CNTs were covalently modified and functionalized with a pyrrolidine ring. By doing this, the authors have obtained more soluble carbon nanotubes, particularly in aqueous solutions. The intracellular delivery was demonstrated in HeLa cells, observed by transmission electron microscopy. Multiwalled carbon nanotubes (MWCNT) were observed crossing the cell membrane as well as present in the cytoplasm, in contact with different organelles.

The use of nanoparticle drug delivery technologies may improve chemotherapeutics distribution along with the reduction of the dose. Different preclinical investigations have focused in two classes of chemotherapeutics, including doxorubicin, paclitaxel, daunorubicin, and docetaxel (Zhang et al. 2008).

Delivery of therapeutic genes has been studied and tested based on viral vectors and constitute a substantial hope to treat cancer, muscular dystrophy, and other genetic-based diseases. However, the use of viruses is a drawback that includes several biosecurity issues. In this way, the use of nonviral gene therapy-based either in antisense or plasmid DNA use is a groundbreaking application of nanotechnology. In this scenario, CNTs have demonstrated remarkable capability in delivery of both DNA and siRNA. Traditional nonviral gene deliver (e.g., Lipofectamine) substances present variable cytotoxicity, which is attributed to the destabilization of the cell membrane. Interestingly, Pantarotto et al. (2004b) observed that DNA carbon nanotube complexes do not increase cell proliferation or develop any toxic effects on cultured lymphocytes.

More recently, it has been shown that nanotubes may spontaneously insert the hydrophilic functional groups into a lipid bilayer so that they may enter cells without any membrane alterations or dysfunctions (Kam et al. 2004). Wang et al. (2008) have demonstrated that targeted RNA interference of cyclin A reduces the proliferation of K562 cells. The siRNA interference was carried out with SWCNTs and indicated a possibly useful therapeutic option for chronic myelogenous leukemia.

Overall, data indicates that the use of nanotechnology, especially the application of carbon nanotubes will contribute to new treatments that will be more efficient and targeted to cancer cells. In this way, therapies will be more efficient and result in less adverse side effects of chemotherapeutic drugs.

2.3.2 Cell Growth and Stem Cell Therapy

Cell culture is mostly based on substrate attachment that is achieved by electric charged surfaces and coating with different molecules. Cell adhesion is an important prerequisite for culture viability, protein expression, and multiplication. Also, certain cell types, such as neurons, are highly dependant on well-defined substrate so that axonal growth and synaptic formation is a function of an adequate growth permissive environment.

The use of CNTs may enhance cell adhesion, differentiation, migration, multiplication, and survival. In this sense, Mattson et al. (2000) have shown that MWCNTs support cultured primary hippocampal neurons and provide permissive substrate for axonal and neurite development and growth. The authors have shown that axons extended across the nanotubes coating, reaching the polyethyleneimine-treated glass surface. For this experiment, 20 nm diameter and 20–100 μm long particles were used and neurons attached to the nanotubes extended up to two neurites that grew to a distance range of 70–90 μm in 3 days. Interestingly, neurons survived and kept viable at least up to 8 days in vitro.

As mentioned before, carbon nanotubes have the remarkable feature of functionalization, by different chemical approaches. Modifications in the CNT structure, by the physical association of 4-hydroxynonenal (4-HNE), led to a remarkable increase in the number of neurites formed by hippocampal neurons, increasing from 1–2 to 4–6 projections. Also, much more elaborated networks were formed. In total, a threefold increase in the number of neuronal branches could be observed (Mattson et al. 2000).

In another work, Zhang et al. (2009) have associated different polysaccharides to single-walled carbon nanotubes in order to produce nanofibrous scaffolds. The results demonstrated that surface properties, regarded as a result of CNT coating,

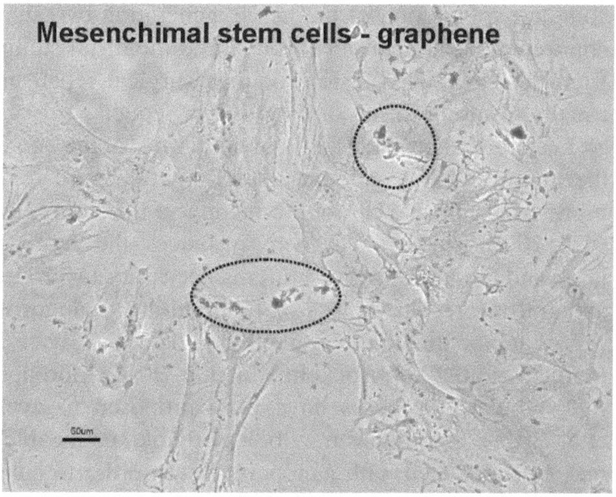

Fig. 2.7 Mesenchymal stem cells cultured in the presence of graphene (1 mg/mL). Observe the accumulation of nanoparticles inside the cells, which appear morphologically normal (picture from Prof. Dr. Alexandre L. R. Oliveira personal archive)

enhanced HeLa cells adhesion and proliferation. In our lab, we have also observed the beneficial effects of graphene and nanotubes of carbon on mesenchymal stem cells. As seen in Fig. 2.7, CNTs enter and accumulate inside the cultured cells without affecting cell proliferation and stability.

Additionally to the intrinsic properties of CNTs, the possibility of generating 2D and 3D organized substrates may further increase their usability in tissue regeneration. In this way, Zang and Yang (2011) have proposed a 3D scaffold produced with multiwalled carbon nanotube-coated polyethylene terephthalate (PET). The substrate was tested for cultivation of mouse embryonic stem cells and provided excellent cell viability and growth. It is possible that similar platforms may in turn facilitate stem cells culture and differentiation, providing new technologies for tissue engineering.

2.3.3 Peripheral Nerve Regeneration

The central nervous system, when compared to the peripheral counterpart, presents a higher regenerative capacity, although a complete recovery is rare (Ide 1996; Pereira Lopes et al. 2006).

The success of the peripheral nerve regeneration depends on the severity of the lesion. Seddon (1943) divided the nerve injuries in three categories according to the seriousness of the lesion: (1) Neuropraxia, which is the milder, when the axons are not ruptured and the continuity of the nerve is preserved. In this case, only a transitory loss of function is observed. (2) Axonotmesis, when the axons are sectioned but the epi-, peri-, and endoneurium are preserved. In this case, loss of function is more persistent, and depending on the proximity to the CNS, several months are necessary for complete recovery. (3) Neurotmesis, which represents the complete transection of the nerve, generating two stumps. The proximal stump is connected to the neuron cell body and the distal stump that remains disconnected from the CNS. This lesion requires surgical repair and the outcome is variable, usually resulting in sensory and motor sequels.

Lesions that produce a loss of continuity of the nerve can be restored by suturing the stumps, what is called neurorrhaphy (Guénard et al. 1992; Grecco et al. 2003). Fibrin glue can also be used to reconnect the stumps (Martins et al. 2005; Sandrini et al. 2007). However, in many instances, the distance between the stumps is too large and the use of nerve grafts is necessary (Moroder et al. 2011). Nevertheless, autografting results in loss of function at the donor nerve site, and usually such nerves are too thin to repair the lesioned nerves (Evans 2000; Nakamura et al. 2004; Pierucci et al. 2008; Bockelmann et al. 2011; Moroder et al. 2011). As an alternative, tubular prostheses have been proposed to overcome longer gaps (Cunha et al. 2011). Such tubes may be resorbable, permeable to cells and molecules, and may be associated with nanoparticles, in order to enhance the nerve regeneration process (Pereira Lopes et al. 2006). The tubulization technique offers orientation and protection to the regenerating axons and has advantages over autografting, including the preservation of the donor nerve area and the potential to modify the microenvironment inside the prosthesis in order to improve the regenerative process (Dubey et al. 1999; Torres et al. 2003).

The main materials used to construct tubular prostheses are poly(α-hydroxy acids), which are considered as the most promising group of bioresorbable polymers (Baraúna et al. 2007). They include the poly(glycolic acid) (PGA), poly (dioxanone) (PDD), poly(L-lactic acid) (PLLA), and the poly(ε-caprolactone) (PCL) (Rezwan et al. 2006; Sun et al. 2010).

Nanoscaled scaffolds have also been used to enhance neuronal growth (Maturana et al. 2011). Results from our laboratory indicate an improved axonal growth and accumulation of nanoparticles in the ECM compartment, as seen in Fig. 2.8.

The nanostructures of carbon are indeed one of the most promising for enhancing neuronal regeneration and include CNTs, graphene sheets, and carbon nanofibers (Armentano et al. 2010). In this sense, Liao et al. (2012) observed that PLLA/PCL membranes, when associated with CNTs, presented enhancement of the mechanical properties, which led to a better biodegradation, cell proliferation, and orientation. Graphene nanocomposites may also improve the resistance to traction as well as to improve the electric conductivity and the thermic stability (Zhu et al. 2010). These are particularly important for biomedical applications, especially for neuronal growth and regeneration, since neurons actively generate and transport action potentials (Kotov et al. 2009). Li et al. (2011) described that films of graphene present excellent biocompatibility for hippocampal primary neuron cultures, leading to the formation of dendrites and axons, especially during the early development stages.

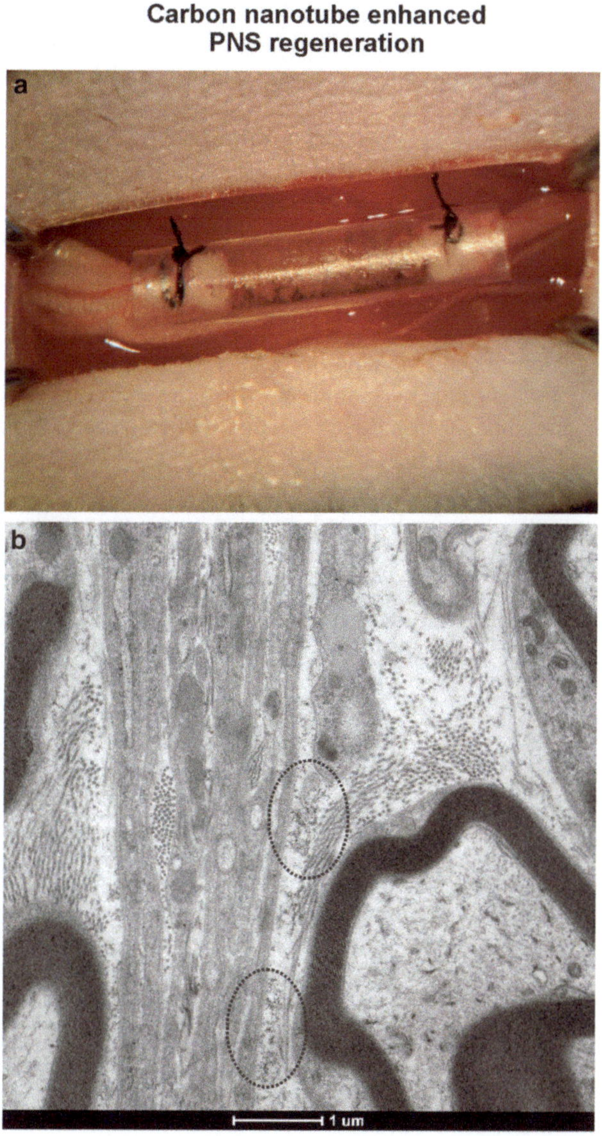

Fig. 2.8 (**a**) Tubulization of the sciatic nerve using polyethylene prosthesis. The gap between the stumps was filled with a saline solution containing CNTs. (**b**) CNT accumulation in between regenerated axons, 6 weeks following sciatic nerve tubulization (pictures from Prof. Dr. Alexandre L. R. Oliveira personal archive)

2.3.4 *Biosensor*

Modified carbon nanotubes and more recently graphene may be used as biosensors due to their unique electronic structure that can allow electrical detection of binding events. Several groups have proposed the use of SWNTs as consistent biosensors. One example is the work by Chen et al. (2012) that investigated carbohydrate-lectin interactions with both SWCNTs and chemically converted graphene. Such nanostructures were obtained in order to develop nanoscaled biosensors to detect several bacteria-derived lectins, what may in turn allow rapid identification of such microorganisms in water systems, soil, and human samples, helping the public health system.

DNA detectors were also proposed by Star et al. (2006). The authors have created nanotube network field-effect transistors that selectively detect tagged DNA sequences. Such approach may help to develop highly effective, low cost, and low complexity systems for molecular diagnostics.

Interestingly, even the enzymatic degradation of starch can be monitored electronically using single-walled carbon nanotubes as semiconductor probes (Star et al. 2004).

More recently, Valentini et al. (2013) have proposed the assembly of amperometric biosensors modified by the use of carbon nanomaterial–hemin nanocomposite for direct electrochemical detection of environmental pollutants, for example, nitrite and H_2O_2.

Overall, it is clear that nanobiotechnology is a fast-growing field that will improve the quality and efficiency of numerous devices, particularly those destinated for the treatment of human diseases and lesions.

Acknowledgement Support from FAPESP, CNPq, and Brazilian Network on Nanotoxicology (MCTI/CNPq) and NanoBioss (MCTI) are acknowledged.

References

Armentano I, Dottori M, Fortunati E et al (2010) Biodegradable polymer matrix nanocomposites for tissue engineering: a review. Polym Degrad Stab 95:2126–2146

Balasubramanian K, Burghard M (2005) Chemically functionalized carbon nanotubes. Small 1:180–192

Baraúna GS, Pierucci A, Oliveira ARL et al (2007) Estudo da degradação "in vivo" de poli (L-co-D, L-ácido láctico) aplicado como prótese para regeneração nervosa periférica. Rev Matéria 12:298–306

Binnig G, Quate CF, Gerber C (1986) Atomic force microscope. Phys Rev Lett 56:930–933

Bockelmann J, Klinkhammer K, Holst A et al (2011) Functionalization of electrospun poly (ε-caprolactone) fibers with the extracellular matrix-derived peptide GRGDS improves guidance of Schwann cell migration and axonal growth. Tissue Eng 17:475–486

Bowie BT, Chase DB, Lewis IR et al (2002) Anomalies and artifacts in Raman spectroscopy. In: Chalmers JC, Griffiths PR (eds) Handbook of vibrational spectroscopy, vol 3. Wiley, Chichester

Bunaciu AA, Aboul-Enein HY, Fleschin S (2010) Application of Fourier transform infrared spectrophotometry in pharmaceutical drugs analysis. Appl Spectrosc Rev 45:206–219

Chan KLA, Kazarian SG, Vassou D et al (2006) In situ highthroughput study of drug polymorphism under controlled temperature and humidity using FT-IR spectroscopic imaging. Vib Spectrosc 43:221–226

Chen J, Hamon MA, Hu H (1988) Solution properties of single-walled carbon nanotubes. Science 282:95–98

Chen J, Rao AM, Lyuksyutov S et al (2001) Dissolution of full-length single-walled carbon nanotubes. J Phys Chem 105:2525–2528

Chen Y, Vedala H, Kotchey GP et al (2012) Electronic detection of lectins using carbohydrate functionalized nanostructures: Graphene versus carbon nanotubes. ACS Nano 6:760–770

Cunha C, Panseri S, Antonini S (2011) Emerging nanotechnology approaches in tissue engineering for peripheral nerve regeneration. Nanomedicine 7:50–59

Das RS, Agrawal YK (2011) Raman spectroscopy: recent advancements, techniques and applications. Vib Spectrosc 57:163–176

Dubey N, Letourneau PC, Tranquillo RT (1999) Guided neurite elongation and Schwann cell invasion into magnetically aligned collagen in simulated peripheral nerve regeneration. Exp Neurol 158:338–350

Esumi K, Ishigami M, Nakajima A et al (1996) Chemical treatment of carbon nanotubes. Carbon 34:279–281

Evans GRD (2000) Challenges to nerve regeneration. Semin Surg Oncol 19:312–318

Fernando KAS, Lin Y, Sun YP (2004) High aqueous solubility of functionalized single-walled carbon nanotubes. Langmuir 20:4777–4778

Georgakilas V, Otyepka M, Bourlinos AB et al (2012) Functionalization of graphene: covalent and non-covalent approaches, derivatives and applications. Chem Rev 112:6156–6214

Grecco MAS, Leite VM, Albertoni WM et al (2003) Estudo da regeneração nervosa em nervos tibiais de ratos Wistar utilizando o Fluoro-Gold® como marcador neuronal. Acta Ortop Bras 11:225–229

Guénard V, Kleitman N, Morrisey TK et al (1992) Syngeneic Schwann cells derived from adult nerves seeded in semipermeable guidance channels enhance peripheral nerve regeneration. J Neurosci 12:3310–3320

Guimarães AE, Pacheco MTT, Silveira L Jr et al (2006) Near infrared Raman spectroscopy (NIRS): a technique for doping control. Spectroscopy 20:185–194

Haguenau F, Hawkes PW, Hutchison JL et al (2003) Key events in the history of electron microscopy. Microsc Microanal 9:96–138

Hamon MA, Chen J, Hu H et al (1999) Dissolution of single-walled carbon nanotubes. Adv Mater 11:834–840

Hamon MA, Hui H, Bhowmik P et al (2002) Ester-functionalized soluble single-walled carbon nanotubes. Appl Phys A 74:333–338

Hanlon EB, Manoharan R, Koo T-W et al (2000) Prospects for in vivo Raman spectroscopy. Phys Med Biol 45:R1–59

Hirsch A (2002) Functionalization of single-walled carbon nanotubes. Angew Chem Int Ed 41:1853–1859

Hirsch A, Vostrowsky O (2005) Functionalization of carbon nanotubes. Top Curr Chem 245:193–237

Ide C (1996) Peripheral nerve regeneration. Neurosci Res 25:101–121

Inahara J, Touhara H, Mizuno T et al (2002) Fluorination of cup-stacked carbon nanotubes, structure and properties. Fluorine Chem 114:181–188

Jeon Y, Chang DW, Kumar NA et al (2011) Functionalization of carbon nanotubes. In: Yellampalli S (ed) Carbon nanotubes—polymer nanocomposites, InTech. http://www.intechopen.com/books/carbon-nanotubes-polymernanocomposites/functionalization-of-carbon-nanotubes

Kam NWS, Jessop TC, Wender PA et al (2004) Nanotube molecular transporters: internalization of carbon nanotube-protein conjugates into Mammalian cells. J Am Chem Soc 126:6850

Katz E, Willner I (2004) Biomolecule-functionalized carbon nanotubes: applications in nanobioelectronics. Chem Phys Chem 5:1085–1104

Kelly KF, Chiang IW, Mickelson ET et al (1999) Insight into the mechanism of sidewall functionalization of single-walled nanotubes: an STM study. Chem Phys Lett 313:445–450

Kotov NA, Winter JO, Clements IP et al (2009) Nanomaterials for neural interfaces. Adv Mater 21:3970–4004

Krafft C, Shapoval L, Sobottka SB et al (2006) Identification of primary tumors of brain metastases by SIMCA classification of IR spectroscopic images. Biochim Biophys Acta Biomembr 1758:883–891

Li N, Zhang X, Song Q et al (2011) The promotion of neurite sprouting and outgrowth of mouse hippocampal cells in culture by graphene substrates. Biomaterials 32:9374–9382

Liao G-Y, Zhou X-P, Chen L et al (2012) Electrospun aligned PLLA/PCL/functionalised multiwalled carbon nanotube composite fibrous membranes and their bio/mechanical properties. Compos Sci Technol 72:248–255

Liu J, Rinzler AG, Dai H et al (1998) Fullerene pipes. Science 280:1253–1256

Marcoux PR, Hapiot P, Batail P et al (2004) Electrochemical functionalization of nanotube films: growth of aryl chains on single-walled carbon nanotubes. New J Chem 28:302–307

Martel R, Shea HR, Avouris P (1999) Rings of single-walled carbon nanotubes. Nature 398:299

Martins RS, Siqueira MG, Silva CF, Plese JPP (2005) Overall assessment of regeneration in peripheral nerve lesion repair using fibrin glue, suture, or a combination of the 2 techniques in a rat model. Which is the ideal choice? Surg Neurol 64:10–16

Mattson MP, Haddon RC, Rao AM (2000) Molecular functionalization of carbon nanotubes and use as substrates for neuronal growth. J Mol Neurosci 14:175–182

Maturana LG, Zanon RG, Pierucci A, et al (2011) Supraorganized collagen enhances Schwann cell reactivity and organization in vitro. Braz J Med Biol Res 44:682–687.

Moreira LM, Silveira L Jr, Santos FV et al (2008) Raman spectroscopy: a powerful technique for biochemical analysis and diagnosis. Spectroscopy 22:1–19

Moroder P, Runge MB, Wang H et al (2011) Material properties and electrical stimulation regimens of polycaprolactone fumarate–polypyrrole scaffolds as potential conductive nerve conduits. Acta Biomater 7:944–953

Movasaghi Z, Rehman S, Rehman I (2008) Fourier transform infrared (FTIR) spectroscopy of biological tissues. Appl Spectrosc Rev 43:134–179

Nakamura T, Inada Y, Fukuda S et al (2004) Experimental study on the regeneration of peripheral nerve gaps through a polyglycolic acid–collagen (PGA–collagen) tube. Brain Res 1027:18–29

Niyogi S, Bekyarova E, Itkis ME et al (2010) Spectroscopy of covalently functionalized graphene. Nano Lett 10:4061–4066

Pantarotto D, Briand J-P, Prato M, Bianco A (2004a) Translocation of bioactive peptides across cell membranes by carbon nanotubes. Chem Commun 1:16–17

Pantarotto D, Singh R, McCarthy D et al (2004b) Functionalized carbon nanotubes for plasmid DNA gene delivery. Angew Chem Int Ed 43:5242–5246

Parker SF (1994) A review of the theory of Fourier-transform Raman spectroscopy. Spectrochim Acta 50A:1841–1856

Pawley J (1997) The development of field-emission scanning electron microscopy for imaging biological surfaces. Scanning 19:324–336

Pereira Lopes FR, Campos LCM, Corrêa JD Jr et al (2006) Bone marrow stromal cells and resorbable collagen guidance tubes enhance sciatic nerve regeneration in mice. Exp Neurol 198:457–468

Pereira RS (2001) Atomic force microscopy as a novel pharmacological tool. Biochem Pharmacol 62:975–983

Pierucci A, Duek EAR, Oliveira ALR (2008) Peripheral nerve regeneration through biodegradable conduits prepared using solvent evaporation. Tissue Eng 14:595–606

Prato M, Kostarelos K, Bianco A (2008) Functionalized carbon nanotubes in drug design and discovery. Acc Chem Res 41:60–68

Rattanaa T, Chaiyakun S, Witit-anun N et al (2012) Preparation and characterization of graphene oxide nanosheets. Procedia Eng 32:759–764

Rezwan K, Chen QZ, Blaker JJ et al (2006) Biodegradable and bioactive porous polymer/inorganic composite scaffolds for bone tissue engineering. Biomaterials 27:3413–3431

Sandrini FAL, Pereira-Júnior ED, Gay-Escoda C (2007) Anastomose do nervo facial de coelhos com cola de fibrina: estudo da velocidade de condução nervosa. Rev Bras Otorrinolaringol 73:196–201

Seddon HJ (1943) Three types of nerve injury. Brain 66:237–288

Stadtländer CTK-H (2007) Scanning electron microscopy and transmission electron microscopy of mollicutes: challenges and opportunities. In: Méndez-Vilas A, Díaz J (eds) Modern research and educational topics in microscopy. Formatex, Badajoz

Star A, Joshi V, Han TR et al (2004) Electronic detection of the enzymatic degradation of starch. Org Lett 24:2089–2092

Star A, Tu E, Niemann J et al (2006) Label-free detection of DNA hybridization using carbon nanotube network field-effect transistors. PNAS 103:921–926

Sun M, Kingham PJ, Reid AJ et al (2010) In vitro and in vivo testing of novel ultrathin PCL and PCL/PLA blend films as peripheral nerve conduit. J Biomed Mater Res A 93(4):1470–1481

Tagmatarchis N, Prato M (2004) Functionalization of carbon nanotubes via 1,3-dipolar cycloadditions. J Mater Chem 14:437–439

Tasis D, Tagmatarchis N, Bianco A, Prato M (2006) Chemistry of carbon nanotubes. Chem Rev 106:1105–1136

Torres MFP, Graça DL, Farias ELP (2003) Reparação microcirúrgica de nervo periférico por meio de sutura, cola de fibrina ou bainha de BioFill® em ratos Wistar. Arq Bras Med Vet Zootec 5:557–561

Unger E, Graham A, Kreupl F et al (2002) Electrochemical functionalization of multi-walled carbon nanotubes for solvation and purification. Curr Appl Phys 2:107–111

Valentini F, Carbone M, Palleschi G (2013) Carbon nanostructured materials for applications in nano-medicine, cultural heritage, and electrochemical biosensors. Anal Bioanal Chem 405:451–465

Voutou B, Stefanaki E-C (2008) Electron microscopy: the basics. physics of advanced materials winter school. Aristotle University of Thessaloniki, Greece

Wang X, Ren J, Qu X (2008) Targeted RNA interference of cyclin A2 mediated by functionalized single-walled carbon nanotubes induces proliferation arrest and apoptosis in chronic myelogenous leukemia K562 cells. ChemMedChem 3:940–945

Wang Y, McCreery RL (1989) Evaluation of a diode laser/charge coupled device spectrometer for near-infrared Raman spectroscopy. Anal Chem 61:2647–2651

Wang ZL (2000) Transmission electron microscopy of shape-controlled nanocrystals and their assemblies. J Phys Chem B 104:1153–1175

Woolley AT, Cheung CL, Hafner JH et al (2000) Structural biology with carbon nanotube AFM probes. Chem Biol 7:R193–R204

Wu X, Sun Z, Meininger GA, Muthuchamy M (2012) Application of atomic force microscopy measurements on cardiovascular cells. Methods Mol Biol 843:229–243

Yao H, Kimura K (2007) Field emission scanning electron microscopy for structural characterization of 3D gold nanoparticle superlattices. In: Méndez-Vilas A, Díaz J (eds) Modern research and educational topics in microscopy. Formatex, Badajoz

Zang X, Yang S-T (2011) High-throughput 3-D cell-based proliferation and cytotoxicity assays for drug screening and bioprocess development. J Biotechnol 151:186–193

Zhang J, Zou H, Qing Q et al (2003) Effect of chemical oxidation on the structure of single-walled carbon nanotubes. J Phys Chem B 107:3712–3718

Zhang L, Gu FX, Chan JM et al (2008) Nanoparticles in medicine: therapeutic applications and developments. Clin Pharmacol Ther 83:761–769

Zhang X, Meng L, Lu Q (2009) Cell behaviors on polysaccharide-wrapped single-wall carbon nanotubes: a quantitative study of the surface properties of biomimetic nanofibrous scaffolds. ACS Nano 3:3200–3206

Zhu Y, Murali S, Cai W et al (2010) Graphene and graphene oxide: synthesis, properties, and applications. Adv Mater 22:3906–3924

Chapter 3
Nanostability

Ana Cauerhff, Yanina N. Martinez, German A. Islan, and Guillermo R. Castro

Abstract Stability of nano-objects should be considered in both ways such as static and dynamic, both directly related to their applications and uses. It can be determined by their physicochemical properties and by the interactions and their effects in a cell, tissue, organ, and/or physiology and structure of an entire organism. Nanotoxicity of a single nanospecies involves not only the interaction of that species but also the interaction of its nanomaterial components with biological systems. In order to analyze the whole picture, studies of nanotoxicity in vitro and in vivo related with the stability are a must. After the physicochemical characteristics of the nano-object, the first relevant factor dictating cellular fate is the size and the second is the surface charge. Other factors as the presence of "bioactive" molecules (e.g., surfactants), surface area, shape, chemistry, crystalline structure, aspect ratio, dimensionality, agglomeration, concentration, and dose are discussed in the present work. The size, structure of nanoparticles (NPs) and their surface characteristics are relevant for protein and other molecules interactions. The binding of NPs with serum proteins influences their biodisponibility, clearance, and toxicity. In addition to traditional methods as optical spectroscopic techniques to perform structural studies, the need of better mechanistic understanding of toxicological pathways is urgent. In this sense, genomics, transcriptomics, proteomics, and metabolomics generically called omics technologies provide a valuable opportunity to refine existing methods and provide information for the so-called integrated testing strategies to study the pathways of toxicity (PoT). Therefore, static and dynamic stability will be taken into account in the future toxicological studies in nanobiotechnology.

G.R. Castro (✉)
Laboratory of Nanobiomaterials, Institute of Applied Biotechnology (CINDEFI), Universidad Nacional de La Plata—CONICET (CCT La Plata), La Plata, Argentina
e-mail: grcastro@gmail.com

N. Durán et al. (eds.), *Nanotoxicology*, Nanomedicine and Nanotoxicology, DOI 10.1007/978-1-4614-8993-1_3,

Abbreviations

AFM	Atomic force microscope
BSA	Bovine serum albumin
CNTs	Carbon nanotubes
CT	Computed tomography
DLS	Dynamic light scattering
ELISA	Enzyme-linked immunosorbent assay
FTIR	Fourier transform infrared spectroscopy
HPMA	*N*-(2-Hydroxypropyl)methacrylamide
HTS	High-throughput screening
ITS	Integrated testing strategies
Iv	Intravenous
MAQC	MicroArray Quality Control
METLIN	Metabolite and tandem MS database
MS	Mass spectrometry
NMR	Nuclear magnetic resonance
NP	Nanoparticle
PAMAM	Polyamidoamine
PEG	Poly(ethylene) glycol
PoT	Pathways of toxicity
QDs	Quantum dots
QELS	Quasi-elastic light scattering
QIVIVE	Quantitative in vitro to in vivo extrapolation
REACH	Registration, Evaluation, Authorization, and Restriction of Chemical substances
RES	Reticulo-endothelial system
SAXS	Small-angle X-ray scattering
sc	Subcutaneous
SEM	Scanning electron microscopies
SLS	Static light scattering
SoT	Signatures of toxicity
SPM	Scanning probe microscope
SPR	Surface plasmon resonance
STM	Scanning tunneling microscope
TEM	Transmission electron microscopy
TOF	Time of flight
XRD	X-ray diffraction
XRR	X-ray reflectometry

3.1 Introduction

Nanotoxicity can be approached from many ways, but considerations related to chemical composition, physicochemical properties and biological activities, and environmental conditions of synthesis, preservation, and degradation, among others, are critical factors to establish the nano-object stability. Additionally, stability of nano-objects should be considered in both ways such as static and dynamic, both directly related to their applications and uses. Stability can be considered as a relationship between the stable or pseudo-stable and non-stable states of the nanospecies in the entity under consideration plus the history of the object and the environmental limits defined by thermodynamic considerations. Therefore, the toxicity of nano-objects is a complex phenomenon depending on their stability and degradation subproducts. This approach comprises a lot of critical factors that must be evaluated in vitro and in vivo models.

Nano-object stability can be determined by their physicochemical properties and by the interactions and their effects in a cell, tissue, organ, and/or physiology and structure of an entire organism. In this sense, time is a crucial parameter that should be taken into account when the traffic and/or the pathway of a nano-object could follow interfering in one or several metabolic pathways or cell structure functions. Static stability of nano-objects can be defined as an intrinsic stability, being a property inherent to the structural and physicochemical aspects of the material from which the object is composed. Also, static stability is related to the environmental conditions such as pH, temperature, ionic strength, as well as complex cell properties like permeability, membrane composition (e.g., interaction with receptors), and binding with circulating molecules (e.g., proteins) which can affect the immune system and produce of toxic compounds, among others.

Dynamic stability of nanospecies comprises a sum of complex interactions with the whole metabolism of an entire organism or tissue or cell producing unpredictable effects that can appear over the time. Those effects can occur after a period of time not defined and/or long lasting after a while. Moreover, stability means that a nanospecies displays resistance to change, to deterioration, and/or to displacement and constant properties, namely, reliability. Besides the intrinsic properties of the nanomaterial type, there are other aspects to take into account: potentially toxic waste products due to a fault in the production of nano-objects or due to poor purification to remove the by-products which are formed inevitably in their production (e.g., residual debris from laser cutting), or due to wear or abrasion of metallic implants, and NPs might be unintentionally released into the body and are known to cause some adverse effects. The properties of these unintentionally released objects have not been qualified/quantified up to now and object-size distribution is expected to be relatively abroad (Hahn et al. 2012). For example, the lifetime and dissolution behavior of implanted NPs might be different depending on the type of NP and the local environment in a body. Many of studies dealing with the cytotoxicity of NPs were reported in the literature, but statements on NP toxicity should be viewed carefully due to the diversity of nano-objects and the complexity of the mechanisms determining the interactions at the bio–nano interface (Nel et al. 2009). Generally, methodological and technical issues may pose a serious problem in studies investigating various nano-products, especially concerning purity and potential contamination of laboratory chemicals and compounds.

In order to analyze the whole picture, studies of nanotoxicity in vitro and in vivo are required. In this regard, the behavior of the nanoparticles in vivo and in vitro must be examined in relation with the stability. In vivo, NPs seem to be continuously transported from organ to organ, whereas, in vitro, the NPs are located within a limited space. Therefore, the differential toxicity between in vivo and in vitro assays at least may be due to exposure concentration and time (kinetic). Many questions remain unanswered regarding the NPs behavior in vivo. For example, how does protein opsonization impact the kinetics of NPs? How are NPs metabolized? What is the long-term fate of NPs? How do the physicochemical properties of nanomaterials affect their biodistribution behavior in vivo? How degradable are the nano-objects? How long nano-objects will take to be degraded? Are the degradation by-products toxic? Where are the main metabolic cells/tissues/organs involved in the detoxification processes? Addressing these questions will be important because the only way to improve the design of biomaterials is to fully elucidate their effects, transformations, and final fate within biological systems (Albanese et al. 2012).

The methods that comprise in vivo and in vitro nanotoxicity must reliably predict and assess the possible spectra of effects, from benefits to potential risks, and health hazards associated with exposure to nanomaterials as they become more widespread, pervasive agents in manufacturing and medicine. Such an assessment regime would best consider traditional pharmacology and toxicology approaches to dose–response, acute, and chronic exposure, as they can correlate to rigorous physicochemical characterization (Warheit 2008) and various levels of physiological reactivity (e.g., molecular, cellular, tissue, organ toxicity) in making risk–benefit analyses of these materials both in various manufactured forms and as incorporated within new technologies (Nel et al. 2006).

Nanotoxicity of a single nanospecies involves not only the interaction of that species but also the interaction of its nanomaterial components with biological systems; in other words, nanotoxicity is related to the complex nano–bio interactions. Then, by analyzing many NPs systematically, it can determine if a single or combination of NP parameters is responsible for a specific biological response. In this regard, a deep observation of behavior of particles present in nature as viruses can help us to understanding the relevance of shape, size, geometry (e.g., symmetry), and topology in the stability and toxicity. By identifying the influence of size, shape, and chemistry in the delivery process, it is possible to redesign the NPs in order to reduce the spread of debris inside the body and accumulate maximally in one tissue reducing the toxic effects in the whole body. The fundamental studies on nano–bio interactions will enable the parameterization of the nano-engineering process by creating specific design normatives and legislation.

In the early years of nanotechnology, various materials have been used to create NPs of various shapes and sizes and even with different symmetries. However, the relevance of nanotoxicity studies was understood much later. For example, noble metal NPs were initially considered as benign carriers because of the well-known bulk properties, but multiple studies have demonstrated their interaction with serum proteins and cell membrane receptors. The NP interaction is essentially determined by the NP design and properties at quantum level, which are in effect influencing cell uptake, gene expression, and toxicity. Also, the manufacturing and consumer utilization then produces multiple different sources of release of these materials into the environment, ecosystem, water and food supplies, and other alternative routes of nonvoluntary entry into the human body. While nanomaterials are very attractive for scientific and technological reasons, increasing human exposure to nanomaterials together with the distinctive properties of these materials all mandate development and validation of accurate nano-device and materials characterization protocols and predictive toxicity and hazard capabilities.

In addition to a nanomaterial's size, shape, and ligand density, surface charge is also important in dictating cellular fate. It has been suggested that the cell membrane possesses a slight negative charge and cell uptake could be driven by electrostatic attractions (Jin et al. 2009; Wang et al. 2010). In such cases, NPs can interact with the cell surface membrane in multiple scenarios. Interactions between NP-bound ligands and cellular receptors depend on the engineered geometry and the ligand density of a nanomaterial.

As it has been mentioned above, NP-based delivery systems improve in vitro and in vivo stability and reduce unwanted side effects. However, NPs are often first picked up by the phagocytic cells of the immune system (e.g., macrophages). Besides, there may be undesirable interactions between NPs and the immune system, such as immunostimulation or immunosuppression, which may promote inflammatory or autoimmune disorders or increase the host's susceptibility to infections and cancer (Zolnik et al. 2010).

By examining one nanostructure parameter at the time, researchers are able to draw correlations between the nanoparticle design and a specific biological response. However, because of the parameters interdependence on properties, a multi-parametric analysis should be required (e.g., Plackett–Burman design for a primary screening of sensitive parameters). The collected data can then be used to populate databases to develop predictive models and software on how NPs behave in that particularly biological system. Such data could guide the engineering of the next generation of nanoscale devices. The next generations of nanostructures design will consist in engineering nanostructures that will be capable of navigating in the body, infecting and transforming cells, with or without undesirable side effects. For this reason, it is required to develop systematic models to predict how the physicochemical properties of a synthetic nanoparticle interact with biological systems, to evaluate its stability, and to predict future toxic effects.

3.2 Static and Dynamic Stability

The toxicity of nano-objects and nanomaterials involves a complex set of phenomena comprising, on the one hand, their physicochemical properties that will dictate their static (or intrinsic) stability. The stability of the nanomaterials can be measured by studying the physical and/or chemical properties and/or the interactions and/or the effect of such nano-objects that can be produced in a cell, tissue, organ, and/or an isolated structure of a whole organism. Furthermore, additional and unexpected interactions of the nano-objects with proteins, ligands, and toxic effects of the products of their metabolism can be evaluated by in vivo methods. In this sense the static stability can be studied by in vitro methods, and dynamic stability can be studied by an in vivo approach.

3.2.1 *In Vitro Stability*

The question of whether or not these nanomaterial properties and/or effects such as intrinsic colloidal instability and aggregation phenomena in aqueous milieu, bioaccumulation in the environment or tissues, and contaminant adsorption and transport would permeate, become persistent in, and influence biological systems remains to be determined. Some tools and methods exist that can adequately track nanomaterial properties and reactivity in biological or physiological systems. Importantly, these methods are primarily in vitro assays to date and, as for many biocompatibility tests conducted for related macrophase materials, could have little, if any, correlation or validation to in vivo materials tolerance (Jones and Grainger 2009). Some broadly used methods are described in the literature to assess nanomaterial surface and bulk properties and biological reactivity in models of in vitro systems. These assays are important to characterizing nanomaterial applications in biotechnology, ecosystems, agri- and aquaculture, biomedical applications, and toxicity screening. Such methods comprise prebiological materials characterization: surface contamination, particle sizing and aggregation, in vitro biological testing (cell types, selection, and use), and cell-based in vitro toxicity assays, among others.

In this regard, the evaluation of in vitro stability can help to predict the behavior in vivo, since the propensity to aggregation of an NP alone or interacting with some molecular component of the organism can activate the immune system or produce obstruction in circulatory system (e.g., thrombosis).

In addition to changes in size, shape, surface area, surface chemistry, and NPs charge, cell suspensions variables and how the particle behaves in this complex environment of proteins, amino acids, and salts are relevant to be considered. In this sense, the composition of serum/media can influence the cytotoxicity of nano-objects. Protein absorption to NPs surface is expected because of the high surface reactivity of NPs, and then this phenomenon may mask NPs from cells (Casey et al. 2007a). For example, carbon nanotubes (CNTs) in the presence of albumin showed differences in dispersion and cytotoxicity compared with controls without the protein (Elgrabli 2007). The presence of other component of media as phenol red also influences cytotoxicity evaluation as the interaction with NPs may influence absorbance during spectroscopic analysis (Casey et al. 2007b). Results from these studies indicate that proper controls of all variables involved are essential to understand fully the cell–NP interactions. Also, mimicking the in vivo environment is crucial to obtaining representative results that may later be applied on living tissues and organs. Finally, the appropriate tests must be performed to measure accurately the cellular response under question taking into account the static and dynamic stability of nano-objects. However, it is difficult to control the individual variables associated with in vivo nanotoxicological studies due to their inherently complex properties (Schulze et al. 2008).

Other relevant aspect is the interaction between the NPs. Many NPs are colloidal and/or show poor water solubility in the presence of attractive forces, e.g., van der Waals; those nano-objects can aggregate, and consequently drastic changes on their properties can be observed. The nano-objects aggregation process has to be checked and controlled considering the potential use. Additionally, in order to prevent NPs aggregation, stabilizing molecules commonly named capping agents are required to reduce agglomeration and increase aqueous dispersion. Capping agents are generally small molecules or polymers (Lowry et al. 2012). However, stabilizing agents like CTAB currently used to synthesize and stabilize gold nanorods are strong detergents known to greatly increase nanorods cytotoxicity (Takahashi et al. 2006). Therefore, the choice of dispersive agents in NP solutions must be considered carefully and proper studies for comparing toxicity in the presence of the capping molecules with and without NPs should be run.

Standardization of NP concentration remains a difficult task; currently, the most common expression of concentration in the literature is mass to volume ratio. Comparison between studies is difficult because variations in particle, size, shape, and surface chemistry produce nonuniform concentrations. In this sense, the publication of size, shape, and concentration in terms of both mass to volume and number of particles per milliliter is suggested (Gormley and Ghandehari 2009). Another significant variable that influences the stability of nano-objects is the incubation time. That variable differs greatly from study to study. The majority of studies only investigate acute toxicity (24 h), but longer incubation times are required to see the influence of NPs on cell cycle and proliferation. The choice of cell lines is an important decision for in vitro experiments design in order to achieve reproducible and representative results that may later be applied in vivo. For example, if a certain NP is expected to be released in aerosol form, then the alveolar epithelial and macrophage cell lines should be chosen to get the best cell model to assess the potential toxicity when inhaled (Gormley and Ghandehari 2009). The most common cell lines and assays used to study nano-objects in vitro are HeLa, macrophages, epithelial cells, fibroblasts, and liver and kidney cells, among others. However, the response of the interaction of nano-objects with perpetual cell lines should be analyzed in details with precaution because of drastic changes of genetic profiles and metabolic pathways with the cell generations.

3.2.2 *In Vivo Stability*

The risk of nanomaterials and/or nanospecies can be evaluated through several different mechanisms in the body. Research involving the use of different types and designs of nanomaterials continues to evolve with the growth of nanotechnology for in vivo applications in such fields as drug delivery, medical imaging, diagnostics, and engineering technology (Aillon et al. 2009).

A summary of the major forms of in vivo nanomaterial toxicity is displayed on a recent review (Aillon et al. 2009), being the most relevant are pulmonary, hematological, splenic, hepatic, and nephritic toxicity, and the main molecular mechanism of in vivo nanotoxicity is the induction of oxidative stress by free radical formation (Soenen et al. 2011).

The majority of nanotoxicity studies have focused on health effects of exposure to ultrafine (unintentionally produced) particles by inhalation, contact through skin, or ingestion (Jones and Grainger 2009). These studies primarily focus on local effects (e.g., lung toxicity after particle inhalation). Characterization of in vivo toxicity has been a daunting task as nanomaterials are quite complex and conflicting studies have led to different views of their use and safety. This makes it difficult to evaluate, generalize, and predict important aspects of toxicity (Warheit 2008). Nanosized particles can enter and penetrate some organs such as the lungs, intestine, and skin. The organs as brain, heart, liver, kidney, spleen, lungs, and lymphatic systems should not be penetrated by NPs. Some can penetrate into the deepest layers of the skin (dermis). Their penetration depends on their size and NP surface features. It must be noted that in vitro tests must be carried out on NP toxicity before in vivo tests are performed.

Acute toxicity resulting from NPs has been investigated in the mouths of rats. The results indicate that toxicity depends on the size, coating, and chemical component of the NPs. Also, the systemic effects of NPs have been shown in different organs and tissues. The effects on inflammatory and immunological systems may include oxidative stress or pre-inflammatory cytotoxin activity in the lungs, liver, heart, and brain. The effects on the circulatory system can include pre-thrombosis effects and paradox effects on heart function. Genotoxicity, carcinogenicity, and teratogenicity may occur as a result of the effects of NPs. Some NPs could pass the blood–brain barrier and cause brain toxicity (Ai et al. 2011).

In the case of epithelial transport of nano-constructs in drug delivery applications such as skin, overcoming the robust barrier may be difficult. Factors that influence penetration include animal model choice, skin structure such as presence of hair follicles, as well as NPs physicochemical properties (Ryman-Rasmussen et al. 2007). It was found that PEGylated quantum dots (QDs) of varying size, shape, and surface chemistry have been applied to porcine skin penetration and toxicity. Diverse studies indicate that skin penetration, toxicity, and inflammation may be influenced by properties such as size, shape, and charge.

Translocation across the GI tract epithelial cell barriers may also be difficult for nanosized constructs due to the tight gap junctions, although direct evidence both in vitro and in vivo indicates that uptake of these particles is possible (Kitchens et al. 2005). Transport of nanoparticles may occur by para-cellular pathways, by trans-cellular transport (endocytosis), or through Peyer patches (Norris et al. 1998) and all these mechanisms are influenced by the physicochemical properties of the nano-objects, so does the transport through intestinal mucus layer and transcytosis across the intestine (Hussain et al. 1997); in this sense, the more important factors are size and charge surface of nano-objects.

The inhalation of particulate matter is known to cause a variety of respiratory diseases including asthma and cancer. Besides, several factors such as physico-chemical properties have to be considered when studying the toxicity of inhaled nano-constructs, including deposition, transport, and clearance (Borm and Kreyling 2004). Anionic dendrimers are transported by caveolae-mediated endocytosis in lung epithelial cells, while cationic and neutral dendrimers are transported by clathrin- and caveolae-independent mechanisms (Perumal et al. 2008).

Upon contact with blood or other biological fluids, nano-objects may absorb proteins that will coat their surface. This phenomenon changes the stability of all the nanomaterials and represents one of the greatest challenges in the study of the in vivo nanomaterials behavior. This item will be described most deeply in the next items.

Slow clearance and tissue accumulation (storage) of potential free radical-producing nanomaterials as well as prevalence of numerous phagocytic cells in the organs of the endothelial reticulum system (ERS) make organs such as the liver and spleen main targets of oxidative stress. Additionally, organs of high blood flow that are exposed to nanomaterials, such as the kidneys and lungs, can also be affected.

Cell culture studies account for the majority of all nanotoxicological research; however, data obtained from in vitro experiments could be misleading for a variety of reasons and will therefore require verification from in vivo (animal) experiments (Fischer and Chan 2007).

3.3 Physicochemical Properties of Nano-objects

Nanotoxicology is a branch of bionanoscience which deals with the study and application of the toxicity of nanomaterials. Nanomaterials, even when made of inert elements such as gold, become highly active at nanometer dimensions. NPs play a remarkable role in toxicity, which is important for toxicologists, especially in respiratory diseases (Ai et al. 2011). Besides, nanotechnology research has focused on understanding the correlation between the optical, electrical, and magnetic properties of nanomaterials with respect to their size, shape, and surface chemistry

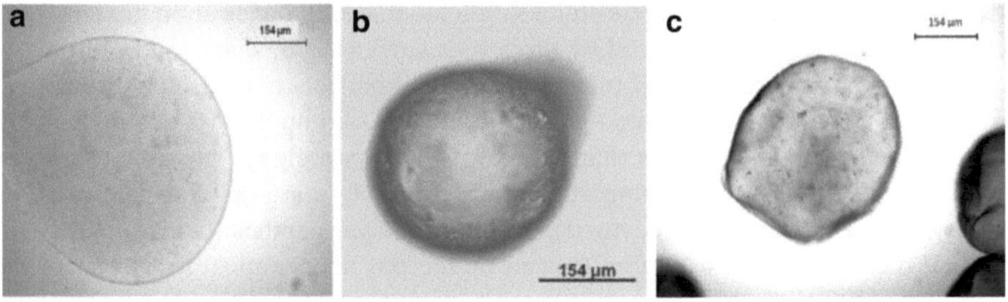

Fig. 3.1 Stability of alginate–pectin microspheres: wet (**a**), freeze dried (**b**), and rehydrated (**c**) (Islan et al., unpublished results)

Table 3.1 Swelling ratio of microspheres under different experimental conditions

		Microsphere swelling (%, ratio)	
Treatment	Time (min)	Alginate	Alginate/HMPectin
Dehydrated	–	88.2 ± 2.4	90.6 ± 0.9
Rehydrated	15	17.8 ± 6.3	207.6 ± 5.9
	30	17.5 ± 1.2	333.4 ± 27.7

(Albanese et al. 2012). Their size is one of the most important factors in the occurrence of disease. From previous knowledge of toxicological properties of fibrous particles, it is believed that the most important parameters in determining the adverse health effects of NPs are dose, dimension, and durability (Oberdörster 2002). However, recent studies show different correlations between various physicochemical properties of NPs and the associated health effects; the most important parameters seem to be mass, number, size, bulk, or surface chemistry, aggregation, hydration, or all together. In the case of solvents, they use to play a crucial role in micro- and nano-particulate systems and their structures and functionalities. For example, changes on gel microspheres gel solvation by dehydration and rehydration showed drastic morphological changes in the bead structure associated with irreversible changes on internal gel structure (Fig. 3.1, Table 3.1).

3.3.1 Size

In the last decade, toxicological studies have demonstrated that small NPs (<100 nm) can cause adverse respiratory health effects, typically producing more inflammation than larger particles made from the same material (Oberdörster et al. 2005; Donaldson and Stone 2003). Rat inhalation and instillation of titanium oxide particles with two sizes, 20- and 250-nm diameter, having the same crystalline structure show that smaller particles led to a persistently high inflammatory reaction in the lungs compared to 12-nm larger-sized particles (Oberdörster et al. 2005). In the postexposure period (up to 1 year), it was observed that the smaller particles had (1) a significantly prolonged retention, (2) increased translocation to the pulmonary *interstitium* and pulmonary persistence of NPs, (3) greater epithelial effects (such as type II cell proliferation), and (4) impairment of alveolar macrophage function (Oberdörster et al. 1994).

For spherical gold/silver NPs, silica NPs, single-walled CNTs, and quantum dots (QDs), a 50-nm diameter is optimal to maximize the rate of uptake and intracellular concentration in certain mammalian cells (Chithrani and Chan 2007).

The size-dependent uptake of NPs is likely related to the membrane-wrapping process. Small NPs have less ligand-to-receptor interaction than do larger NPs. A 5-nm NP coated with 50-kDa proteins may interact with only one or two cell receptors. By contrast, a 100-nm NP has many more ligand–receptor interactions per particle. Larger NPs can act as a cross-linking agent to cluster receptors and induce uptake. Thermodynamically, a 40–50-nm NP is capable of recruiting and binding enough receptors to successfully produce membrane wrapping. NPs larger than 50 nm bind with high affinity to a great number of receptors and may limit the binding of additional NPs. Studies about the effect of NP diameter on uptake using mathematical modeling demonstrated that when NPs present diameters of 30–50 nm where ligand density is optimal, optimal endocytosis occurs when there is no ligand shortage on the NP and no localized receptor shortage on the cell surface (Yuan et al. 2010). However, because each cell type possesses a unique phenotype, optimal NP uptake size may depend on the cell being assayed. Once inside the cell, additional evidence suggests that compartmentalization of CdTe QD into different subcellular organelles depends on size and cell type: sub-2.1-nm QDs enter the nucleus, whereas 4.4-nm QDs are found in the cytoplasm (Williams et al. 2009).

Once in the cytosol, NPs can elicit biological responses by disrupting mitochondrial function, eliciting production of reactive oxygen species and activation of the oxidative stress-mediated signaling cascade (AshaRani et al. 2008); however, there is no consensus on the toxicity and properties of NPs inside the cytoplasm. When compared with other sizes, 30-nm amorphous TiO_2 and 15-nm silver NPs induce the highest generation of reactive oxygen species (Carlson et al. 2008). However, the uptake of silver NPs and QDs into macrophages induces the expression of inflammatory mediators independent of size (Fischer et al. 2010). For intravenously administered NPs, diameter is an important determinant of pharmacokinetics and biodistribution owing to the variable size of inter-endothelial pores lining the blood vessels. NPs with diameters smaller than 6 nm are quickly eliminated from the body because they can be excreted by the kidneys (Choi et al. 2007). Unless a nanomaterial consists of degradable materials such as polymers, lipids, or hydrogels, it cannot be eliminated by the kidneys when the diameter is greater than 6 nm. NPs with diameters larger than 200 nm accumulate in the spleen and liver where they are processed by the MPS cells. Tumors possess large fenestrations between the endothelial cells of blood vessels produced by angiogenesis and can retain particulates found in the blood. This response, termed enhanced permeation and retention, allows NPs to accumulate inside the tumor if they are not cleared by the liver or spleen or excreted through the kidney. To produce long-circulating NPs that can accumulate inside tumor tissues, a diameter between 30 and 200 nm is desired (Jain and Stylianopoulos 2010). The capacity of NPs to navigate between the tumor *interstitia* after extravasation increased with decreasing size. By contrast, larger NPs (100 nm) do not extravasate far beyond the blood vessel because they remain trapped in the extracellular matrix between cells. Thus, the smallest NPs (20 nm) penetrate deep into the tumor tissue but are not retained beyond 24 h.

3.3.2 Surface Charge

In addition to a nanomaterial's size, shape, and ligand density, surface charge is also important in dictating cellular fate. Compared with NPs with a neutral or negative charge, positively charged NPs are taken up at a faster rate (Thorek and Tsourkas 2008). It has been suggested that the cell membrane possesses a slight negative charge and cell uptake is driven by electrostatic attractions (Jin et al. 2009). A recent study demonstrated that this electrostatic attraction between membrane and positively charged NPs favors adhesion onto the cell's surface, leading to uptake. For small NPs (2 nm), a positive charge can perturb the cell membrane potential causing Ca^{2+} influx into cells and the inhibition of cell proliferation (Mukherjee et al. 2010). For larger NPs (4–20 nm), surface charge induces the reconstruction of lipid bilayers (Wang et al. 2008). Binding of negatively charged NPs to a lipid bilayer causes local gelation, whereas binding of positively charged NPs induces fluidity. Several studies have confirmed the pivotal role surface charge plays in downstream biological responses to NPs. It is important to remember that, in the presence of serum or other biological environments, the nanomaterial's surface charge is quickly covered by a corona made up of multiple proteins.

In vivo models' blood half-life is highest for neutral NPs. Positively charged NPs are cleared most quickly from the blood and cause several complications such as hemolytic and platelet aggregation. The unequal half-lives of different charges are a result of the interactions of NPs with serum proteins such as immunoglobulin, lipoproteins, complement and coagulation factors, acute phase proteins, as well as metal-binding and sugar-binding proteins (Cedervall et al. 2007). These proteins instantaneously bind onto the NP surface and dictate the long-term fate, metabolism, clearance, and immune response. The NP surface chemistry appears to determine the type of proteins adsorbed onto the surface and the strength of the interaction (Lynch and Dawson 2008). Positive NPs are quickly adsorbed by serum proteins onto their surface that "tag" them for removal by the mononuclear phagocyte system (MPS) inside the liver and spleen. Rapid clearance by the phagocytic cells of the MPS can be avoided by adding poly(ethylene) glycol (PEG) to the surface of nanomaterials. By preventing opsonization, the addition of PEG drastically increases the blood half-life of all nanomaterials regardless of surface charge.

3.3.2.1 Surface Coating and Functionalization

Due to the possibility of chemical interactions, the combined effects of inhalation, ingestion, or dermal application of NPs with other NPs, chemicals, and gases are largely unknown. The estimated risk of two or more pollutants is not a simple additive process. Particle surface plays a critical role in toxicity as it makes contact with cells and biological material (Buzea et al. 2007).

Surfactants can drastically change the physicochemical properties of NPs, such as magnetic, electric, and optical properties and chemical reactivity (Oberdörster et al. 2005) affecting their cytotoxicity. Surface coatings can render noxious particles nontoxic while less harmful particles can be made highly toxic. The presence of oxygen, ozone, oxygen radicals, and transition metals on NP surfaces leads to the creation of reactive oxygen species and the induction of inflammation. The cytotoxicity of C_{60} molecules systematically correlates with their chemical functionality in human (skin and liver) carcinoma cells with cell death occurring due to lipid oxidation caused by the generation of oxygen radicals. Spherical gold NPs with various surface coatings are not toxic to human cells, despite the fact that they are internalized (Connor et al. 2005). QD of CdSe can be rendered nontoxic when appropriately coated (Derfus et al. 2004).

3.3.3 Surface Area

For the same mass of particles with the same chemical composition and crystalline structure, a greater toxicity was found from NPs than from their larger counterparts. Thus, the inflammatory effect may be dependent on the surface area of NPs, suggesting a need for changes in definitions and regulations related to dose and exposure limits. Indeed, smaller NPs have higher surface area and particle number per unit mass compared to larger particles. The body will react differently to the same mass dose consisting of billions of NPs compared to several microparticles. Larger surface area leads to increased reactivity (Roduner 2006) and is an increased source of reactive oxygen species, as demonstrated by in vitro experiments (Donaldson and Stone 2003). Intratracheal instillation studies on mice with TiO_2 anatase show that small NPs (20 nm) induce a much greater inflammatory response than larger NPs (250 nm) for the same mass dose (Oberdörster et al. 2005). The higher surface area of NPs causes a dose-dependent increase in oxidation and DNA damage (Risom et al. 2005), much higher than larger particles with the same mass dose (Donaldson and Stone 2003). In a simplified calculation, for a total surface area of the human lung alveolar region of 75 m^2, from which 3 % are type II epithelial cells (target for cancer development), this dose is equivalent to about 4 years of exposure at the highest ambient particle concentration (Risom et al. 2005). However, mathematical modeling of particle deposition in the airways indicates that some cells may receive 100-fold more particles depending on their orientation geometry (Balashazy et al. 2003). Other studies suggested a threshold of 20-cm^2 surface area of instilled NPs below which there is no significant inflammatory response in mice. Extrapolating these findings to humans and environmental pollution, the critical surface area of NPs becomes 30,000 cm^2 (Stoeger et al. 2006). Also, subjects with respiratory or cardiovascular diseases may have a lower threshold. In addition, cardiovascular consequences may appear at a lower pollution threshold. We must emphasize that epidemiological studies do not indicate the existence of a threshold below which there are no adverse health effects (Stoeger et al. 2006).

3.3.4 Morphology

Morphological characteristics to be taken into account are flatness, sphericity, and aspect ratio. A general classification exists between high and low aspect ratio particles. High aspect ratio NPs include nanotubes and nanowires, with various shapes, such as helices, zigzags, and belts, or perhaps nanowires with diameter that varies with length. Small aspect ratio morphologies include spherical, oval, cubic, prism, helical, or pillar. Collections of many particles exist as powders, suspension, or colloids (Buzea et al. 2007).

3.3.4.1 Shape

An NP's shape directly influences uptake into cells: rods show the highest uptake, followed by spheres, cylinders, and cubes. Gratton et al. (2008) determined the uptake cell priorities by the synthesis of NPs larger than 100 nm. In studies with sub-100-nm NPs, spheres show an appreciable advantage over rods. In fact, at this size range, increasing the aspect ratio of nanorods seems to decrease total cell uptake. Ligand-coated rod-shaped NPs may present to the cell with two different orientations. Compared with the short axis, the long axis will interact with many more cell surface receptors (Chithrani and Chan 2007). For spiky nanostructures such as gold nano-urchins, whether the ligand is located on or between the spikes affects how it is presented to the target cell receptors (Hutter et al. 2010). For the engineering process, asymmetrical NPs may provide another level of control in presenting ligands to the target receptors.

Blood half-life also depends on an NP's shape, size, and surface chemistry. For example, rod-shaped micelles have a circulation lifetime ten times longer than that of spherical micelles (Geng et al. 2007).

3.3.4.2 Particle Chemistry and Crystalline Structure

In toxicological studies it is good to remark that similar composition does not necessarily imply similar chemistry, chemical bonds. The best example is carbon, whose allotropes are graphite, diamond, CNTs, and fullerenes, each with distinct physical and biological characteristics.

Although there have been suggestions that size may be more important than chemical composition in deciding NPs toxicity (Risom et al. 2005), one cannot generally extrapolate the results of studies showing similar extent of inflammation for different NPs chemistries. Particle chemistry is critical in determining NPs toxicity, and it is especially relevant from the point of view of cell molecular chemistry and oxidative stress. In consequence depending on their chemistry, NPs can show different cellular uptake, subcellular localization, and ability to catalyze the production of reactive oxygen species (Xia et al. 2006). Distinction between composition and chemistry should be done in order to better understand the stability of nano-objects. It is good to remark that though particles may have the same composition, they may have different chemical or crystalline structure. The toxicity of a material depends on its type of crystalline form (Gurr et al. 2005). For example rutile and anatase, both allotropes of TiO_2, i.e., polymorphs sharing the chemical composition, but with different crystalline structure, and hence different chemical and physical properties. Rutile NPs (200 nm) were found to induce oxidative DNA damage in the absence of light, but anatase NPs of the same size did not (Gurr et al. 2005). Another aspect is that NPs can change crystal structure after interaction with water or liquids. NPs often exhibit unexpected crystal structures due to surface effects. Condensation dynamics dictate that gold under these conditions will form these crystalline particles, which form equilibrium-seeking quasi-spheres as the condensing atoms jostle each other in random walks on the surface towards final resting places within the crystal. The effects of crystallinity on condensation are clearly observed in the faceting and fine (nano) structure of the crystal faces. Incidentally interesting are the dendritic patterns on the (111) faces where the condensation forms a classic diffusion-limited aggregation structure.

3.3.4.3 Aspect Ratio

It was found that the higher the aspect ratio, the more toxic the particle is. More exactly, lung cancer was associated with the presence of asbestos fibers longer than 10 μm in the lungs, mesothelioma with fibers longer than 5 μm, and asbestosis with fibers longer than 2 μm. All of these fibers had a minimum thickness of about 150 nm (Lippmann 1990). Long fibers (longer than 20 μm for humans) will not be effectively cleared from the respiratory tract due to the inability of macrophages to phagocytize them. The toxicity of long aspect fibers is closely related to their bio-durability. The bio-durability of a fiber depends on its dissolution and mechanical properties (breaking). Longer fibers that break perpendicular to their long axis become shorter and can be removed by macrophages. Asbestos fibers break longitudinally, resulting in more fibers with smaller diameter, being harder to clear (Hoet et al. 2004). Fibers that are sufficiently soluble in lung fluid can disappear in a matter of months, while the insoluble fibers are likely to remain in the lungs indefinitely. Even short insoluble fibers that are efficiently phagocytized by alveolar macrophages may induce biochemical reactions (release of cytokines, reactive oxygen species, and other mediators). Long aspect ratio engineered NPs, such as CNTs, are new materials of emerging technological relevance and have recently attracted a lot of attention due to their possible negative health effects (Muller et al. 2005) as suggested by their morphological similarities with asbestos.

3.3.4.4 Dimensionality

As shape, or morphology, of NPs plays an important role in their toxicity, it is useful to classify them based on their number of dimensions. This is a generalization of the concept of aspect ratio. 1D nanomaterials are materials with one dimension in the nanometer scale; they are typically thin films or surface coatings and include the circuitry of computer chips and the antireflection and hard coatings on eyeglasses. Thin films have been developed and used for decades in various fields, such as electronics, chemistry, and engineering. 2D nanomaterials are two-dimensional nanomaterials having two dimensions in the nanometer scale; they include 2D nanostructured films, with nanostructures firmly attached to a substrate or nanopore filters used for small particle separation and filtration. Also, free particles with a high aspect ratio, with dimensions in the nanoscale range, are also considered 2D nanomaterials. Asbestos fibers are an example of 2D NPs. 3D nanomaterials are those materials that are considered nanoscaled in all three dimensions. These include thin films deposited under conditions that generate atomic-scale porosity, colloids, and free NPs with various morphologies (Rosi and Mirkin 2007).

3.3.5 Composition of Nano-objects

Nano-objects can be composed of a single constituent material or be a composite of several materials. The nano-objects found in nature are often coacervates, structures with undefined and variable compositions. Man-made gel coacervates have many potential uses, from foods, cosmetics, medicine, and pharmacology, but predictive analysis of their properties is often complex (Table 3.1). Meanwhile pure single-composition materials can be easily synthesized today by a variety of methods, but the presence of potential impurities at nano- or pico-scale brings the attention to the sensitivity limits of analytical methods. The composition of the nanomaterials also affects uptake because single-walled CNTs and gold NPs, each 50 nm in diameter, possess endocytosis rates of 10^{-3} and 10^{-6} per minute, respectively. This 1,000-fold difference may be due to the intrinsic properties of carbon versus gold (Albanese et al. 2012).

3.3.6 *Other Important Aspects*

3.3.6.1 Uniformity and Agglomeration

Based on their chemistry and electromagnetic properties, NPs can exist as dispersed aerosols, as suspensions/colloids, or in an agglomerate state. For example, magnetic NPs tend to cluster, forming an agglomerate state, unless their surfaces are coated with a nonmagnetic material. In an agglomerate state, NPs may behave as larger particles, depending on the size of the agglomerate. Hence, it is evident that NP agglomeration, size, and surface reactivity, along with shape and size, must be taken into account when deciding considering health and environmental regulation of new materials.

3.3.6.2 Concentration

There are many contradictory results related to the toxic effects of NPs at different concentrations (Buzea et al. 2007). When comparing the results of different studies, one must take into account that there are differences in the aggregation properties of NPs in air and water, resulting in inherent discrepancies between inhalation studies and instillation or in vitro experiments. The aggregation may depend on surface charge, material type, and size, among others. One must stress the fact that aggregation of NPs is essential in determining their toxicity, due to a more effective macrophage clearance for larger particles compared to smaller ones (that seem to evade easier this defense mechanism), leading to reduced toxicity of NP aggregates larger than 100–200 nm (Oberdörster et al. 2005; Takenaka et al. 2001). It has been demonstrated that a high concentration of NPs would promote particle aggregation and therefore reduce toxic effects compared to lower concentrations (Takenaka et al. 2001). Most aggregates are observed to be larger than 100 nm, a size that seems to be a threshold for many of the adverse health effects of small particles. Therefore, experiments performed with high concentrations of NPs will lead to the formation of NP aggregates that may not be as toxic as lower concentrations of the same NPs.

3.3.6.3 Dose

Dose is defined as the amount or quantity of substance that will reach a biological system. The dose is directly related to exposure or the concentration of substance in the relevant medium (air, food, water) multiplied by the duration of contact. Generally, the negative health effects of NPs do not correlate with NP mass dose (Oberdörster et al. 2005; Donaldson and Stone 2003). Comparing the health effects of inhaled TiO_2 NPs with different sizes, it is remarkable that the low-dose (10 mg/m^3) exposure to 20-nm-diameter particles resulted in a greater lung tumor incidence than the high-dose (250 mg/m^3) exposure of 300-nm-diameter particles (Hoet et al. 2004).

3.4 Relationship Between Binding and Stability

Multiple studies have demonstrated that NPs interact with serum proteins (or other biological fluids) and cell membrane receptors and that interaction is determined by the NP design, in effect influencing cell uptake, gene expression, and toxicity (Albanese et al. 2012).

The size, the structure of NPs and their surface characteristics are important for protein interactions; the higher size NP the higher interactions with a specific protein. It means that high-volume NP can interact with several molecules of the specific protein. This characteristic can improve the beneficial effect of the NPs but at the same time can produce unexpected and undesirable effects. When NPs enter in the body, some events can take place. The immune system can react against NPs recognizing them as foreign antigens and an immune response can be arisen (Dobrovolskaia and McNeil 2007). The particles will be targeted and removed very quickly. Another undesirable event is produced with the NPs aggregation which can form clots that can obstruct arteries and veins. At the same time, nanotoxicity can be produced for the nonspecific interactions; thus, NPs can interact with different proteins in an unexpected and uncontrolled ways provoking undesirable side effects. The architecture of these adsorbed proteins on the NP surface is complex but can be described as hard and soft corona layers. The hard corona layer contains proteins that are strongly adsorbed to the NP surface ($K_d \approx 10^{-6}$ to 10^{-8} M) (Lacerda et al. 2009), whereas the soft corona layer contains serum proteins that weakly interact with the hard corona layer. This outermost layer is likely to be dynamic and could vary during the course of the NP's life in vivo. Proteins as fibronectin and albumin have high affinity for many material surfaces and after those reactions convert their surface from nonbiologically recognized material to one that is. In this sense once absorbed, those NPs acquired new surface characteristics that will influence their biodisponibility, clearance, and toxicity. Thus, the static and dynamic stability of nano-objects will change as they will absorb biological proteins. Complete clearance from the body may not be possible once phagocytosed due to the resistance of NPs to degradation. The capacity of absorption of protein depends of the charge and hydrophobicity of the nano-object surface (Lück et al. 1998). It has been found that latex NPs with high surface charge density adsorb proteins to a greater extent than those with similar size and hydrophobicity (Gessner et al. 2002). At the same time the decrease in hydrophobicity of NPs results in a decrease in protein adsorption. Other material parameters as size and shape influence protein adsorption. Thus, the curvature of silica NPs was found to greatly influence the secondary structure of adsorbed proteins; this aspect dictates the overall stability and hence adsorption of these proteins onto the NPs surface. It means that the dynamic stability of those silica NPs is even more difficult to predict. Analyzing the exposed results, the study of protein–nano-objects interactions is indispensable as one aspect of toxicological effects. In this regard, nano-object surface characterization before and after serum protein exposure is necessary. Studies of adsorption kinetics either in single-protein solutions or in a complex mixture will provide affinity data in the protein–NPs reactions. Some efforts were made to stabilize the surface of NPs to avoid protein adsorption. One strategy is to coat the NP surface with long polymer chain as PEG (Otsuka et al. 2003). PEG hydrophilicity and steric repulsion of other molecules prevent proteins as fibronectin from adsorbing to biomaterial surfaces. Similarly, other surface-immobilized biomolecules as saccharides, lipids, natural proteins, and nucleic acids may be used to reduce nonspecific protein adsorption and therefore unwanted bio-recognition of nano-objects.

Analyzing the factors that can influence NP–cell interactions at the nano–bio interface is diverse: size, shape, and charge of NPs; ligand density; receptor expression levels; internalization mechanism; and cell properties (phenotypes location). The ligand-coated NPs bind to receptors on the membrane and induce a signaling cascade without entering the cell. The ligand-coated NPs can also be internalized and exocytosed by the cell, without ever leaving the vesicle. They bind to the membrane receptor, enter the cells, and then exit from the cell (Albanese et al. 2012). Besides, NPs can interact nonspecifically with the cell surface membrane; in this sense, unpredictable effects of NPs are expected and therefore changes in the dynamic stability. Interactions between NP-bound ligands and cellular receptors depend on the engineered geometry and the ligand density of a nanomaterial. The NP acts as a scaffold whose design dictates the number of ligands that interact with the receptor target. A multivalent effect occurs when multiple ligands on the NP interact with multiple receptors on the cell. The binding strength of complexed ligands is more than the sum of individual affinities and is measured as the avidity for the entire complex. This phenomenon can be exemplified by analyzing the binding affinity of Herceptin to the ErbB2 receptor, being 10^{-10} M in solution, 5.5×10^{-12} M on a 10-nm nanoparticle, and 1.5×10^{-13} M on a 70-nm nanoparticle (Jiang et al. 2008). This example illustrates how a ligand's binding affinity increases proportionally to the size of an NP owing to a higher protein density on the NP surface. Similarly, receptor-specific peptides improved their ability to induce angiogenesis when conjugated to an NP surface (Kanaras et al. 2011). Presentation of the peptide on a structured scaffold increased angiogenesis, which is dependent on receptor-mediated signaling. These findings highlight the advantages of having a ligand bound to an NP as opposed to its being free in solution. The NP surface creates a region of highly concentrated ligand, which increases avidity and, potentially, cell signaling.

An additional concern with NP–ligand complexes is the potential denaturation of proteins when bound to the engineered surface. The denaturation of a protein can affect binding to its receptor, increase nonspecific interactions, or provoke inflammation. Lysozyme, for example, when bound onto gold NPs will denature and interact with other lysozyme molecules and produce protein–NP aggregates (Zhang et al. 2009). Fibrinogen also unfolds when bound to the surface of polyacrylic acid-coated gold NPs. The denatured fibrinogen can then bind to the integrin receptor Mac-1 and lead to inflammation (Deng et al. 2010). Which ligand is used to coat the nanomaterial will also affect downstream biological responses. Once bound to their receptor, NPs will typically enter the cell via receptor-mediated endocytosis (Albanese et al. 2012). The binding of the NP–ligand conjugate to the receptor produces a localized decrease in the Gibbs free energy, which induces the membrane to wrap around the NP to form a closed-vesicle structure (Gao et al. 2005). For example, the uptake and cytotoxicity of NPs were significantly altered when the NPs were coated with two different proteins targeting the same receptor (Wang et al. 2010). The localization of NPs in the intracellular space may be directed using peptides such as the mitochondrial localization sequence. If an NP is engineered to escape the endolysosomal system, it can enter the cytosol, where it is free to interact with a wide number of organelles and can affect cell behavior. Once in the cytosol, NPs can elicit biological responses by disrupting mitochondrial function, eliciting production of reactive oxygen species and activation of the oxidative stress-mediated signaling cascade (AshaRani et al. 2008).

Some examples about the role of binding of nano-objects can be described. The binding of amphiphilic lipids with dendrimers may disturb the structure of the cell membrane, causing the leakage of intracellular molecules. Daily administration of dendrimer-based nanomedicines may decrease the protein concentration and lead to acute hypoalbuminemia. The interactions between dendrimer and nucleic acids may induce interferences in the cell cycle and protein synthesis. The molecular toxicity of a dendrimer depends a lot on its surface functionality (Cheng et al. 2011). Some enzymes can exhibit conformational changes upon binding with polyamidoamine (PAMAM) dendrimers, such as the case of acetylcholinesterase (Shcharbin et al. 2006). Glucose oxidase lost almost 50 % of its catalytic activity because of amine-terminated PAMAM dendrimers (Miller et al. 2006), while streptokinase retained 80 % of the protein activity when the enzyme was bound to a carboxylate-terminated PAMAM dendrimer (Wang et al. 2007). Lipase conjugated to a carboxylate-terminated poly(phenylene sulfide) dendrimer retained 90 % of its original activity even after 20 times of recycling (Yemul and Imae 2005). However, in a recent study, a dendritic polymer consisting of poly(benzyl ether) and poly(ethylene glycol) (PEG) represented a friendly environment for the immobilization of laccase and even improved the catalytic activity of the bound enzyme (Gitsov et al. 2008). Though limited data are available in the references on molecular toxicity of dendrimers, we should pay more attention to such issues when evaluating the long-term safety of dendrimer-based nanomedicines.

3.5 Immune System and Stability

NPs can be engineered to either avoid immune system recognition or specifically inhibit or enhance the immune responses (Zolnik et al. 2010). Therefore, all new chemical and biological entities require adequate investigations into their interactions with the immune system before their use in industry, biology, and medicine. An interaction between an NP and the immune system is considered desirable when it may lead to various beneficial medical applications, such as vaccines or therapeutics for inflammatory and autoimmune disorders. NP-based delivery systems offer the following potential advantages: (1) site-specific delivery of drugs, peptides, and genes; (2) improved in vitro and in vivo stability; and (3) reduced side effect profile. However, because NPs are often first picked up by the phagocytic cells of the immune system (e.g., macrophages), there may be undesirable interactions between NPs and the immune system, such as immunostimulation or immunosuppression, which may promote inflammatory or autoimmune disorders or increase the host susceptibility to infections and cancer. Consequently, inadvertent recognition of NPs as foreign by the immune cells may result in a multilevel immune response against NPs and eventually lead to toxicity in the host and/or lack of therapeutic efficacy. For example, granuloma formation was observed in the lungs, skin, and pleural lining of the animals exposed to CNTs (Poland et al. 2008). However, when NPs are recognized as self or there is an absence of immune recognition, this represents a major area of interest in the field of drug delivery. Some polymeric nano-objects as PAMAM dendrimers, PEGylated dendrimers, and *N*-(2-hydroxypropyl)methacrylamide (HPMA), among others, present high stability producing minimal toxicity, reduced immunogenicity, and excellent solubility in aqueous and most organic solutions (Yang and Lopina 2006). As many other aspects of the biodisponibility of nano-objects, the degree to which immune responses occur upon NP exposure is governed by NP physicochemical properties (Vonarbourg et al. 2006). It is now well accepted that properties such as NP size, surface charge, hydrophobicity/hydrophilicity, and the steric effects of particle coating can dictate NP compatibility with the immune system (Dobrovolskaia and McNeil 2007). The immune system can be divided into two different and interconnected systems termed innate and adaptive immunity. The first line of protection comprises the innate immunity, but one important aspect of the immune response of nano-objects involves the humoral adaptive immunity (antibodies). Opsonization of antibodies as IgG may activate humoral immune response against NPs by stimulating natural killer cells (NK) and the complement system (Salvador-Morales et al. 2006). Finally, those NPs will be phagocyted by antigen-presenting cells and removed by reticulo-endothelial system (RES). Both opsonization (a process in which an antigen is marked for modification and destruction by immune system cells) and phagocytosis (a form of endocytosis in which the foreign body is ingested into the immune system cell called phagocyte) may pose a significant problem in therapeutic effectiveness of engineered biomaterials (Pantic 2011). In some cases, such as in vaccine development strategies, these processes may be beneficial, because they enable the NP to reach certain phagocytic routes and thereby increase vaccine efficiency. However, when NPs are used as drug delivery mediums, opsonization and phagocytosis may significantly decrease concentrations and their therapeutic load in target tissues, limit the half-life, as well as their availability.

Phagocytosis of NPs usually occurs in the liver (by the Kupffer cells), although spleen and bone marrow macrophages may also play a significant role. Some researchers have found that the opsonization of hydrophobic, negatively charged NPs, occurs more quickly and efficiently than in hydrophilic, neutral NPs, due to the enhanced adsorption of plasma proteins on the NP surface (Shan et al. 2009). One commonly used technique is conjugation of NPs with PEG, a polyether compound prepared by polymerization of ethylene oxide. PEGylation procedure results in prolonged half-life, higher stability, water solubility, and lower antigenicity and immunogenicity, making specific cell targeting much easier and effective (Ryan et al. 2008). PEG chains create a barrier layer resulting in prevention of the adhesion of the blood serum opsonins, making the particles camouflaged or invisible to the cells of the mononuclear phagocytic system (Owens and Peppas 2006). Although PEG or other polymers shield the NPs from recognition by the immune system, there are data suggesting the formation of PEG-specific antibodies after administration of PEG-coated liposomes (Ishida et al. 2007). Consequently, these antibodies resulted in accelerated clearance of the PEG liposomes from blood and contributed to the change in pharmacokinetic profile of subsequent injected doses of PEG liposome (Ishida et al. 2006). Therefore, it should be noted that generation of particle-specific antibodies may affect the efficacy and the safety of NP-based therapeutics. NPs can also be designed to elicit an immune response by either direct immunostimulation of antigen-presenting cells or delivering antigen to specific cellular compartments (Kalkanidis et al. 2006). In the past few years, polysaccharides have been reported as possible alternative to PEG, as examples we can mention dextrin vinyl acrylate SC16NPs; they are noncytotoxic and do not elicit a reactive response when making contact with macrophages. Folic acid (FA) could decrease the protein adsorption and avoid the NP recognition by macrophage cells, facilitating the nanoparticle uptake to specific cancer cells (Zhang et al. 2002).

NP immunogenicity is drawing interest because NPs have been shown to improve antigenicity of conjugated weak antigens and thus serve as adjuvants and because some NPs have been shown to be antigenic themselves. The former property has been shown to depend on particle size and surface charge (Manolova et al. 2008) and can significantly contribute to the development of improved vaccine formulations. Particle size has been reported as a major factor in determining whether antigens loaded into NPs induce type I (interferon-γ) or type II (IL-4) cytokines, thereby contributing to the type of immune response (Mottram et al. 2007). A leading hypothesis on why nanotechnology-driven formulations are effective in vaccine development is that non-soluble NPs provide controlled, slow release of antigens, creating a depot at the site of injection and providing protection in the destabilizing in vivo environment (O'Hagan and De Gregorio

2009). On the other hand, antigenicity of the NPs is less predictable. Some studies demonstrated the generation of particle-specific antibodies when C_{60} fullerene derivatives conjugated to a protein carrier, BSA, were used for immunization (Chen et al. 1998). Other studies using different fullerene derivatives, gold colloids, and cationic polyamidoamine and polypropylenimine dendrimers have reported no particle-specific immune response, even in the presence of strong adjuvants (Agashe et al. 2006). However, PAMAM dendrimers conjugated to bovine serum albumin (BSA) showed increased antigenic properties and as a result dendrimer-specific antibody was observed in vivo (Lee et al. 2004). It is important to remark that all of these studies in which NP-specific antibodies were generated, the antigen consists in NPs conjugated a protein carrier BSA. Therefore, these limited data may suggest that some water-soluble NPs may behave as haptens, i.e., they are not antigenic until they bind to protein carrier, possibly as a result of their small size. Another factor that may enhance the immune response towards nano-objects is the capacity of those objects or particles to bind serum protein and acts as an antigen because of the change in protein structure.

Nanoemulsions are emulsions with droplet size in the nanometer scale. An emulsion is a thermodynamically unstable system, unless stabilized by the presence of emulsifying agent, consisting of at least two immiscible liquid phases, one of which is dispersed as globules (the dispersed phase) in the other liquid phase (the continued phase) (Matrin 1993). Emulsions are traditionally used in vaccines. It has been reported that the nanoscale-sized emulsions are able to permeate the nasal mucosa and carry the antigen to the antigen-presenting cells more efficiently than larger-sized emulsions. Nanoscale emulsion-based intranasal vaccines have been investigated for hepatitis B, HIV, influenza, and anthrax (Makidon et al. 2008). One example of such an application is a nanoemulsion-based seasonal influenza vaccine NB-1008, currently undergoing phase 1 clinical trials in the United States. Many immunostimulatory reactions initiated by NPs are mediated by the production of inflammatory cytokines. Several studies have reported cytokine induction by different types of nanomaterials (gold colloids, dendrimers, polymers, lipid NPs, etc.). NP size has been suggested as a leading parameter that determines an NP's potential to induce cytokine responses (Fifis et al. 2004; Mottram et al. 2007). For example, the length of CNTs was shown to correlate with subcutaneous (sc) inflammation induced by CNT in vivo. However, a few studies have shown that cytokines were induced not by NPs per se but by surfactants or bacterial endotoxins present in the formulation (Schöler et al. 2001). Therefore, it is important to quantify the presence of chemical (formation of by-products) and biological (endotoxin) contaminants before analysis of inflammatory properties of NPs. There are few studies about the elicitation of allergic reactions by NP-mediated immunostimulation. For example, both single-walled and multiwalled CNTs enhanced the allergenicity of egg albumin when administered via intranasal or sc routes in mice. The mechanism of this enhanced allergenicity is thought to be due to the CNT-mediated induction of the acute inflammatory response (Nygaard et al. 2009). Occupational exposure during the manufacture of nanomaterials has been linked to allergic reactions. Toxic epidermal necrolysis-like dermatitis was observed in a chemist exposed to high levels of intermediate or final products of dendrimers while performing dendrimer synthesis (Toyama et al. 2008). However, it is unclear whether the dermatitis was caused by NPs or reactive species used in their synthesis. It should be noted that these studies dealt with raw nanomaterials, which were not intended for biological or medicinal use. More studies involving well-characterized NPs, relevant animal models, and routes of NP administration are required to understand whether NPs can cause allergy in humans.

NPs are evaluated for their immunostimulatory potential based on their ability to stimulate innate or adaptive immune responses. Activation of the complement cascade can be harmful if particles inadvertently, or by design, enter the systemic circulation because this may lead to hypersensitivity reactions and anaphylaxis (Chanan-Khan et al. 2003). Recent data also suggested that activation of the complement system at tumor sites stimulates tumor-associated immune cells and promotes their conversion into a tumor-supportive phenotype, thereby stimulating cancer progression (Markiewski et al. 2008). This type of response may impact the therapeutic efficacy of NP formulations intended for cancer diagnosis or therapy. On the other hand, if particles are intended for sc or intradermal administration, activation of the complement by the particles can benefit vaccine efficacy. Whereas many questions still remain, it is important to elucidate how complement activation relates to NP toxicity and the development of nanotechnology-based formulations for medical applications.

3.6 Classical Methods for Evaluation of Nano-object Stability

Since static stability is related to the intrinsic properties of nano-objects, structural characterization should be performed to analyze the homogeneity/heterogeneity of the system under study. At present, some traditional methods are used to structural characterization. The most useful techniques to determine structural properties can be grouped into microscopies and spectroscopies.

Microscopies involved traditional techniques like optical and confocal microscopies, SEM (scanning electron microscopies), TEM (transmission electron microscopy), and their variants. Additionally, during the last decades the developments in the microscopy field culminate with scanning probe microscope (SPM) extensively used in two main versions, the STM (scanning tunneling microscope) and the AFM (atomic force microscope). Also, spectroscopic techniques are widely used from X-ray crystallography, UV–visible spectroscopy, DLS and SLS, Raman–FTIR, and NMR. The developments of synchrotrons in where both spectroscopies and microscopies can be simultaneously used on samples are allowing to establish the physicochemical and topological properties of samples at nanoscale (e.g., synchrotron Soleil, France).

SEM, TEM, and their variants together with electron diffraction have been commonly used in nano-characterization of objects from materials to cells. In both cases, electron beans under different experimental vacuum conditions interact with the sample determining some of the properties. In SEM sample analysis information on surface and topography (like 3D image) can be determined with a resolution close to 1 nm. TEM analysis involves the use of electrons under high acceleration that can impact and across the surface of ultrathin samples producing 2D images but very useful in crystallography analysis. By these microscopy techniques alone or combined, it is possible to establish morphological and intrinsic properties of the samples with definition at nanoscale.

SPM is a relatively new characterization technique and has found wide spread applications in nanotechnology. The two major members of the SPM family are scanning tunneling microscopy (STM) and atomic force microscopy (AFM). Although both STM and AFM are true surface image techniques that can produce topographic images of a surface with atomic resolution in all three dimensions, combining with appropriately designed attachments, both the STM and AFM have found a much broadened range of applications, such as nano-indentation, nanolithography (as discussed in the previous chapter), and patterned self-assembly. Almost all solid surfaces, whether hard or soft, electrically conductive or not, can all be studied with STM and AFM. Surfaces can be studied in gas such as air or vacuum or in liquid. In the following, we will briefly discuss the aforementioned characterization techniques and their applications in nanotechnology.

Optical spectroscopies are commonly used to determine the presence of metal NPs typically gold or silver, through the localized surface plasmon resonance (SPR). The SPR is the frequency of excited valence electrons in solids mediated by photons. SPR is able to provide label-free biomolecular analysis on kinetic processes of interactions (association and dissociation), equilibrium binding (affinity and enthalpy), analyte concentration, and real-time molecule detection. This technology can also be applied to determine the size of NPs and QD and also for lab-on-a-chip sensors. In studies of static stability related to protein–nano-objects interactions, some simple methods as shifts in SPR peak measurements in gold NPs and QD after protein exposure may provide evidence of nonspecific protein adsorption. Another method as dynamic light scattering (DLS) may help to predict if protein adsorption would be possible when NPs interact with biological proteins. It is important to reproduce as possible the in vivo conditions to extrapolate results.

DLS is a powerful technique commonly used when NP size and size distribution are required, since the latter parameter is necessary to define the homogeneity in one sample of nano-object. DLS, sometimes referred to as quasi-elastic light scattering (QELS), is a noninvasive, well-established technique for measuring the size and size distribution of molecules and particles typically in the submicron region and with the latest technology lower than 1 nm (Zetasizer-nano 2007). Typical applications of DLS are the characterization of particles, emulsions, or molecules, which have been dispersed or dissolved in a liquid. The Brownian motion of particles or molecules in suspension causes laser light to be scattered at different intensities. Analysis of these intensity fluctuations yields the velocity of

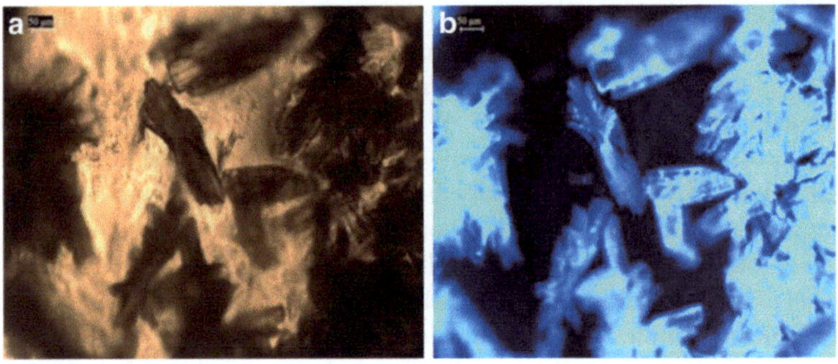

Fig. 3.2 Determination of enrofloxacin contained in PVA–pectin films by optical (**a**) and fluorescent (**b**) microscopies (Martinez et al., unpublished results)

the Brownian motion and hence the particle size using the Stokes–Einstein relationship.

Another technique that utilizes light scattering is static light scattering (SLS) which is performed to measure molecular weight without the requirement of column calibration. The lack of calibration required means that the result is an absolute value. There are two available methods of performing this measurement for polymers and proteins in solution. One is based on the use of size exclusion chromatography detectors and the other one is based on the batch measurements. In this technique a Debye plot is constructed (Zetasizer-nano 2007). This requires the measurement of the light scattered from a number of known concentrations of the analyte, followed by a linear extrapolation to determine the excess scattering at zero concentration.

Fluorescence applied at nanoscale is a powerful tool to synthesize, detect, and characterize nanomaterials. One of the main projected applications of nanofluorescence is in medicine for diagnostic but also to detect molecular impurities. An example is the inclusion of the antibiotic enrofloxacin into PVA–pectin films (Fig. 3.2). Fluorescence microscopy revealed the presence of antibiotic microcrystals inside the PVA–pectin matrix. The main advantage of fluorescence techniques is the high sensitivity, but also among the disadvantages are the unspecific interactions between the probe and molecules present in the environment. Several companies are currently running to develop equipment for specific nanoapplications.

Raman–FTIR and MNR spectroscopies are classical tools used alone or combined with other methods to characterize molecular system and their interactions at region and/or atomic level. However, in the case of low-energy spectroscopies, i.e., Raman or FTIR, the most serious limitation is the low sensitivity of these techniques because of red zone-energy beams employed. The use of synchrotron light allows increasing the sensitivity signals in about 1,000 times compared to benchtop equipments. In addition, synchrotron facilities are becoming more useful to characterize nano-objects since more than one technique can be used simultaneously (Bosio et al. 2013).

X-ray spectroscopy is currently used for characterization of different biological and chemical nanomaterials and moreover applied to physicochemical, biological, and engineering studies. X-ray diffraction techniques (XRD) and associated techniques are powerful techniques for nanotechnology because of the nano-objects dimensions are in the same magnitude as X-ray wavelengths. XRD can be employed to determine chemical composition, layer thickness, roughness, lattice spacing, and relaxation. Recently, XRD microscopy was used to determine the 3D internal structure of a whole yeast cell (Jiang et al. 2010). X-ray diffraction (XRD) crystallography has been used for the determination of crystallinity, crystal structures, lattice constants and geometry, identification of unknown materials, orientation of single crystals, and preferred orientation of polycrystals, defects, stresses, etc., of NPS, nanowires, and thin films. It was demonstrated that stability is related with crystallinity, for example, in InP (indium phosphide) QD to determine the crystal sizes (Guzelian et al. 1996). In XRD, a collimated beam of X-rays, with a wavelength typically ranging from 0.7 to 2 Å, is incident on a specimen and is diffracted by the crystalline phases in the specimen according to Bragg's law. The intensity of the diffracted X-rays is measured as a function of the diffraction angle (2θ) and the specimen's orientation. This diffraction pattern is used to identify the specimen's crystalline phases and to measure its structural properties. Additionally, X-ray diffuse scattering can be used to determine lateral and transversal correlations, distortions, density, and porosity of nano-objects. As well in-plane grazing incidence diffraction is used to analyze lateral correlations of thinnest organic and inorganic structural layers as well as depth profiling.

The small-angle X-ray scattering (SAXS) technique allows establishing size, shape, distribution, orientation, and correlation of NPs present in solids and solutions. Strong diffraction peaks result from constructive interference of X-rays scattered from ordered arrays of atoms and molecules. A lot of information can be obtained from the angular distribution of scattered intensity at low angles. Fluctuations in electron density over lengths on the order of 10 nm or larger can be sufficient to produce an appreciable scattered X-ray intensities at angles $2\theta > 5°$ (Masadeh 2008). These variations can be from differences in density, from differences in composition, or from both and do not need to be periodic. The amount and angular distribution of scattered intensity provide information, such as the size of very small particles or their surface area per unit volume, regardless of whether the sample or particles are crystalline or amorphous. SAXS is the scattering due to the existence of heterogeneous regions of sizes of several nanometers to several tens of nanometers, whereas XRD is useful to determine atomic structures. For example, SAXS is a powerful technique to determine the homogeneity of CdSe nanocrystals: nanocrystal sample size and size distribution (Murray et al. 2000). Recently, changes of myelin sheaths in a rat brain all the way down to the molecular level without surgery were observed by combining SAXS and CT scanning (computed tomography scanning) (Nielsen et al. 2012).

Another useful technique is X-ray reflectometry (XRR) which allows analyzing layer thickness, surface and interface roughness, surface density gradients, and layer density of nano-objects.

3.6.1 In Vitro Assays Interferences

Certain nanomaterials may interfere with the readout systems of commonly used assays for cell viability and/or mitochondrial function (Lewinski et al. 2008). Monteiro-Riviere et al. (2009) recently conducted a comprehensive study of the cytotoxic effects of carbonaceous nanomaterials on human epidermal keratinocytes using nine widely accepted cytotoxicity assays and found that classical dye-based assays produce invalid results with some nanomaterials due to nanomaterial/dye interactions and/or nanomaterial adsorption of the dye/dye products. Furthermore, Laaksonen et al. (2007) reported that the MTT assay failed to report toxicity of certain porous silica microparticles due to spontaneous redox reactions. However, for other completely oxidized particles, the assay yielded the expected results. Consequently, more than one assay may be required when determining nanoparticle toxicity for risk assessment (Fadeel and Garcia-Bennett 2010).

3.6.2 Lack of Correlation Between In Vitro and In Vivo Assays

Furthermore, recent studies have indicated a lack of correlation between in vivo and in vitro effects of TiO_2 NPs (Sayes et al. 2007). Indeed, while in vitro assays may provide useful mechanistic information, revealing cell type-specific responses, such assays may fail to capture intercellular effects such as cross talk between inflammatory cells that requires that the cells and their signaling occur in its natural state (Fischer and Chan 2007). Moreover, and for obvious reasons, the influence of transport of NPs via blood, lymph, or bile cannot be readily incorporated into in vitro studies. For instance, in a study of the biodistribution of QD in isolated perfused porcine skin, Lee et al. (2007) revealed a phenomenon of skin–QD interaction not observable in vitro. One obvious conclusion is that better cell culture and coculture systems are required and a better understanding of the relevant endpoints to test for in vitro toxicity is needed (Warheit et al. 2008). The emerging field of nanotoxicogenomics (Ding et al. 2005; Waters et al. 2009) which attempts to correlate global gene expression profiles of cells or tissues exposed to NPs with biological/toxicological responses using cDNA microarray technologies may provide useful information in this regard. Potentially, the application of two-dimensional electrophoresis and mass spectrometry (MS) methods (proteomics) could also enhance our understanding of the biological responses induced by NPs (Sheehan 2007).

3.7 New Trends in Toxicological Evaluation

The study of toxicological effects of nano-objects could lead to the development of predictive and simulation tools to assist in the engineering process to get less harmful nanomaterials and nano-objects to be used in human and animals. Already, Monte Carlo simulations have been used to model the effect of NP size and ligand density on cellular uptake and tumor targeting to improve the NP design for optimal tumor accumulation for diagnostic and drug delivery applications (Wang and Dormidontova 2010). Ideally, the outcomes of these studies would also be entered into a database, which would further enable practitioners to use computer simulation programs to identify quickly the most appropriate nanostructure design for a specific application. All fundamental studies on nano–bio interactions at the systemic, cellular, and molecular levels could populate this database, making it useful in generating correlations between the physicochemical properties of nanostructures and biological responses. Ultimately, this database could create blueprints for constructing nanosystems (Albanese et al. 2012).

Toxicology testing is likely to witness a significant influx of new testing methodologies over the next few years to study the impact of long-term exposure to a wide range of chemical and biological substances and by the need to gain a deeper understanding of the mechanisms by which these substances lead to toxicity (Hartung 2010). The need for a more mechanistic understanding of the ways in which chemicals modulate biological pathways is urgent if we are to identify and better assess safety issues relating to a wide range of substances developed by the pharmaceutical, chemical, agri–bio, and cosmetic industries. Omics technologies provide a valuable opportunity to refine existing methods and provide information for the so-called integrated testing strategies (ITS) via the creation of signatures of toxicity. By mapping these signatures to underlying pathways of toxicity (PoT), some of which have been identified by toxicologists over the last few decades, and bringing them together with pathway information determined from biochemistry and molecular biology. A "systems toxicology" approach will enable virtual experiments to be conducted that can improve the prediction of hazard and the assessment of compound toxicity. Diverse approaches as modern high-content and high-throughput technologies for predictive in vitro systems have received a great support from major funding programs. Those technologies are used to phenotype substance effects (signatures of toxicity, SoT). Increasingly, attempts are made to link these to specific PoT.

Another approach is based largely on "off the shelf" pathway assays for high-throughput testing, and other approaches arose to identify PoT starting from metabolomics and transcriptomics (Hartung and McBride 2011). All these approaches are complementary and strongly overlapping. New technologies will improve the toxicology system attempting to model potential pathologies and the physiology of the body with computational tools. The modeling is necessary both to identify putative PoT and to make predictions about the effect of a substance in humans.

Traditional in vivo tests and in vitro assays typically provide only limited mechanistic information, while new comprehensive screening studies could reveal the interactions of chemicals with biochemical pathways that control cell function, communication, and adaptation to environmental changes. They are based on a broad biological (initially) phenotypic profiling by either omics or high-throughput screening (HTS) technologies. The latter requires large batteries of tests to characterize an individual substance, as each and every one reflects only a limited amount of information regarding biological characteristics. A qualitative hazard identification requires the analysis of the dose–response characteristics of the perturbations of these pathways, together with in vitro to in vivo extrapolation of the pharmacokinetics (Wetmore et al. 2012), could ultimately provide the basis for the so-called pathway-based hazard assessment.

In 2007 a new paradigm was proposed in contrast to animal-based testing towards the application of emerging technologies, including genomic, metabolomic, and proteomic approaches, as well as systems biology. This new paradigm would provide greater mechanistic insight into the ways in which many compounds, including pharmaceuticals, affect human health (Hartung 2001). In order to reduce time, costs, and animal numbers, REACH (the European Community Regulation on chemicals and their safe use which deals with *R*egistration, *E*valuation, *A*uthorization and Restriction of *Ch*emical substances) foresees increased use of in vitro and in silico testing methods. However, it is clear those simple in vitro test systems or in silico approaches will not be able to reflect the complexity of the organism and its derangement by substances. To answer this question combinations of information sources in ITS that together map the hazard of interest have been developed, among other approaches. Other aspects such as toxico-kinetics (Basketter et al. 2012) and exposure (Wetmore et al. 2012) need to be integrated in order to make them useful for risk assessment. In systems toxicology sequencing of the genome, annotation of genes, and gene chip developments, especially for transcriptomics, clearly has been the starting point for elucidation of toxic effects. Multi-omics research, including data integration and modeling, using both in vivo and cell-based assays, is beginning to gain attraction as a viable approach in the development of a pathway-based mechanistic view of chemically induced bioactivity. For instance, both medium- and high-throughput genomics, proteomics, and metabolomics measurements have been applied to identify and to understand molecular and biochemical pathways that control homeostasis (Heijne et al. 2005).

Transcriptomics is the study of transcriptome based on the fact that gene expression can rapidly change in response to chemical exposure providing a sensitive endpoint for toxicity. In addition, advances in microarray technology now include the capability to survey the entire transcriptome, thus providing a global evaluation of cellular changes. The technology also has become more reliable and reproducible (Shippy et al. 2006), and extensive validation studies performed by the MicroArray Quality Control (MAQC) project demonstrated consistent results across multiple platforms and research sites (Shi et al. 2006). For systems toxicology applications, transcriptomic measures typically are performed across both dose and time. The dose–response characteristics are important for inferring dose-dependent transitions in what pathways are perturbed, while the time course measurements provide a better understanding of the pathways involved in the temporal transitions from the molecular initiating event to secondary and tertiary responses (Andersen et al. 2010). For chemicals or their metabolites that interact weakly with cellular macromolecules, multiple pathways may initially be perturbed, and for chemicals that interact with higher affinity to a single cellular macromolecule, a single pathway would be perturbed initially. In both cases, a systems-level analysis provides an understanding of the key events in the mode of action that ultimately leads to toxicity. Transcriptomic studies often are complemented by proteomic and metabolomic approaches in order to functionally validate the observed changes in the transcriptome.

Proteomics studies in the field of toxicology have focused mainly on identifying biomarkers and refining a mechanistic understanding of molecular mechanisms of toxicity (Kennedy 2002). Currently, the human blood plasma proteome is being mapped to gain more insight into disease and toxicity. Proteomics analysis is increasingly applied in vitro, including applications based on stem cells, embryotoxicity in embryonic stem cells, and mechanisms of toxicity in primary cell cultures. In vitro proteomics is proving to be a valuable method of identifying protein biomarkers for mechanisms of toxicity (e.g., oxidative stress, cell death, or energy metabolism). Proteomics is still less standardized than transcriptomics or metabolomics, however, and thus is more difficult to apply for PoT identification. The sensitivity of proteomics methods for identification of single proteins is lower than immunohistochemistry, enzyme-linked immunosorbent assay (ELISA), or Western blotting (Heijne et al. 2005). However, in contrast to those methods, proteomics allows the detection of changes in unexpected and unknown proteins and often subsequent identification with MS. Typically, only water-soluble proteins within a specified mass and isoelectric point range will be analyzed.

The number of metabolites in a biological system is estimated to be a few thousand, which is relatively small compared to the number of genes and proteins. Metabolomics, therefore, is considered a more approachable methodology than proteomics, particularly in translating results to phenotypic changes. At the same time, while transcriptomics and proteomics only indicate possible derangements of physiology, a shift in metabolites indicates actual change in physiologic chemistry. The principal technologies for metabolomics studies are NMR spectroscopy and MS. NMR examines the proton spectrum of a sample, which represents a robust and quantitative measurement. Advantages of NMR are that it is noninvasive, requires no metabolite extraction procedure, and allows a relatively easy structural identification of metabolites. In the case of MS, signals are characterized by mass, time of flight (TOF), and, if coupled with chromatography, a retention time. To confirm the identities of metabolites relevant to a specific toxic perturbation, their accurate mass, retention time parameters, or NMR spectra can be compared to a database with annotated metabolites, such as METLIN containing thousands of metabolites (Smith et al. 2005). These two techniques can be used to form an SoT, even without identification of the substance. Metabolomics toxicity studies have been performed both in vivo and in vitro. Metabolomics applications in cell cultures have been initiated only recently, but they are highly promising, as they could provide insight into human toxicity pathways (Kleinstreuer et al. 2011).

The underlying structure of systems biology and toxicology is a network. Networks can vary in their functionalities. Some are undirected graphs that enable only the study of structure; others, like the biochemical network, are characterized by interactions of varying strengths, strongly nonlinear dynamics, and saturating response to inputs. Omics technologies offer a prime opportunity to refine existing tests and to provide information for ITS via the creation of SoT. Finally, the components of quantitative in vitro to in vivo extrapolation (QIVIVE) and exposure move us to the risk paradigm that ultimately is required. Besides the techniques described above, some ultimate researchers have reported other new proposes for how to overcome the acknowledged scientific gaps for the full replacement of systemic toxicity testing using animals. In this sense, researches as Basketter et al. (2012) have reviewed new alternative methods which address toxico-kinetics, skin sensitization, repeated-dose toxicity, carcinogenicity, and reproductive toxicity testing. In summary, systems toxicology, with its new datasets and large-scale data integration, will enable the exploration of properties of biological systems beyond what is currently possible.

3.8 Final Considerations

In summary, existing studies have demonstrated that nanotechnology offers many advantages, such as improved stability, favorable biodistribution profiles, slower drug release kinetics, lower immunotoxicity, and targeting to specific cell populations.

NP toxicology is a relatively young field, and the bulk of reports have focused on acute toxicity. Long-term toxicity of the materials and examination of chronic exposure are critical to understanding the toxicology of nanomaterials in vivo. Evaluation of toxicity has proven to be challenging as several factors may be working in tandem to cause NP toxicity. Moreover, as nanomaterials are inherently quite complex, many unexpected interactions (based on bulk properties) with biological components may arise. However, with appropriately validated analytical methods and carefully designed experimentation, the mechanisms of toxicity may become clearer so that nanomaterials can safely be used as therapeutics and as diagnostic tools. For a good understanding about toxicological effects of nano-objects, it will be important to characterize the effect of nanomaterials on biological systems. In the next decade, it will be crucial to elucidate how the physicochemical properties of nanomaterials and their by-products interact with subcellular organelles, cells, tissues, and organisms. This will greatly affect our ability to engineer new generations of nontoxic products containing NPs. The publication of well-executed fundamental studies will provide the design criteria for successful NP-based strategies in vivo. The progress in nanobiotechnology, material synthesis, and computer simulation studies can potentially change how nanostructures can be engineered that could lead to novel applications. Static and dynamic stability of nano-objects in relation to their structure, binding to proteins and other subcellular and cellular components, tissues, and organs, will be crucial to avoid undesirable toxic effects of nano-objects from new generations. In this regard, advanced analysis of the physical and chemical characteristics of NPs will continue to be essential in revealing the relationship between their size, composition, crystallinity, and morphology and their electromagnetic response properties, reactivity, aggregation, and kinetics. It is important to remark that fundamental properties of NPs are still being discovered. A systematic scientific approach to the study of NP toxicity requires correlation of the physicochemical properties of NPs with their toxicity. It is a big challenge in the engineered nano-objects for applications in nanomedicine and changes in research areas are expected since existing research on nanotoxicity has concentrated on empirical evaluation of the toxicity of various NPs, with less regard given to the relationship between NP physicochemical properties and toxicity. At present time, even much less importance in the consideration of static and dynamic stability as crucial concepts in nanotoxicity. Further studies on kinetics and biochemical interactions of NPs within organisms are imperative. These studies must include, at least, research on NPs translocation pathways, accumulation, short- and long-term toxicity, their interactions with cells, the receptors and signaling pathways involved, cytotoxicity, and their surface functionalization for an effective phagocytosis. A relatively new field of research is nano-immunology should be developed in the near future to prevent or enhance immunomodulatory effects product of the administration of nano-objects as carriers or adjuvants in vaccines or as therapeutic agents in the treatment of diverse immunological diseases.

The extent of nano-immunological studies is vast since lessons learned from previous studies include the importance of detection and prevention of potential particle contamination with such things as bacterial endotoxins and/or toxic synthesis by-products and the importance of understanding how route of administration and particle biodistribution in the body may result in either desirable or undesirable immunomodulation (e.g., complement activation on Iv administration is not desirable, whereas on sc administration, it is beneficial for vaccinations). Nanotechnology platforms are being investigated as vaccine carriers, adjuvants, and drug delivery systems to target inflammatory and inflammation-associated disorders. Some formulations are already in clinical trials, whereas many others are in various phases of preclinical development. Although in recent years our understanding of NP interaction with components of the immune system has improved, many questions still require more thorough investigation and deeper understanding. Further mechanistic studies investigating particle immunomodulatory effects (immunostimulatory and immunosuppression) are required to improve our understanding of the physicochemical parameters of NPs that define their effects on the immune system.

Therefore, we can expect medical nanotechnology will continue to be a scientific area that will most certainly draw a lot of attention from both basic medical researchers and clinical practitioners, providing that issues concerning toxicity of NPs are resolved.

The challenge in new ways to explore the nanotoxicity also reaches the methods to study toxicity; in this way, traditional in vivo tests and in vitro assays typically provide only limited mechanistic information, while new comprehensive screening studies are needed to reveal the interactions of chemicals with biochemical pathways that control cell function, communication, and adaptation to environmental changes. The whole comprehension of all the aspects of biology and toxicity concerning the use of nano-object should be done from another perspective; it means with new approaches as the use of omics such as transcriptomics, proteomics, and metabolomics. These approaches will imply more collaborative work between research groups with different scopes in a multidisciplinary environment to give answers to complex matters in nanotoxicity contextualized into nanobiotechnology. All these new aspects to consider in the study of nano-objects taking into account the physical chemistry not only in structure but in the interactions that they may have in a whole organism offer the possibility of a quantum leap in science of nanobiomaterials. This quantum leap will bring a significant advance in applied science to biomedicine, agriculture, environmental, etc.

References

Agashe HB, Dutta T, Garg M et al (2006) Investigations on the toxicological profile of functionalized fifth-generation poly (propylene imine) dendrimer. J Pharm Pharmacol 58:1491–1498

Ai J, Biazar E, Jafarpour M et al (2011) Nanotoxicology and nanoparticle safety in biomedical designs. Int J Nanomedicine 6:1117–1127

Aillon KL, Xie Y, El-Gendy N et al (2009) Effects of nanomaterial physicochemical properties on *in vivo* toxicity. Adv Drug Deliv Rev 61(6):457–466

Albanese A, Tang PS, Chan WC (2012) The effect of nanoparticle size, shape, and surface chemistry on biological systems. Annu Rev Biomed Eng 14:1–16

Andersen ME, Clewell HJ III, Bermudez E et al (2010) Formaldehyde: integrating dosimetry, cytotoxicity, and genomics to understand dose-dependent transitions for an endogenous compound. Toxicol Sci 118:716–731

AshaRani PV, Low Kah Mun G, Hande MP et al (2008) Cytotoxicity and genotoxicity of silver nanoparticles in human cells. ACS Nano 3:279–290

Balashazy I, Hofmann W, Heistracher T (2003) Local particle deposition patterns may play a key role in the developing of lung cancer. J Appl Physiol 94:1719–1725

Basketter DA, Clewell H, Kimber I et al (2012) A roadmap for the development of alternative (non-animal) methods for systemic toxicity testing. ALTEX 29:3–91

Borm PJ, Kreyling W (2004) Toxicological hazards of inhaled nanoparticles-potential implications for drug delivery. J Nanosci Nanotechnol 4:521–531

Bosio VE, Cacicedo M, Calvignac B et al (2013) Synthesis and characterization of $CaCO_3$-biopolymers hybrid nanoporous microparticles for cancer damaged bone therapy. Submitted

Buzea C, Pacheco I, Robbie K (2007) Nanomaterials and nanoparticles: sources and toxicity. Biointerphases 2:MR17–MR71

Carlson C, Hussain SM, Schrand AM et al (2008) Unique cellular interaction of silver nanoparticles: size-dependent generation of reactive oxygen species. J Phys Chem B 112:13608–13619
Casey A, Davoren M, Herzog E et al (2007a) Probing the interaction of single walled carbon nanotubes within cell culture medium as a precursor to toxicity testing. Carbon 45:34–40
Casey A, Herzog E, Davoren M et al (2007b) Spectroscopic analysis confirms the interactions between single walled carbon nanotubes and various dyes commonly used to assess cytotoxicity. Carbon 45:1425–1432
Cedervall T, Lynch I, Lindman S et al (2007) Understanding the nanoparticle protein corona using methods to quantify exchange rates and affinities of proteins for nanoparticles. Proc Natl Acad Sci U S A 104:2050–2055
Chanan-Khan A, Szebeni J, Savay S (2003) Complement activation following first exposure to pegylated liposomal doxorubicin (Doxil): possible role in hypersensitivity reactions. Ann Oncol 14:1430–1437
Chen BX, Wilson SR, Das M et al (1998) Antigenicity of fullerenes: antibodies specific for fullerenes and their characteristics. Proc Natl Acad Sci U S A 95:10809–10813
Cheng Y, Zhao L, Li Y et al (2011) Design of biocompatible dendrimers for cancer diagnosis and therapy: current status and future perspectives. Chem Soc Rev 40:2673–2703
Chithrani BD, Chan WC (2007) Elucidating the mechanism of cellular uptake and removal of protein-coated gold nanoparticles of different sizes and shapes. Nano Lett 7:1542–1550
Choi HS, Liu W, Misra P et al (2007) Renal clearance of quantum dots. Nat Biotechnol 25:1165–1170
Connor EE, Mwamuka J, Gole A et al (2005) Gold nanoparticles are taken up by human cells but do not cause acute cytotoxicity. Small 1:325–327
Deng ZJ, Liang M, Monteiro M et al (2010) Nanoparticle-induced unfolding of fibrinogen promotes Mac-1 receptor activation and inflammation. Nat Nanotechnol 6:39–44
Derfus AM, Chan WCW, Bhatia SN (2004) Probing the cytotoxicity of semiconductor quantum dots. Nano Lett 4:11–18
Ding L, Stilwell J, Zhang T et al (2005) Molecular characterization of the cytotoxic mechanism of multiwall carbon nanotubes and nano-onions on human skin fibroblast. Nano Lett 5:2448–2464
Dobrovolskaia MA, McNeil SE (2007) Immunological properties of engineered nanomaterials. Nat Nanotechnol 2:469–478
Donaldson K, Stone V (2003) Current hypotheses on the mechanisms of toxicity of ultrafine particles. Ann Ist Super Sanita 39:405–410
Elgrabli D (2007) Effect of BSA on carbon nanotube dispersion for in vivo and in vitro studies. Nanotoxicology 1:266–278
Fadeel B, Garcia-Bennett AE (2010) Better safe than sorry: understanding the toxicological properties of inorganic nanoparticles manufactured for biomedical applications. Adv Drug Deliv Rev 62:362–374
Fifis T, Gamvrellis A, Crimeen-Irwin B et al (2004) Size-dependent immunogenicity: therapeutic and protective properties of nano-vaccines against tumors. J Immunol 173:3148–3154
Fischer HC, Chan WC (2007) Nanotoxicity: the growing need for in vivo study. Curr Opin Biotechnol 18:565–571
Fischer HC, Hauck TS, Gomez-Aristizabal A et al (2010) Exploring primary liver macrophages for studying quantum dot interactions with biological systems. Adv Mater 22:2520–2524
Gao H, Shi W, Freund LB (2005) Mechanics of receptor-mediated endocytosis. Proc Natl Acad Sci U S A 102:9469–9474
Geng Y, Dalhaimer P, Cai SS et al (2007) Shape effects of filaments versus spherical particles in flow and drug delivery. Nat Nanotechnol 2:249–255
Gessner A, Lieske A, Paulke B et al (2002) Influence of surface charge density on protein adsorption on polymeric nanoparticles: analysis by two-dimensional electrophoresis. Eur J Pharm Biopharm 54:165–170

Gitsov I, Hamzik J, Ryan J et al (2008) Enzymatic nanoreactors for environmentally benign biotransformations. 1. Formation and catalytic activity of supramolecular complexes of laccase and linear-dendritic block copolymers. Biomacromolecules 9:804–811

Gormley AJ, Ghandehari H (2009) Evaluation of toxicity of nanostructures in biological systems. In: Sahu SC, Casciano DA (eds) Nanotoxicity. Wiley, Chichester

Gratton SE, Ropp PA, Pohlhaus PD et al (2008) The effect of particle design on cellular internalization pathways. Proc Natl Acad Sci U S A 105:11613–11618

Gurr JR, Wang ASS, Chen CH et al (2005) Ultrafine titanium dioxide particles in the absence of photoactivation can induce oxidative damage to human bronchial epithelial cells. Toxicology 213:66–73

Guzelian AA, Katari JEB, Kadavanich AV et al (1996) Synthesis of size-selected, surface-passivated InP nanocrystals. J Phys Chem 100:7212–7219

Hahn A, Fuhlrott J, Loos A, Barcikowski S (2012) Cytotoxicity and ion release of alloy nanoparticles. J Nanopart Res 14:1–10

Hartung T (2001) Three Rs potential in the development and quality control of pharmaceuticals. ALTEX 18(suppl 1):3–13

Hartung T (2010) Food for thought…on alternative methods for chemical safety testing. ALTEX 27:3–14

Hartung T, McBride M (2011) Food for thought…on mapping the human toxome. ALTEX 28:83–93

Heijne WH, Kienhuis AS, van Ommen B et al (2005) Systems toxicology: applications of toxicogenomics, transcriptomics, proteomics and metabolomics in toxicology. Expert Rev Proteomics 2:767–780

Hoet PHM, Bruske-Hohlfeld I, Salata OV (2004) Nanoparticles-known and unknown health risks. J Nanobiotechnol 2:12–27

Hussain N, Jani PU, Florence AT (1997) Enhanced oral uptake of tomato lectin-conjugated nanoparticles in the rat. Pharm Res 14:613–618

Hutter E, Boridy S, Labrecque S et al (2010) Microglial response to gold nanoparticles. ACS Nano 4(5):2595–2606

Ishida T, Atobe K, Wang X et al (2006) Accelerated blood clearance of PEGylated liposomes upon repeated injections: effect of doxorubicin-encapsulation and high-dose first injection. J Control Release 115:251–258

Ishida T, Wang X, Shimizu T et al (2007) PEGylated liposomes elicit an anti-PEG IgM response in a T cell-independent manner. J Control Release 122:349–355

Jain RK, Stylianopoulos T (2010) Delivering nanomedicine to solid tumors. Nat Rev Clin Oncol 7:653–664

Jiang W, Kim BY, Rutka JT et al (2008) Nanoparticle-mediated cellular response is size-dependent. Nat Nanotechnol 3:145–150

Jiang H, Song C, Chen CC et al (2010) Quantitative 3D imaging of whole, unstained cells by using X-ray diffraction microscopy. Proc Natl Acad Sci U S A 107:11234–11239

Jin H, Heller DA, Sharma R et al (2009) Size-dependent cellular uptake and expulsion of single-walled carbon nanotubes: single particle tracking and a generic uptake model for nanoparticles. ACS Nano 3:149–158

Jones C, Grainger DW (2009) In vitro assessments of nanomaterial toxicity. Adv Drug Deliv Rev 61:438–456

Kalkanidis M, Pietersz GA, Xiang SD et al (2006) Methods for nano-particle based vaccine formulation and evaluation of their immunogenicity. Methods 40:20–29

Kanaras AG, Bartczak D, Sanchez-Elsner T et al (2011) Receptor-mediated interactions between colloidal gold nanoparticles and human umbilical vein endothelial cells. Small 7:388–394

Kennedy S (2002) The role of proteomics in toxicology: identification of biomarkers of toxicity by protein expression analysis. Biomarkers 7:269–290

Kitchens KM, El-Sayed ME, Ghandehari H (2005) Transepithelial and endothelial transport of poly (amidoamine) dendrimers. Adv Drug Deliv Rev 57:2163–2176

Kleinstreuer NC, Smith AM, West PR et al (2011) Identifying developmental toxicity pathways for a subset of ToxCast chemicals using human embryonic stem cells and metabolomics. Toxicol Appl Pharmacol 15:111–121
Laaksonen T, Santos H, Vihola H et al (2007) Failure of MTT as a toxicity testing agent for mesoporous silicon microparticles. Chem Res Toxicol 20:1913–1918
Lacerda SH, Park JJ, Meuse C et al (2009) Interaction of gold nanoparticles with common human blood proteins. ACS Nano 4:365–379
Lee SC, Parthasarathy R, Botwin K et al (2004) Biochemical and immunological properties of cytokines conjugated to dendritic polymers. Biomed Microdevices 6:191–202
Lee HA, Imran M, Monteiro-Riviere NA et al (2007) Biodistribution of quantum dot nanoparticles in perfused skin: evidence of coating dependency and periodicity in arterial extraction. Nano Lett 7:2865–2870
Lewinski N, Colvin V, Drezek R (2008) Cytotoxicity of nanoparticles. Small 4:26–49
Lippmann M (1990) Effects of fiber characteristics on lung deposition, retention, and disease. Environ Health Perspect 88:311–317
Lowry GV, Gregory KB, Apte SC et al (2012) Transformations of nanomaterials in the environment. Environ Sci Technol 46:6893–6899
Lück M, Pistel KF, Li YX et al (1998) Plasma protein adsorption on biodegradable microspheres consisting of poly(D, L-lactide-co-glycolide), poly(L-lactide) or ABA triblock copolymers containing poly(oxyethylene). Influence of production method and polymer composition. J Control Release 55:107–120
Lynch I, Dawson KA (2008) Protein-nanoparticle interactions. Nano Today 104:2050–2055
Makidon PE, Bielinska AU, Nigavekar SS et al (2008) Pre-clinical evaluation of a novel nanoemulsion-based hepatitis B mucosal vaccine. PLoS One 3:e2954
Manolova V, Flace A, Bauer M et al (2008) Nanoparticles target distinct dendritic cell populations according to their size. Eur J Immunol 38:1404–1413
Markiewski MM, DeAngelis RA, Benencia F (2008) Modulation of the antitumor immune response by complement. Nat Immunol 9:1225–1235
Masadeh AS (2008) Quantitative structure determination of nanostructured materials using the atomic pair distribution function analysis. ProQuest LLC, Ann Arbor, MI
Matrin A (1993) Physical pharmacy: physical chemical principles in the pharmaceutical sciences, 4th edn. Williams & Wilkins, Baltimore, MD
Miller SA, Hong ED, Wright D (2006) Rapid and efficient enzyme encapsulation in a dendrimer silica nanocomposite. Macromol Biosci 6:839–845
Monteiro-Riviere NA, Inman AO, Zhang LW (2009) Limitations and relative utility of screening assays to assess engineered nanoparticle toxicity in a human cell line. Toxicol Appl Pharmacol 234:222–235
Mottram PL, Leong D, Crimeen-Irwin B et al (2007) Type 1 and 2 immunity following vaccination is influenced by nanoparticle size: formulation of a model vaccine for respiratory syncytial virus. Mol Pharm 4:73–84
Mukherjee P, Arvizo RR, Miranda OR et al (2010) Effect of nanoparticle surface charge at the plasma membrane and beyond. Nano Lett 10:2543–2548
Muller J, Huaux F, Moreau N et al (2005) Respiratory toxicity of multi-wall carbon nanotubes. Toxicol Appl Pharmacol 207:221–231
Murray CB, Kagan CR, Bawendi MG (2000) Synthesis and characterization of monodisperse nanocrystals and close-packed nanocrystal assemblies. Ann Rev Mater Sci 30:545–610
Nel A, Xia T, Mädler L, Li N (2006) Toxic potential of materials at the nanolevel. Science 311 (5761):622–627
Nel AE, Mädler L, Velegol D et al (2009) Understanding biophysicochemical interactions at the nano-bio interface. Nat Mater 8:543–557
Nielsen MS, Lauridsen T, Thomsen M et al (2012) X-ray tomography using the full complex index of refraction. Phys Med Biol 57:5971–5979

Norris DA, Puri N, Sinko PJ (1998) The effect of physical barriers and properties on the oral absorption of particulates. Adv Drug Deliv Rev 34:135–154

Nygaard UC, Hansen JS, Samuelsen M (2009) Single-walled and multi-walled carbon nanotubes promote allergic immune responses in mice. Toxicol Sci 109:113–123

Oberdörster G (2002) Toxicokinetics and effects of fibrous and nonfibrous particles. Inhal Toxicol 14:29–56

Oberdörster G, Ferin J, Lehnert BE (1994) Correlation between particle size, *in vivo* particle persistence, and lung injury. Environ Health Perspect 102(suppl 5):173–179

Oberdörster G, Oberdörster E, Oberdörster J (2005) Nanotoxicology: an emerging discipline evolving from studies of ultrafine particles. Environ Health Perspect 113:823–839

O'Hagan DT, De Gregorio E (2009) The path to a successful vaccine adjuvant—'the long and winding road'. Drug Discov Today 14:541–551

Otsuka H, Nagasaki Y, Kataoka K (2003) PEGylated nanoparticles for biological and pharmaceutical applications. Adv Drug Deliv Rev 55:403–419

Owens DE III, Peppas NA (2006) Opsonization, biodistribution, and pharmacokinetics of polymeric nanoparticles. Int J Pharm 307:93–102

Pantic I (2011) Nanoparticles and modulation of immune response. Sci Prog 94:97–107

Perumal OP, Inapagolla R, Kannan S, Kannan RM (2008) The effect of surface functionality on cellular trafficking of dendrimers. Biomaterials 29:3469–3476

Poland CA, Duffin R, Kinloch I et al (2008) Carbon nanotubes introduced into the abdominal cavity of mice show asbestos-like pathogenicity in a pilot study. Nat Nanotechnol 3:423–428

Risom L, Moller P, Loft S (2005) Oxidative stress-induced DNA damage by particulate air pollution. Mutat Res 592:119–137

Rosi NL, Mirkin CA (2005) Nanostructures in Biodiagnostics. Chem Rev 105:1547–1562

Roduner E (2006) Size matters: why nanomaterials are different. Chem Soc Rev 35:583–592

Ryan SM, Mantovani G, Wang X et al (2008) Advances in PEGylation of important biotech molecules: delivery aspects. Expert Opin Drug Deliv 5:371–383

Ryman-Rasmussen JP, Riviere JE, Monteiro-Riviere NA (2007) Surface coatings determine cytotoxicity and irritation potential of quantum dot nanoparticles in epidermal keratinocytes. J Invest Dermatol 127:143–153

Salvador-Morales C, Flahaut E, Sim E et al (2006) Complement activation and protein adsorption by carbon nanotubes. Mol Immunol 43:193–201

Sayes CM, Reed KL, Warheit DB (2007) Assessing toxicity of fine and nanoparticles: comparing *in vitro* measurements to in vivo pulmonary toxicity profiles. Toxicol Sci 97:163–180

Schöler N, Olbrich C, Tabatt K et al (2001) Surfactant, but not the size of solid lipid nanoparticles (SLN) influences viability and cytokine production of macrophages. Int J Pharm 221:57–67

Schulze C, Kroll A, Lehr C-M et al (2008) Not ready to use-over coming pitfalls when dispersing nanoparticles in physiological media. Nanotoxicology 2:51–61

Shan X, Liu C, Yuan Y et al (2009) In vitro macrophage uptake and in vivo biodistribution of long-circulation nanoparticles with poly(ethylene-glycol)-modified PLA (BAB type) triblock copolymer. Colloids Surf B Biointerfaces 72:303–311

Shcharbin D, Jokiel M, Klajnert B, Bryszewska M (2006) Effect of dendrimers on pure acetylcholinesterase activity and structure. Bioelectrochemistry 68:56–59

Sheehan D (2007) The potential of proteomics for providing new insights into environmental impacts on human health. Rev Environ Health 22:175–194

Shi L, Reid LH, Jones WD et al (2006) The MicroArray Quality Control (MAQC) project shows inter- and intraplatform reproducibility of gene expression measurements. Nat Biotechnol 24:1151–1161

Shippy R, Fulmer-Smentek S, Jensen RV et al (2006) Using RNA sample titrations to assess microarray platform performance and normalization techniques. Nat Biotechnol 24:1123–1131

Smith CA, O'Maille G, Want EJ et al (2005) METLIN: a metabolite mass spectral database. Ther Drug Monit 27:747–751

Soenen SJ, Rivera-Gil P, Montenegro JM et al (2011) Cellular toxicity of inorganic nanoparticles: Common aspects and guidelines for improved nanotoxicity evaluation. Nano Today 6:446–465

Stoeger T, Reinhard C, Takenaka S et al (2006) Instillation of six different ultrafine carbon particles indicates a surface area threshold dose for acute lung inflammation in mice. Environ Health Perspect 114:328–333

Takenaka S, Karg E, Roth C et al (2001) Pulmonary and systemic distribution of inhaled ultrafine silver particles in rats. Environ Health Perspect 109(suppl 4):547–551

Takahashi H, Niidome Y, Niidome T et al (2006) Modification of gold nanorods using phosphatidylcholine to reduce cytotoxicity. Langmuir 22:2–5

Thorek DLJ, Tsourkas A (2008) Size, charge and concentration dependent uptake of iron oxide particles by non-phagocytic cells. Biomaterials 29:3583–3590

Toyama T, Matsuda H, Ishida I et al (2008) A case of toxic epidermal necrolysis-like dermatitis evolving from contact dermatitis of the hands associated with exposure to dendrimers. Contact Dermatitis 59:122–123

Vonarbourg A, Passirani C, Saulnier P et al (2006) Parameters influencing the stealthiness of colloidal drug delivery systems. Biomaterials 27:4356–4373

Wang S, Dormidontova EE (2010) Nanoparticle design optimization for enhanced targeting: Monte Carlo simulations. Biomacromolecules 11:1785–1795

Wang XT, Inapagolla R, Kannan S et al (2007) Characterization and *in vitro* activity of dendrimer-streptokinase conjugates. Bioconjug Chem 18:791–799

Wang B, Zhang L, Bae SC et al (2008) Nanoparticle-induced surface reconstruction of phospholipid membranes. Proc Natl Acad Sci U S A 105:18171–18175

Wang J, Tian S, Petros RA et al (2010) The complex role of multivalency in nanoparticles targeting the transferrin receptor for cancer therapies. J Am Chem Soc 132:11306–11313

Warheit DB (2008) How meaningful are the results of nanotoxicity studies in the absence of adequate material characterization? Toxicol Sci 101:183–185

Warheit DB, Sayes CM, Reed KL et al (2008) Health effects related to nanoparticle exposures: environmental, health and safety considerations for assessing hazards and risks. Pharmacol Ther 120(1):35–42

Waters KM, Masiello LM, Zangar RC et al (2009) Macrophage responses to silica nanoparticles are highly conserved across particle sizes. Toxicol Sci 107:553–569

Wetmore BA, Wambaugh JF, Ferguson SS et al (2012) Integration of dosimetry, exposure, and high-throughput screening data in chemical toxicity assessment. Toxicol Sci 125:157–174

Williams Y, Sukhanova A, Nowostawska M et al (2009) Probing cell-type-specific intracellular nanoscale barriers using size-tuned quantum dots. Small 5:2581–2588

Xia T, Kovochich M, Brant J et al (2006) Comparison of the abilities of ambient and manufactured nanoparticles to induce cellular toxicity according to an oxidative stress paradigm. Nano Lett 6:1794–1807

Yang H, Lopina ST (2006) In vitro enzymatic stability of dendritic peptides. J Biomed Mater Res A 76:398–407

Yemul O, Imae T (2005) Covalent-bonded immobilization of lipase on poly(phenylene sulfide) dendrimers and their hydrolysis ability. Biomacromolecules 6:2809–2814

Yuan HY, Li J, Bao G et al (2010) Variable nanoparticle-cell adhesion strength regulates cellular uptake. Phys Rev Lett 105:138101–138104

Zetasizer-nano (2007) Zetasizer nano user manual. http://www.malvern.com

Zhang Y, Kohler N, Zhang M (2002) Surface modification of superparamagnetic magnetite nanoparticles and their intracellular uptake. Biomaterials 23:1553–1561

Zhang D, Neumann O, Wang H et al (2009) Gold nanoparticles can induce the formation of protein-based aggregates at physiological pH. Nano Lett 9:666–671

Zolnik BS, González-Fernández A, Sadrieh N et al (2010) Nanoparticles and the immune system. Endocrinology 151:458–465

Chapter 4
Pharmacokinetics and Pharmacodynamics of Nanomaterials

Priscyla D. Marcato

Abstract Nanotechnology has provided a means to generate new and more effective ways to deliver drugs into body. Nanoparticles as drug delivery system are a promising approach to obtain a drug formulation with pharmacokinetic and pharmacodynamic properties modified. Nanoparticles can be used for delivering drug in a site-specific, alleviating unwanted toxicity due to nonspecific distribution, increasing the extent of tissue-specific accumulation, improving patient compliance, and providing favorable clinical outcomes. Furthermore, the nanocarrier systems can increase the drug bioavailability, sustain drug/gene effect in the target tissue, solubilize drugs for intravascular delivery, and/or improve the stability of therapeutic agents against enzymatic degradation. However, the understanding about how the physicochemical properties of nanoparticles interact with biological systems is fundamental in order to help to design "smarter" nanostructures. The particle size, surface charge, surface modifications (e.g., targeting ligand and pegylation functionalization), and composition can influence the pharmacokinetics and pharmacodynamics of nanostructures, thus influencing in its efficacy. Particles with specific characteristics can exhibit long circulation time in the bloodstream due to delayed opsonization and increase the cellular uptake and organ accumulation, thereby enhancing the therapy effect.

P.D. Marcato (✉)
Faculty of Pharmaceutical Sciences of Riberão Preto, Universidade de São Paulo, Ribeirão Preto, Sao Paulo, Brazil
e-mail: pmarcato@gmail.com

N. Durán et al. (eds.), *Nanotoxicology*, Nanomedicine and Nanotoxicology, 97
DOI 10.1007/978-1-4614-8993-1_4, © Springer Science+Business Media New York 2014

4.1 The Pharmacokinetics and Pharmacodynamics Profiles and Overview

Pharmacokinetics (PK) can be defined as the study of the time course of drug including absorption, biodistribution, metabolism, and excretion of an administered drug in the body in order to describe and predict the time course of drug concentration in body fluids (Fig. 4.1). In vivo experiments to determine the pharmacokinetics of a drug refer to time series measurements of drug concentration in body fluid (e.g., plasma or blood) or tissue (Benet 1984). An important parameter measured in PK is the bioavailability, which evaluates the fraction of an administered dose of unchanged drug that reaches the systemic circulation when the drug is not administrated by intravenous route. When the drug is administrated by intravenous route, the bioavailability is 100 %, and by other administration routes, the bioavailability decreases due to incomplete absorption, first-pass metabolism, or can vary from patient to patient. The studies of pharmacokinetics have been used in clinical medicine for many years in order to optimize the efficacy of medications administered to treat disease (Jenkins 2007). Pharmacodynamics, on the other hand, refers to the relationship between drug concentration at the site of action and the resulting effect, including the time course and intensity of therapeutic and adverse effects (Fig. 4.1).

So, pharmacodynamics establishes a relationship between drug exposure and its effect. The effect of a drug present at the site of action is determined by that drug's binding with a receptor. For most drugs, the concentration at the site of the receptor determines the intensity of a drug's effect. In vivo experiments to determine the pharmacodynamics include a direct or indirect measure of drug effect, such as pain relief or the concentration of a certain hormone. There is a relationship between the concentration of drug at the receptor site and the pharmacologic effect. In pharmacodynamic studies, if enough concentrations are tested, a maximum effect (E_{max}) and the drug potency (EC_{50}) (50 % effective concentration) can be determined, plotting the plasma drug concentration versus effect. Another parameter that is possible to determine is the tolerance that refers to a decrease in drug effectiveness with its continued use. The tolerance can be due to pharmacokinetic or pharmacodynamic factors. So, integrating pharmacokinetic and pharmacodynamic modeling, the time course of drug effect in the patient can be described and predicted.

Mathematical modeling has been used to predict the pharmacokinetics and pharmacodynamics of drugs. Different models were developed based on parameters derived from the experimental measurements obtained parameter estimation forms that a part of model development. The models may be used to predict concentration vs. time profiles for different dosing patterns (Jenkins 2007). In some case, in the mathematical model, it is the unique possibility to predict the pharmacokinetics/pharmacodynamics, for example, in studies in children and pregnant women. Thus, mathematical models can be used to predict drug concentration and drug effect prior to in vivo experiments, and when in vivo study becomes available,

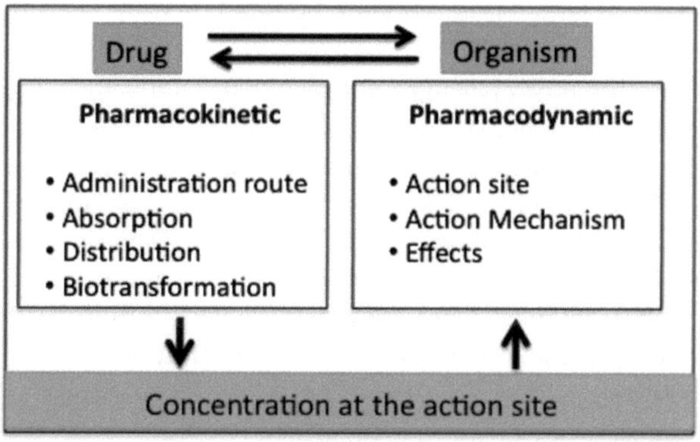

Fig. 4.1 Relation of pharmacokinetics and pharmacodynamics

the comparison of the data to the model predictions can be used to assess the reliability of the model assumptions.

In pharmacokinetics (PK) studies, the levels of drug and its metabolites (products of biotransformation) in the body are determined, allowing to obtain important data on these substances, such as:

- Conditions for its proper use.
- Dosage (dose and dosing interval).
- Side effects, for example, in the case of accumulation of drug in a given compartment (organotropism) or those from drug interactions at the level of the absorption, distribution, biotransformation, and excretion.
- Determination of the major sites of biotransformation.
- Determination of the excretion routes.
- Determination of drug frequency of administration.

The parameters available in PK studies are clearance, volume of distribution, bioavailability, and half-life. A clearance rate is defined as the volume of blood or plasma completely cleared of drug per unit time:

$$\text{Clearance (CL)} = \text{Rate of elimination}/\text{Concentration}$$

The clearance is the ability of the body to eliminate drug but is not a measure of the amount of drug removed; however, this parameter measures the volume of biological fluid that would have to have the drug removed to account for drug elimination (Bourne 2002).

The volume of distribution, also known as apparent volume of distribution (V), is an important indicator of the extent of drug distribution into body fluids and tissues. V is defined as a proportionality constant that relates the amount of drug in the body to the measured concentration in body fluid (e.g., plasma). This parameter has the dimensions of volume (e.g., liters):

$$V = \text{Amount in body/Plasma drug concentration}$$

Volume of distribution does not normally reflect a physiological volume. This parameter indicates the apparent volume of fluid required to contain all the drug in the body at the same concentration as in the plasma or blood.

Bioavailability describes the fraction of an administered dose that is absorbed and becomes available at the site of drug action. The bioavailability of a drug by various routes may also be determined by comparing the area under the curve (AUC) obtained from the plasma concentration vs. time curve after intravenous (AUC_{IV}) and other routes of administration (Jenkis and Cone 2007; Chereson 2009):

$$\text{Bioavailability} = \text{AUC}/\text{AUC}_{\text{IV}}$$

The half-life ($t_{1/2}$) is the time it takes for the plasma drug concentration to decrease by 50 % and is typically calculated from the following equation:

$$t_{1/2} = 0.693/k$$

where k is the elimination rate constant ($K = CL/V$).

4.2 Comparison of Pharmacokinetics (PK) and Pharmacodynamics (PD) of Free and Encapsulated Drug in Nanoparticles

The pharmacokinetics (PK) and pharmacodynamics (PD) of duloxetine hydrochloride (DLX) free and encapsulated in nanostructured lipid carriers (NLC) were evaluated after intranasal administration. DLX is the first-line drug in the treatment of depression (major depressive disorder, MDD) because this drug is a potent reuptake inhibitor of serotonin and norepinephrine. Thus, its efficacy depends on their continued presence at the site of action (brain) over a prolonged period of time. The amount of DLX in brain and blood was analyzed, and the pharmacodynamics, through efficacy, was evaluated using locomotor activity test (LAT) and forced swimming test (FST). The DLX was encapsulated in NLC composite by solid lipid glyceryl monostearate and Capryol PGMC as liquid lipid, and Pluronic F-68 and sodium taurocholate were used as stabilizer. The particles were prepared by sonication method. The animals received intranasal administration of free and encapsulated drug through sample instillation in each nostril. These groups were compared with other groups that received the drug free or encapsulated by oral gavage. Intranasal DLX-NLC treatment significantly increased the total swimming and climbing time and significantly reduced the immobility period when compared with control and with DLX solution administrated by intranasal or oral pathway.

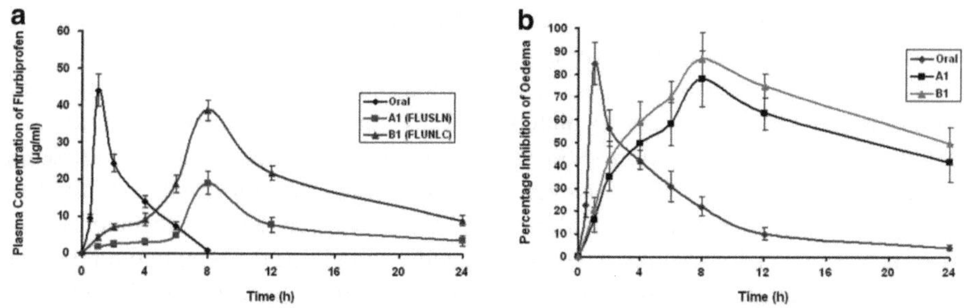

Fig. 4.2 (**a**) Plasma concentrations of flurbiprofen (FLU) after oral and transdermal administration ($n = 6$), (**b**) anti-inflammatory activity of FLU encapsulated in SLN (A1)- and NLC-enriched hydrogels (B1) after transdermal application in comparison to oral administration in carrageenan-induced rat paw edema ($n = 6$) (from Bhaskar et al. 2009, by permission of Biomed Central)

The DLX-NLC intranasally administrated increased the total swimming in fourfold in relation to control, twofold than DLX solution orally, and 1.25-fold than DLX solution intranasal pathway. Furthermore, the DLX concentration in brain was much higher with intranasal DLX-NLC than DLX solution formulation (Fig. 4.2b). Thus, the NLC enhanced the nose-to-brain delivery compared to drug solution formulations probably due to lipidic nature of NLC and due to enhancement of the permeability through nasal mucosa induced by sodium taurocholate which is used to prepare the particles (Alam et al. 2012).

The PK and PD of flurbiprofen (FLU) free and encapsulated in solid lipid nanoparticles (SLN) and in nanostructured lipid carriers (NLC) were studied after the transdermal application. Flurbiprofen is an inhibitor of platelet aggregation used to treat gout, osteoarthritis, rheumatoid arthritis, and others. This drug exhibits side effects as abdominal discomfort along with other gastrointestinal effects when it is administrated orally. An alternative to avoid these side effects is transdermal administration, but its skin permeation is very low due to efficient lipid barrier of the stratum corneum. In this way, FLU was encapsulated in SLN composite by solid lipid Dynasan 114, and phosphatidylcholine and polysorbate 80 were used as stabilizer. The NLC was composite by the same reagent and with liquid lipid Captex 355 EP/NF (caprylic/capric triglycerides). The FLU concentration in blood and the pharmacokinetic (PK) parameters (C_{max}, T_{max}, $t_{1/2}$, AUC_{0-24}, and $AUC_{0-\infty}$) were estimated. For pharmacodynamic (PD) assay, the paw edema method was used to determine the anti-inflammatory activity of FLU free and encapsulated. The C_{max} and AUC of FLU-NLC were, respectively, 1.8- and 2.5-fold higher than the FLU-SLN, showing the influence of particle type in PK. Another interesting point was the sustained FLU release. The FLU free showed a C_{max} in 1 h and its concentration decreases quickly until zero in 8 h. However, FLU encapsulated in SLN and NLC showed a C_{max} in 8 h and after 24 h there was still FLU in the plasma. Furthermore, the $AUC_{0-\infty}$ of FLU encapsulated and administrated transdermal route was higher than FLU orally administered.

The FLU bioavailability increased 4.4-fold when transdermal gel formulations were applied (Fig. 4.2a) (Bhaskar et al. 2009).

The same behavior was verified in pharmacodynamic (PD) assay. The formulations with FLU encapsulated in SLN and NLC showed a significant decrease of the inflammation (~60 % in 8 h), and this effect was sustained for more than 24 h (Fig. 4.2b). However, FLU free orally administrated exhibited an inhibition effect of below 40 % after 4 h. This difference can be attributed to the slow release of FLU from the nanoparticles (SLN and NLC), which maintained the FLU concentration in plasma for a longer period of time (Bhaskar et al. 2009).

Chlorambucil (CHL) was encapsulated in long-circulating nanoemulsion (LNE) composite by soybean oil, egg lecithin, cholesterol, and PEG_{2000}DSPE (Ganta et al. 2010). The nanoemulsion was modified with poly(ethylene glycol) (PEG) to obtain an LNE to improve the in vivo pharmacokinetics and enhance the therapeutic effect of the encapsulated chlorambucil. Chlorambucil is used clinically against chronic lymphocytic leukemia, lymphomas, and advanced ovarian and breast cancers. This lipophilic drug binds with DNA to prevent cell replication. The PK and PD were studied after intravenous administration of chlorambucil (CHL) free and encapsulated in LNE and encapsulated in nanoemulsion (NE) without PEG modification (NE). The encapsulation of chlorambucil in LNE improved the PK parameters. Six hours after the IV administration of CHL-LNE, the drug in the plasma was detected, whereas the CHL-NE and CHL free were measurable after 4 and 2 h, respectively. The $AUC_{0-\infty}$ for CHL-LNE was 1.4-fold higher than CHL-NE and 2.7-fold higher than that of CHL free. Furthermore, the CHL-LNE increased the half-life of drug to 1.3-fold compared to a nanoemulsion without PEG modification (CHL-NE) and showed lower clearance than either CHL-NE or CHL free. Another parameter modified was the volume of distribution of CHL that was increased after the CHL-LNE administrations compared to CHL free and CHL-NE (Table 4.1). Additionally, CHL-LNE decreased drug uptake in the reticuloendothelial system compared to nanoemulsion without PEG modification. This difference can be due to PEG modification that may play an important role in inhibiting the adsorption of opsonins onto lipid surface, thus preventing the clearance by the reticuloendothelial system (Ganta et al. 2010).

To evaluate the PD of chlorambucil (CHL), the formulations (CHL free, CHL-NE, and CHL-LNE) were IV administrated in mice with subcutaneous colon-38 adenocarcinoma tumor. This administration was repeated after 7 days and the tumor growth delay was determined. The CHL-LNE exhibited greater antitumor effect, increasing significantly the tumor growth delay and the tumor volume doubling time. In CHL free, the tumor growth delay time was 7.5 days whereas in CHL-NE and CHL-LNE it was 18.7 and 23.5 days, respectively. In relation to tumor volume doubling time, CHL-LNE and CHL-NE showed, respectively, 10.6 and 7.9 days while CHL free exhibited 4.1 days. Furthermore, CHL-LNE prolonged the life span of the tumor-bearing mice compared to the CHL-NE and CHL free—mice surviving over 48 days compared to 40 days with CHL-NE and 26 days with CHL free (Ganta et al. 2010).

Table 4.1 Pharmacokinetic (PK) parameter after IV administration of CHL free and in nanoemulsion formulations

PK parameters	CHL free	CHL-NE	CHL-LNE
$AUC_{0-\infty}$	17	32.4	45
$t_{1/2}$ β (h)	0.3	1.8	2.3
V_{ss} (mL/kg)	178	409	339

Modified from Ganta et al. (2010)

The pharmacodynamics and pharmacokinetics of artemether free and encapsulated in NLC were studied. Artemether (ARM) is a potent antimalarial drug, which is used for the treatment of severe multiresistant malaria. The NLC, with 63 nm of size, composite by glyceryl dilaurate, Capmul® MCM, tween 80, and solutol HS15 was produced by microemulsion method obtained particles with 63 nm of size. The ARM free and encapsulated in NLC was parenterally administrated in mice infected with *P. berghei*. The NLC-ARM showed a long circulation time in the bloodstream (more than 20 days) and exhibited a greater in vivo antimalarial efficacy than ARM solution. The encapsulation of ARM increased 20-fold the antimalarial effect than ARM solution and 2.5-fold than marketed formulation (Larither®). Furthermore, a complete clearance of parasite of the animals was observed in the groups treated with NLC-ARM, whereas marketed formulation (Larither 206®) showed around 45 % of activity on the eighth day. Thus, NLC-ARM showed significant improvement in the antimalarial activity and duration of action of ARM as compared to the conventional formulation, indicating an interesting system as a viable alternative to the current injectable intramuscular formulation (Joshi et al. 2008).

All these studies showed that the pharmacokinetics and pharmacodynamics were enhanced when the drug was encapsulated, probably because the nanocarrier can increase the drug bioavailability, its concentration at the site of action, and its uptake to the cells, improving the drug effects. Thus, the drug encapsulation in different nanoparticles is an interesting strategy to enhance drug activity.

4.3 Factors That Influence Pharmacokinetics of Nanoparticles

Nanostructures with a variety of sizes, shapes, composition, and surface modification can be produced. The interaction of these particles with biological system is depending on the nanoparticle's characteristics, such as size, superficial charge (zeta potential), and surface modification. Thus, the understanding about how the physicochemical properties of a synthetic nanoparticle interact with biological systems can help to design nanostructures more specific (Albanese et al. 2012; Elsaesser and Howard 2012; Mohanraj and Chen 2006). However, the analysis of pharmacokinetics of nanoparticles is not easy to do due to analytical methods that are inadequate to identify the particles in different organs or in the blood. Thus, several studies evaluated the PK of drug released from the nanoparticles and not of the nanostructure. In this topic, studies that evaluated the PK of different nanoparticles and evaluated the influence of particle's characteristics in the PK will be discussed.

4.3.1 Surface Modifications

Interactions between nanoparticles with surface modification, e.g., bound ligands and cellular receptors, depend on the nanoparticle geometry and the ligand density of a nanomaterial (Albanese et al. 2012). Nanoparticles with surface functionalization can be utilized to increase residence time of the particles in the blood, reduce nonspecific distribution, as well as to target tissues or specific cell. One of the problems with intravenous administration of nanoparticles is its rapid clearance from the bloodstream due to opsonization process. In this process, proteins are adsorbed on the particle surface leading to its aggregation and its rapid phagocytosis by the mononuclear phagocyte system (MPS) in the liver and spleen (Alex et al. 2008). Thus, several particles have a short circulation half-life and, consequently, exhibit low efficacy. To avoid this process, modification on nanoparticle surface can be done with hydrophilic polymers, e.g., poly(ethylene glycol) (PEG). This polymer can be grafted, conjugated, or absorbed to the nanoparticle surface to form the corona, which provides steric stabilization and prevents the protein absorption in the particle surface. In this case, a long-circulating nanoparticle is obtained (Ganta et al. 2010; Alexis et al. 2008).

Another point is to direct the particles to specific tissue or cells. For this goal nanoparticles are functionalized with antigens with a targeting ligand such as a peptide, aptamer, antibody/antibody fragment, small molecule, and others. Folic acid and biotin are most extensively studied as targeting molecules because of their small size and low cost of production (Jiao et al. 2012). Due to overexpression of the folate receptor on cancer cells, an attractive strategy for nanoparticle delivery in tumoral cells is to add acid folic on particle surface. In this case folate receptor-mediated endocytosis is an alternative interestingly because simultaneously it enhances drug's cell penetration and improves tumor specificity (Wang et al. 2011; Kularatne and Low 2010).

Superparamagnetic iron oxide nanoparticles were functionalized with poly (glycidyl methacrylate) (PGMA) and folic acid (FA) for cancer cell targeting (SPIONs–PGMA–FA) (Fig. 4.3a). These particles were highly stable in an aqueous phase and exhibited selective internalization of these nanoparticles in human epidermoid carcinoma cells (three and five times higher than normal cells) while concomitantly minimized its uptake by macrophages and normal cells (Fig. 4.3b). This study demonstrated interesting nanoparticles with a great potential to be used to selectively target cancer cells for imaging, drug delivery, as well as hyperthermia (Huang et al. 2011).

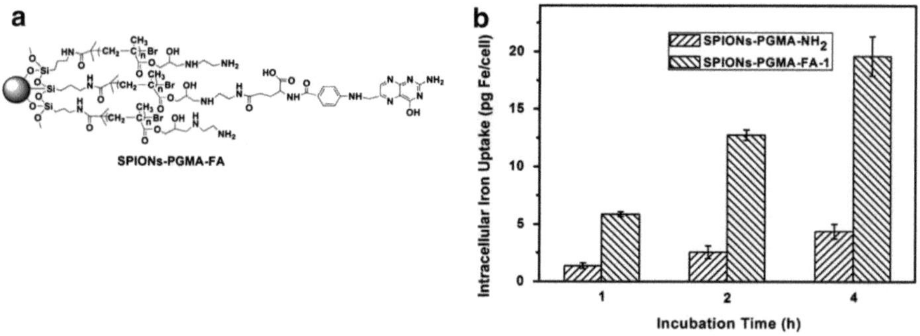

Fig. 4.3 (**a**) SPIONs–PGMA–FA structure; (**b**) intracellular uptake of SPIONs–PGMA–NH_2 and SPIONs–PGMA–FA-1 by KB cells as a function of the incubation time (modified from Huang et al. 2011)

4.3.2 Sizes

The size of nanoparticles is an important parameter to pharmacokinetics. This parameter also plays an important role as it influences surface pressure and adhesion forces (Elsaesser and Howard 2012). Particles with high specific surface area have a high interfacial chemical and physical reactivity of many types that translates to biological reactivity. Furthermore, particle size and plasma binding profile are a key factor in the biodistribution of long-circulating nanoparticles and achieving therapeutic efficacy (Hirn et al. 2011; Alexis et al. 2008). In general, nanoparticles with sizes smaller than 6 nm are quickly excreted by the kidneys. In the case of particles composite by degradable materials (e.g., polymers, lipids, or hydrogels), it cannot be eliminated by the kidneys when the diameter is greater than 6 nm. Particles with size larger than 200 nm accumulate in the spleen and liver, where they are processed by the mononuclear phagocyte system cells (Albanese et al. 2012).

Brinkhuis et al. (2012), using polymersome made of block copolymer polybutadiene-block-poly(ethylene glycol) with different sizes (90–250 nm) (Fig. 4.4a), verified a relation between blood circulation times and particle size. Particles larger than 120 nm were rapidly cleared in 4 h, whereas more than 30 % of particles with size around 90 nm were detected in blood pool after 24 h. Thus, smaller particles showed long circulating time with an estimated blood half-life of 24 h and the large particles accumulated in liver and spleen, probably due to recognition of these particles by the reticuloendothelial system (Fig. 4.4b).

As mentioned above, the particle surface modification with PEG improves the circulation time of the particles. In this line, Brinkhuis et al. (2012) studied the influence of the PEG molecular weight in the circulation time of polymersome. No influence in blood circulation half-life of the particles was verified when PEG's molecular weight (MW) was varied (Fig. 4.4b). Similar results were obtained with liposomes (Litzinger et al. 1994). Liposomes larger than 200 nm accumulated in the

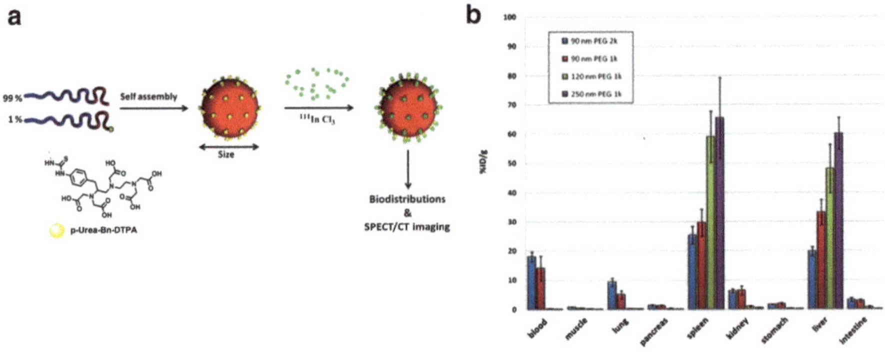

Fig. 4.4 (**a**) Schematic presentation of polymersome formation, ^{111}In labeling, and in vivo SPECT/CT imaging; (**b**) biodistribution of differently sized ^{111}In-labeled polymersomes in Balb/C mice (n = 4/group). (a) 4 h postinjection and (b) 24 h postinjection (from Brinkhuis et al. 2012 by permission of American Chemical Society)

spleen and liver, whereas liposomes of less than 70 nm accumulated predominantly in the liver. However, liposomes with 90 and 150 nm of diameter showed long blood circulation times. These studies show that the particle size influences the blood circulation kinetics and biodistribution of the particles. Thus the evaluation of size is a fundamental parameter in pharmacokinetic study of the nanoparticles.

The particle size influences the interaction between nanoparticles and serum proteins too. When particles are administrated into the bloodstream, serum proteins quickly coat the particles. The properties of the protein capping may define the biological response to the nanoparticle, such as influence on cellular uptake, organ accumulation, immune response, metabolism, and route of clearance (Albanese et al. 2012). In a recent study it was verified that particle size also plays a role in the interaction between gold nanoparticles with serum. Gold nanoparticle with 30 nm of size bound almost twofold more protein mass than the gold nanoparticles with size of 50 nm. This result can be explained by the difference in the surface area. The total surface area of nanoparticles with 30 nm of diameter is larger than the total surface area of the particles with 50 nm of size. This data was confirmed by 2D PAGE analysis that showed significantly more proteins detected in samples isolated from nanoparticles with 30 nm than from sample of gold nanoparticles with 50 nm (Dobrovolskaia et al. 2009).

4.3.3 Zeta Potential

Surface charge affects the biodistribution of nanoparticles and influences the adsorption of serum protein on particle surface too. Surface charge of particles plays a critical role in downstream intracellular events; thus, the understanding about interactions between cells and nanoparticles with different surface charges is very important to determine intracellular uptake and localization of the nanoparticles and their biological functions as well as their relative cytotoxicity (Albanese et al. 2012; Arvizo et al. 2010).

Nanoparticles with positive surface charge, in in vitro assay, are taken up at a faster rate than nanoparticles with a neutral or negative surface charge. This difference can be explained by difference between surface charge of cell membrane, that is negative, and nanoparticles. Thus, the interaction of positive particles with cell membrane leads to particle uptake by electrostatic attractions and by adhesion of the particles in the membrane. On the other hand, in in vivo assay, particles with positive surface charge are cleared most quickly from the blood and cause several complications such as hemolysis and platelet aggregation than negative or neutral particles, and this last one exhibits higher blood half-life than negative or positive particles. This difference can be due to the effect of particle charges and the interactions of nanoparticles with serum proteins. Thus, the particle surface charge is an important parameter to pharmacokinetics and biodistribution of the nanostructures (Albanese et al. 2012; Arvizo et al. 2010; Thorek and Tsourkas 2008; Slowing et al. 2006).

The influence of particle surface charge in a cellular membrane potential was studied using gold nanoparticles (~10 nm) with positive or negative charge. Positive particles depolarized more the membrane potential, causing an increased [Ca^{2+}] and, consequently, inhibiting normal cell proliferation (Arvizo et al. 2010). In other study, gold nanoparticles with 2.8 nm of diameter, prepared with ^{198}Au radioactivity, and with different surface charges (positive and negative) showed different accumulations in several organs. The quantification of gold nanoparticles in different organs, urine, and in the blood was made with γ-spectroscopy in either a lead-shielded 10 mL or a lead-shielded 1 L well-type NaI(Tl) scintillation detector (Hirn et al. 2011). After 24 h most of the administrated gold nanoparticles, with negative or positive surface charge, were found in the liver. However, the amount of particles in the liver was significantly higher for negative particles (81 %) than for positive particles (72 %). On the other hand, in the spleen the positive particles showed a significantly higher accumulation (11.4 %) than the negative particles (8.6 %). But in the urine, more amount of negative particles (0.71 %) than positive particles (0.51 %) were detected. Furthermore, more positive gold nanoparticles than negative gold nanoparticles were cleared through the hepatobiliary pathway, indicating the influence of surface charge in the pharmacokinetics of nanoparticles (Hirn et al. 2011).

4.4 Pharmacokinetics and Pharmacodynamics Issues in Nanotechnology

The method by which drugs are delivered to the body can have a very large effect on its pharmacokinetics (PK) and pharmacodynamics (PD). The encapsulation or association of a drug with nanostructures has been used as a physical approach to modify and/or improve the PK and PD properties of various drugs. Nanotechnology is widely used for delivery of various drugs to the body by increasing the bioavailability, reducing toxicity, sustaining drug/gene effect in the target tissue, solubilizing drugs for intravascular delivery, and/or improving the stability of therapeutic agents against enzymatic degradation (Arvizo et al. 2010; Marcato and Durán 2008). Through this technology, it is possible to protect the drug from degradation, overcome physical barriers within the body, as well as improve drug penetration into the target cell. Due to these advantages, nanomedicine is a fast-growing area. Improvements in many areas of drug delivery are only made possible by nanotechnology (Mohanraj and Chen 2006). For example, drug systemically administered is diluted by blood and is adsorbed on various tissue surfaces. Thus, significantly large doses of the drug need to be administered in order to compensate the dilution. If the drug is encapsulated in the nanostructure and this drug is delivered in a specific part of the body, the dilution effect will be low. In this case lower doses could be administered in order to achieve the same therapeutic effect (Devalapally et al. 2007).

A recent trend in nanotechnology has been to investigate the interactions of nanomaterials with biological systems. The understanding of how the physicochemical properties of nanostructures are related to biological interactions and functions is very important in order to adequately materialize the potential of nanotechnology (Albanese et al. 2012). For example, the study of effect of size, surface particle modification, and surface charge in the interaction with their target cells and in the PK/PD is fundamental to determinate the toxicity of these particles and its efficacy. The drug biodistribution can be modified when the drug is encapsulated or associated in a nanostructure. For example, nanoparticles with surface modification with PEG show low concentration in the organ of reticuloendothelial system indicating an interesting long-circulation formulation. Furthermore, the surface can be also modified with a targeting ligand to a target cell, obtaining a most specific formulation, which is able to improve the drug efficacy, increase the PK parameters, and decrease the drug toxicity.

Several nanostructures have been studied as drug delivery carriers such as liposomes, polymeric nanoparticles, dendrimers, solid lipid nanoparticles or lipid carrier systems, metallic nanoparticles (e.g., gold nanoparticles), carbon nanostructures (e.g., carbon nanotube, fullerenes, and graphene), micelles, nanoemulsion, and others. In general, the goal, in all cases, is to modify the PK/PD, improve the efficacy, and reduce the drug toxicity. However, in spite of the advantages of nanomedicine, nanoparticles have limitations. The small particle size and large surface area can lead to particle aggregation, making physical handling of nanoparticles difficult in liquid and dry forms. Furthermore, small particles with large surface area show, in general, limited drug loading, and fast drug release. These problems have to be overcome before nanoparticles can be used clinically or formulated in a commercial product. Additionally, nanostructure can induce hemolysis and long-circulating nanoparticles in the bloodstream increase the duration of the nanoparticle's contact with blood components including the coagulation system, amplifying activation of the coagulation cascade and blood clotting (Dobrovolskaia et al. 2008). Thus, nowadays, novel engineered nanoparticles have been developed to make these nanostructures increasingly "smarter," in order to obtain a better formulation with high stability in biological fluids, high drug loading and sustained release of drug, as well as a more specific nanocarrier to target cells.

4.5 Conclusions

Bioavailability, pharmacokinetic, pharmacodynamic, and biodistribution analyses are required for pharmacological understanding of nanoparticles. Particle's characteristics have a great influence in its interaction with biological medium modifying, consequently, the PK and PD of these particles. Thus, modification in the particle's characteristic can be done to enhance its interaction with biological system, such as the production of particles with long circulation time in the bloodstream or more specific particles for a targeting cell. Large particles tend to be quickly recognized by the reticuloendothelial system, thus accumulating in the liver and spleen. In relation to surface charge, particles with positive charge exhibit good adhesion on cellular membrane and are taken up at a faster rate than nanoparticles with a neutral or negative surface charge. On the other hand, particles with positive surface charge are quickly cleared from the blood and cause several complications (e.g., hemolysis and platelet aggregation) than negative or neutral particles.

Nanostructures can also be used as a drug delivery system for several drugs in order to modify the PK, PD, and biodistribution of the drug, improve the therapeutic efficiency, and reduce the side effects. Many interesting results have been obtained as shown in this chapter demonstrating that the nanotechnology could play a potential role in optimizing drug delivery and thus enhancing the therapeutic efficacy of several drugs. Nowadays, more "smarter" particles can be produced enhancing significantly the effect of these particles on the body obtaining a better formulation, which reduce the unwanted toxicity, enhance the therapy efficacy, improve patient compliance, and, consequently, improve clinical outcomes.

Acknowledgment The author acknowledges the FAPESP, INOMAT (MCT/CNPq), NanoBioss (MCTI), and Brazilian Network of Nanotoxicology (MCT/CNPq).

References

Alam MI, Baboota S, Ahuja A et al (2012) Intranasal administration of nanostructured lipid carriers containing CNS acting drug: pharmacodynamic studies and estimation in blood and brain. J Psychiatr Res 46:1133–1138

Albanese A, Tang PS, Chan WCW (2012) The effect of nanoparticle size, shape, and surface chemistry on biological systems. Annu Rev Biomed Eng 14:1–16

Alexis F, Pridgen E, Molnar LK et al (2008) Factors affecting the clearance and biodistribution of polymeric nanoparticles. Mol Pharm 5:505–515

Arvizo RR, Miranda OR, Thompson MA et al (2010) Effect of nanoparticle surface charge at the plasma membrane and beyond. Nano Lett 10:2543–2548

Benet LZ (1984) Pharmacokinetics: basic principles and its use as a tool in drug metabolism. In: Mitchell JR, Horning MG (eds) Drug metabolism and drug toxicity. Raven Press, New York

Bhaskar K, Anbu J, Ravichandiran V et al (2009) Lipid nanoparticles for transdermal delivery of flurbiprofen: formulation, in vitro, ex vivo and in vivo studies. Lipids Health Dis 8:1–15

Bourne DWA (2002) Pharmacokinetics. In: Schoenwald RD (ed) Pharmacokinetics in drug discovery and development, 1st edn. CRC Press, Boca Raton

Brinkhuis RP, Stojanov K, Laverman P et al (2012) Size dependent biodistribution and SPECT imaging of ^{111}In-labeled polymersomes. Bioconjug Chem 23:958–965

Chereson R (2009) Bioavailability, bioequivalence, and drug selection. In: Makoid CM, Vuchetich PJ, Banakar UV (eds) Basic pharmacokinetics, 1st edn. Pharmaceutical Press, London

Devalapally H, Chakilam A, Amiji MM (2007) Role of nanotechnology in pharmaceutical product development. J Pharm Sci 96:2547–2565

Dobrovolskaia MA, Aggarwal P, Hall JB et al (2008) Preclinical studies to understand nanoparticle interaction with the immune system and its potential effects on nanoparticle biodistribution. Mol Pharm 5:487–495

Dobrovolskaia MA, Patri AK, Zheng J et al (2009) Interaction of colloidal gold nanoparticles with human blood: effects on particle size and analysis of plasma protein binding profiles. Nanomedicine 5:106–117

Elsaesser A, Howard CV (2012) Toxicology of nanoparticles. Adv Drug Deliv Rev 64:129–137

Ganta S, Sharma P, Paxton JW et al (2010) Pharmacokinetics and pharmacodynamics of chlorambucil delivered in long-circulating nanoemulsion. J Drug Target 18:125–133

Hirn S, Semmler-Behnke M, Schleh C et al (2011) Particle size-dependent and surface charge-dependent biodistribution of gold nanoparticles after intravenous administration. Eur J Pharm Biopharm 77:407–416

Huang C, Neoh KG, Kang E-T et al (2011) Surface modified superparamagnetic iron oxide nanoparticles (SPIONs) for high efficiency folate-receptor targeting with low uptake by macrophages. J Mater Chem 21:16094–16102

Jenkins AJ (2007) Pharmacokinetics: basic concepts and models. In: Chang LW (ed) Pharmacokinetics: basic concepts and models, 1st edn. CRC Press/Taylor & Francis Group, Boca Raton

Jenkis AJ, Cone EJ (2007) Pharmacokinetics: drug absorption, distribution, and elimination. In: Karch SB (ed) Drug abused handbook, 2nd edn. CRC Press, Boca Raton

Jiao PF, Zhou HY, Chen LX et al (2012) Cancer-targeting multifunctionalized gold nanoparticles in imaging and therapy. Curr Med Chem 18:2086–2102

Joshi M, Pathak S, Sharma S, Patraval V (2008) Design and in vivo pharmacodynamic evaluation of nanostructured lipid carriers for parenteral delivery of artemether: nanoject. Int J Pharm 364:119–126

Kularatne SA, Low PS (2010) Targeting of nanoparticles: folate receptor. Methods Mol Biol 624:249–265

Litzinger DC, Buiting AMJ, Vanrooijen N, Huang L (1994) Effect of liposome size on the circulation time and intraorgan distribution of amphipathic poly(ethylene glycol)-containing liposomes. Biochim Biophys Acta 1190:99–107

Marcato PD, Durán N (2008) New aspects of nanopharmaceutical delivery systems. J Nanosci Nanotechnol 8:2216–2229

Mohanraj VJ, Chen Y (2006) Nanoparticles: a review. Trop J Pharm Res 5:561–573

Slowing I, Trewyn BG, Lin VSY (2006) Effect of surface functionalization of MCM-41-type mesoporous silica nanoparticles on the endocytosis by human cancer cells. J Am Chem Soc 128:14792–14793

Thorek DLJ, Tsourkas A (2008) Size, charge and concentration dependent uptake of iron oxide particles by non-phagocytic cells. Biomaterials 29:3583–3590

Wang X, Li J, Wang Y et al (2011) A folate receptor-targeting nanoparticle minimizes drug resistance in a human cancer model. ACS Nano 5:6184–6194

Chapter 5
In Vitro Cytotoxicity Assays of Nanoparticles on Different Cell Lines

Patricia S. Melo, Priscyla D. Marcato, Daniele R. de Araújo, and Nelson Durán

Abstract One of the greatest interests in the pharmaceutical and cosmetic industries today is the promising of new substances without adverse effects. Preclinical animal safety investigations and clinical trials increase the time necessary to bring a new candidate compound to the market and increase development costs and time dispensed in this process. Cell culture models are adequate for screening toxicity of several substances including nanomaterials. The data from those models can provide an indication of the safety use of these particles in humans since they are proven hazardous.

5.1 Introduction of Cell Culture in the Toxicity Study of Nanomaterials

The initial experiments in animal cell culture as a method for investigating the behavior of these cells in the absence of systemic variations were first devised by the pioneering work of Ross Harrison, Alexis Carrel, and others in the beginning of the twentieth century (Russell 1969). The use of cell culture has as principal advantage to minimize the changes that might have in vivo experiments mainly under the stress of an experiment. This research tool was elaborated first with fragments of tissue. Cell growth was limited to the migration of cells from the tissue piece or by surgical subdivision of the culture. The development of cell culture technology was possible due to the expansion of various undefined and defined culture media, as well the use of trypsin for subculturing. Adjustments in the atmosphere in which cells were maintained involved changes in the concentration of carbon dioxide in conjunction with sodium bicarbonate to maintain

P.S. Melo (✉)
METROCAMP, Grupo Ibmec, Campinas, Sao Paulo 13035-270, Brazil
e-mail: patricia.melo@metrocamp.edu.br

N. Durán et al. (eds.), *Nanotoxicology*, Nanomedicine and Nanotoxicology,
DOI 10.1007/978-1-4614-8993-1_5,

physiological pH. Indeed, the introduction of antibiotics in the defined media allowed the long-term cell line propagation (Freshney 2005).

Cell culture has several applications such as investigation of pharmacology (biological activity of a drug, drug metabolism and resistance, receptor interactions, etc.), cell–cell interaction (cell proliferation, cell adhesion, invasion and metastasis), toxicology (cytotoxicity, carcinogenesis, inflammation, etc.), genomics, and tissue engineering (Olabisi et al. 2010; Yang et al. 2011; Polchow et al 2012; Ayaki et al. 2012). The progress of cell culture allowed the production of antiviral vaccines and advances in the understanding of neoplasia expansion. In addition to these areas, genetic manipulation and advances in cell fusion techniques can be cited as areas of research that are heavily dependent on cell culture procedures (Harris 2012). In summary, cultured cells are widely used for studying biochemistry and functional cell biology, the effects of pharmaceuticals and drugs on cell physiology, the pathways and processes triggering cell death and aging, nutritional studies, and the correlations between disease-causing substances and cells (Freshney 1994, 2005; Machana et al. 2011; Pintus et al 2012; Fukazawa et al. 2012; Acosta et al. 2013).

In recent years, the idea of replacing animals in experimental tests with cell culture is being supported by the researchers around the world not only for humanitarian reasons embraced by animal rights communities but mainly because the advantages of tissue culture (Breheny et al 2011; Costin et al. 2011; Balls 2012). The principal benefits of cell culture are the control of the physicochemical environment (pH, temperature, osmotic pressure, and O_2 and CO_2 tension) and physiological conditions (maintenance of hormones and concentrations of nutrients, cell line homogeneity, assays replicable, etc.) (Freshney 2005).

Tissue culture technology has been used in medicine and in general industries, mainly in pharmaceutical and cosmetics areas. In the cosmetics development the public opinion is concerned about the extensive use of experimental animals and consider this kind of activity without moral justification (Combes and Balls 2011). In medical use, chromosomal analysis by cultivation of amniotic fluid cells in the handling of pregnancies can reveal genetic disorders; moreover, the toxic effects of pharmaceutical, cosmetics, and environmental pollutants can be also verified by cell culture assays.

Nanotechnology is used to break new insights in the fight and prevention of diseases using materials at the nanoscale. These nanomaterials gain new properties due to their small size and can be used in many areas, including medicine. Predicted applications in biomedical area include drug delivery, in vitro and in vivo diagnosis, nutraceuticals, and generation of biocompatible materials (De Jong and Borm 2008). However, nanoparticles could also provoke new types of effects that are not seen with major particles, mainly in in vitro assays (e.g., mitochondrial damage, oxidative stress, alterations in macrophage phagocytosis, platelet aggregation) (Moller et al. 2010; Frohlich 2012; You et al. 2012). The cytotoxicity of nanoparticles depends on several parameters as morphology, surface charges, and size. Some characteristics as hydrophobicity and porosity induce the production of reactive oxygen species and ascertain binding sites for receptors contributing to the toxic effects (Frohlich 2012). However, biological aspects such as cell type used in the cytotoxic assays and exposure protocols (e.g., cell density, particle concentration, medium composition including the addition of fetal serum) also interfere in the cytotoxicity (De Jong and Borm 2008).

Biocompatibility and toxicity are necessary to be investigated before a new material is employed as biomedical application. Cell culture models which make adequate systems for screening of nanomaterials toxicity can provide an indication about the secure use of these particles in humans since they are proven hazardous. Several nanoparticles show toxic effects in some cell types and these activities can be exploited as inhibitory agents in cancer cells (Gao et al. 2012).

Stem cell research has expanded dizzily in the biomedical area, but its applicability as therapeutic agent has progressed much more slowly. The prospects for using normal cells of adult or fetal tissue donors have generated perspectives in the tissue engineering mainly the possibility of a pluripotent stem cell providing new organs. The controversy over human cloning procedures brings the ethical questions about the use of stem cell limiting the research (Brunt et al. 2012). The technical feasibility of stem cell therapy and the several possibilities in this research area has generated great expectations in regenerating damaged organs and in the treatment of chronic and neuronal diseases, such as diabetes and Parkinson (Jang et al. 2012; Clover et al. 2012).

The major limitations of cell culture are the stability of cell lines, sample preparation, and assay evaluated. Culture procedures must be carried out under austere aseptic conditions and the cost of obtaining cells in culture is superior compared to using animal tissue. However, the advantage of microscale method is the low cost associated toreduced manipulation times, volumes and easily automated (Harris 2012).

Variability can be found from one passage to the next due to the cell capacity to differentiate inside a population. To reduce those variables, several criteria are investigated in the characterization of cell lines such as karotype, cell surface-specific antigen, aalysis of gene expression, and cell morphology. Observation of cell morphology is easy and is the most direct technique used to recognize cells (Freshney 2005), it is necessary to avoid cross-contamination, one common problem found in cell culture. In the late 1960s it was demonstrated that the majority of cell lines in the United States was cross-contaminated with HeLa cells, due to this authentication of a cell line is crucial to provide its characterization as well to determine its functionally (Capes-Davis et al. 2010).

5.2 Rationale for Cytotoxicity Screening

The discovery and development of a new therapeutic drug needs a substantial financial support and can take more than 10 years of research effort. The drug development process can be divided as early discovery until clinical studies. New pharmaceutical drugs have to be approved in terms of efficacy and toxicity because of the high cost in terms of money; the ability to identify and reduce stages can increase the process of drug development by improving efficiency, reducing time in the research, and reducing the costs (McKim 2010). Cell-based tests are getting increasingly essential for the pharmaceutical industry, not only for cytotoxicity assays but also for high-throughput screening of several substances that may have possible use as therapeutic agents (Chandler et al. 2011).

Cell culture models are fundamentals in the investigation of the toxicological characteristics in relation to pharmacological and environmental characteristics, constituting a strong tool in the evaluation of biological and/or therapeutic versus toxicological effects (Pernot et al. 2011; Fang and Al-Suwayeh 2012). In this way, the primary goal of in vitro models in the toxicity evaluation could be to predict the toxicity in vivo, specially human toxicity. In vitro screening permits investigation of the metabolism and biochemical reactions of different substances in order to obtain knowledge about the pharmacokinetics and bioavailability of the drugs (Mahato et al. 2011; Fang and Al-Suwayeh 2012; Shahbazi and Santos 2013). Studies about bioavailability, chemical and metabolic stability, and permeability are necessary in vitro screening models that help predicting the human toxicity (Xin et al. 2011).

Several studies have been presented with interesting results about the good bioavailability, the biochemical (physicochemical and metabolic) stability, and also the controlled release parameters for different formulations containing nanoparticles (Xin et al. 2011). However, the development of in vitro screening models would be of great interest to predict the toxicity of the human body.

Toxicity is a complex event showing several effects, even cell death until metabolic aberrations. Neuro-, liver, and/or kidney toxicity are examples of functional changes not necessarily events linked to cell death. Due to these evidences, the in vitro cytotoxicity assays need to exploit different parameters in the cell biochemistry. In vitro assays are important tools to amplify the knowledge about the cytotoxic effects triggered by chemical substances and to predict the toxicity in humans (Roguet et al. 1993; Melo et al. 2002).

Several pharmacological substances show instability in biological systems that can make their use in therapeutic treatments difficult. Others have collateral effects in normal cells. Toxicity is a limiting factor in pharmaceuticals use; therefore, the toxicity evaluation versus therapeutic activity of a compound is crucial to determine its applicability as a therapeutic substance. Ethical and financial questions corroborate in vitro assays, instead using animals in initial toxicological studies (Costin et al. 2011; Combes and Balls 2011; Balls 2012). In vitro cell culture model establishment and the development of cell lines increased the reproducibility of the assays since the use of the same cellular population during different times is possible. The cell culture assays have as advantages the fast results in the research, reduced cost, and a few quantities of substance to be used (Melo et al. 2002). Besides that, those models can predict toxicity in animals including humans to get more precise results to establish multi-tiered in vitro screening models (McKim 2010). These models must be fully characterized and predictive of in vivo effects with fewer incidences of false-positive/false-negative results. In vitro cell-based toxicity assays can be used to assess potential safety use in the investigation of new potential drugs and to study the correlation between structure, toxicity, and biological activity, permitting changes in the structure chemical or in the formulation with the aim to improve drug-like properties (Yang et al. 2011).

Previous studies demonstrated that classic cytotoxicity assays may not be adequate to investigate nanomaterials toxicity since they can interfere with reagents inducing false-positive or false-negative results, such as inducing changes on membrane permeability and/or changing oxidation–reduction processes, since these properties are essential to determine cell viability using neutral red uptake and MTT reduction test (Kroll et al. 2011).

In addition, contradictory effects on cells have been determined when the nanoparticle concentration is much higher than tissue concentrations in in vivo models. Due to this, the comparisons of in vitro versus in vivo studies showed little correlation (L'Azou et al. 2008; Jonhston et al. 2009; Sayes et al. 2009; Warheit et al. 2009). For this reason, the development of more appropriate in vitro methods, such as cell coculture models for determining nanoparticle permeability parameters, in combination with cell viability assays is of interest for nanotoxicology fields.

5.3 In Vitro Assay System

There are several cell models and in vitro cytotoxicity assays available for in vitro toxicity assays. It is important to understand the positive and negative aspects of each model to obtain reliable results (Freshney 2005).

5.3.1 Primary Cultures

Definition of a primary culture is that stage of the culture before the first subculture and after the isolation. The cells can be obtained by dissection or disaggregation of the tissue seeding into the culture flask. This disaggregation of the tissue can be processed mechanically or enzymatically producing a suspension which will attach to the substrate, an event that appears to be essential for most untransformed cells, with the exception of hematopoietic cells (Freshney 2005; Harris 2012). Several enzymes are used for tissue disaggregation such as trypsin, hyaluronidase, collagenase, elastase, and DNase. It can be used as a crude preparation alone or in combination. Trypsin is mostly used because it is promoting a great disaggregation but has the inconvenience of triggering damage to the cells. The cell injury can be reduced minimizing the exposure of cells to trypsin in order to obtain maximum viability. Collagenase promotes incomplete disaggregation, but it is less prejudicial to the cells (Freshney 2005; Harris 2012).

Primary cell cultures maintain biochemical and morphological technical features that are more similar to the primary tissue, but the results obtained are not always reproducible. The principal advantage of primary culture is its use for comparative studies of some specialized cells removed from different organs and animal species.

Previous studies demonstrated that cytotoxic responses after treatment with nanomaterials differ in individual cell lines and immortalized cells are usually more sensitive than primary cells (Kroll et al 2011). Primary cell culture can be used to research microglial biology in Alzheimer's disease since primary microglia cultures are very predominant in neuro-inflammatory studies due to the similarities in phenotype compared to in vivo cells. Potential of compounds to be involved in collateral or adverse interactions by the inhibition or induction of liver enzymes can be investigated in primary cultures (LeCluyse et al. 2012). Kim et al. (2010) investigated the zinc oxide nanoparticle-induced injury to alveolar epithelium by exposing primary cultured rat alveolar epithelial cell to these nanoparticles and observed a severe injury to these cells. The cytotoxic effects were dose- and time-dependent inducing mitochondrial dysfunction and formation of reactive oxygen species.

5.3.2 Cell Lines

Immediately when a primary culture had a passage or is subcultured, it is named a cell line. After selection, by cloning, physical separation, or by another technique, the cell line showing specific characteristics in the culture is known as a cell strain. There are several parameters to be analyzed in choosing a cell line: finite versus continuous, normal or transformed, species, growth characteristics, availability, validation, phenotype expression, and stability (Freshney 2005).

A substantial number of literature gives some insights in relation to cytotoxicity induced by different nanomaterials in several cell lines. Lanone et al. (2009) studied comparatively 24 nanoparticles in the same experimental protocol. They reported toxicity of some nanomaterials, but not all, of the nanostructures tested in human alveolar epithelial and macrophage cell lines (A549 and THP-I) by MTT reduction and neutral red uptake assays. These two cell lines were chosen because they are the major targets of nanoparticles in vivo after inhalation. With this experimental model, copper- and zinc-based nanoparticles were the most toxic and no toxicity was observed for tungsten carbide.

Studies conducted by Sayes et al. (2007) and Lanone et al. (2009) demonstrated that rat lung epithelial (L2 cell line) and primary alveolar macrophages treated with different nanoparticles showed different cytotoxicity evaluated by viability assays and inflammatory response.

Kroll et al. (2011) evaluated 23 engineered nanomaterials in ten different standard cell lines. Six cell lines were of animal origin: lung epithelial cells (RLE-6TN), fibroblasts (NIH-3T3), macrophages (RAW264.7), and three different epithelial cell lines (MDCK, MDCK II, NRK52E) with kidney origin, which is representative as a secondary target organ. Four of the standard cell lines were of human origin and derived from organs affected directly after exposure to nanoparticles upon inhalation (lung: A549, CaLu3), dermal exposure (skin: HaCaT), or ingestion (colon: CaCo2). According to the results they showed that sensitivity towards nanoparticles treatment is not only cell type-specific but also is related to the particle type used as well as on the cytotoxic endpoint analyzed. Furthermore, other biomarkers like inflammatory response or genotoxicity may be evaluated depending on the pathological pathways of interest.

Pujalté et al. (2011) investigated the cytotoxicity and oxidative stress triggered by different nanoparticles on human kidney cells: IP15 (glomerular mesangial) and HK-2 (epithelial proximal) cell lines. The cytotoxic effects measured were correlated with the physicochemical parameters of nanoparticles assayed and the cell type used (Liang et al. 2008). Therefore the authors showed that nanoparticles of ZnO and CdS induced toxic effects related to metal composition, particle size, and metal solubility. The generation of reactive oxygen species induced oxidative stress provoking nephrotoxic responses.

5.4 Distinctive Methodologies for Cytotoxic Studies on Nanoparticles

Numerous assays have been developed for assessing cytotoxicity in vitro investigating the cell viability through ATP content, enzyme release or activities (lactate dehydrogenase and phosphatase, respectively), cell morphology, plating efficiency, viable cell count using trypan blue exclusion test, dye uptake, protein and acid nucleic content, etc. The goal of these studies is reducing animal usage by using in vitro cytotoxicity results to predict the optimal starting doses for in vivo acute toxicity test or to improve the process of drug development by the investigation of desired drug attributes in a large scale (Spielmann et al. 1999; Martin et al. 2009).

Cytotoxicity can be defined as the adverse effects deriving from reactions with structures and/or processes crucial for cell maintenance such as proliferation, survival, and normal biochemical/physiology. These adverse effects may involve changes in the cellular metabolism; in the synthesis, degradation, or release of cellular constituents; in the integrity of membrane; and/or in the cytoskeleton, as many alterations as in the cell death. Commonly, some principal mechanisms for toxicity have been investigated. These include basal, selective, and cell-specific function toxicities. Basal or general cytotoxicity means changes in processes or structures that are intrinsic to all cell sorts such as alterations in organelle functions or in the membrane integrity. Selective cytotoxicity depends on the existence of cell specificities such as different metabolic reactions (e.g., in hepatocytes) and receptors (in neurons). Those features can affect the cytotoxic response in terms of sensibility to the different in vitro methods.

Cell-specific function cytotoxicity may not affect directly the cells, but some toxic substances can interfere with the organism as a whole, for example, the toxicity can be due to cell signaling changes induced by hormones and neurotransmitters. In vitro cytotoxicity assays have been standardized and used by researchers around the world since the middle of the last century. Nevertheless, in vitro assays are not completely reliable in predicting in vivo toxicity. Some factors fully not addressed in vitro assays comprise binding to specific proteins, metabolic reactions, metabolites production, and metabolic stability. Numerous assays have been developed for assessing cytotoxicity in vitro investigating the cell viability through ATP content, enzyme release or activities (lactate dehydrogenase and phosphatase, respectively), cell morphology, plating efficiency, viable cell count using trypan blue exclusion test, dye uptake, protein and acid nucleis content, etc. (Sittampalam et al. 2007). The goals of these studies are to reduce animal usage by using in vitro cytotoxicity results, to predict the optimal starting doses for in vivo acute toxicity test, or to improve the process of drug development by the investigation of desired drug attributes in a large scale (Spielmann et al. 1999; Martin et al. 2009).

5.4.1 Cell Viability

The first and most easily visualized effect after exposure of cells to a chemical toxicant is the observation of changes in cell morphology. Thus, morphological alterations are useful as a toxicity biomarker. Rude changes in cell morphology can include formation of blebs in the membrane, loss or increase in the cellular volume, vacuolization, cytoplasmatic and/or mitochondrial swelling, and plasmatic membrane rupture (Freshney 2005; Harris 2012).

Changes in the cellular growth also can be determined as a toxic effect. Chemical toxicities can be evaluated by an assay called plating efficiency, which evaluates the ability of cells to form colonies after 15 days of culture in the presence of a substance, signaling cell survival and ability to reproduce. Another assay which evaluates the capability of cells to replicate and determines the concentration of the substance at which 50 % of the cells do not multiply is named the median inhibitory doses (ID_{50}).

Viability tests are used to measure the percentage of viable cells after an immediate- or short-term response, such as changes in membrane permeability or alterations in specific metabolic pathways linked to cell survival or cell proliferation. The viability assay named neutral red uptake does not measure the total number of cells but living cells that can take up the dye and sequester in the lysosomes. Thus, it shows a reduction in the absorbance correlated to loss of viable cells. It is based on the ability of viable cells to incorporate and bind the supravital dye neutral red in the lysosomes. Particles are known as lysosomotropic substances and these kinds of agents can cause lysosomal dysfunction. Changes in the neutral red uptake by the cells need to be evaluated since lysosomal disorders are associated with toxicities of lung, liver, and kidney (De Duve et al. 1974; Schneider et al. 1997; Stern et al. 2012). Lysosome membrane permeabilization has been detected in mussels treated with nanoscale glass wool and gold nanoparticles. In the assay to study the neutral red retention in the hemolymph, a loss of dye from the lysosomes to cytosol was observed (Tedesco et al. 2010). the same results in the RTG-2 cells (rainbow trout gonadal tissue) treated with titanium dioxide nanoparticles were observed (Vevers and Jha 2008). Another cell viability assay is MTT reduction, which is based on the reduction of a yellow water-soluble tetrazolium salt by live cells to an aqueous purple insoluble product.

5.4.2 Cell Death

Cell death is a crucial cellular mechanism and is divided into several types, including apoptosis, necrosis, and autophagy. Specific morphological features and biochemical characteristics can be determined with each death pathway (Zhivotovsky 2004).

Apoptosis or programmed cell death is a genetically controlled and evolutionary conserved form of cell death that occurs in many physiological and pathological processes in multicellular organisms. Cell death by apoptosis can be triggered by intrinsic or extrinsic pathways, receptor-mediated or mitochondria-mediated, respectively. The morphological events that occur in apoptosis are well organized and triggered in sequence. The unique events include the degradation of chromatin in internucleosomal fragments, a reduction in cellular volume, and the formation of bleb in the plasma membrane. It then breaks up into membrane-enclosed fragments, named as apoptotic bodies, which are recognized due to externalization of phosphatidylserine in the plasma membrane and are engulfed by neighboring cells or macrophages. In addition to morphological alterations, apoptosis is also characterized by DNA fragmentation induced by endonuclease activation and activation of the caspase enzyme family. Caspases are aspartate-specific cysteinyl proteases which mediate the fragmentation of several hundred proteins and also activate DNase, inducing chromatin degradation into a characteristic ladder at nucleosomes. According to structure and function, caspases can be divided into two groups: initiator such as caspase-8 or -9 and effector caspases such as caspase-3, -6, and -7 (Wlodkowick et al. 2011). Galluzzi et al. (2012) proposed cell death subroutines applied to in vitro and in vivo situations and are defined by measurable biochemical markers.

Several in vitro assays have demonstrated that increased reactive oxygen species generation as a first factor of toxicity in nanoparticle-treated cells. Cells exposure to TiO_2 nanoparticles leads to lipid peroxidation, DNA injury, caspase activation triggering micronuclei formation, chromatin condensation, and cell death by apoptosis (Park et al. 2008; Yang et al. 2011; Sohaebuddin et al. 2010).

Necrosis, on the other hand, is characterized by loss of membrane integrity due to irreversible swelling of the cytoplasm and organelles. Cell death by necrosis generally occurs due to a severe injury to the cells provoking cell lysis and the release of noxious cellular constituents promoting an inflammatory response, which trigger the cell death of neighboring cells.

Several other forms of cell death are known; one of these is autophagic cell death, which is thought to have evolved before the apoptosis. The term autophagic cell death has been used to indicate cell death that is accompanied by a massive cytoplasmatic vacuolization indicating cellular self-digestion, involving the sequestration of load such as cellular organelle, proteins, and pathogens into double-membrane structures named autophagosomes. Microtubule-associated protein 1 light chain 3 (LC3), which is a biomarker of the early steps of autophagy, is cleaved during the process, forming a membrane-bound 16 kDa protein (LC3-II) from the 18 kDa protein (LC3-I). The ratio between LC3-II and LC3-1 characterizes the activity of autophagy or an increase on autophagy flux (Mizushima 2004; Tsai et al. 2012). Literature data suggest that intracellular nanoparticles may suffer autophagic sequestration, and autophagy dysfunction can play a crucial role in nanoparticle toxicity (Stern et al. 2012). Several nanoparticles have also been shown to induce changes in the autophagy pathway and this may be involved in their mechanism of toxicity (Stern and Johnson 2008; Johnson-Lyles et al. 2010).

Apoptosis and autophagy are modalities of cell death that play critical functions in regulating cell death and survival; thus, they are strongly regulated by key proteins and enzymes. Autophagy as apoptosis has both pro- and anti-survival mediators and this is dependent on the cell's ability to dye by apoptosis. In this way, cells with defective apoptosis and autophagy may serve as a strategy for cellular demise or as a mechanism to trigger other forms of death such as necrosis (Frohlich 2012).

5.5 Conclusions

Bio- or toxic effects of nanoparticles in the cell depend not only on the amount of nanoparticles but also in its size, morphology, and surface area of nanoparticles. Nano–bio interface investigates the interface between nanostructured materials and biological systems initially with proteins, then membrane lipids, and finally with the cell. Nanotoxicology examines the bioeffects of nanomaterials concerning the toxic activities and depends on the in vitro system investigated. Then, validation of in vitro assays will be valuable for safety or hazard investigation in an analysis against standard nanostructures. We are in agreement with the hypothesis of Han et al. (2012) in which they concluded that the use of the steepest slope analysis offers an excellent correlation in vitro with in vivo results of nanoparticle toxicity and for ranking their toxic potency.

Acknowledgments Supports from FAPESP, CNPq, INOMAT (MCTI/CNPq), NanoBioss (MCTI), and Brazilian Network on Nanotoxicology (MCTI/CNPq) are acknowledged.

References

Acosta SA, Tajiri N, Shinozuka K et al (2013) Long-term upregulation of inflammation and suppression of cell proliferation in the brain of adult rats exposed to traumatic brain injury using the controlled cortical impact model. PLoS One 8:e53376

Ayaki M, Wasawa A, Niwano Y (2012) Cell viability score as an integrated indicator for cytotoxicity of benzalkonium chloride-containing antiglaucoma eyedrops. Biocontrol Sci 17:121–128

Balls M (2012) The conflict over animal experimentation: is the field of battle changing? Altern Lab Anim 40:189–191

Breheny D, Oke O, Faux SP (2011) The use of in vitro systems to assess cancer mechanisms and the carcinogenic potential of chemicals. Altern Lab Anim 39:233–255

Brunt KR, Weisel RD, Li RK (2012) Stem cells and regenerative medicine: future perspectives. Can J Physiol Pharmacol 90:327–335

Capes-Davis A, Theodosopoulos G, Atkin I et al (2010) Check your cultures! A list of cross-contaminated or misidentified cell lines. Int J Cancer 127:1–8

Chandler KJ, Barrier M, Jeffay S et al (2011) Evaluation of 309 environmental chemicals using a mouse stem cell adherent cell differentiation and cytotoxicity assay. PLoS One 6:1–11

Clover AJ, O'Neil BL, Kumar AH (2012) Analysis of attitudes toward the source of progenitor cells in tissue-engineered products for use in burns compared with other disease states. Wound Repair Regen 20:311–316

Combes RD, Balls M (2011) Integrated testing strategies for toxicity employing new and existing technologies. Altern Lab Anim 39:213–225

Costin GE, Raabe HA, Priston R et al (2011) Vaginal irritation models: the current status of available alternative and in vitro tests. Altern Lab Anim 39:317–337

De Duve C, De Barsy T, Poole B et al (1974) Commentary. Lysosomotropic agents. Biochem Pharmacol 23:2495–2531

De Jong WH, Borm PJA (2008) Drug delivery and nanoparticles: applications and hazards. Int J Nanomedicine 3:133–149

Fang JY, Al-Suwayeh SA (2012) Nanoparticles as delivery carriers for anticancer prodrugs. Expert Opin Drug Deliv 9:657–669

Freshney RI (ed) (1994) Culture of animal cells: a manual of basic technique. Wiley, Hoboken, NJ

Freshney RI (2005) Culture of animal cells: a manual of basic technique. Wiley, Hoboken, NJ, p 649

Frohlich E (2012) The role of surface charge in cellular uptake and cytotoxicity of medical nanoparticles. Int J Nanomedicine 7:5577–5591

Fukazawa H, Suzuki T, Wakita T et al (2012) A cell-based, microplate colorimetric screen identifies7,8-benzoflavone and green tea gallate catechins as inhibitors of the hepatitis C virus. Biol Pharm Bull 35:1320–1327

Galluzzi L, Vitale I, Abrams JM et al (2012) Molecular definitions of cell death subroutines: recommendations on the Nomenclature Committee on Cell Death 2012. Cell Death Differ 19:107–120

Gao W, Lai JCK, Leung SW (2012) Functional enhancement of chitosan and nanoparticles in cell culture, tissue engineering, and pharmaceutical applications. Front Physiol 3:1–12

Han X, Corson N, Wade-Mercer P et al (2012) Assessing the relevance of in vitro studies in nanotoxicology by examining correlations between in vitro and in vivo data. Toxicology 297:1–9

Harris C (2012) Overview of in vitro models in developmental toxicology. Methods Mol Biol 889:105–113

Jang J, Yoo JE, Lee JA, Lee DR et al (2012) Disease-specific induced pluripotent stem cells: a platform for human disease modeling and drug discovery. Exp Mol Med 44:202–213

Johnson-Lyles DN, Peifley K, Lockett S et al (2010) Fullerenol cytotoxicity in kidney cells is associated with cytoskeleton disruption, autophagic vacuole accumulation, and mitochondrial dysfunction. Toxicol Appl Pharmacol 248:249–258

Jonhston HJ, Hutchison GR, Christensen FM et al (2009) Identification of the mechanisms that drive the toxicity of TiO_2 particulates: the contribution of physicochemical characteristics. Part Fibre Toxicol 6:33–60

Kim YH, Fazlollahi F, Kennedy IM et al (2010) Alveolar epithelial cell injury due to zinc oxide nanoparticle exposure. Am J Respir Crit Care Med 182:1398–13409

Kroll A, Dierker C, Rommel C et al (2011) Cytotoxicity screening of 23 engineered nanomaterials using a test matrix of ten cell lines and three different assays. Part Fibre Toxicol 8:1–19

L'Azou B, Jorly J, On D et al (2008) In vitro effects of nanoparticles on renal cells. Part Fibre Toxicol 5:22–36

Lanone S, Rogerieux F, Geys J et al (2009) Comparative toxicity of 24 manufactured nanoparticles in human alveolar epithelial and macrophage cell lines. Part Fibre Toxicol 6(14):26

LeCluyse EL, Witek RP, Andersen ME et al (2012) Organotypic liver culture models: meeting current challenges in toxicity testing. Crit Rev Toxicol 42:501–548

Liang XJ, Chen C, Zhao Y et al (2008) Biopharmaceutics and therapeutic potential of engineered nanomaterials. Curr Drug Metab 9:697–709

Machana S, Weerapreeyakul N, Barusrux S (2011) Cytotoxic and apoptotic effects of six herbal plants against the human hepatocarcinoma (HepG2) cell line. Chin Med 6:39

Mahato R, Tai W, Cheng K (2011) Prodrugs for improving tumor targetability and efficiency. Adv Drug Deliv Rev 63:659–670

Martin MT, Judson RS, Reif DM et al (2009) Profiling chemicals based on chronic toxicity results from the U.S. EPA ToxRefDatabase. Environ Health Perspect 117:392–399

McKim JM Jr (2010) Building a tiered approach to in vitro predictive toxicity screening: a focus on assays with in vivo relevance. Comb Chem High Throughput Screen 13:188–206

Melo PS, Durán N, Haun M (2002) Derivatives of dehydrocrotonin, a diterpene lactone isolated from Croton cajucara: cytotoxicity in rat cultured hepatocytes and V79 cells. Hum Exp Toxicol 21:273–280

Mizushima N (2004) Methods for monitoring autophagy. Int J Biochem Cell Biol 36:2491–2502

Moller P, Jacobsen NR, Folkmann JK et al (2010) Role of oxidative damage in toxicity of particulates. Free Radic Res 44:1–46

Olabisi RM, Lazard ZW, Franco CL et al (2010) Hydrogel microsphere encapsulation of a cell-based gene therapy system increases cell survival of injected cells, transgene expression, and bone volume in a model of heterotropic ossification. Tissue Eng 16:3727–3736

Park EJ, Yi L, Chung KH et al (2008) Oxidative stress and apoptosis induced by titanium dioxide nanoparticles in cultured BEAS-2B cells. Toxicol Lett 180:222–229

Pernot M, Vanderesse R, Frochot C et al (2011) Stability of peptides and therapeutic success in cancer. Expert Opin Drug Metab Toxicol 7:793–802

Pintus F, Floris G, Rufini A (2012) Nutrient availability links mitochondria, apoptosis, and obesity. Aging 4:734–741

Polchow B, Kebbel K, Schmiedeknecht G et al (2012) Cryopreservation of human vascular umbilical cord cells under good manufacturing practice conditions for culture cell banks. J Transl Med 10:98

Pujalté I, Passagne I, Brouillaud B et al (2011) Cytotoxicity and oxidative stress induced by different metallic nanoparticles on human kidney cells. Part Fibre Toxicol 8:10–26

Roguet R, Cotovio J, Gaetani Q et al (1993) Cytotoxicity of 28 MEIC chemicals to rat hepatocytes using two viability endpoints: correlation with acute toxicity data in rat and man. Altern Lab Anim 21:216–224

Russell K (1969) Tissue culture: a brief historical review. Clio Med 4:109–119

Sayes CM, Reed KL, Warheit DB (2007) Assessing toxicity of fine and nanoparticles: comparing in vitro measurements to in vivo pulmonary toxicity profiles. Toxicol Sci 97:163–180

Sayes C, Reed K, Subramoney S et al (2009) Can in vitro assays substitute for in vivo studies in assessing the pulmonary hazards of fine and nanoscale materials? J Nanopart Res 11:421–431
Schneider P, Korolenko TA, Busch U (1997) A review of drug-induced lysosomal disorders of the liver in man and laboratory animals. Microsc Res Tech 36:253–275
Shahbazi MA, Santos HA (2013) Improving oral absorption via drug-loaded nanocarriers: absorption mechanisms, intestinal models and rational fabrication. Curr Drug Metab 14(1):28–56
Sittampalam GS, Gal-Edd N, Arkin M et al (eds) (2007) Assay guidance manual. Eli Lilly & Company, Bethesda, MD
Sohaebuddin SK, Thevenot PT, Baker D et al (2010) Nanomaterial cytotoxicity is composition, size, and cell type dependent. Part Fibre Toxicol 7:22–39
Spielmann H, Genschow E, Leibsch M et al (1999) Determination of the starting dose for acute oral toxicity (LD50) testing in the up and down procedure (UDP) from cytotoxicity data. ATLA 27:957–966
Stern ST, Johnson DN (2008) Role for nanomaterial-autophagy interaction in neurodegenerative disease. Autophagy 4:1097–1100
Stern ST, Adiseshaiah PP, Crist RM (2012) Autophagy and lysosomal dysfunction as emerging mechanisms of nanomaterial toxicity. Part Fibre Toxicol 9:20–35
Tedesco S, Doyle H, Blasco J, Redmond G et al (2010) Oxidative stress and toxicity of gold nanoparticles in Mytilus edulis. Aquat Toxicol 100:178–186
Tsai TL, Hou CC, Wang HC et al (2012) Nucleocytoplasmic transport blockage by SV40 peptide-modified gold nanoparticles induces cellular autophagy. Int J Nanomedicine 7:5215–5234
Vevers WF, Jha AN (2008) Genotoxic and cytotoxic potential of titanium dioxide (TiO_2) nanoparticles on fish cells in vitro. Ecotoxicology 17:410–420
Warheit DB, Sayes CM, Reed KL (2009) Nanoscale and fine zinc oxide particles: can in vitro assays accurately forecast lung hazards following inhalation exposures? Environ Sci Technol 43:7939–7945
Wlodkowick D, Telford W, Skommer J et al (2011) Apoptosis and beyond: cytometry in studies of programmed cell death. Methods Cell Biol 03:55–98
Xin GZ, Qi LW, Shi ZQ et al (2011) Strategies for integral metabolism profile of multiple compounds in herbal medicines: pharmacokinetics, metabolites characterization and metabolic interactions. Curr Drug Metab 12:809–817
Yang D, Van S, Liu J et al (2011) Physicochemical properties and biocompatibility of a polymer-paclitaxel conjugate for cancer treatment. Int J Nanomedicine 6:2557–2566
You C, Han C, Wang X et al (2012) The progress of silver nanoparticles in the anti-bacterial mechanism, clinical application and cytotoxicity. Mol Biol Rep 39:9193–9201
Zhivotovsky B (2004) Apoptosis, necrosis and between. Cell Cycle 3:64–66

[illegible]

Chapter 6
Carbon Nanotubes: From Synthesis to Genotoxicity

Diego Stéfani T. Martinez, Leonardo P. Franchi, Camila M. Freria, Odair P. Ferreira, Antonio G. Souza Filho, Oswaldo L. Alves, and Catarina S. Takahashi

Abstract Massive industrial production of carbon nanotubes (CNTs) is increasing year after year, and it is urgent to address their safety-related issues. Due to their morphological similarities with asbestos fibers, which are classical carcinogenic materials, these CNTs have been considered as hazardous manufactured products by regulatory agencies. In this context, genotoxic effects of CNTs and the mechanisms proposed in current literature are reviewed and discussed in this chapter. Relevant aspects of preparation and physicochemical characterization of CNTs in toxicological context as well as the recent perspectives involving cytotoxicity assessment are also highlighted. Finally, this chapter aims to contribute

D.S.T. Martinez (✉) • O.L. Alves
Solid State Chemistry Laboratory, Institute of Chemistry, Universidade Estadual de Campinas—UNICAMP, Cidade Universitária "Zeferino Vaz"-Barão Geraldo, Campinas, São Paulo 13083-970, Brazil
e-mail: diegostefani.br@gmail.com; oalves@iqm.unicamp.br

L.P. Franchi • C.S. Takahashi
Department of Genetics, Faculty of Medicine of Ribeirão Preto, Universidade de São Paulo–USP, Ribeirão Preto, São Paulo 14049-900, Brazil
e-mail: leonardofranchi@yahoo.com.br; cstakaha@usp.br

C.M. Freria
Department of Structural and Functional Biology, Institute of Biology, Universidade Estadual de Campinas—UNICAMP, Cidade Universitária "Zeferino Vaz"- Barão Geraldo, Campinas, São Paulo 13083-862, Brazil
e-mail: camilafreria@yahoo.com.br

O.P. Ferreira
Advanced Functional Materials Laboratory, Department of Physics, Universidade Federal do Ceará–UFC, Fortaleza, Ceará, 60440-554, Brazil
e-mail: opferreira@fisica.ufc.br

A.G. Souza Filho
Departamento de Física, Universidade Federal do Ceará–UFC, Fortaleza, Ceara 60440-554, Brazil
e-mail: agsf@fisica.ufc.br

N. Durán et al. (eds.), *Nanotoxicology*, Nanomedicine and Nanotoxicology,
DOI 10.1007/978-1-4614-8993-1_6,

to point out to a proactive discussion towards a responsible and sustainable development of nanotechnology lined up with environmental, health, and safety (EH&S) requirements.

6.1 Introduction

Carbon nanotubes (CNTs) are very interesting materials for technological innovation owing to their striking physical and chemical properties. The deep understanding of new phenomena emerging from these nanomaterials is leading to important breakthroughs in materials, electronics, and biotechnology areas (Ajayan and Tour 2007; Schnorr and Swager 2011; Heister et al. 2013). Today, there are many companies manufacturing and commercializing CNTs and the massive industrial production of these nanocarbons has been increasing year after year (Thayer 2007; Kumar and Ando 2010). However, the safety-related aspects of nanomaterials to human health and environment are not well understood so far (Nel et al. 2006; Krug and Wick 2011) and the toxicity assessment of CNTs is an important issue to be addressed towards the development of a safe nanotechnology. Although considerable experimental toxicity data related to CNTs have been published so far, the results are often conflicting and a systematic understanding of CNT toxicity is needed (Liu et al. 2013; Li et al. 2013). Moreover, interactions of CNTs with different biological entities (biomolecules, cells, organisms, etc.) can be tuned by their physicochemical properties (Senanayake et al. 2008; Jain et al. 2011; Bussy et al. 2012). In this sense, understand how these properties affect the overall bio-toxicological responses is a key step towards the sustainable exploration of nanotube technological potential lined up with environment, health, and safety (EH&S) issues (Hussain et al. 2009; Lehman et al. 2011; Lowry et al. 2012).

Different methodologies are used for accessing the toxicity of nanomaterials at different hierarchical levels of biological systems. Genotoxicity is a destructive effect which affects the DNA integrity through the action of damage-inducing agents (genotoxins) such as chemicals and radiation. The identification of genotoxic compounds is extremely important since an increasing in the genomic instability is associated with the raising of cancer (Shahab et al. 2013). There are many agents that can act as genotoxins and they can be classified as physical (e.g., UV, X-rays), chemical (e.g., benzopyrene, ethidium bromide), and biological (e.g., virus, transposons). These agents can also be described according to their action mechanism such as oxidants (e.g., hydrogen peroxide), alkylating (e.g., methyl methanesulfonate), inductors of DNA breaking (e.g., ionizing radiation), and aneugenic agents which affect chromosome division (e.g., taxanes) (Parry and Parry 2012). As DNA damage may either initiate and promote carcinogenesis or impact fertility, the genotoxicology science became a vital area governing regulatory health risk assessment.

Experimental studies have demonstrated that CNTs induce genotoxic effects (Kato et al. 2013). Intrapulmonary administration of CNTs induced mesothelial proliferation that is potentially associated with mesothelioma cancer development in rats (Xu et al. 2012). Therefore, these carbon-based nanomaterials have been considered to be hazardous to human beings, unless evidence shows otherwise following the precaution principle (Schulte et al. 2012).

In this chapter, we discuss some aspects of preparation and characterization of CNTs in toxicological context as well as their interactions with biological systems. Genotoxic effects of CNTs and mechanisms proposed in current literature are reviewed and discussed as these are central elements of risk assessment of chemicals and materials. By considering the importance and growing of nanotoxicology and nanosafety fields, some relevant considerations to cytotoxicity assessment of these nanomaterials are also discussed.

6.2 Carbon Nanotubes: Preparation, Functionalization, and Characterization

CNTs are made of single layers of graphite (graphene) that are rolled up to produce structures with tubular or cylindrical morphology (Endo 1988; Iijima 1991). In terms of the number of layers, these nanomaterials are classified as single-walled carbon nanotubes (SWCNTs), which are a cylinder formed by only one wrapped graphene sheet, and multiwalled carbon nanotubes (MWCNTs), which are formed by various concentric graphitic cylinders with an interwall spacing ranging from 0.34 to 0.39 nm. While SWCNTs have diameters varying from 0.5 to 1.5 nm and length of few nanometer-to-micrometers range, MWCNT counterparts have diameters varying from 2 to 100 nm and lengths from some tens of nanometers to millimeters (Jorio et al. 2008).

The syntheses of CNTs are well documented in literature (Ying et al. 2011) and the most commonly used techniques are chemical vapor deposition (CVD), high-pressure carbon monoxide (HiPCo), laser ablation (LA), and arc discharge (AD). Therefore, samples with different features and purities can be obtained, depending either on the synthetic method or purification process employed (Dai 2002; Awasthi et al. 2005; Nessim 2010). As-synthesized CNTs are strongly agglomerated and have low dispersibility in any solvent media. Indeed, the agglomeration or bundling effect is due to van der Waals interactions among the nanotubes, and this effect is more pronunciated for SWCNTs than for MWCNTs. Depending on the applications required, the agglomerated state of CNTs need to be destroyed in order to obtain the CNTs in an isolated state (Ma et al. 2010; Kim et al. 2012b). A way to perform the aforesaid process is via functionalization and dispersion in a liquid media. Functionalization process consists in the insertion of chemical groups of interest in the tips and outer and inner walls of the CNTs. This process is crucial for developing applications of these nanomaterials in different areas because it allows creating and engineering new functionalities. Additionally, the functionalization allows one to exploit and modulate the intrinsic physicochemical properties of the CNTs leading to development of multifunctional hybrid systems (Balasubramanian and Burghard 2010; Paula et al. 2011). In this sense, several approaches to functionalize CNTs have been proposed, and they can be classified in two

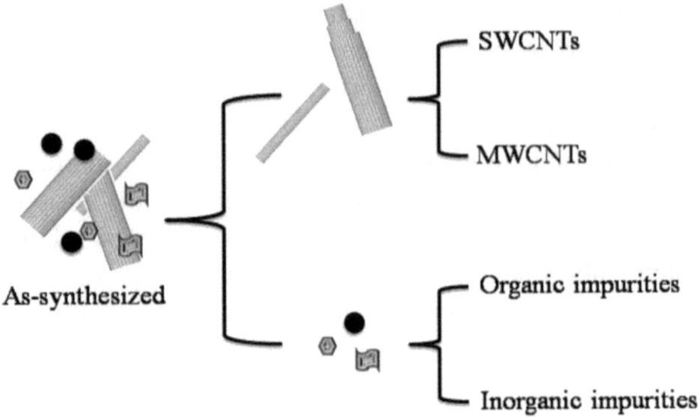

Fig. 6.1 Carbon nanotube samples in toxicological context. *SWCNTs* single-walled carbon nanotubes, *MWCNTs* multiwalled carbon nanotubes

categories, non-covalent and covalent, depending on the chemical bond nature between the functionalizing moieties and CNT wall (Akbar and Taimoor 2009; Karousis et al. 2010).

As-synthesized CNTs have inherent impurities such as metallic catalysts (e.g., Fe, Ni, or Co) carbonaceous species like amorphous and graphitic carbons and fullerenes, among others. Thus, the purification process may be extremely important depending on the application (Hou et al. 2008). In general, biological applications require CNT samples with high purity, because it is well known the deleterious effects of metals to biological systems (Kagan et al. 2006; Hull et al. 2009; Berhanu et al. 2009; Manshian et al. 2013). The main purification routes of the as-synthesized CNTs involve the oxidation of the impurities by either acid attack or thermal treatment. The former uses inorganic acids (i.e., HNO_3 and H_2SO_4) to eliminate the metal catalysts, and as side effects, oxygenated groups on the nanotube surface (e.g., OH, COOH, and C=O) are created (Osswald et al. 2007; Datsyuk et al. 2008). Furthermore, this treatment can also generate carbon-based by-products such as carboxylated carbonaceous fragments (CCFs) or oxidation debris (Salzmann et al. 2007; Stéfani et al. 2011). The mechanism of formation and influence of these CCFs during the purification have received little attention of the researchers so far (Del Canto et al. 2011). Nevertheless, one possible consequence of the presence of these carbon-based by-products along with the CNTs could be the interference on the subsequent steps of functionalization because these fragments are more reactive than the CNTs themselves (Worsley et al. 2009; Faria et al. 2012). The thermal treatment can also be used to eliminate the carbonaceous impurities, since the decomposition temperatures of these species are lower than those of CNTs. Thermal annealing also contributes to generate oxygenated functional groups on the nanotube surface (Datsyuk et al. 2008; Romanos et al. 2011).

Regarding toxicity assessment, the CNT samples must be considered as very heterogenic class of nanomaterials (Fig. 6.1). This is due to the fact that the current synthetic methods are not yet developed in order to have a controlled CNT production with respect to their morphological, structural, and surface properties as well as impurities. In addition, the purification and functionalization steps would change the nature and quality of CNTs as well. Grobert (2007) has pointed out the importance of having clean CNTs to be successful in some applications. In fact, no currently available CNT application (e.g., electronics, gas sensors, composites, or biotechnology) makes use of the properties of individual nanotubes, and to date, the majority of commercial applications of CNTs rely on the bulk properties of the CNT materials as a whole. Moreover, in toxicity assessment, this situation is more critical due to this key limiting factor which is the unavailability of clean samples. Furthermore, toxicological studies require a precise classification and specification of the CNTs evaluated and the set of a standard sample is very difficult (Lacerda et al. 2006; Firme and Bandaru 2010).

In this context, it is urgent to develop methods for an integrated physicochemical characterization of CNT samples. Although it is a very difficult task, it could be reached by using different measurements and techniques as briefly described (Gommes et al. 2004; Belin and Epron 2005; Wepasnick et al. 2010, 2011; Lehman et al. 2011). The main parameters or properties to be addressed for CNT samples in bio-toxicological context are the following:

(a) Morphology and size distribution.

To access the CNTs's morphological aspects (straight, bent, tangled, etc.) and size distribution (e.g., diameters and length), the following electronic microscopies techniques can be used: scanning electron microscopy (SEM) and (high-resolution) transmission electron microscopy (TEM or HRTEM) (Thien-Nga et al. 2002; Lehman et al. 2011). Additionally, these techniques could be used to distinguish between SWCNTs and MWCNTs as well as determine the inner and outer diameters, number of shells, and if the nanotube ends are opened or capped. In some cases, atomic force microscopy (AFM) could also be used to study the size distribution (Li et al. 2001; Hayashida and Umemura 2013).

(b) Atomic structure and structural defects.

The atomic structure, e.g., carbon atoms arrangement, can be evaluated using X-ray diffraction (XRD) for MWCNTs and Raman spectroscopy and high-resolution transmission microscopy (HRTEM) for both SWCNTs and MWCNTs (Saito et al. 1993; Zhang et al. 1999; Dresselhaus et al. 2002). By combining these techniques, it is possible to evaluate the internal order in nanotube walls (crystallinity degree), tube alignment, chirality, and diameters. There are indeed several structural defects catalogued for CNTs. Nevertheless, structural defects are herein being understood only as vacancies or imperfections on the walls like heptagon–pentagon pairs. The identification and relative quantification of these defects can be evaluated by combining HRTEM and Raman spectroscopy (Dresselhaus et al. 2010).

(c) Surface area.

The most popular technique used to measure CNT surface areas is N_2 adsorption–desorption isotherms at 77 K by applying the BET (Brunauer–Emmett–Teller) model. Moreover, information about pore volume and size distribution is also obtained from the BJH (Barrett–Joyner–Halenda) model of the isotherms data (Raymundo-Pinero et al. 2002; Li et al. 2004).

(d) Surface functional groups.

As discussed above, functional groups at the ends and along with the walls of CNTs arise from the purification process or functionalization reactions. Fourier transform infrared spectroscopy (FTIR) is the technique most used to identify functional groups attached to nanotube surface such as hydroxyl, carboxylate acid, amines, amides, and ketones, among others. Additionally, X-ray photoelectron spectroscopy (XPS) can be used with the same goal and it is very important for providing information about the carbon bonds at the surface (Kundu et al. 2008; Kingston et al. 2010; Wepasnick et al. 2011).

(e) Surface charge.

The surface charge is determined by using zeta potential (ZP) measurements. These charges would be positive or negative as function of the chemical groups attached to CNTs (Jiang et al. 2003). In a selective way, potentiometric titration (PT) is used to obtain the number of oxygenated and amine groups on the nanotubes (Samori et al. 2010).

(f) Organic impurities.

Organic impurities are considered as particles/molecules that are non-nanotube carbon (e.g., fullerenes, amorphous or graphitic carbons, onion-like particles, among others) produced during the synthesis processes or particles/molecules obtained after purification process such as oxidation debris (Verdejo et al. 2007; Tobia et al. 2009; Stéfani et al. 2011). It is very difficult to identify and quantify the organic impurities in CNT samples, and there is no single technique which is able to determine both features. For as-synthesized CNTs, the percentage of undesired materials possessing other morphologies per unit area can be quantified by using SEM (Lehman et al. 2011). HRTEM enables one to evaluate the presence of amorphous carbon coating the outer layers of CNTs (Endo et al. 1997). Near-infrared spectroscopy (NIR) can be a useful technique to access the content of SWCNTs after purification process (Itkis et al. 2005). The most used technique to estimate the mass percentage of carbon nanoparticles containing other morphologies in the CNT samples is the thermogravimetric analysis (TGA). It is well known that these organic impurities decompose at temperatures lower than CNTs and the derivate of the weight loss curve can be used to determine these percentages. Also, derivate peak width indicates the relative purity of CNTs (Bom et al. 2002; Trigueiro et al. 2007). Mass spectrometry would be used to study the chemical nature of the oxidation debris, after the purification process (Wang et al. 2009; Stéfani et al. 2011).

(g) Inorganic impurities.

Inorganic impurities are commonly the metal catalysts that remain in the CNT samples after synthetic process. In order to identify these impurities, energy-dispersive X-ray spectroscopy (EDS) is generally used whereas TGA analyses are employed to quantify the inorganic impurities as by following the method proposed by NIST (Freiman et al. 2008). Additionally, inductively coupled plasma optical emission spectroscopy (ICP-OES) technique can be used to identify and quantify these metallic residues in CNT samples (Chattopadhyay et al. 2002; Itkis et al. 2005; Lima et al. 2009; Datsyuk et al. 2008; Mello et al. 2011).

The abovementioned techniques give substantial information about the quality of the CNT samples in which it will be submitted to biological interactions. This is a key issue because any bio-related study to be compared with each other needs to have the samples with the same characteristics. Indeed, a precise definition of the term purity for CNTs has to be made, since it may depend on the point of view (Schweinberger and Meyer-Plath 2011). In this way, it is easy to find in the literature the use of the terms "High Purity CNTs" and these terms should be interpreted with caution. As already described, even the purified samples of CNTs (washed with acids and/or heat treated, etc.) may still contain carbon nanostructures such as oxidation debris and trace metals, and thus CNTs always will be a complex mixture of components. So to obtain toxicological comprehensive results, it is necessary to stratify samples of CNTs according to the kind of treatments performed before toxicological assays. For example, a sample of CNTs would be classified as acid-CNT (acid treated), thermo-CNT (heat treatment), etc. keeping in mind which impurities are being "eliminated" from the samples. Comparing these types of samples with raw-CNT (from same origin) would be possible to identify degrees of risk associated to CNTs' preparation.

6.3 Interaction of Carbon Nanotubes with Biological Systems

Biological systems are a complex group of interconnected elements derived from the living world. They were self-organized in hierarchical levels (from biomolecules to biosphere) during evolutionary process of life in the Earth (Campbell et al. 1999). Nowadays, CNTs emerge as promising materials to interact with biological systems regarding practical implications in medicine, agriculture, and environment. Due to their nanometric size, cell-penetrating capability, and physicochemical versatility, CNTs have been considered a key material for the development of carrier systems for bioactive molecules such as drugs, peptides, enzymes, antibodies, and nucleic acids (Pantarotto et al. 2004a, b; Liu et al. 2009; Benincasa et al. 2011); modulation of specific biochemical cellular functions (Menard-Moyon et al. 2010); cell and tissue engineering (Edwards et al. 2009); biosensors and bioconjugates (Yang et al. 2010a; Feng and Ji 2001; Wang et al. 2011); environmental remediation of pollutants (Mauter and Elimelech 2008); and in silico, in vitro, and in vivo toxicity studies (Johnston et al. 2010).

By analyzing the hierarchical levels of biological systems, we can observe that CNTs could impact the living world in different ways. For example, at molecular level, nanotubes interact with proteins and DNA changing the native structures of these biomolecules (Karajanagi et al. 2004; Wu et al. 2012). Complement proteins from the immune system have been reported as effective during the opsonization process of CNTs (Moghimi and Hunter 2010). At cellular level, nanotubes have negatively impacted on the growth rates of prokaryotic and eukaryotic cells, thus showing antimicrobial activities and cellular toxicity (Kang et al. 2009; Cui et al. 2010). However, nanotubes have improved the growth of neuronal and bone cells, thus opening up exciting perspectives to neuroscience and tissue engineering fields, respectively (Cellot et al. 2009; Nayak et al. 2010). At organism level, water-soluble MWCNTs have showed to enhance the growth rate of plants, thereby suggesting better water absorption and retention related to enhanced growth (Tripathi et al. 2011). On the other hand, nanotubes have showed adverse effects on the survival and growth of invertebrates and vertebrates organisms (Mouchet et al. 2008, 2010; Scott-Fordsmand et al. 2008; Cheng et al. 2009a; Zhu et al. 2009; Li and Huang 2011). Finally, at ecological level, there are many research gaps to be filled before laying down conclusive statements about the impacts of nanotubes on the ecosystems (Petersen and Henry 2012). Therefore, it is important to take into account that the nanotube presence in the environment does not mean that they will always deliver deleterious or adverse effects to biological systems; their use in electronics, in batteries, and other equipments does not necessarily lead to consumer exposition. It is more rational to consider the exposure conditions employed towards realistic scenarios to these nanomaterials, for example, occupational exposure to CNTs can occur during laboratory research, product development (synthesis), and waste recycling/disposal (Aschberger et al. 2010; Castranova et al. 2013).

In fact, more detailed studies on CNT toxicity are required (Donaldson 2012) and the evaluation of cytotoxicity and genotoxicity of CNT materials is a hot topic. Due to the similarities of CNTs with asbestos fibers (classical human genotoxic and carcinogenic agents), namely, biopersistence and needlelike structure, there are many reports drawing attention to nanotube carcinogenic potential (Palomaki et al. 2011; Stella 2011; Nagai and Toyokuni 2012). In next section, we are going to discuss the current studies describing the adverse effects of CNTs on the genetic material (DNA) by using classical and modern biological models (in vitro and in vivo). Furthermore, the mechanisms involved in the genotoxicity of CNTs are also discussed focusing the positive responses found in the literature.

6.4 Genotoxic Effects of Carbon Nanotubes

The cell interactions with CNTs could lead to different toxic routes. The CNTs can affect cell signaling and induce membrane perturbations, production of cytokines, chemokines, DNA damage, and oxidative stress (Lelimousin and Sansom 2013; Shvedova et al. 2008). These injuries, if not correctly repaired/recovered, would culminate in different kinds of cell death such as apoptosis, necrosis, and autophagy (Cheng et al. 2013; Liu et al. 2011; Wang et al. 2012b). Shvedova et al. (2012) proposed that an important molecular mechanism of nanotoxicity was excessive raise of the oxidative stress. It is well known that DNA damage and oxidative stress are tightly associated (Moller et al. 2012). Here, we propose genotoxicity as the primary mechanism of toxicity induction by CNTs, thus the main focus is DNA damage events. It is worth mentioning that genotoxicology represents a regulatory science within OECD test guidelines and the field is a strategic area in regulation of nanomaterials (Doak et al. 2012).

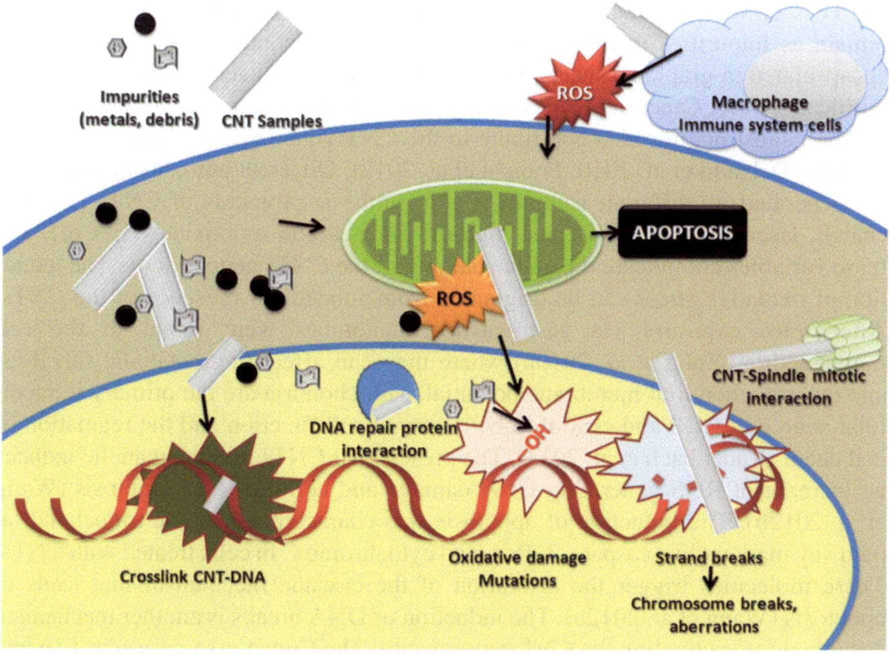

Fig. 6.2 Proposed mechanisms of CNT genotoxicity. CNTs can induce primary genotoxicity (directly and indirectly) and also secondary genotoxicity mediated by inflammation

What makes the genotoxicity studies of CNTs more intriguing is the fact that nanotubes exhibit multiple mechanisms of action that differentiate them from the classical genotoxins (Singh et al. 2009; van Berlo et al. 2012) and they present more than one DNA damage pathway (Fig. 6.2). This multiple mechanisms scenario is very clear when we analyze the results presented by Zhu et al. (2007). The authors observed that MWCNTs increased the expression of two isoforms of base excision repair protein 8-oxoguanine-DNA glycosylase 1 (OGG1), induced double-strand break repair protein Rad 51, phosphorylation of H2AX histone at serine 139, and SUMO modification of XRCC4, and finally increased the mutation frequency by twofold compared with the spontaneous mutation frequency in mouse embryonic stem cells.

Among the various toxicological mechanisms described, oxidative stress is the main pathway associated to CNTs genotoxicity (Shvedova et al. 2012). Oxidative stress is characterized by cellular redox imbalance, usually as a result of increased intracellular reactive oxygen species (ROS) and decreased antioxidants molecules and enzymes. ROS are highly reactive molecules that can disrupt the intracellular homeostasis reacting unfavorably with cellular macromolecules, including lipids, proteins, and DNA. The presence of metal catalysts that can generate ROS by Fenton reaction is the main feature responsible for this biological response. These catalysts are used in the process of synthesis of CNTs; and among them, we could cite Fe, Co, Mn, Mo, and Ni, which even after CNT purification process may still remain as impurities in the samples. These metallic contaminants can be found encapsulated in graphene sheets, which protect them from acid washes used in the purification step. Other impurities such as amorphous carbon, oxidation debris, and carbon nanostructures also contribute to the CNT genotoxicity (Silva et al. 2009; Campos-Delgado et al. 2010; Franchi et al. 2012). Different amounts of impurities are generated by different techniques and synthesis protocols of CNTs [arc discharge, laser ablation, CVD, and high-pressure carbon monoxide (HiPCo)], and these variables complicate the understanding of the CNTs genotoxicity. The induction of oxidative stress can be originated from subcellular presence of the CNTs. After mice exposure via gastrogavage, nanotubes were found in the cell mitochondria (Yang et al. 2010b) where they can affect the metabolic functions and the mitochondrial membrane potential. Mitochondria are the primary intracellular source of ROS and central to both normal cell function and the regulation of cell death (Kadenbach et al. 2013). The presence of CNTs in this organelle induces an increase of ROS generating DNA damage and cell death by apoptosis (Wang et al. 2012b). The induction of apoptosis was characterized by the mitochondrial pathway markers like caspase-3, Bax, and cytochrome C in cells treated with CNTs. These molecules trigger the activation of the cascade mechanism that leads to apoptosis (Wang et al. 2012b). The induction of DNA breaks is another mechanism described for accounting the CNT genotoxicity. The Comet assay is a sensitive and reproducible method able to detect DNA breaks. Especially, the alkaline version of the assay detects double strand breaks, single strand breaks, and alkali labile sites. The use of enzymes (OGG1, oxoguanine glycosylase; Endo III, endonuclease III; FPG, formamidopyrimidine-DNA glycosylase) also allows the oxidative damage detection. This methodology is the most used in the detection of genotoxicity of CNTs and it has been very sensitive to probe the damage induced by these nanomaterials (Karlsson 2010). Positive responses of DNA breaks were detected in in vitro systems using different cell lines for both SWCNTs and MWCNTs using the Comet assay (Table 6.1). In some circumstances, oxidative DNA damages were not detected in cells exposed to CNT and this may be associated to samples of high purity and with few metal residues, but DNA breaks were also detected in these cases using Comet assay. The evidences for supporting this mechanism are unveiled by results that point out that oxidative stress is not the unique pathway of DNA damage induction (Karlsson et al. 2008; Kisin et al. 2011; Cavallo et al. 2012; Ursini et al. 2012; Pelka et al. 2013). The DNA breaks generated in this context would be a direct consequence between CNTs and DNA. Furthermore, DNA strand break can also be assessed by measuring H2AX phosphorylation and γ-H2AX foci markers. These markers were found after exposition of cells to SWCNTs and MWCNTs (Zhu et al. 2007; Pacurari et al. 2008a, b; Cveticanin et al. 2010; Guo et al. 2011).

Other evidences also indicate that CNTs are able to directly interact with DNA. Ghosh et al. (2011) described that a cross-link DNA-CNT can occur and they observed an inhibition of DNA migration after exposition to MWCNTs using Comet assay. Moreover, SWCNTs have been also reported as a kind of telomeric

Table 6.1 Summary of oxidative and DNA strand breaks induced by CNTs reported in the literature

CNTs	Oxidative DNA damage	DNA breaks (Comet assay)	Conc.[a] (μg/mL)	References
MWCNTs	ND	Negative	40	Patlolla et al. (2010)
MWCNTs	Positive (FPG-Comet assay)	Negative	20	McShan and Yu (2012)
MWCNTs	Negative (FPG-Comet assay)	Positive	40	Ursini et al. (2012)
MWCNTs-OH	Negative (FPG-Comet assay)	Positive	5	Ursini et al. (2012)
MWCNTs-OH	Negative (FPG-Comet assay)	Positive	10	Ursini et al. (2012)
MWCNTs	Negative (OGG1, MTH1, MYH; HPLC 8-OhdG)	Positive	20	Ogasawara et al. (2012)
MWCNTs	Negative (FPG-Comet assay)	Positive	5	Cavallo et al. (2012)
MWCNTs	ND	Positive	300	Lindberg et al. (2012)
MWCNTs	Positive (ENDOIII- and FPG-Comet assay)	Positive	1	Migliore et al. (2010)
MWCNTs	Negative (FPG-Comet assay)	Positive	2	Karlsson et al. (2008)
MWCNTs	ND	Positive	2	Ghosh et al. (2011)
MWCNTs	Positive (DCF fluorescence)	Positive	25	Pacurari et al. (2008a, b)
MWCNTs	Positive (DCF fluorescence)	Positive	3	Di Giorgio et al. (2011)
SWCNTs	ND	Negative	10	Zeni et al. (2008)
SWCNTs	Positive (FPG-Comet assay)	Negative	100	Jacobsen et al. (2008)
SWCNTs	Negative (FPG-Comet assay)	Positive	0.00005	Pelka et al. (2013)
SWCNTs	ND	Positive	3.8	Lindberg et al. (2009)
SWCNTs	Positive (ESR detection of free radicals)	Positive	25	Pacurari et al. (2008a, b)
SWCNTs	Positive (DCF fluorescence)	Positive	5	Yang et al. (2009)
SWCNTs	ND	Positive	152	Lindberg et al. (2012)
SWCNTs	Positive (DCF fluorescence)	Positive	75	Cicchetti et al. (2011)
SWCNTs	ND	Positive	8	Nam et al. (2011)
SWCNTs	Negative (ESR detection of free radicals)	Positive	240	Kisin et al. (2011)
SWCNTs	Positive (ENDOIII- and FPG-Comet assay)	Positive	10	Migliore et al. (2010)
SWCNTs	Negative (OGG1, MTH1, MYH; HPLC 8-OhdG)	Positive	20	Ogasawara et al. (2012)
SWCNTs	Positive (DCF fluorescence)	Positive	10	Di Giorgio et al. (2011)

[a]Concentration stands for the lowest value that provided a positive response, and for the negative results, highest tested concentration is informed

Table 6.2 Summary of chromosome breaks induced by CNTs according to literature

CNTs	MN test	Conc.[a] (μg/mL)	References
MWCNTs	Negative	100	Ponti et al. (2013)
MWCNTs-COOH	Negative	100	Ponti et al. (2013)
MWCNTs-NH_2	Negative	100	Ponti et al. (2013)
MWCNTs-OH	Negative	100	Ponti et al. (2013)
MWCNTs	Negative	1	Szendi and Varga (2008)
MWCNTs	Negative	5	Asakura et al. (2010)
MWCNTs	Positive	10	Srivastava et al. (2011)
MWCNTs	Positive	1	Migliore et al. (2010)
MWCNTs	Positive	25	Cveticanin et al. (2010)
MWCNTs	Positive	25	Muller et al. (2008)
MWCNTs	Positive	10	Muller et al. (2008)
MWCNTs	Positive	20	Kato et al. (2013)
MWCNTs	Positive	1	Di Giorgio et al. (2011)
SWCNTs (~50 %)	Negative	360	Lindberg et al. (2009)
SWCNTs	Negative	480	Kisin et al. (2007)
SWCNTs (400–800 nm)	Positive	20	Manshian et al. (2013)
SWCNTs (1–3 μm)	Positive	20	Manshian et al. (2013)
SWCNTs (5–30 μm)	Positive	20	Manshian et al. (2013)
SWCNTs	Positive	75	Cicchetti et al. (2011)
SWCNTs	Positive	60	Kisin et al. (2011)
SWCNTs	Positive	0.1	Migliore et al. (2010)
SWCNTs	Positive	25	Cveticanin et al. (2010)
Amide-SWCNTs	Positive	25	Cveticanin et al. (2010)
SWCNTs	Positive	1	Di Giorgio et al. (2011)

[a]Concentration stands for the lowest value that provided a positive response, and for the negative results, highest tested concentration is informed

DNA–ligand. This structure stabilization can inhibit telomerase activity and leads to telomere uncapping and displaces telomere-binding proteins from telomere. The dysfunctional telomere triggers DNA damage response and elicits upregulation of p16 and p21 proteins (Li et al. 2006a, b; Chen et al. 2012). DNA breaks may result in the formation of chromosome breaks or rearrangements (clastogenesis), which can be evaluated by the micronucleus (MN) test, which also detects loss of chromosomes (aneugenesis). This technique has been an effective tool to detect such events induced by CNTs (Gonzalez et al. 2011). The data of MN tests pointing out a positive effect to SWCNTs while to MWCNTs conflicting results are found in the literature (Table 6.2). The use of probes (FISH) allows one to identify the origin of the micronuclei: if the micronucleus containing a chromosomal fragment is characterized as clastogenic event or if an entire chromosome is found inside of the micronucleus, an aneugenic event has happened. Muller et al. (2008) made use of FISH and found that MWCNTs significantly increases both events. Therefore, these findings reinforce the idea that CNTs can induce DNA chromosome breaks and also act as aneugenic agent.

Aneugenic agents can act on different cellular targets, but mainly on mitotic spindle (kinetochores, centrosomes, microtubules, and anaphase promoter complex). These kinds of agent alter chromosome segregation between the two daughter cells. This missegregation leads to wrong formation of cells with abnormal number of chromosomes (aneuploidy). The aneuploidy is another event detected after CNT exposure. The size and physical properties of CNTs are similar to cellular microtubules, thus suggesting that nanotubes could replace or associate to microtubules. This association creates hybrid-form nanotubes/microtubules that subsequently associate with the mitotic apparatus. This hybrid-form nanotubes/microtubule alters the delicate balance of assembly/disassembly of microtubules, with in turn has critical implications such as the number of chromosomes, mitotic spindle multipolarity, division, and cell morphology and consequently may induce cell death (Sargent et al. 2009, 2012; Rodriguez-Fernandez et al. 2012).

Gene (point) mutations where only a single base is modified, or one or a relatively few bases are inserted or deleted from DNA, are readily measured in bacteria and other cell systems. The *Salmonella*/microsome assay (Salmonella test; Ames test) is a widely accepted short-term bacterial assay for identifying substances that can produce genetic damage that leads to gene mutations. Negative results for CNTs were observed by using this methodology (Table 6.3). However, a detail is very important for interpreting these results correctly. The Ames test uses bacteria, and there is one crucial difference between the membranes of mammalian cells and bacteria. The bacteria possesses a cell wall that can prevent the entry of CNTs in these cells (Fig. 6.3). Umbuzeiro et al. (2011) demonstrated that MWCNTs did not penetrate the bacteria cell in the Ames test. Unlike the in vitro gene mutation assays using mammalian cells showed positive results when treated with MWCNTs and SWCNTs (Zhu et al. 2007; Manshian et al. 2013) (Table 6.3). Therefore, this difference might be related to the degree of CNT uptake by the bacterial cells, which is likely to be less than in human cells for two reasons. Firstly, prokaryotes cannot perform endocytosis as well as mammalian cells and, secondly, their cell wall forms a barrier against simple diffusion of CNTs (particularly those in agglomerated form) into the bacterial cell—this lack of uptake could potentially lead to false-negative results (Doak et al. 2012). However, one positive result was detected by using TA1538 strain (Table 6.3) and it is possible that well-dispersed CNTs of approximately 20 nm diameter could enter the bacteria, and therefore the system would report genotoxicity appropriately. Another explanation is related to differences in the strains used in Ames test, which also carry mutations in the genes responsible for lipopolysaccharide synthesis (LPS mutant), thus making the cell wall of the bacteria more permeable to chemicals. In this sense, the identification or engineering of strains more sensitive to the entry of CNTs inside the bacteria would be a strategy against false results using this system as model.

The adsorption capability of CNTs is another property that would also impact their genotoxicity evaluation. This adsorptive nature may potentially lead to an indirect genotoxicity due to the binding of essential nutrients, DNA repair proteins, enzymes, cytokines, and/or growth factors (Casey et al. 2007; Umbuzeiro et al. 2011). However, this property also could lead to adsorption of mutagenic

Table 6.3 Gene mutations induced according to literature.

CNTs	Strains	Gene mutations	Conc.[a] (μg/mL)	References
MWCNTs	TA100, TA1535, WP2uvrA, TA98, TA1537	Negative	100	Ema et al. (2012a)
MWCNTs	TA98, TA100, WP2uvrA	Negative	9	Di Sotto et al. (2009)
Baytubes (agglomerated MWCNTs)	TA1535, TA100, TA1537, TA98, TA102	Negative	5,000	Wirnitzer et al. (2009)
MWCNTs	hprt mutation assay (mammalian cells)	Negative	100	Asakura et al. (2010)
MWCNTs	TA1538	Positive	10^{-5}	Taylor et al. (2012)
MWCNTs	aprt mutation assay (mammalian cells)	Positive	5	Zhu et al. (2007)
SWCNTs	TA100, TA1535, WP2uvrA, TA98, TA1537	Negative	100	Ema et al. (2012b)
SWCNTs (400–800 nm)	hprt mutation assay (mammalian cells)	Negative	100	Manshian et al. (2013)
SWCNTs (5–30 mm)	hprt mutation assay (mammalian cells)	Negative	100	Manshian et al. (2013)
SWCNTs	TA100, TA1535, WP2 uvrA, TA98, TA97	Negative	500	Naya et al. (2011)
SWCNTs	cII mutant frequency (mammalian cells)	Negative	100	Jacobsen et al. (2008)
SWCNTs	YG1024, YG1029	Negative	240	Kisin et al. (2007)
SWCNTs (1–3 μM)	hprt mutation assay (mammalian cells)	Positive	25	Manshian et al. (2013)

[a]Concentration stands for the lowest value that provided a positive response, and for the negative results, highest tested concentration is informed

molecules by CNTs thus either avoiding or enhancing their effects inside of the cells. Moreover, it was reported that SWCNTs act as electron scavengers and thereby decrease the chance of DNA lesion by OH-adduct in an ultrasonication in vitro environment. Thus, CNTs may mitigate the impacts of certain oxidative stresses and mutagenic molecules in cells. However, this activity will be dependent of the cellular environment and more studies are necessary to understand these effects (Dolash et al. 2013). Oxidative DNA damage, DNA strand breakage, structural chromosomal aberrations, and micronucleus formation and mutations were also detected by in vivo genotoxicity experiments (Muller et al. 2008; Kisin et al. 2007; Folkmann et al. 2009; Jacobsen et al. 2009; Patlolla et al. 2010; Kato et al. 2013; Kim et al. 2012a). An in vivo approach indicates that CNTs might induce genotoxicity by secondary pathway. In this scenario, ROS generated indirectly by target cells involved in an inflammatory response affect neighboring cells. Macrophages and neutrophils are probably the first cells to combat CNTs after an exposure by inhalation or injection. The activation of macrophages is accompanied

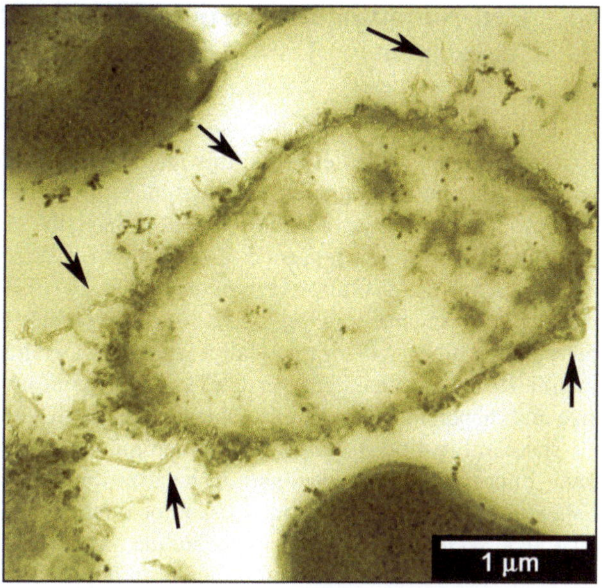

Fig. 6.3 Bacteria cells (*Salmonella* TA98 strain) exposed to acid-treated multiwalled carbon nanotubes (CNT-LQES$_1$) under the transmission electronic microscope. *Arrows* indicate the nanotubes outside the bacterial cell. Reproduced from Umbuzeiro et al. (2011) by permission of Elsevier Ltd.

by production and release of ROS. Moreover, if the CNTs are long enough, an incomplete or "frustrated" phagocytosis happens (Brown et al. 2007). This leads to prolonged production of ROS, cytokines, and inflammatory mediators. The important mediators of inflammation induced by CNTs are the IL-1 family (i.e., IL-33), IL-6, and TNF-α. The IL-33 has a potential role in alerting the immune system of tissue injury or trauma (Katwa et al. 2012). Thus, CNTs seem to be implicated in physical damage to cells and tissues and cell proliferation in this context of chronic inflammation may result in accumulation of DNA damage and if not correctly repaired in mutations.

Thus, we discussed here direct genotoxicity of CNTs that results from physical interactions of materials with the genomic DNA and microtubules, and the indirect genotoxicity that can result from increased ROS generation upon interaction with other cellular components (e.g., mitochondria, endoplasmic reticulum, cell membrane), and it also can arise when CNTs elicit depletion of intracellular antioxidants or even by the presence of metal contaminants. Finally, the secondary genotoxicity was also discussed above and happens when an inflammatory action induced by CNTs results in increased ROS and DNA damage. However, it is important to highlight that many controversies regarding the genotoxicity of CNTs exist. The misunderstandings are probably associated with physicochemical and dispersion issues of diferent nanotube samples and the state of the art in the field is moving knowledge towards to answer these questions (Nagai and Toyokuni 2012; Murray et al. 2012 and Wang et al. 2012a).

6.5 Cell-Nanotube Interaction: Considerations for Toxicity Assessment

CNTs have hydrophobic nature and their existence in the form of aggregated/agglomerated and bundles make this material inadequately soluble or dispersible in most of common aqueous solvents and biological buffers, which is crucial to their processing in (geno)toxicity studies. Indeed, the mixture of CNTs and water or organic solvents in the presence of surfactants or (bio)polymers is just a colloidal dispersion and not a solution (Premkumar et al. 2012). The differences in dispersion ranging from macroscopic aggregates to micrometer bundles or individually dispersed nanotubes will dramatically affect the absolute size and, hence, the cell internalization mechanisms for CNTs. Thus, (geno)toxicity depends critically on size and dispersibility of CNTs (Manshian et al. 2013).

The combination of microscopy techniques with cell viability assays will have great importance for understanding localized effects of nanomaterials and establishment of their potential toxicity. Confocal laser scanning microscopy (CLSM) is an optical sectioning microscopy technique that offers fluorescence imaging of thin slices of specimens several micrometers in depth. Studies have demonstrated that CNTs may be visualized in cell culture directly by using the reflection mode of CLSM (Zhang and Monteiro-Riviere 2010). Moreover, the combination of SEM and CLSM is especially well appropriate to the task of establishing the spatial localization of nanomaterials within living cells (Cheng et al. 2009b). To better detect the accurate trajectory of nanomaterials, spinning-disk confocal microscopy technique can be used to monitor the trafficking of nanomaterials within cells (Ruan et al. 2007). Regarding to achieve the mapping of nanomaterials within a cell, conventional dark-field (DF) microscopy is a detection method, which allows observation, but does not measure the exact size or amount of nanomaterials (Gibbs-Flournoy et al. 2011). On the other hand, the photoacoustic microscopy (PAM) is suitable for quantifying, detecting, and mapping CNTs in histological specimens (Avti et al. 2012). Transmission electronic microscopy (TEM) can provide the most detailed information regarding nanoparticle localization by allowing visualization and location of nanoparticle within a cell or tissue. Porter et al. (2007) observed cellular uptake of SWCNTs by TEM, the majority of SWCNTs were located inside of phagosomes and lysosomes of human monocyte-derived macrophages, suggesting uptake by phagocytosis. The diffusion through cell membrane was also characterized, and in this investigations, nanotube length clearly influences uptake and shorter (sub-1 μm) MWCNTs are readily internalized by cells. Short nanotubes (e.g., hundreds of nanometers in length) can act as tiny straight "nano-needles" able to penetrate the cell membrane more efficiently than longer nanotubes (e.g., >1 μm length), which are often arranged in a "ball" or bundled shape (Pantarotto et al. 2004a; Raffa et al. 2008). Endocytosis as an internalization mechanism for CNTs was also proved, but in this case the presence of biomolecules (protein, DNA) conjugated to the nanotubes is pivotal (Kam et al. 2006; Kam and Dai 2006; Raffa et al. 2008, 2010). More interesting question arises from the uptake mechanism of CNTs. As mentioned before, endocytosis happens when biomolecules such as proteins are conjugated to CNTs. Layer of proteins spontaneously and stably adsorb on CNT when they are exposed to physiological solutions forming a "protein corona" (Dutta et al. 2007; Lundqvist et al. 2008). Thus, the protein corona of a nanomaterial is a complex system with several proteins on their surface. But the interesting factor on protein corona is that the binding efficiency of proteins is not merely related to their abundance in the given sample but there is specificity in the binding, and it may be due to 3D arrangement of carbon atoms in CNTs, protein size, or charge. Thus, proteins with a high affinity to bind at CNTs are potential targets after the latter enters the cell. The conformation and functions of these specific proteins may be affected, thus resulting in disruptions to cellular structure and cell toxicity (Monopoli et al. 2012; Cai et al. 2013).

"Cell vision" is a complementary concept to protein corona in nanotoxicology (Mahmoudi et al. 2012). It was described that what the cell is seeing is a crucial feature in the internalization and hence in the toxic effects of nanomaterials. For example, consider nanotube with a protein corona approaching the surface of a cell. From the cell's standpoint, the interactions driven by receptor–ligand which are being presented to the cells are proteins and other biomolecules. In this point of view, proteins that are not recognized by any receptors on the cell, which will be specific for each type of cell/tissue, could bind to the nanomaterial without presenting a relevant receptor-binding sequence, and it will make the particle less tasty to the cell. After all, cells associate and take up only the proteins they need. A naked particle surface has a much greater nonspecific affinity to the cell surface than a particle hiding behind a corona of bystander proteins (Lynch et al. 2009). Therefore, proteins have influences in the amount of uptake of CNTs into cells as well as their fate in the intracellular environment, since membrane transport greatly depends on the composition of cellular membranes. In this perspective, the impact of exactly the same CNTs on various cells is significantly different and could not be assumed for other cells. The cellular response is also related to the numerous detoxification strategies that any particular cell can utilize in response to nanoparticles. Thus, what the cell really "sees", when it is faced with CNTs, is most likely dependent on the cell type and their mechanisms to deal with these nanomaterials (Lynch et al. 2009; Laurent et al. 2012; Mahmoudi et al. 2012).

There are about 200 types of differentiated cells in the human body (Alberts et al. 2007), all of which will have a significant variability in the interaction with CNTs as mentioned earlier as "cell vision". The toxicity evaluation at all these kinds of cells to detect potential cell and tissue target associated to physicochemical characteristics of CNTs is a complex task by using standard in vitro and in vivo systems. Thereby, in silico approach and high-throughput omics screening can speed up toxicology evaluations and might be an important tool in nanotoxicology (George et al. 2011; Thomas et al. 2011). Thus, a great potential to predict properties, reactivity, and mechanisms of actions of CNTs can be explored by using these modern methodologies. Computational studies have been used to elucidate possible toxic effects of various carbon nanostructures and nanoparticles on peptide/protein structure and dynamics. The π–π interactions between the aromatic residues of the peptide and the carbon surfaces have been reported to be responsible for stabilization of the complexes. The surface curvature of the carbon nanostructure has the influence on the conformational changes in the peptide (Makarucha et al. 2011). Recent examples of global omics studies (e.g., transcriptomics, proteomics, and lipidomics) of CNT-induced tissue damage were employed to achieve a comprehensive understanding of the complex and interrelated molecular changes in cells and tissues exposed to SWCNTs. The oxidative lipidomics demonstrated a nonrandom profile of peroxidation in the lungs of mice exposed to SWCNT with high content of transition metals. The peroxidation products were detected and identified in three relatively minor classes of anionic phospholipids: cardiolipin, phosphatidylserine, and phosphatidylinositol (Tyurina et al. 2011; Shvedova et al. 2012). Proteomics approaches were also employed to demonstrate the influence of metal impurities on CNT cytotoxicity (Haniu et al. 2010) and to identify the protein corona associated with carbon nanomaterials after interaction with cell culture media containing fetal bovine serum (Cai et al. 2013).

6.6 Comments and Future Perspectives

The interaction of CNTs with biological systems is a key point towards a responsible and sustainable development of nanotechnology in broad sense. However, integrated methods for appropriately characterizing CNTs in bio-toxicological context are not yet available. At present, precise influence of physicochemical features in the genotoxicity of CNTs is still unclear and because some results are conflictuos, and in part, this is due to the differences in samples processing. More studies are required with adapted and harmonized protocols and better characterized CNT samples to properly evaluate the genotoxic potential of these nanomaterials. Moreover, protein "corona" and cell "vision" are emerging concepts in nanotoxicology that should be addressed. By considering some structural similarities of nanotubes with classical carcinogenic agents (asbestos fibers), and the absence of reliable genotoxicity studies, these materials have been considered as hazardous products by regulatory agencies and the *precaution principle* is recommended. We have highlighted the potential use of different approaches (omics, informatics, and microscopies) that can be used to understanding the interactions of CNTs with cells/biological systems. These approaches may not only lead to the generation of new hypotheses regarding CNT-biointeractions but could also yield models and biomarkers of human and environmental exposure to engineered nanomaterials. Finally, the most important issue is the need of a multidisciplinary approach with researchers with different backgrounds working together to holistically solve the complex problems of nanotoxicology.

Acknowledgments D.S.T.M. and O.L.A. thank INCT-Inomat and Brazilian Nanotoxicology Network—CIGENANOTOX; L.P.F. and C.M.F. acknowledge the FAPESP for doctorate fellowships; O.P.F. and A.G.S.F. acknowledge CNPq, FUNCAP, PRONEX, INCT-NanoBioSimes, and Núcleo de Estudos em Nanoestruturas de Carbono; D.S.T.M. and A.G.S.F. thank the European Union Seventh Framework Programme (FP7) Small Collaborative project, Neuronano (NMP4-SL-2008-214547); A.G.S.F. and O.L.A. thank PROCAD-CAPES program. The authors thank Prof Dr Gisela A. Umbuzeiro for discussions and valuable suggestions.

References

Ajayan PM, Tour JM (2007) Materials science: nanotube composites. Nature 447(7148): 1066–1068

Akbar S, Taimoor AA (2009) Functionalization of carbon nanotubes: manufacturing techniques and properties of customized nanocomponents for molecular-level technology. Recent Pat Nanotechnol 3(2):154–161

Alberts B, Lewis J, Roberts RM, Walter P (eds) (2007) Molecular biology of the cell, 5th edn. Garland Science, New York, NY

Asakura M, Sasaki T, Sugiyama T et al (2010) Genotoxicity and cytotoxicity of multi-wall carbon nanotubes in cultured chinese hamster lung cells in comparison with chrysotile A fibers. J Occup Health 52(3):9–20

Aschberger K, Johnston HJ, Stone V et al (2010) Review of carbon nanotubes toxicity and exposure- appraisal of human health risk assessment based on open literature. Crit Rev Toxicol 40(9):759–790

Avti PK, Hu S, Favazza C et al (2012) Detection, mapping, and quantification of single walled carbon nanotubes in histological specimens with photoacoustic microscopy. PLoS One 7(4): e35064

Awasthi K, Srivastava A, Srivastava ON (2005) Synthesis of carbon nanotubes. J Nanosci Nanotechnol 5(10):1616–1636

Balasubramanian K, Burghard M (eds) (2010) Carbon nanotubes: methods and protocols. In: Methods in molecular biology, vol 625. Humana Press, New York, NY

Belin T, Epron F (2005) Characterization methods of carbon nanotubes: a review. Mat Sci Eng B Solid 119(2):105–118

Benincasa M, Pacor S, Wu W et al (2011) Antifungal activity of amphotericin B conjugated to carbon nanotubes. ACS Nano 5(1):199–208

Berhanu D, Dybowska A, Misra SK et al (2009) Characterisation of carbon nanotubes in the context of toxicity studies. Environ Health 8(suppl 1):S3. doi:10.1186/1476-069X

Bom D, Andrews R, Jacques D et al (2002) Thermogravimetric analysis of the oxidation of multiwalled carbon nanotubes: evidence for the role of defect sites in carbon nanotube chemistry. Nano Lett 2(6):615–619

Brown DM, Kinloch IA, Bangert U et al (2007) An in vitro study of the potential of carbon nanotubes and nanofibres to induce inflammatory mediators and frustrated phagocytosis. Carbon 45(9):1743–1756

Bussy C, Pinault M, Cambedouzou J et al (2012) Critical role of surface chemical modifications induced by length shortening on multi-walled carbon nanotubes-induced toxicity. Part Fibre Toxicol 9:46. doi:10.1186/1743-8977-9-46

Cai X, Ramalingam R, Wong HS et al (2013) Characterization of carbon nanotube protein corona by using quantitative proteomics. Nanomedicine 9(5):583–593. doi:10.1016/j.nano.2012.09.004

Campbell NA, Reece JB, Mitchell LG (1999) Biology, 5th edn. Benjamin Cummings, Menlo Park, CA

Campos-Delgado J, Maciel IO, Cullem DA et al (2010) Chemical vapor deposition synthesis of N-, P-, and Si-doped single-walled carbon nanotubes. ACS Nano 4(3):1696–1702
Casey A, Davoren M, Herzog E et al (2007) Probing the interaction of single walled carbon nanotubes within cell culture medium as a precursor to toxicity testing. Carbon 45(1):34–40
Castranova V, Schulte PA, Zumwalde RD (2013) Occupational nanosafety considerations for carbon nanotubes and carbon nanofibers. Acc Chem Res 46(3):642–649
Cavallo D, Fanizza C, Ursini CL et al (2012) Multi-walled carbon nanotubes induce cytotoxicity and genotoxicity in human lung epithelial cells. J Appl Toxicol 32(6):454–464
Cellot G, Cilia E, Cipollone S et al (2009) Carbon nanotubes might improve neuronal performance by favouring electrical shortcuts. Nat Nanotechnol 4(2):126–133
Chattopadhyay D, Galeska I, Papadimitrakopoulos F et al (2002) Complete elimination of metal catalysts from single wall carbon nanotubes. Carbon 40(7):985–988
Chen Y, Qu K, Zhao C et al (2012) Insights into the biomedical effects of carboxylated single-wall carbon nanotubes on telomerase and telomeres. Nat Commun 3:1074. doi:10.1038/ncomms2091
Cheng J, Chan CM, Veca LM et al (2009a) Acute and long-term effects after single loading of functionalized multi-walled carbon nanotubes into zebrafish (Danio rerio). Toxicol Appl Pharmacol 235(2):216–225
Cheng C, Müller KH, Koziol KK (2009b) Toxicity and imaging of multi-walled carbon nanotubes in human macrophage cells. Biomaterials 30(25):4152–4160
Cheng W-W, Lin Z-Q, Wei B-F et al (2013) Single-walled carbon nanotube induction of rat aortic endothelial cell apoptosis: reactive oxygen species are involved in the mitochondrial pathway. Int J Biochem Cell Biol 43(4):564–572
Cicchetti R, Divizia M, Valentini F et al (2011) Effects of single-wall carbon nanotubes in human cells of the oral cavity: geno-cytotoxic risk. Toxicol In Vitro 25(8):1811–1819
Cui HF, Vashist SK, Al-Rubeaan K et al (2010) Interfacing carbon nanotubes with living mammalian cells and cytotoxicity issues. Chem Res Toxicol 23(7):1131–1147
Cveticanin J, Joksic G, Leskovac A et al (2010) Using carbon nanotubes to induce micronuclei and double strand breaks of the DNA in human cells. Nanotechnology 21(1):015102. doi:10.1088/0957
Dai HJ (2002) Carbon nanotubes: synthesis, integration, and properties. Acc Chem Res 35(12): 1035–1044
Datsyuk V, Kalyva M, Papagelis K et al (2008) Chemical oxidation of multiwalled carbon nanotubes. Carbon 46(6):833–840
Del Canto E, Flavin K, Movia D et al (2011) Critical investigation of defect site functionalization on single-walled carbon nanotubes. Chem Mater 23(1):67–74
Di Giorgio ML, Di Bucchianico S, Ragnelli AM et al (2011) Effects of single and multi walled carbon nanotubes on macrophages: cyto and genotoxicity and electron microscopy. Mutat Res 722(1):20–31
Di Sotto A, Chiaretti M, Carru GA et al (2009) Multi-walled carbon nanotubes: Lack of mutagenic activity in the bacterial reverse mutation assay. Toxicol Lett 184:192–197
Doak SH, Manshian B, Jenkins GJ et al (2012) In vitro genotoxicity testing strategy for nanomaterials and the adaptation of current OECD guidelines. Mutat Res 745(1–2):104–111
Dolash BD, Lahiji RR, Zemlyanov DY et al (2013) Sonication mediated covalent cross-linking of DNA to single-walled carbon nanotubes. Chem Phys 413:11–19. doi:10.1016/j.chemphys.2012.07. 004
Donaldson K (2012) The toxicology of carbon nanotubes. Cambridge University Press, New York
Dresselhaus MS, Dresselhaus G, Jorio A et al (2002) Single nanotube Raman spectroscopy. Acc Chem Res 35(12):1070–1078
Dresselhaus MS, Jorio A, Souza Filho AG et al (2010) Defect characterization in graphene and carbon nanotubes using Raman spectroscopy. Phil Trans A Math Phys Eng Sci 368(1932): 5355–5377
Dutta D, Sundaram SK, Teeguarden JG et al (2007) Adsorbed proteins influence the biological activity and molecular targeting of nanomaterials. Toxicol Sci 100(1):303–315

Edwards SL, Werkmeister JA, Ramshaw JA et al (2009) Carbon nanotubes in scaffolds for tissue engineering. Expert Rev Med Devices 6(5):499–505

Ema M, Imamura T, Suzuki H et al (2012a) Evaluation of genotoxicity of multi-walled carbon nanotubes in a battery of in vitro and in vivo assays. Regul Toxicol Pharmacol 63:188–95

Ema M, Imamura T, Suzuki H et al (2012b) Genotoxicity evaluation for single-walled carbon nanotubes in a battery of in vitro and in vivo assays. J Appl Toxicol 33:933–9

Endo M (1988) Grow carbon fibers in the vapor phase. Chem Technol 18:568–576

Endo M, Takeuchi K, Hiraoka T et al (1997) Stacking nature of graphene layers in carbon nanotubes and nanofibres. J Phys Chem Solids 58(11):1707–1712

Faria AF, Martinez DST, Moraes ACM et al (2012) Unveiling the role of oxidation debris on the surface chemistry of graphene through the anchoring of Ag nanoparticles. Chem Mater 24(21): 4080–4087

Feng W, Ji P (2001) Enzymes immobilized on carbon nanotubes. Biotechnol Adv 29(6):889–895

Firme CP, Bandaru PR (2010) Toxicity issues in the application of carbon nanotubes to biological systems. Nanomedicine 6(2):245–256

Folkmann JK, Risom L, Jacobsen NR et al (2009) Oxidatively damaged DNA in rats exposed by oral gavage to C-60 fullerenes and single-walled carbon nanotubes. Environ Health Perspect 117(5):703–708

Franchi LP, Santos RA, Matsubara EY et al (2012) Citotoxicidade e genotoxicidade de nanotubos de carbono. Quím Nova 35:571–580

Freiman S, Hooker S, Migler K et al., (2008) Messurent issues in single wall carbon nanotubes. National Institute of Standards and Technology (NIST). Special Publication, 960–019

George S, Xia T, Rallo R et al (2011) Use of a high-throughput screening approach coupled with In vivo zebrafish embryo screening to develop hazard ranking for engineered nanomaterials. ACS Nano 5(3):1805–1817

Ghosh M, Chakraborty A, Bandyopadhyay M et al (2011) Multi-walled carbon nanotubes (MWCNT): induction of DNA damage in plant and mammalian cells. J Hazard Mater 197: 327–336

Gibbs-Flournoy EA, Bromberg PA, Hofer TP et al (2011) Darkfield-confocal microscopy detection of nanoscale particle internalization by human lung cells. Part Fibre Toxicol 8(1):2. doi:10.1186/1743-8977-8-2

Gommes C, Blacher S, Dupont- Pavlovsky N et al (2004) Comparison of different methods for characterizing multi-walled carbon nanotubes. Colloids Surf A Phys Eng Aspects 241(1–3): 155–164

Gonzalez L, Sanderson BJS, Kirsch-Volders M (2011) Adaptations of the in vitro MN assay for the genotoxicity assessment of nanomaterials. Mutagenesis 26(1):185–191

Grobert N (2007) Carbon nanotubes—becoming clean. Mater Today 10(1–2):28–35

Guo YY, Zhang J, Zheng YF et al (2011) Cytotoxic and genotoxic effects of multi-wall carbon nanotubes on human umbilical vein endothelial cells in vitro. Mutat Res 721(2):184–191

Haniu H, Matsuda Y, Takeuchi K et al (2010) Proteomics-based safety evaluation of multi-walled carbon nanotubes. Toxicol Appl Pharmacol 242(3):256–262

Hayashida T, Umemura K (2013) Surface morphology of hybrids of double-stranded DNA and single-walled carbon nanotubes studied by atomic force microscopy. Colloids Surf B Biointerfaces 101:49–54

Heister E, Brunner EW, Gregg R et al (2013) Are carbon nanotubes a natural solution? Applications in Biology and Medicine. ACS Appl Mater Interfaces 5(6):1870–1891

Hou PX, Liu C, Cheng HM (2008) Purification of carbon nanotubes. Carbon 46(15):2003–2025

Hull MS, Kennedy AJ, Steevens JA et al (2009) Release of metal impurities from carbon nanomaterials influences aquatic toxicity. Environ Sci Technol 43(11):4169–4174

Hussain SM, Braydich-Stolle LK, Schrand AM et al (2009) Toxicity evaluation for safe use of nanomaterials: recent achievements and technical challenges. Adv Mater 21(16):1549–1559

Iijima S (1991) Helical microtubules of graphitic carbon. Nature 354(6348):56–58

Itkis ME, Perea DE, Jung R et al (2005) Comparison of analytical techniques for purity evaluation of single-walled carbon nanotubes. J Am Chem Soc 127:3439–3448

Jacobsen NR, Pojana G, White P et al (2008) Genotoxicity, cytotoxicity, and reactive oxygen species induced by single-walled carbon nanotubes and C60 fullerenes in the FE1-Muta™Mouse lung epithelial cells. Environ Mol Mutagen 49(6):476–487

Jacobsen NR, Moller P, Jersen KA et al (2009) Lung inflammation and genotoxicity following pulmonary exposure to nanoparticles in ApoE(-/-) mice. Part Fibre Toxicol 6:2. doi:10.1186/1743-8977-6-2

Jain S, Thakare VS, Das M et al (2011) Toxicity of multiwalled carbon nanotubes with end defects critically depends on their functionalization density. Chem Res Toxicol 24(11):2028–2039

Jiang LQ, Gao L, Sun J (2003) Production of aqueous colloidal dispersions of carbon nanotubes. J Colloid Interface Sci 260(1):89–94

Johnston HJ, Hutchison GR, Christensen FM et al (2010) A critical review of the biological mechanisms underlying the in vivo and in vitro toxicity of carbon nanotubes: the contribution of physico-chemical characteristics. Nanotoxicology 4(2):207–246

Jorio A, Dresselhaus G, Dresselhaus MS (2008) Carbon nanotubes: advanced topics in the synthesis, structure, properties, and applications. Springer, Berlin

Kadenbach B, Ramzan R, Vogt S (2013) High efficiency versus maximal performance—the cause of oxidative stress in eukaryotes: a hypothesis. Mitochondrion 13(1):1–6

Kagan VE, Tyurina YY, Tyurin VA et al (2006) Direct and indirect effects of single walled carbon nanotubes on RAW 264.7 macrophages: role of iron. Toxicol Lett 165(1):88–100

Kam NWS, Dai H (2006) Single walled carbon nanotubes for transport and delivery of biological cargos. Phys Status Solidi 243(13):3561

Kam NWS, Liu ZA, Dai H (2006) Carbon nanotubes as intracellular transporters for proteins and DNA: an investigation of the uptake mechanism and pathway. Angew Chem Int Ed Engl 45(4):577–581

Kang S, Mauter MS, Elimelech M (2009) Microbial cytotoxicity of carbon-based nanomaterials: implications for river water and wastewater effluent. Environ Sci Technol 43(7):2648–2653

Karajanagi SS, Vertegel AA, Kane RS et al (2004) Structure and function of enzymes adsorbed onto single-walled carbon nanotubes. Langmuir 20(26):11594–11599

Karlsson H (2010) The comet assay in nanotoxicology research. Anal Bioanal Chem 398(2):651–666

Karlsson HL, Cronholm P, Gustafsson J et al (2008) Copper oxide nanoparticles are highly toxic: a comparison between metal oxide nanoparticles and carbon nanotubes. Chem Res Toxicol 21(9):1726–1732

Karousis N, Tagmatarchis N, Tasis D (2010) Current progress on the chemical modification of carbon nanotubes. Chem Rev 110(9):5366–5397

Kato T, Totsuka Y, Ishino K et al (2013) Genotoxicity of multi-walled carbon nanotubes in both in vitro and in vivo assay systems. Nanotoxicology 7(4):452–461. doi:10.3109/17435390.2012.674571

Katwa P, Wang X, Urankar RN et al (2012) A carbon nanotube toxicity paradigm driven by mast cells and the IL-33/ST2 axis. Small 8(18):2904–2912

Kim JS, Sung JH, Song KS et al (2012a) Persistent DNA damage measured by comet assay of sprague dawley rat lung cells after five days of inhalation exposure and 1 month post-exposure to dispersed multi-wall carbon nanotubes (MWCNTs) generated by new MWCNT aerosol generation system. Toxicol Sci 128(2):439–448

Kim SW, Kim T, Kim YS et al (2012b) Surface modifications for the effective dispersion of carbon nanotubes in solvents and polymers. Carbon 50(1):3–33

Kingston CT, Martinez-Rubi Y, Guan J et al (2010) Coupled thermogravimetry, mass spectrometry, and infrared spectroscopy for quantification of surface functionality on single-walled carbon nanotubes. Anal Bioanal Chem 396(3):1037–1044

Kisin ER, Murray AR, Keane MJ et al (2007) Single-walled carbon nanotubes: geno- and cytotoxic effects in lung fibroblast V79 cells. J Toxicol Environ Health A 70:2071–2079

Kisin ER, Murray AR, Sargent L et al (2011) Genotoxicity of carbon nanofibers: are they potentially more or less dangerous than carbon nanotubes or asbestos? Toxicol Appl Pharmacol 252(1):1–10

Krug HF, Wick P (2011) Nanotoxicology: an interdisciplinary challenge. Angew Chem Int Ed 50 (6):1260–1278

Kumar M, Ando Y (2010) Chemical vapor deposition of carbon nanotubes: a review on growth mechanism and mass production. J Nanosci Nanotechnol 10(6):3739–3758

Kundu S, Wang YM, Xia M et al (2008) Thermal stability and reducibility of oxygen-containing functional groups on multiwalled carbon nanotube surfaces: a quantitative high-resolution XPS and TPD/TPR study. J Phys Chem C 112(43):16869–16878

Lacerda L, Bianco A, Prato M et al (2006) Carbon nanotubes as nanomedicines: from toxicology to pharmacology. Adv Drug Deliv Rev 58(14):1460–1470

Laurent S, Burtea C, Thirifays C et al (2012) Crucial ignored parameters on nanotoxicology: the importance of toxicity assay modifications and "Cell Vision". PLoS One 7(1):306–314

Lehman JH, Terrones M, Mansfield E et al (2011) Evaluating the characteristics of multiwall carbon nanotubes. Carbon 49(8):2581–2602

Lelimousin M, Sansom MS (2013) Membrane pertubation by carbon nanotube insertion: pathways to internalization. Small, http://onlinelibrary.wiley.com/doi/10.1002/smll.201202640/citedby

Li MH, Huang CP (2011) The responses of Ceriodaphnia dubia toward multi-walled carbon nanotubes: effect of physical-chemical treatment. Carbon 49(5):1672–1679

Li YM, Kim W, Zhang Y et al (2001) Growth of single-walled carbon nanotubes from discrete catalytic nanoparticles of various sizes. J Phys Chem B 105(46):11424–11431

Li ZJ, Pan ZW, Dai S (2004) Nitrogen adsorption characterization of aligned multiwalled carbon nanotubes and their acid modification. J Colloid Interface Sci 277(1):35–42

Li X, Peng YH, Qu X (2006a) Carbon nanotubes selective destabilization of duplex and triplex DNA and inducing B-A transition in solution. Nucleic Acids Res 34(13):3670–3676

Li X, Peng YH, Ren J et al (2006b) Carboxyl-modified single-walled carbon nanotubes selectively induce human telomeric i-motif formation. Proc Natl Acad Sci U S A 103(52):19658–19663

Li R, Wang X, Ji Z et al (2013) The surface charge and cellular processing of covalently functionalized multiwall carbon nanotubes determine pulmonary toxicity. ACS Nano 7(3): 2352–2368. doi:10.1021/nn305567s

Lima AMF, Musumeci AW, Waclawik ER et al (2009) Purity evaluation and influence of carbon nanotube on carbon nanotube/graphite thermal stability. J Therm Anal Calorim 97(1):257–263

Lindberg HK, Falck GC, Suhonen S et al (2009) Genotoxicity of nanomaterials: DNA damage and micronuclei induced by carbon nanotubes and graphite nanofibres in human bronchial epithelial cells in vitro. Toxicol Lett 186(3):166–173

Lindberg HK, Falck GC, Singh R et al (2012) Genotoxicity of short single-wall and multi-wall carbon nanotubes in human bronchial epithelial and mesothelial cells in vitro. Toxicology http://www.sciencedirect.com/science/article/pii/S0300483X12004271#. doi:10.1016/j.tox.2012.12.008

Liu Z, Tabakman S, Welsher K et al (2009) Carbon nanotubes in biology and medicine: in vitro and in vivo detection, imaging and drug delivery. Nano Res 2(2):85–120

Liu HL, Zhang YL, Yang N et al (2011) A functionalized single-walled carbon nanotube-induced autophagic cell death in human lung cells through Akt-TSC2-mTOR signaling. Cell Death Dis 2:E159. doi:10.1038/cddis.2011.27

Liu Y, Zhao Y, Sun B et al (2013) Understanding the toxicity of carbon nanotubes. Acc Chem Res 46(3):702–713

Lowry GV, Gregory KB, Apte SC et al (2012) Transformations of nanomaterials in the environment. Environ Sci Technol 46:6893–6899

Lundqvist M, Stigler J, Elia G et al (2008) Nanoparticle size and surface properties determine the protein corona with possible implications for biological impacts. Proc Natl Acad Sci U S A 105 (38):14265–14270

Lynch I, Salvati A, Dawson KA (2009) Protein-nanoparticle interactions: what does the cell see? Nat Nanotechnol 4(9):546–547
Ma PC, Siddiqui NA, Marom G et al (2010) Dispersion and functionalization of carbon nanotubes for polymer-based nanocomposites: a review. Composites A 41(10):1345–1367
Mahmoudi M, Saeedi-Eslami SN, Shokrgozar MA et al (2012) Cell "vision": complementary factor of protein corona in nanotoxicology. Nanoscale 4(17):5461–5468
Makarucha AJ, Todorova N, Yarovsky I (2011) Nanomaterials in biological environment: a review of computer modelling studies. Eur Biophys J 40(2):103–115
Manshian BB, Jenkins GJ, Williams PM et al (2013) Single-walled carbon nanotubes: differential genotoxic potential associated with physico-chemical properties. Nanotoxicology 7:144–156
Mauter MS, Elimelech M (2008) Environmental applications of carbon-based nanomaterials. Environ Sci Technol 42(16):5843–5859
McShan D, Yu H (2012) DNA damage in human skin keratinocytes caused by multiwalled carbon nanotubes with carboxylate functionalization. Toxicol Ind Health 25:1–10
Mello PA, Rodrigues LF, Nunes et al (2011) Determination of metal impurities in carbon nanotubes by direct solid sampling electrothermal atomic absorption spectrometry. J Braz Chem Soc 22(6):1040–1049
Menard-Moyon C, Kostarelos K, Prato M et al (2010) Functionalized carbon nanotubes for probing and modulating molecular functions. Chem Biol 17(2):107–115
Migliore L, Saracino D, Bonelli A et al (2010) Carbon nanotubes induce oxidative DNA damage in RAW 264.7 cells. Environ Mol Mutagen 51(4):294–303
Moghimi SM, Hunter AC (2010) Complement monitoring of carbon nanotubes. Nat Nanotechnol 5(6):382–383
Moller P, Folkmann JK, Danielsen PH et al (2012) Oxidative stress generated damage to DNA by gastrointestinal exposure to insoluble particles. Curr Mol Med 12(6):732–745
Monopoli MP, Aberg C, Salvati A et al (2012) Biomolecular coronas provide the biological identity of nanosized materials. Nat Nanotechnol 7(12):779–786
Mouchet F, Landois P, Sarremejean E et al (2008) Characterisation and in vivo ecotoxicity evaluation of double-wall carbon nanotubes in larvae of the amphibian Xenopus laevis. Aquat Toxicol 87(2):127–137
Mouchet F, Landois P, Puech P et al (2010) Carbon nanotube ecotoxicity in amphibians: assessment of multiwalled carbon nanotubes and comparison with double-walled carbon nanotubes. Nanomedicine 5(6):963–974
Muller J, Decordier I, Hoet PH et al (2008) Clastogenic and aneugenic effects of multi-wall carbon nanotubes in epithelial cells. Carcinogenesis 29(2):427–433
Murray AR, Kisin ER, Tkach AV et al (2012) Factoring-in agglomeration of carbon nanotubes and nanofibers for better prediction of their toxicity versus asbestos. Part Fibre Toxicol 9(10):1–19
Nagai H, Toyokuni S (2012) Differences and similarities between carbon nanotubes and asbestos fibers during mesothelial carcinogenesis: shedding light on fiber entry mechanism. Cancer Sci 103(8):1378–1390
Nam CW, Kang SJ, Kang YK et al (2011) Cell growth inhibition and apoptosis by SDS-solubilized single-walled carbon nanotubes in normal rat kidney epithelial cells. Arch Pharm Res 34(4): 661–669
Naya M, Kobayashi N, Mizuno K et al (2011) Evaluation of the genotoxic potential of single-wall carbon nanotubes by using a battery of in vitro and in vivo genotoxicity assays. Regul Toxicol Pharmacol 61: 192-8
Nayak TR, Jian L, Phua LC et al (2010) Thin films of functionalized multiwalled carbon nanotubes as suitable scaffold materials for stem cells proliferation and bone formation. ACS Nano 4(12): 7717–7725
Nel A, Xia T, Madler L et al (2006) Toxic potential of materials at the nanolevel. Science 311 (5761):622–627
Nessim GD (2010) Properties, synthesis, and growth mechanisms of carbon nanotubes with special focus on thermal chemical vapor deposition. Nanoscale 2(8):1306–1323

Ogasawara Y, Umezu N, Ishii K (2012) DNA damage in human pleural mesothelial cells induced by exposure to carbon nanotubes. Nihon Eiseigaku Zasshi 67(1):76–83

Osswald S, Havel M, Gogotsi Y (2007) Monitoring oxidation of multiwalled carbon nanotubes by Raman spectroscopy. J Raman Spectrosc 38(6):728–736

Pacurari M, Yin XJ, Ding M et al (2008a) Oxidative and molecular interactions of multi-wall carbon nanotubes (MWCNT) in normal and malignant human mesothelial cells. Nanotoxicology 2(3):155–170

Pacurari M, Yin XJ, Zhao J et al (2008b) Raw single-wall carbon nanotubes induce oxidative stress and activate MAPKs, AP-1, NF-kappaB, and Akt in normal and malignant human mesothelial cells. Environ Health Perspect 116(9):1211–1217

Palomaki J, Valimaki E, Sund J et al (2011) Long, needle-like carbon nanotubes and asbestos activate the NLRP3 inflammasome through a similar mechanism. ACS Nano 5(9):6861–6870

Pantarotto D, Briand JP, Prato M et al (2004a) Translocation of bioactive peptides across cell membranes by carbon nanotubes. Chem Commun (Camb) 1:16–17

Pantarotto D, Singh R, McCarthy D et al (2004b) Functionalized carbon nanotubes for plasmid DNA gene delivery. Angew Chem Int Ed 43(39):5242–5246

Parry JM, Parry EM (2012) Genetic toxicology: principles and methods. Humana Press, New York, NY

Patlolla AK, Hussain SM, Schlager JJ et al (2010) Comparative study of the clastogenicity of functionalized and nonfunctionalized multiwalled carbon nanotubes in bone marrow cells of swiss-webster mice. Environ Toxicol 25(6):608–621

Paula AJ, Stefani D, Souza Filho AG et al (2011) Surface chemistry in the process of coating mesoporous SiO(2) onto carbon nanotubes driven by the formation of SiOC bonds. Chemistry 17(11):3228–3237

Pelka J, Gehrke H, Rechel A et al (2013) DNA damaging properties of single walled carbon nanotubes in human colon carcinoma cells. Nanotoxicology 7:2–20

Petersen EJ, Henry TB (2012) Methodological considerations for testing the ecotoxicity of carbon nanotubes and fullerenes: review. Environ Toxicol Chem 31(1):60–72

Ponti J, Broggi F, Mariani V et al (2013) Morphological transformation induced by multiwall carbon nanotubes on Balb/3T3 cell model as an in vitro end point of carcinogenic potential. Nanotoxicology 7:221–233

Porter AE, Gass M, Muller K et al (2007) Direct imaging of single-walled carbon nanotubes in cells. Nat Nanotechnol 2:713–717

Premkumar T, Mezzenga R, Geckeler KE (2012) Carbon nanotubes in the liquid phase: addressing the issue of dispersion. Small 8(9):1299–1313

Raffa V, Ciofani G, Nitoda S et al (2008) Can the properties of carbon nanotubes influence their internalization by living cells? Carbon 46(12):1600–1610

Raffa V, Ciofani G, Vittorio O et al (2010) Physicochemical properties affecting cellular uptake of carbon nanotubes. Nanomedicine 5(1):89–97

Raymundo-Pinero E, Cazorla-Amoros A, Linares- Solano A et al (2002) High surface area carbon nanotubes prepared by chemical activation. Carbon 40(9):1614–1617

Rodriguez-Fernandez L, Valiente R, Gonzalez J et al (2012) Multiwalled carbon nanotubes display microtubule biomimetic properties in vivo, enhancing microtubule assembly and stabilization. ACS Nano 6(8):6614–6625

Romanos GE, Likodimos V, Marques RRN et al (2011) Controlling and quantifying oxygen functionalities on hydrothermally and thermally treated single-wall carbon nanotubes. J Phys Chem C 115(17):8534–8546

Ruan G, Agrawal A, Marcus AI et al (2007) Imaging and tracking of tat peptide-conjugated quantum dots in living cells: new insights into nanoparticle uptake, intracellular transport, and vesicle shedding. J Am Chem Soc 129(47):14759–14766

Saito Y, Yoshikawa T, Bandow S et al (1993) Interlayer spacings in carbon nanotubes. Phys Rev B 48(3):1907–1909

Salzmann CG, Llewellyn SA, Tobias G et al (2007) The role of carboxylated carbonaceous fragments in the functionalization and spectroscopy of a single-walled carbon-nanotube material. Adv Mater 19(6):883–887

Samori C, Sainz R, Ménard- Moyon C et al (2010) Potentiometric titration as a straightforward method to assess the number of functional groups on shortened carbon nanotubes. Carbon 48 (9):2447–2454

Sargent LM, Shvedova AA, Hubbs AF et al (2009) Induction of aneuploidy by single-walled carbon nanotubes. Environ Mol Mutagen 50(8):708–717

Sargent LM, Hubbs AF, Young SH et al (2012) Single-walled carbon nanotube-induced mitotic disruption. Mutat Res 745(1–2):28–37

Schnorr JM, Swager TM (2011) Emerging applications of carbon nanotubes. Chem Mater 23(3): 646–657

Schulte PA, Kuempel ED, Zumwalde RD et al (2012) Focused actions to protect carbon nanotube workers. Am J Ind Med 55(5):395–411

Schweinberger FF, Meyer-Plath A (2011) Status of characterization techniques for carbon nanotubes and suggestions towards standards suitable for toxicological assessment. J Phys Conf Ser 304:1–10

Scott-Fordsmand JJ, Krogh PH, Schaefer M et al (2008) The toxicity testing of double-walled nanotubes-contaminated food to *Eisenia veneta* earthworms. Ecotoxicol Environ Saf 71(3): 616–619

Senanayake V, Juurlink BH, Zhang C et al (2008) Do surface defects and modification determine the observed toxicity of carbon nanotubes? J Biomed Nanotechnol 4(4):515–523

Shahab U, Moinuddin, Ahmad S et al (2013) Genotoxic effect of N-hydroxy-4-acetylamino-biphenyl on human DNA: implications in bladder cancer. PLoS One 8(1):e53205

Shvedova AA, Kisin E, Murray AR et al (2008) Inhalation vs. aspiration of single-walled carbon nanotubes in C57BL/6 mice: inflammation, fibrosis, oxidative stress, and mutagenesis. Am J Physiol Lung Cell Mol Physiol 295(4):L552–L565

Shvedova AA, Pietroiusti A, Fadeel D et al (2012) Mechanisms of carbon nanotube-induced toxicity: focus on oxidative stress. Toxicol Appl Pharmacol 261(2):121–133

Silva G, Musumeci A, Gome A et al (2009) Characterization of commercial double-walled carbon nanotube material: composition, structure, and heat capacity. J Mater Sci 44(13):3498–3503

Singh N, Manshian BB, Jenkins GJ et al (2009) Nano genotoxicology: the DNA damaging potential of engineered nanomaterials. Biomaterials 30(23–24):3891–3914

Srivastava RK, Rahman Q, Kashyap MP et al (2011) Ameliorative effects of dimetylthiourea and N-acetylcysteine on nanoparticles induced cyto-genotoxicity in human lung cancer cells-A549. PLoS One 6(9):e25767

Stéfani D, Paula AJ, Vaz BG et al (2011) Structural and proactive safety aspects of oxidation debris from multiwalled carbon nanotubes. J Hazard Mater 189(1–2):391–396

Stella GM (2011) Carbon nanotubes and pleural damage: perspectives of nanosafety in the light of asbestos experience. Biointerphases 6(2):P1–P17

Szendi K, Varga C (2008) Lack of genotoxicity of carbon nanotubes in a pilot study. Anticancer Res 28(1A):349–352

Taylor AA, Aron GM, Beall GW et al (2012) Carbon and clay nanoparticles induce minimal stress responses in gram negative bacteria and eukaryotic fish cells. Environ Toxicol. doi:10.1002/tox.21824

Thayer AM (2007) Carbon nanotubes by the metricton. Chem Eng News 85(46):29–30

Thien-Nga L, Bonard JM, Richard G et al (2002) Comparison of catalytically grown and arc-discharge carbon nanotube tips. Appl Phys Lett 80(5):850–852

Thomas CR, George S, Host AM et al (2011) Nanomaterials in the environment: from materials to high-throughput screening to organisms. ACS Nano 5(1):13–20

Tobia G, Shao L, Ballesteros B et al (2009) Enhanced sidewall functionalization of single-wall carbon nanotubes using nitric acid. J Nanosci Nanotechnol 9(10):6072–6077

Trigueiro JPC, Silva GG, Laval RL (2007) Purity evaluation of carbon nanotube materials by thermogravimetric, TEM, and SEM methods. J Nanosci Nanotechnol 7(10):3477–3486
Tripathi S, Sonkar SK, Sarkar S (2011) Growth stimulation of gram (Cicer arietinum) plant by water soluble carbon nanotubes. Nanoscale 3(3):1176–1181
Tyurina YY, Kisin ER, Murray A et al (2011) Global phospholipidomics analysis reveals selective pulmonary peroxidation profiles upon inhalation of single-walled carbon nanotubes. ACS Nano 5(9):7342–7353
Umbuzeiro GA, Coluci VR, Honório JG et al (2011) Understanding the interaction of multi-walled carbon nanotubes with mutagenic organic pollutants using computational modeling and biological experiments. TRAC Trend Anal Chem 30(3):437–446
Ursini CL, Cavallo D, Fresegna AM et al (2012) Comparative cyto-genotoxicity assessment of functionalized and pristine multiwalled carbon nanotubes on human lung epithelial cells. Toxicol In Vitro 26(6):831–840
van Berlo D, Clift M, Albrecht C et al (2012) Carbon nanotubes: an insight into the mechanisms of their potential genotoxicity. Swiss Med Wkly 142:w13698. doi:10.4414/smw.2012.13698
Verdejo R, Lamoriniere S, Cottam B et al (2007) Removal of oxidation debris from multi-walled carbon nanotubes. Chem Commun 5:513–515
Wang ZW, Shirley MD, Meikle ST et al (2009) The surface acidity of acid oxidised multi-walled carbon nanotubes and the influence of in-situ generated fulvic acids on their stability in aqueous dispersions. Carbon 47(1):73–79
Wang Q, Zhou LY, Jiang et al (2011) Improved stability of the carbon nanotubes-enzyme bioconjugates by biomimetic silicification. Enzyme Microb Technol 49(1):11–16
Wang J, Sun P, Bao Y et al (2012a) Vitamin E renders protection to PC12 cells against oxidative damage and apoptosis induced by single-walled carbon nanotubes. Toxicol In Vitro 26(1): 32–41
Wang X, Guo J, Chen T et al (2012b) Multi-walled carbon nanotubes induce apoptosis via mitocondrial pathway and scavenger receptor. Toxicol In Vitro 26(6):799–806
Wepasnick KA, Smith BA, Bitter JL et al (2010) Chemical and structural characterization of carbon nanotube surfaces. Anal Bioanal Chem 396(3):1003–1014
Wepasnick KA, Smith BA, Schrote KE et al (2011) Surface and structural characterization of multi-walled carbon nanotubes following different oxidative treatments. Carbon 49(1):24–36
Wirnitzer U, Herbold B, Voetz M et al (2009) Studies on the in vitro genotoxicity of baytubes®, agglomerates of engineered multi-walled carbon-nanotubes (MWCNT). Toxicol Lett 186: 160–165
Worsley KA, Kalinina I, Bekyarova E et al (2009) Functionalization and dissolution of nitric acid treated single-walled carbon nanotubes. J Am Chem Soc 131(50):18153–18158
Wu N, Wang Q, Arash B (2012) Ejection of DNA molecules from carbon nanotubes. Carbon 50 (13):4945–4952
Xu J, Futakuchi M, Shimizu H et al (2012) Multi-walled carbon nanotubes translocate into the pleural cavity and induce visceral mesothelial proliferation in rats. Cancer Sci 103(12): 2045–2050
Yang H, Liu C, Yang D et al (2009) Comparative study of cytotoxicity, oxidative stress and genotoxicity induced by four typical nanomaterials: the role of particle size, shape and composition. J Appl Toxicol 29(1):69–78
Yang W, Ratinac KR, Ringer SP et al (2010a) Carbon nanomaterials in biosensors: should you use nanotubes or graphene? Angew Chem Int Ed 49(12):2114–2138
Yang Z, Zhang Y, Yang Y et al (2010b) Pharmacological and toxicological target organelles and safe use of single-walled carbon nanotubes as drug carriers in treating Alzheimer disease. Nanomedicine 6(3):427–441
Ying LS, Salleh MAM, Hamdan BM et al (2011) Continuous production of carbon nanotubes—a review. J Ind Eng Chem 17(3):367–376
Zeni O, Palumbo R, Bernini R et al (2008) Cytotoxicity investigation on cultured human blood cells treated with single-wall carbon nanotubes. Sensors 8(1):488–499

Zhang LW, Monteiro-Riviere NA (2010) Lectins modulate multi-walled carbon nanotubes cellular uptake in human epidermal keratinocytes. Toxicol In Vitro 24(2):546–551

Zhang Y, Iijima S, Shi Z et al (1999) Defects in arc-discharge-produced single-walled carbon nanotubes. Philos Mag Lett 79(7):473–479

Zhu L, Chang DW, Dai L et al (2007) DNA damage induced by multiwalled carbon nanotubes in mouse embryonic stem cells. Nano Lett 7(12):3592–3597

Zhu X, Zhu L, Chen Y et al (2009) Acute toxicities of six manufactured nanomaterial suspensions to Daphnia magna. J Nanopart Res 11:67–75

Chapter 7
From Combinatorial Display Techniques to Microarray Technology: New Approaches to the Development and Toxicological Profiling of Targeted Nanomedicines

Giselle Z. Justo, Eloah R. Suarez, Carina Melo, Marcelo A. Lima, Helena B. Nader, and Maria Aparecida S. Pinhal

Abstract Advances in genome scale analysis provide an alternative tool to combine large-scale screening of chemicals with detailed mechanistic studies, in which expression signatures for specific factors can be used as predictors of response or serve as new therapeutic targets in vitro and in vivo. Recently, the possibility of analyzing the effects of nanomaterials on a large number of genes has led to toxicogenomic studies in nanotechnology. Furthermore, expression signatures are providing potential targets for the design of ligand-mediated tumor targeting. In this respect, the integration of phage display technique with nanotechnology has been explored as a new approach to generate cancer-targeted nanomedicines. In this chapter we outline advances in microarray technology and in the selection of peptides and antibodies by display techniques. Furthermore, we add insights into their application in the field of toxicogenomics and in the development of targeted nanomedicines.

7.1 Introduction

With the advent of nanotechnology, the exposure to nanomaterials is inevitable and concerns about the potential risk to humans and the environment have emerged. This is increasingly more relevant when considering that the diversity of nanoparticles and their applications are still increasing, whereas limited methods are available in comparison, to

G.Z. Justo (✉)
Departamento de Bioquímica (Campus São Paulo) and Departamento de Ciências Biológicas (Campus Diadema), Universidade Federal de São Paulo, Três de Maio, 100, São Paulo, São Paulo 04044-020, Brazil
e-mail: giselle.zenker@unifesp.br

M.A.S. Pinhal (✉)
Departamento de Bioquímica (Campus São Paulo), Universidade Federal de São Paulo, Três de Maio, 100, São Paulo, São Paulo 04044-020, Brazil
e-mail: maspinhal@yahoo.com.br

N. Durán et al. (eds.), *Nanotoxicology*, Nanomedicine and Nanotoxicology, DOI 10.1007/978-1-4614-8993-1_7,

predict and characterize their potential toxicity. As a consequence, high-throughput cell-based approaches to conduct the initial screening of several samples of nanomaterials have been employed (Damoiseaux et al. 2011). A major drawback is that these studies are generally conducted in in vitro systems, making difficult to determine if the effects observed are physiologically relevant (Lewinski et al. 2008). In addition, environmental concentrations are likely to be low, thus limiting the study of the mechanisms of action by most toxicological methods (Fisichella et al. 2012).

Recent advances in genome scale analysis provide expression signatures for specific factors to be used as predictors of response or to serve as new therapeutic targets in vitro and in vivo. The application of this methodology in toxicological studies also provides information for networking genes modulated by toxicants, even at low doses (Lettieri 2006). Recently, the possibility of analyzing the effects of nanomaterials on a large number of genes in a unique experiment and identifying the mechanism of action or the altered cellular pathway has led to toxicogenomic studies in nanotechnology (Poynton and Vulpe 2009).

Another consequence of expression signatures is that they may serve as a target for the design of site-directed nanocarriers. In the case of cancer, this is particularly important, since a molecular signature in the cancer cell surface may distinguish it from normal healthy cells (Allen 2002; Lammers et al. 2008; Torchilin 2010; Hanahan and Weinberg 2011). A recent approach has reported the integration of nanotechnology with the development of cancer-specific targets selected by combinatorial phage display technique (Wang et al. 2011a). Of interest, the discovery of new targets is based on the ability of phage display peptide libraries to interact with specific targets on cancer cells. This approach represents an evolution in the synthesis of targeted cancer nanomedicines.

In the following sections we add insights into microarray technology and the selection of peptides and antibodies by display techniques. Furthermore, we discuss some applications in the field of toxicogenomics and in the development of targeted nanomedicines.

7.2 Specific Selection of Peptides and Antibodies with Pharmacological Application Using Display Techniques

7.2.1 Phage Display and In Vivo Display

Different peptides and proteins can be displayed on the bacteriophage capsid which generates phage display libraries or simply phage display. Phage display system was established in 1985 by George Smith (1985) and improved significantly immunology, biochemistry, physiology, pharmacology, cell biology, and medical fields. Bacteriophages used to construct the library are generally non-lytic and have a single-strand DNA encapsulated by a capsid approximately 7 nm wide and 1,000 nm in length. They belong to the Inoviridae family called filamentous bacteriophages (M13, Fd, fl) or simply Ff (Van Den Hondel and Schoenmakers 1976). Ff phages are able to infect a variety of gram-negative bacteria, such as *Escherichia coli*, depending on the F plasmid for infection. They present 11 different coat proteins, defined in Roman numbers. Each coat protein has a specific function: protein II (pII) and pX are involved with DNA replication; pI, pIV, and pXI are essential for phage assembly; and pV binds to single-strand DNA. Next group of proteins (III, VI, VII, VIII, and XI) constitute the capsid proteins of the phage particle.

Ff phages are excellent cloning vehicles, due to the fact that replication or assembly is not constrained by the size of the DNA. Genetic manipulation of coding genes, such as peptides, proteins, or antibody fragments, is cloned by insertion of specific oligonucleotides or entire coding regions into genes encoding specific capsid proteins, the minor coat protein (pIII) or the major coat protein (pVIII) presenting, respectively, 5 and 2,700 copies in each phage particle.

The first fused protein on the surface of pIII coat protein of bacteriophage M13 was the EcoRI endonuclease (Smith 1985), demonstrating more than 1,000-fold enrichment for the fusion protein compared to the yield of the same endonuclease produced by the wild-type M13. Yet, it is important to point out that only short peptides can be displayed at the N-terminus of pIII or pVIII. Although intolerance is still unclear, two different systems, using phagemids and phage vectors, have been developed to overcome cloning difficulties.

Phagemid is a normal plasmid vector, used to clone DNA fragments, which additionally has M13 origin (f1 origin), essential for single-strand replication, as well as double-strand replication origin. Phagemid vectors carry a copy of pIII or pVIII (fusion proteins) and can grow as a plasmid and be packaged as single-strand DNA. When bacterial cells are transformed by a phagemid, together with a helper phage (defective in replication), which carry the full capsid-encoding genes, secreted phage particles are able to express the encoded phagemid proteins and a mixture of wild-type and fusion pIII or pVIII. Besides phagemid system, a phage vector that carries two copies of pIII or pVIII can also be used to compensate cloning difficulties. Phage display technology can be used as vehicles for vaccines, to detect specific bacterial strain, display proteins and antibodies (Haq et al. 2012; Prisco and De Berardinis 2012). Also, phage display allows manageable selection of high affinity reagents to be used in biomarkers detection, an essential tool in the health-care field improving early and accurate diagnosis, to discover target molecules for innovative types of treatment of many diseases and to better understand human physiological conditions. Furthermore, phages are promising vehicle candidate because they can be easily engineered both to display appropriate ligands for targeting molecules in specific cells (Bar et al. 2008), parasites (Tonelli et al. 2013), or virus (Hashemi et al. 2012) and to carry a cytotoxic load that is delivered inside the target.

As demonstrated in Fig. 7.1, usually three to five rounds of panning are enough to obtain phage clones leading to specific target. In order to amplify selected bacteriophages after panning assay, an *E. coli* strain is infected by the pool of phages, submitted to growth in Luria Broth medium (LB medium), containing specific antibiotics (Fig. 7.1). The pool of selected phages obtained after each

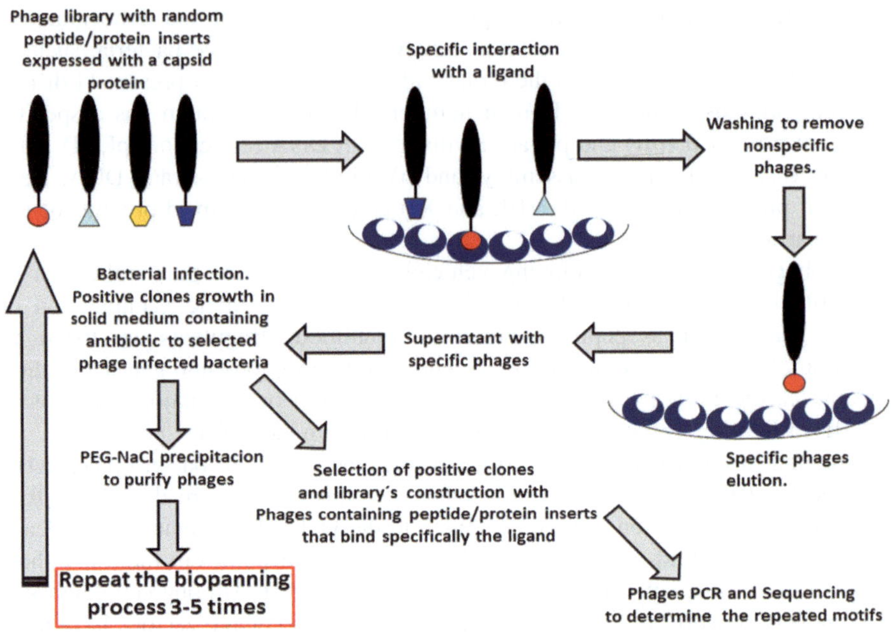

Fig. 7.1 Schematic steps of phage display panning

panning round is subjected to precipitation with polyethylene glycol (PEG6000) and NaCl (Fig. 7.1). An aliquot of the LB-growing phages is used to obtain isolated colonies on LB-agar plates. Usually, 20 different colonies are submitted to DNA sequencing analysis in order to define selected peptides (Fig. 7.1). At the end of the third or fifth panning assay, approximately 90–100 % of selected phages must present the same amino acid sequence at the surface of the capsid, confirming the specificity of this technology.

A faster method for probing target cell surface molecules was developed with a broad range of potential applications. Biopanning and rapid analysis of selective interactive ligands, defined as BRASIL, is another approach for screening, selection, and sorting of cell surface-binding peptides from phage libraries (Giordano et al. 2001). In BRASIL method, an aqueous upper phase containing cell suspension and selected phages is centrifuged through a nonmiscible organic lower phase, thus avoiding washing or limiting dilution steps of panning assays.

In cancer research, for example, the screening of peptide phage library in vivo allows the identification of different ligand receptors that can be distinguished from normal and cancer tissues. In vivo screening using a peptide phage library demonstrates that systemic tissue targeting is feasible and highlights specific vascular molecular markers that can be attractive as therapy targets (Staquicini et al. 2011; Deramchia et al. 2012). The use of selective peptide phage library addressed to tumor cells by displaying specific molecule on the phage coat associated with chemotherapy drug is a great promise as a novel way of not only delivering new drugs but also avoiding the toxic effects for healthy cells (Porto et al. 2011). Thus, phage display is a very interesting tool to destroy tumor cells, bypassing healthy cells. It is also possible to design a drug-release mechanism with genetic engineering of the phage particle (Chen et al. 2004).

The use of a pool of phage preparation against a range of drug-resistant bacteria allows to isolate resistant bacterial strains, identify matching phages, and promote mechanisms to destroy specific bacteria (Gu et al. 2012). A specific clinical trial in the Institute of Immunology and Experimental Therapy of Polish Academy of Sciences is currently recruiting patients for the use of phage preparations against a range of drug-resistant bacteria such as *methicillin-resistant Staphylococcus aureus, Escherichia, Enterococcus, Enterobacter, Klebsiella, Shigella,* and *Salmonella.*

Phage display of natural peptides can be used to determine the antigenic epitope relatively quickly. Peptides displayed on M13 virus can be used to elicit antibodies against specific coat proteins of certain parasites, viruses, and bacteria as previously described. Random peptide libraries are often used to target specific molecules of interest and the affinity selection is known as biopanning. Random peptide libraries can also be used to map the specificity of antibodies and search peptide ligands for a variety of receptors. It is also a sensitive and fast assay to define substrate specificity or specific inhibitors for enzymes (Gesteira et al. 2011).

The selection of carbohydrate-mimetic peptides using a phage display libraries is also a powerful tool, since glycoconjugates play essential roles in the surface of animal cells involved with cell–cell interaction and virus and parasites infection. Consequently, the inhibition of carbohydrate–protein interaction in such biological processes using a mimetic peptide selected by phage display has a potential effect on the mechanisms, such as cell infection and cell adhesion.

Recombinant antibodies production offers many advantages compared to the traditional production of polyclonal and monoclonal antibodies. Using a phage displayed antibody, the selection procedure is fast and the sequential rounds enrich significantly the target specific antibody binders (Dantas-Barbosa et al. 2012). Such approach involves an *E. coli* transformation, with a repertoire of random peptides or light chain genes encoded on a plasmid and the infection with another repertoire of random peptides or heavy chain sequences encoded on a phage or phagemid. In each infected bacteria, fusion proteins are expressed simultaneously from each vector and assembled at the surface of the phage particles (Clementi et al. 2012).

Additionally, a mammalian-cell-based antibody can also be applied to display full-length functional glycosylated antibodies (IgGs), on the cell surface, instead of just small fragments of the antibodies. Mammalian-cell display platform coupled with fluorescence-activated cell sorting (FACS) is throughput to distinguish antibodies with high binding affinities, enriched by more than 500-fold (Zhou et al. 2010; Bowers et al. 2011).

Despite the use of mammalian-cell display, as mentioned before, bacteria or lower eukaryotic organisms such as yeast can also be used to screen, select, and characterize antibody fragments. A peptide library using bacteria offers some advantages as a display platform compared to the phage display system, like it is easier to be generated due to the straightforward plasmid manipulations, does not require bacteriophage production or purification, does not use viral infection steps for amplification during panning selection, and can readily be analyzed by simple plasmid isolation and sequencing.

Yeast display system also offers advantages of larger library sizes and significant control over reaction conditions because displayed peptides, proteins, or antibodies are displayed extracellularly or expressed in the absence of a host cell. In addition, posttranslational modifications such as glycosylation can be present on displayed libraries using yeast platform.

The use of bacterial or yeast cells enables up to 100,000 copies of a library member linked to one copy of the gene, increasing significantly sensitivity during selection steps. Finally, cell surface displayed libraries are compatible with powerful FACS analysis that allows precise and quantitative control during screening assay. Multicolor FACS analysis also enables the simultaneous selection of positive and negative screens, which is difficult to be implemented in phage display system.

All display methods are based on the ability to physically link the protein produced by a library clone to its corresponding genotype. Phage display technique was the first display technique to be developed and, as a consequence, it is the most well established and used. However, more recent display strategies have been developed, since the random protein expression in a simple bacteriophage surface limits the use of large size libraries. These alternative methods are able to keep the phenotype associated with the genotype without the use of a bacteriophage.

7.2.2 Ribosomal Display

Ribosomal display was one of the first cell-free in vitro display methods developed (Mattheakis et al. 1994). This method is based in a complex, constituted by a correctly folded protein and its encoding RNA, both attached to the ribosome. The in vitro ribosome display for screening protein libraries for ligand binding starts by a DNA library that codifies peptide or antibody fragments that are first amplified by PCR, whereby a T7 RNA polymerase promoter, ribosome-binding site, and stem loops are introduced. As outlined in Fig. 7.2a, the DNA library is fused with a DNA spacer, without a stop codon, using a DNA ligase, and then transcribed to RNA (Baugh et al. 2001; Peano et al. 2006). The RNA is then purified and translated in vitro in an *E. coli* S-30 system (Zawada 2012) or a pure translation system (Shimizu et al. 2005) in the presence of different factors, enhancing the stability of ribosomal complexes and improving the folding of protein fragments on the ribosomes. Translation is stopped by chloramphenicol addition, once this molecule reversibly binds to the L16 protein of the 50S subunit of bacterial ribosomes, where transfer of amino acids to growing peptide chains is prevented by suppression of peptidyl transferase activity, thus inhibiting peptide bond formation and subsequent protein synthesis. The translation can also be stopped by cooling on ice. The ribosome complexes are stabilized by increasing the magnesium concentration. When a DNA spacer without stop codon is used during

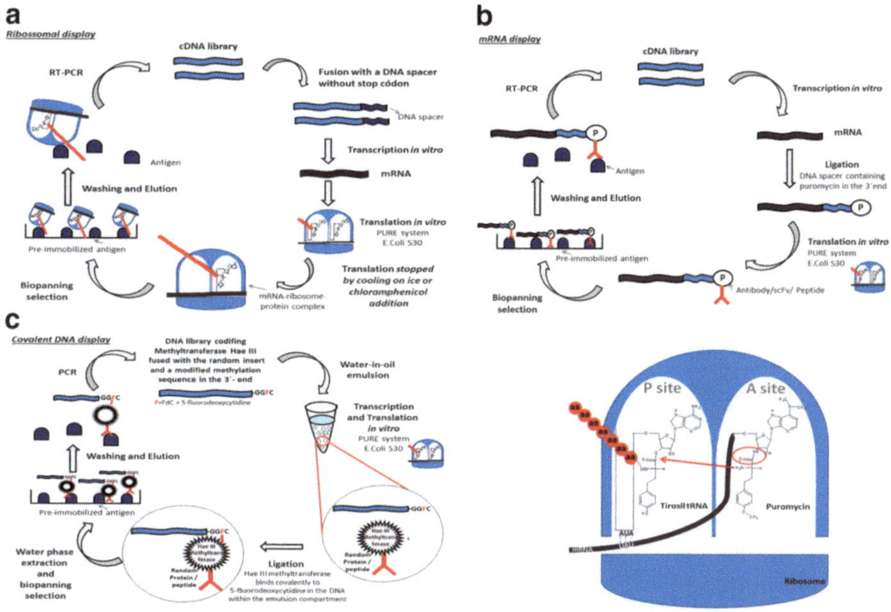

Fig. 7.2 In vitro display methods. (**a**) mRNA display. (**b**) Ribosomal display. (**c**) Covalent DNA display

translation, it occurs normally; however, the protein remains bound to the peptidyl tRNA and, consequently, to the ribosomal tunnel and the mRNA (Yabe et al. 2007). The complexes are selected by affinity from the translation mixture by binding of the protein to the immobilized target. Unspecific ribosome complexes are removed by intensive washing. The bound ribosome complexes can be dissociated by EDTA, or whole complexes can be specifically eluted with a specific ligand. After that, the RNA is isolated from the complexes and it is reverse transcribed to cDNA. This cDNA is then amplified by PCR and used for the next cycle of enrichment, and a portion can be analyzed by cloning and sequencing or by ELISA or RIA, in the case of antibodies/antibodies fragment selection (Hanes and Pluckthun 1997).

The affinity maturation of phage display populations by using ribosome display can generate high affinity ligands with K_d in the picomolar range, without the use of error-prone mutagenesis, since the affinity of the proteins produced by phage display presents constants in the nanomolar range (Groves et al. 2006). The error-prone mutagenesis approach uses a PCR in which the polymerase has a reasonably high error rate to amplify the wild-type sequence. This PCR can be made increasing the $MgCl_2$ concentration in the reaction, adding $MnCl_2$, or using unequal concentrations of each nucleotide. After amplification, the library of mutant coding sequences must be cloned into a suitable vector. The drawback of this method is that the size of the library is limited by the efficiency of the cloning (Pritchard et al. 2005).

The main advantage of the ribosomal display system is found when it is compared to the systems using microorganisms because the library diversity

(10^{13}–10^{15}) is not limited by the efficiency of the transformation systems in vivo. A second diversification of the protein sequence selected can be easily performed by PCR. The insertion of nonnatural amino acids that could enhance the resistance to proteases, for example, is also possible. Besides, the whole assay can be performed in only 1 day. Some disadvantages founded in this system are that the ribosome–mRNA–protein complex is maintained by non-covalent interactions, so the steps to avoid the unspecific ligand binding are impaired. Due to the complex size, impairment in the selection cycle is found, with loose of potential ligands. Besides, the requirement of specific conditions, like low temperature, is also a negative point when this technique is applied (Lipovsek and Pluckthun 2004; Sergeeva et al. 2006).

7.2.3 mRNA Display

The libraries applied in the selection of ligands against a potential target using mRNA display are constituted by peptides/proteins associated with its progenitor mRNA.

A cDNA library containing a T7 promoter and a ribosome-binding sequence in the 5′ end (Shine–Dalgarno sequence) is firstly transcribed to mRNA in vitro (Fig. 7.2b). This mRNA is bound to a DNA spacer containing puromycin in the 3′ end using T4 DNA ligase and a splint. Normally during the translation process, when the mRNA is bound to the ribosome, the tRNA containing the polypeptide chain in formation is localized at the peptidyl (P) site of the ribosome. The next amino acid to be incorporated in the polypeptide chain in formation is brought by the aminoacyl tRNA that binds to the aminoacyl (A) site by pairing of mRNA codon with the tRNA anticodon. When this pairing occurs, peptidyl transferase catalyzes a peptide bond formation between the polypeptide chain and the amino acid present in the A site, transferring the polypeptide chain to the A site. At this moment, the ribosome is translocated, allowing the entrance of a new aminoacyl tRNA in the A site. The puromycin mimics the 3′ end of a tyrosyl tRNA; however, this molecule differs from the tyrosyl tRNA due to the presence of an amide group that is non-hydrolyzed when compared to the hydrolyzed ester bound found in the original tyrosyl tRNA, as outlined in Fig. 7.2b. This chemical difference allows the release of the peptide fused with its encoding mRNA when puromycin enters in the ribosomal A site during translation in vitro (using *E. coli* S30 or PURE system methods) (Tabata et al. 2009). The product of these steps can now be used to the ligand selection assays using a defined target and after the elution, the mRNA complexes are submitted to RT-PCR and the cDNA library obtained are subjected to a new round of selection that begin with the transcription step explained above (Fig. 7.2b).

In this technique, the library size is also not limited by the transformation efficiency, since it is not dependent on a microorganism. The system can be constituted by $>10^{15}$ different members and one round can be performed in

3 days. In addition, RNA display allows mutagenesis and recombination in vitro, and if it is associated with the PURE system of translation, the insertion of nonnatural amino acids is also possible. Some important disadvantages of this system are the use of RNA and the need of cloning the selected sequences in a vector to determine their structure. This method differs from phage display, which contains its polypeptide sequence expressed in the microorganism surface, allowing the recovery of the sequence from its genome (Lipovsek and Pluckthun 2004; Sergeeva et al. 2006).

7.2.4 Covalent DNA Display

The covalent DNA display consists in a linear DNA library which is packaged with an in vitro transcription/translation mix into a water-in-oil emulsion. Ideally, each emulsion compartment contains only one DNA molecule. These constructs codify for an N-terminal Hae III DNA methyltransferase fusion protein associated with a C-terminal random polypeptidic sequence, with potential to bind in a predetermined target. The Hae III DNA methyltransferase sequence and the potential binding domain are flanked by a T7 promoter and T7 terminator for efficient expression. At the downstream end of the DNA T7 terminator, the modified methylation sequence 5′-GGFC-3′ (F = FdC = 5-fluorodeoxycytidine) is introduced by PCR. When the construct is expressed within the emulsion compartment, the Hae III DNA methyltransferase normally catalyzes a methylation that is initiated by nucleophile attack of an active site thiol group (Cys71) at C6 of cytosine. *S*-Adenosylmethionine (AdoMet) acts as the methyl donor at the C5. The methylated dihydrocytosine intermediate undergoes beta-elimination to generate a 5-methylcytosine and an active enzyme. If the nucleotide 5-fluorodeoxycytidine is inserted in the methylation sequence (5′-GGFC-3′), the beta-elimination is blocked and the enzyme associated with the random polypeptide remains covalently bound to the DNA substrate, which is present in the same compartment of the water-in-oil emulsion (Bertschinger and Neri 2004), as shown in Fig. 7.2c.

After in vitro expression and formation of the DNA–protein complexes, the water phase is extracted from the emulsion and DNA molecules displaying a protein with desired binding properties are selected from the pool of DNA–protein fusions by affinity panning against a predefined target. The nucleotide sequence of the selected DNA–protein fusions is amplified by PCR and can either be used for a further round of selection or for cloning and characterization of the potential ligands (Fig. 7.2c).

This technique has the advantage of not using the RNA in its final structure, avoiding the use of specific conditions to avoid contamination. In addition covalent DNA display does not depend on a microorganism structure in the selection process. However, more than one DNA molecule can be present in an aqueous compartment, which consists of a bias. It is necessary to improve the encapsulation techniques. Another disadvantage is the necessity of cloning the DNA in the end of the biopanning, in order to determine the sequences that present affinity against the target (Bertschinger and Neri 2004; Sergeeva et al. 2006).

Another covalent display technology found exploits an unusual property of a replication initiator protein from the *E. coli* bacteriophage P2, which replicates by attaching the P2A endonuclease to its own DNA. P2A binds to the viral origin and introduces a single-strand discontinuity in the DNA, beginning the DNA synthesis by using the host replication machinery. Finally, this protein attaches to the molecule of DNA from which it has been expressed, a property called cis-activity that is still poorly understood. These features enable pools of polypeptides that are linked to P2A to be synthesized in vitro, so that they also become covalently attached to their own coding DNA (FitzGerald 2000; Sergeeva et al. 2006). Another DNA replication initiator protein, called RepA, has also been used in covalent DNA techniques, since it binds exclusively to the template DNA from which it has been expressed. A diverse peptide library is created by ligation of DNA fragments of random sequence to a DNA fragment that encodes RepA. After in vitro transcription and translation, a pool of protein–DNA complexes is formed, where each protein is stably associated with the DNA that encodes it (Odegrip et al. 2004).

Independently of the covalent DNA method used, the library pool is incubated with an immobilized target, non-binding polypeptides are washed away, and the retained DNA that encodes the target-binding peptides is eluted and amplified by PCR, to form a DNA library ready for the next round of selection. After three to five rounds of selection, the recovered DNA is cloned into an appropriate expression vector for the identification of individual target-binding peptide sequences.

Despite the DNA and RNA complexes vulnerability for in vitro display methods, in the future, these techniques could become useful for directed evolution of proteins and for in vivo selections, as well as for other applications currently limited to phage display (Sergeeva et al. 2006).

7.2.5 Display Technology in the Development of Targeted Nanomedicines

It is known that systemic chemotherapy results in indiscriminated drug distribution and severe toxicity. The approach to manipulate filamentous bacteriophages with the ability to display a host specificity ligand, genetically modified, loaded with a large payload of a cytotoxic drug prepared by chemical modification, presents a powerful technology to target specific tumor cells.

It is only in the past years that regulatory guidelines for phage products approval, both in therapy and food safety, have been introduced. The first phage product, ListShield™, was approved by U.S. Food and Drug Administration (FDA) in 2006, for using in food safety. This product is a cocktail of several phages that target *Listeria monocytogenes* contaminants in meat and poultry products. Moreover, in 2008, the first phase 1 clinical trial using phages was approved by FDA. This trial uses a cocktail of eight phages that target various bacteria (*Staphylococcus aureus*, *Pseudomonas aeruginosa*, and *E. coli*) involved in venous leg ulcers (www.clinicaltrials.gov).

Biotechnology industries are currently investing in display techniques in order to provide solutions to many of modern medicine's unsolved problems. Among the most important biotechnology industries from the world, around 75 % use phage display, 19 % use ribosomal display, 3 % use mRNA display, and 3 % use covalent DNA display to select new biomolecules with therapeutic potential (Rothe et al. 2006).

Industries have an important focus in the developing monoclonal antibodies against specific targets. Some of these antibodies, which were developed by display techniques and were approved by FDA will be discussed as follows.

Adalimumab is an anti-TNF alpha antibody completely developed by phage display. This antibody is able to bind and neutralize soluble TNF alpha. The FDA approved adalimumab in 2002, and this is the first human monoclonal antibody approved by this agency. This antibody is approved to the treatment of many autoimmune diseases, like rheumatoid arthritis, psoriatic arthritis, ankylosing spondylitis, Crohn's disease, psoriasis, and juvenile idiopathic arthritis, presenting good results (Benucci et al. 2012).

Another antibody developed by phage display that was recently approved by the FDA was belimumab. This antibody is in use for the treatment of systemic lupus erythematosus (SLE), a chronic connective tissue disease of unknown origin. The hallmark of SLE is B cell activation and the production of many autoantibodies. The most effective treatment to this disease is corticosteroid use, which is responsible for many collateral effects (Miller and Ranatunga 2012). Belimumab is an anti-BLyS monoclonal antibody. BLyS is a cell membrane protein produced by a wide variety of cells that is overexpressed in the SLE and correlates with disease activity. BLyS molecules are processed by cleavage and released into the circulation as a biologically active soluble protein that stimulates B cells by interaction with some specific receptors. Belimumab has presented beneficial effects when compared to placebo in the treatment of SLE; however, these effects were quite modest (Bezalel et al. 2012).

Raxibacumab is another phage display-derived antibody developed against the *Bacillus anthracis* protective antigen, and it has been developed to be used in prevention and treatment of inhalation anthrax. It was approved by the FDA in December 2012. Inhalation anthrax manifests initially with fever, nausea, fatigue, and chest discomfort. Most patients show pleural effusions, hemorrhagic thoracic lymphadenitis, and, in some cases, hemorrhagic meningitis. Antibiotics might control the bacterial infection, but they are not able to block released toxins from the bloodstream, and antibiotic-resistant strains may not be effectively treated. Vaccines can be more effective only in the long term and require booster doses to maintain immunity. Postexposure studies have demonstrated raxibacumab efficacy even after toxins were released into the bloodstream, representing a significant advance against anthrax (Mazumdar 2009).

Nanocarriers are emerging as efficient ways to enhance the selectivity and specificity of anticancer therapeutics toward their targets, and several nanocarriers have been used for the delivery not only of therapeutic but also of diagnostic agents (Torchilin 2010). Among the several delivery systems, liposomes have attracted a lot of interest as versatile carriers to be developed for clinical application. Among its characteristics is the ability of passive targeting of drugs by the enhanced permeability and retention effect, a physiological phenomenon of increased accumulation of the drug in tumors as a result of active angiogenesis (Torchilin 2010).

At present, multifunctional liposomes carrying targeting ligands in their surfaces are envisioned. In this direction, Wang et al. (2011a) performed an elegant study in which phage display technology was integrated with nanotechnology to direct synthesize the assembly of phage-derived binding peptides to liposomes for tumor cell targeting. This methodology contrast with the general synthesis used to combine phage-derived targeting ligands with liposomal surfaces, which is well known to occur in multiple steps. Unfortunately, these chemical schemes are complex and in most cases difficult to achieve (Noble et al. 2004). With the use of the alternative methodology proposed by Petrenko (2008), named landscape phage protein-based approach (Jayanna et al. 2010), the chemical modification steps are bypassed by the binding peptide fused to the phage pVIII coat protein, which has the ability of spontaneous insertion into the liposomal surface and also allows for site-directed targeting of the disease area. As a consequence, this system can be used to the design of targeted therapies against different pathologies. A series of landscape phage fusion proteins were integrated into liposomal membranes and evaluated against prostate and breast tumor cells. The results indicated high degree of affinity and specificity toward the targets (Jayanna et al. 2010; Wang et al. 2011b).

However, the characteristics of the phage pVIII, such as its amphiphilic transmembrane motif, which facilitates interaction with biological membranes, may prompt its interaction with cells and organelles, interfering with the selective targeting. In this respect, these authors also evaluated the contribution of both the phage pVIII and the peptide to the targeting specificity of phage–liposome interaction. The results indicated that only the binding peptide is involved in the targeting efficiency of this mechanism (Wang et al. 2011a).

The application of display techniques in nanotechnology is not limited to the field of targeted nanomedicines. Indeed, combinatorial phage display approach has been proven to directly screen peptides able to catalyze the growth of semiconductor nanocrystals (Wei et al. 2011), showing that a number of opportunities for its application in the future might be expected.

7.3 Toxicogenomics of Nanomaterials: Expanding the Scope of Microarray Technology

7.3.1 Microarray Technology

Man-made nanomaterials possess a wide range of potential applications; hence, animal exposure to such structures is certain. Exposure may be achieved directly, i.e., via lungs or via food, drinks, and medicines affecting a wide range of tissues, which may cause undesirable cytotoxic effects leading to various diseases. Owing their complex structures/composition, their effects on biological systems and toxicity can be, sometimes, hitherto unknown and depend upon several properties, such as size, concentration, solubility, chemical and biological properties, and stability (Singh and Nalwa 2007).

Alteration in gene expression of specific markers is a suitable technique to address nanoparticles genotoxicity. Such analysis can be accomplished by microarray assay, serial analysis of gene expression (SAGE), and RNA-sequencing (RNA-seq) (Fig. 7.3).

7.3.1.1 Microarray

The microarray technology is a fluorescence hybridization-based process, which allows verifying mRNA concentration. Microarray is a quantitative method since the intensity of fluorescence is measured by the brightness signals. The first step of microarray assay involves gene selection. The next step is the immobilization of specific sequences of the selected genes in a DNA chip. Finally, gene expression is detected by the hybridization of synthetic complementary DNA (cDNA) labeled with a fluorophore. The cDNA is synthesized by reverse transcription of mRNA obtained from biological samples, such as tissues, culture cells, smears, peripheral blood, and urine (Yauk and Berndt 2007). The mRNA expression is determined by the ratio between the levels of two different samples of cDNA hybridization. Specific wavelengths are used to excite the fluorophore of labeled cDNA and such excitation promotes an emission wave which is further detected by scanning devices that generate an image, which is finally analyzed by software (Allison et al. 2006).

7.3.1.2 Serial Analysis of Gene Expression

Serial analysis of gene expression, or SAGE, is a cDNA sequence-based analysis. Briefly, eukaryotic mRNA is isolated by a poly-thymidine oligonucleotide (oligo-dT)$_{18–20}$ column. cDNA is synthesized by reverse transcription, using biotinylated primers. Obtained cDNA is digested by specific endonucleases and

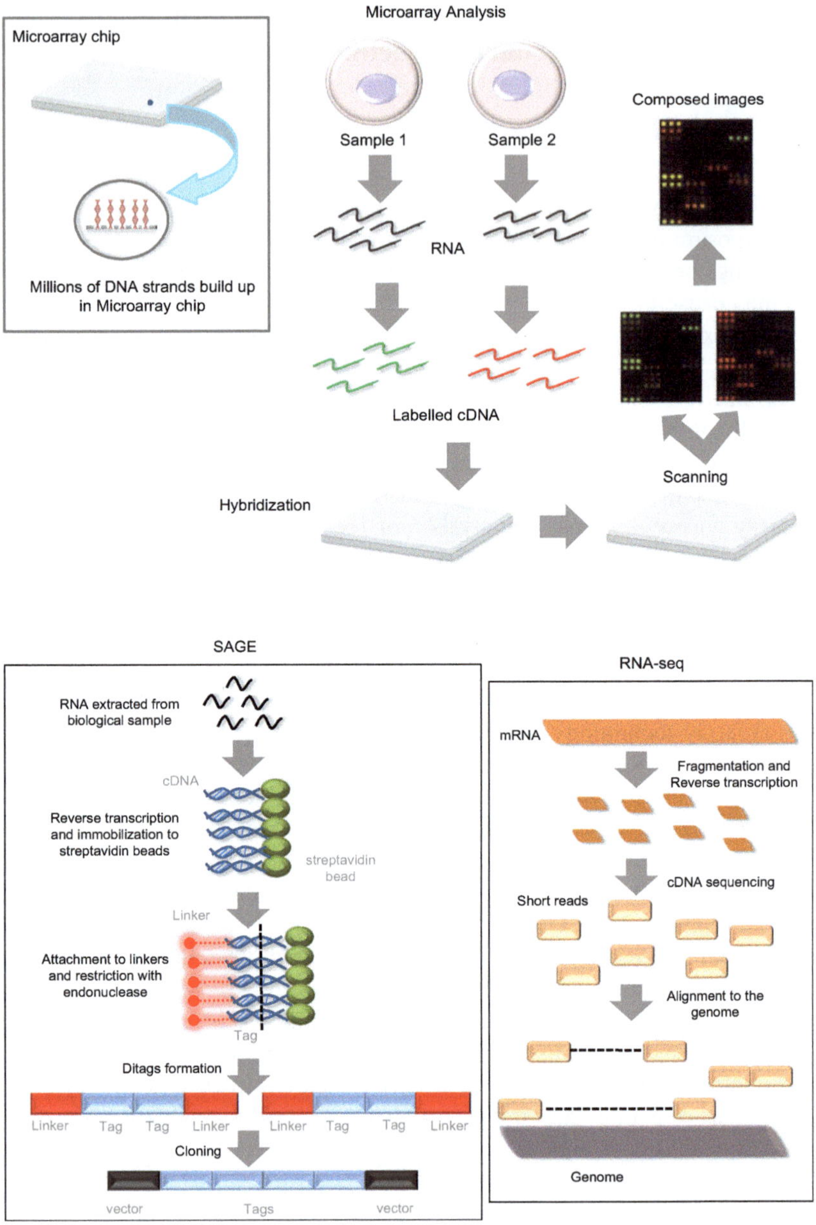

Fig. 7.3 Microarray analysis. DNA strands are printed onto treated glass slides. RNA is reverse transcribed and labeled with fluorophore. Samples are hybridized to DNA in the treated glass and laser excitation is applied. The emissions are measured and fluorescence intensity values to each spot are analyze by specialized software. *SAGE*. Gene expression profiles are determined from cells by cDNA. The endonuclease cleaves cDNA that was attached to the beads. Linkers are ligated to the cDNA. The tags formed are paired into ditags and cloned into a sequencing vector for

immobilized in streptavidin-coated beads. It is important to observe that the cDNA fragment obtained by digestions generates a specific tag (9–10 bp). An oligonucleotide linker is attached at tagged-cDNA. The oligonucleotides linkers are essential to clone the cDNA into a specific vector that is used to perform the sequencing analysis. This method allows that all tagged-cDNA sequences are cloned together at the same vector. Finally, the sequence analysis enables quantification of each specific tag, which delineates the identification and the abundance of the transcripts (Yamamoto et al. 2001; Tuteja and Tuteja 2004).

7.3.1.3 RNA-Sequencing

RNA-seq is obtained by deep-sequencing method. Initially, mRNA obtained from biological sample is converted into a cDNA library. Total obtained cDNA is sequenced using a high-throughput manner to obtain short sequences from one end (single-end sequencing) or both ends (pair-end sequencing). Defined sequences present typically 30–400 bp (Wang et al. 2009).

The ideal method for transcriptomics should be able to directly identify and quantify all small or large cDNA. Long cDNA fragments are converted into small cDNA fragments by endonucleases digestion. Sequencing adaptors are subsequently added to each short cDNA fragment. Finally, short cDNA fragments containing adaptors are sequenced using high-throughput technology. Resulting sequences are aligned using specific software that allows blast analysis in the genome or transcriptome bank. In addition, the alignment can be performed using three different genome regions: exons, exon junction region, or 3′-terminal region. These three different random analyses generate a base-resolution expression profile for each gene (Wang et al. 2009; Haas and Zody 2010).

Although RNA-seq technology is still under development, it offers several key advantages over existing hybridization-based approaches, since transcripts detection is not limited to detect known genes that have already been included in existing genomic database (Wang et al. 2009).

Tag-based sequencing approaches correspond to high throughput and can provide precise gene expression levels. However, it is important to observe that frequently Tag-based method uses an expensive sequencing technology. Another disadvantage that must be considered is the fact that isoforms frequently cannot be distinguished (Wang et al. 2009).

Fig. 7.3 (continued) cDNA sequencing. *RNA-seq*. Long RNAs or cDNAs are fragmented and short sequences are obtained from each cDNA. The resulting of cDNA sequencing is aligned to the genome

7.3.2 Microarray Gene Expression Profiling in Nanotoxicology

Nanomaterials hold promising applications for biomedical and cosmetic purposes. Several products currently in the market contain nanoparticles, such as metal oxide nanoparticles, polymeric nanoparticles, and liposomes, among others, and toxicity is a critical factor to consider when progressively more are expected to emerge. As a consequence, nanotoxicology has been regarded as a subdiscipline of nanotechnology, aiming at studying the toxic potential of nanostructures to humans and the environment (Fischer and Chan 2007; Jones and Grainger 2009; Damoiseaux et al. 2011). In this way, studies have highlighted the relevance of toxicogenomic analysis in nanotoxicology (Poynton and Vulpe 2009).

Recently, microarray technology has been recognized as a powerful tool to elucidate the mechanism of action of chemicals on living organisms, leading to the establishment of potential biomarkers at the population level and risk assessment.

Microarray technology has been extensively used for characterizing the effects of several stressors, such as temperature, pH, nutrients, oxidation, and even metals, in yeast (Lettieri 2006). These studies have stimulated the application of this technology for other organisms and for the evaluation of more complex stressors, including nanoparticles. In this regard, Niazi et al. (2011) reported the use of *Saccharomyces cerevisiae* to compare the effects of Ag nanoparticles (NP) and Ag ions on gene global expression. While no differences were found after longer period of exposure to Ag-NP and Ag ions, differences in the expression profiles of the stressors were detected after a shorter time of treatment. This study revealed that 234 genes were up- and downregulated by Ag ions, in contrast to 17 genes differentially induced by Ag-NP (Niazi et al. 2011).

Griffitt et al. (2009, 2011) have used microarrays to study responses of zebrafish to metallic nanoparticles. This popular aquarium fish presents several advantages for high-throughput genetic screening of small substances, including nanoparticles (Froehlicher et al. 2009; Ferreira et al. 2011). The aim of their studies is to characterize the effects of metallic nanoparticles from those of metal ions released from the particles on zebrafish gills. In a recent study, these authors demonstrated little differences in gene response to Al-NP, whereas a higher number of genes were affected by soluble aluminum. In addition, a significant differential gene expression was observed between soluble and particulate aluminum. This study suggests that Al-NP is unlikely to induce acute toxicity to freshwater fish (Griffitt et al. 2011). However, it is important to note that depending on the experimental conditions, other potential biological effects may be observed.

In this same direction, Gagné et al. (2012) have employed rainbow trout to examine the effects of nano-Ag and dissolved Ag that may be released from Ag-NP used as antimicrobial adjuvant in medical devices. Analysis of the results obtained in gene expression and biochemical assays indicated distinct effects for both forms of Ag. Nano-Ag affected genes involved in inflammatory responses, while dissolved Ag involved oxidative stress and protein stability. *Daphnia magna* is a freshwater crustacean commonly used as an indicator species for toxicity that has also been proved useful in toxicogenomic studies of nanotoxicity. Studies of Poynton et al. (2011) identified specific biomarkers for NP that can be developed for using in environment. Furthermore, new evidences about the effects of different coatings on the toxicity and gene expression profiles of Ag-NP were provided using this model system (Poynton et al. 2012). The effects of Ag-NP and Ag ions were also investigated on the transcriptome of mammalian cells (Foldbjerg et al. 2012). Microarray analysis of human lung epithelial cells (A549 cell line) exposed to Ag-NP and Ag ions revealed that less genes were altered after Ag ions treatment (133 genes) compared with Ag-NP (>1,000 genes). Among these genes, metallothionein, heat shock protein, and histone family genes were associated with oxidative stress and cell cycle arrest (Foldbjerg et al. 2012). Silver-NP-hydrogel is available for medical application and its potential toxicity has been studied in recent years. Interestingly, global gene expression analysis of HeLa cells exposed to Ag-NP-hydrogel showed that metallothionein and oxidative stress response genes were upregulated, suggesting a protective response of cells against oxygen reactive species-induced toxicity. In contrast, genes associated with DNA damage, cell cycle, and apoptosis were upregulated, which is consistent with chromosome damage and cell fate (Xu et al. 2012).

The development of microarrays for other species is contributing to improve toxicogenomic studies in ecosystems. Recently, microarray for enchytraeids *Enchytraeus albidus*, one of the most abundant members of the soil fauna, which is responsible for the degradation of organic matter, has been developed (Amorim et al. 2011). Using this approach, Gomes et al. (2012) evaluated the worm gene expression profile after exposure to increased concentrations of Cu-NP and Cu ions. The number of genes differently expressed increased with the increase in Cu ions concentration. In contrast, no differences were observed for Cu-NP. An important finding was that the effects were caused directly by the nanoparticle and not by possible released ions. Recently, these authors have used this worm model and microarray analysis to compare the effects of Ag-NP and $AgNO_3$ in soils (Gomes et al. 2013). The differences in altered genes indicated that the dissimilarities observed in toxicity are probably due to a release of Ag ions from the nanoparticles. However, the authors also suggested that similarities might become apparent at a longer exposure time (Gomes et al. 2013). Together, these reports indicate an important contribution of microarray technology in studies of ecotoxicology.

Another potential area of research includes the application of toxicogenomics to in vitro and animal studies of nanotoxicity. For example, titanium dioxide (TiO_2) surface-treated nanoparticles (STNP) are widely used in sunscreens. Although the toxicity of unmodified TiO_2-NP is known, alterations in the toxicological parameters of STNP caused by environmental degradation are largely unknown. Studies conducted by Fisichella et al. (2012) in Caco-2 cells indicated that the surrounding layer of these NP protects cells from a potentially toxic effect of TiO_2 core. Conversely, whole-genome microarray analysis of the livers of mice submitted to 90 days of exposure to TiO_2-NP revealed significant alterations in several genes associated with liver toxicity, suggesting the reduction in complement factor D as a biomarker (Cui et al. 2012). The application of TiO_2-NP in food, toothpastes, cosmetics, and medicine led Gui et al. to investigate the ability of TiO_2-NP to cause nephrotoxicity when administered intragastrically to mice for 90 days (Gui et al. 2013). This study revealed that long-lasting exposure to TiO_2-NP caused severe pathological changes and apoptosis in the kidney, concomitant with oxidative stress and alterations in mineral element distribution. In addition, analysis of kidney gene expression profile by microarray indicated changes in genes associated with the immune and inflammatory responses, apoptosis, oxidative stress, cell cycle, ion transport, signal transduction, transcription, translation, and cellular differentiation. Furthermore, from this analysis ten genes were proposed as potential biomarkers of kidney toxicity induced by TiO_2-NP (Gui et al. 2013). Importantly, these findings support previous work describing the role of inflammatory cytokines in TiO_2-NP-induced kidney inflammation and necrosis (Gui et al. 2011). The work of Liao and Liu (2012) also reported changes in gene expression profile of rat kidney exposed to nanocopper, which might explain the widespread renal proximal tubule necrosis and the alterations in blood urea nitrogen.

Recently, Simon at al. applied the RNA-seq methodology to characterize the transcriptomic effects of four metal-based NPs commonly used in the manufacturing (nano-Ag, nano-ZnO, nano-TiO_2) and biomedical imaging (quantum dots, CdTe/CdS) in the eukaryotic green algae *Chlamydomonas reinhardtii* (Simon et al. 2013). Specific toxicity was inferred by analysis of the increased or decreased gene expression. Alterations in the expression of several marker genes of stress were observed after exposure to nano-Ag and nano-TiO_2, while nano-ZnO induced nonspecific cellular stress. Nano-TiO_2 also induced a decrease in the levels of transcripts related to photosynthesis, suggesting toxicity to chloroplasts or a response that simulates light transition. Nano-Ag decreased the expression of genes associated with cell wall and flagella. Finally, an increase in genes related to proteasome inhibition, which is important in some neurological diseases, was induced by nano-TiO_2, nano-ZnO, and QD (Simon et al. 2013). This article points out a new high-throughput method to provide information in toxicogenomic studies of nanomaterials.

Microarray analysis can be used with cells from several histological origins and different species. For instance, Lee et al. investigated the effects of cerium oxide nanoparticles (nanoceria) on global gene expression and cellular functions in mouse neuronal cells (Lee et al. 2012). Nanoceria are engineered nanoparticles with potential application in neurodegenerative diseases due to their redox properties. In contrast to other studies that focused primarily on the beneficial effects of nanoceria, the study of Lee et al. described that nanoceria exposure led to downregulation of the huntingtin gene, with consequences for various developmental processes, as well as modulation of several genes associated with cell cycle arrest, growth inhibition, and apoptosis (Lee et al. 2012). Even though more tests are necessary, this study highlights the importance of not neglecting the potential side effects of nanoparticles in cellular systems. Ze et al. (2013) used microarray technology to investigate the mechanisms involved in mice brain cognition and behavioral injury caused by nasal administration of TiO_2-NP. The authors verified significant alterations in 249 genes, which were associated with oxidative stress, immune response, apoptosis, memory and learning, brain development, signal transduction, metabolic process, DNA repair, response to stimulus, and cellular process. Furthermore, some of the studied genes were proposed as potential biomarkers of brain injury (Ze et al. 2013).

Recently, the cytotoxicity of different-sized hematite nanoparticles in intestinal epithelia was evaluated at the cellular and genetic levels by Kalive et al. (2012). Using microarray analysis, a differential expression of genes related to cell junction maintenance was observed after treatment with the different-sized nanoparticles, indicating that the epithelial integrity is affected by the size of the nanoparticle. Thus, among their physical characteristics, the size is crucial to determine their toxicity, implying the need for particle characterization before assessing toxicity.

The potential of microarray technology to point out biomarkers of toxicity was explored by Fujita et al. to characterize time-dependent changes in rat lung gene expression after intratracheal instillation with C_{60} fullerenes at different doses and to identify the candidate expressed genes as potential biomarkers (Fujita et al. 2010). In this study ten genes involved in inflammatory response and cell migration were found to positively correlate with the dose of C_{60} fullerenes at both times, which correspond to the acute-phase and the persistent responses to C_{60} fullerenes. These genes were proposed as potential biomarkers for identifying C_{60} fullerenes responses in the lung tissue (Fujita et al. 2010).

Shim et al. have investigated gene expression and metabolic changes induced by silica-coated magnetic nanoparticles containing chemically bound rhodamine B isothiocyanate (RITC)—MNPs@SiO_2(RITC)—using a combined approach of transcriptome and metabolic gas chromatography–mass spectrometry analyses in human embryo kidney (HEK) 293 cells (Shim et al. 2012). The results indicated that MNPs@SiO_2(RITC) at high concentration induced changes associated with the metabolism of glutamic acid and its precursors. Furthermore, in this condition, gene expression and cell function analyses revealed mitochondrial dysfunction through oxidative stress induction (Shim et al. 2012).

Taken together, these studies illustrate how toxicogenomics can provide important information on the molecular mechanisms involved in the toxic response of biological systems to nanomaterials, thus improving risk assessment.

7.4 Conclusion

The application of toxicogenomics in nanotoxicology is rapidly progressing. However, further studies are needed to optimize the established model and non-model organisms and to improve interlaboratory reproducibility, especially when ecotoxicological studies are envisioned. Furthermore, the characterization of gene expression signatures will certainly open new possibilities for the development of more refined and specific future therapies. Indeed, molecular signatures in cancer cells allow for the design of site-directed nanocarriers. Novel approaches based on the integration of different technologies are desired, bringing new challenges to the development of nanomedicines in the forthcoming years.

Acknowledgments Our research in this field is supported by the government agencies Fundação de Amparo à pesquisa do Estado de São Paulo (FAPESP), Coordenação de Aperfeiçoamento de Pessoal de Nível Superior (CAPES), Conselho Nacional de Desenvolvimento Científico e Tecnológico (CNPq), and Brazilian Network on Nanotoxicology (MCTI/CNPq).

References

Allen TM (2002) Ligand-targeted therapeutics in anticancer therapy. Nat Rev Cancer 2 (10):750–763

Allison DB, Cui X, Page GP, Sabripour M (2006) Microarray data analysis: from disarray to consolidation and consensus. Nat Rev Genet 7(1):55–65

Amorim MJ, Novais SC, Van Der Ven K et al (2011) Development of a microarray for Enchytraeus albidus (Oligochaeta): preliminary tool with diverse applications. Environ Toxicol Chem 30(6):1395–1402

Bar H, Yacoby I, Benhar I (2008) Killing cancer cells by targeted drug-carrying phage nanomedicines. BMC Biotechnol 8:37

Baugh LR, Hill AA, Brown EL et al (2001) Quantitative analysis of mRNA amplification by in vitro transcription. Nucleic Acids Res 29(5):E29

Benucci M, Saviola G, Manfredi M et al (2012) Tumor necrosis factors blocking agents: analogies and differences. Acta Biomed 83(1):72–80

Bertschinger J, Neri D (2004) Covalent DNA display as a novel tool for directed evolution of proteins in vitro. Protein Eng Des Sel 17(9):699–707

Bezalel S, Asher I, Elbirt D et al (2012) Novel biological treatments for systemic lupus erythematosus: current and future modalities. Isr Med Assoc J 14(8):508–514

Bowers PM, Horlick RA, Neben TY et al (2011) Coupling mammalian cell surface display with somatic hypermutation for the discovery and maturation of human antibodies. Proc Natl Acad Sci U S A 108(51):20455–20460

Chen L, Zurita AJ, Ardelt PU et al (2004) Design and validation of a bifunctional ligand display system for receptor targeting. Chem Biol 11(8):1081–1091

Clementi N, Mancini N, Solforosi L et al (2012) Phage display-based strategies for cloning and optimization of monoclonal antibodies directed against human pathogens. Int J Mol Sci 13 (7):8273–8292

Cui Y, Liu H, Ze Y et al (2012) Gene expression in liver injury caused by long-term exposure to titanium dioxide nanoparticles in mice. Toxicol Sci 128(1):171–185

Damoiseaux R, George S, Li M et al (2011) No time to lose—high throughput screening to assess nanomaterial safety. Nanoscale 3(4):1345–1360

Dantas-Barbosa C, de Macedo Brigido M, Maranhao AQ (2012) Antibody phage display libraries: contributions to oncology. Int J Mol Sci 13(5):5420–5440

Deramchia K, Jacobin-Valat MJ, Vallet A et al (2012) *In vivo* phage display to identify new human antibody fragments homing to atherosclerotic endothelial and subendothelial tissues [corrected]. Am J Pathol 180(6):2576–2589

Ferreira C, Sartori-da-Silva M, Justo G (2011) Zebrafish as a suitable model for evaluating nanocosmetics and nanomedicines. In: Beck R, Guterres S, Pohlmann A (eds) Nanocosmetics and nanomedicines. Springer, Berlin, pp 239–251

Fischer HC, Chan WC (2007) Nanotoxicity: the growing need for in vivo study. Curr Opin Biotechnol 18(6):565–571

Fisichella M, Berenguer F, Steinmetz G et al (2012) Intestinal toxicity evaluation of TiO_2 degraded surface-treated nanoparticles: a combined physico-chemical and toxicogenomics approach in caco-2 cells. Part Fibre Toxicol 9:18

FitzGerald K (2000) In vitro display technologies—new tools for drug discovery. Drug Discov Today 5(6):253–258

Foldbjerg R, Irving ES, Hayashi Y et al (2012) Global gene expression profiling of human lung epithelial cells after exposure to nanosilver. Toxicol Sci 130:145–157

Froehlicher M, Liedtke A, Groh KJ et al (2009) Zebrafish (Danio rerio) neuromast: promising biological endpoint linking developmental and toxicological studies. Aquat Toxicol 95 (4):307–319

Fujita K, Morimoto Y, Endoh S et al (2010) Identification of potential biomarkers from gene expression profiles in rat lungs intratracheally instilled with C(60) fullerenes. Toxicology 274:34–41

Gagne F, Andre C, Skirrow R et al (2012) Toxicity of silver nanoparticles to rainbow trout: a toxicogenomic approach. Chemosphere 89(5):615–622

Gesteira TF, Coulson-Thomas VJ, Taunay-Rodrigues A et al (2011) Inhibitory peptides of the sulfotransferase domain of the heparan sulfate enzyme, N-deacetylase-N-sulfotransferase-1. J Biol Chem 286(7):5338–5346

Giordano RJ, Cardo-Vila M, Lahdenranta J et al (2001) Biopanning and rapid analysis of selective interactive ligands. Nat Med 7(11):1249–1253

Gomes SI, Novais SC, Scott-Fordsmand JJ et al (2012) Effect of Cu-nanoparticles versus Cu-salt in Enchytraeus albidus (Oligochaeta): differential gene expression through microarray analysis. Comp Biochem Physiol C Toxicol Pharmacol 155(2):219–227

Gomes SI, Soares AM, Scott-Fordsmand JJ et al (2013) Mechanisms of response to silver nanoparticles on Enchytraeus albidus (Oligochaeta): survival, reproduction and gene expression profile. J Hazard Mater 254–255:336–344

Griffitt RJ, Hyndman K, Denslow ND et al (2009) Comparison of molecular and histological changes in zebrafish gills exposed to metallic nanoparticles. Toxicol Sci 107(2):404–415

Griffitt RJ, Feswick A, Weil R et al (2011) Investigation of acute nanoparticulate aluminum toxicity in zebrafish. Environ Toxicol 26(5):541–551

Groves M, Lane S, Douthwaite J et al (2006) Affinity maturation of phage display antibody populations using ribosome display. J Immunol Methods 313(1–2):129–139

Gu J, Liu X, Li Y et al (2012) A method for generation phage cocktail with great therapeutic potential. PLoS One 7(3):e31698

Gui S, Zhang Z, Zheng L et al (2011) Molecular mechanism of kidney injury of mice caused by exposure to titanium dioxide nanoparticles. J Hazard Mater 195:365–370

Gui S, Sang X, Zheng L et al (2013) Intragastric exposure to titanium dioxide nanoparticles induced nephrotoxicity in mice, assessed by physiological and gene expression modifications. Part Fibre Toxicol 10:4

Haas BJ, Zody MC (2010) Advancing RNA-Seq analysis. Nat Biotechnol 28(5):421–423

Hanahan D, Weinberg RA (2011) Hallmarks of cancer: the next generation. Cell 144(5):646–674

Hanes J, Pluckthun A (1997) In vitro selection and evolution of functional proteins by using ribosome display. Proc Natl Acad Sci U S A 94(10):4937–4942

Haq IU, Chaudhry WN, Akhtar MN et al (2012) Bacteriophages and their implications on future biotechnology: a review. Virol J 9:9. doi:10.1186/1743-422X-9-9

Hashemi H, Pouyanfard S, Bandehpour M et al (2012) Immunization with M2e-displaying T7 bacteriophage nanoparticles protects against influenza A virus challenge. PLoS One 7(9): e45765

Jayanna PK, Bedi D, Gillespie JW et al (2010) Landscape phage fusion protein-mediated targeting of nanomedicines enhances their prostate tumor cell association and cytotoxic efficiency. Nanomedicine 6(4):538–546

Jones CF, Grainger DW (2009) In vitro assessments of nanomaterial toxicity. Adv Drug Deliv Rev 61(6):438–456

Kalive M, Zhang W, Chen Y, Capco DG (2012) Human intestinal epithelial cells exhibit a cellular response indicating a potential toxicity upon exposure to hematite nanoparticles. Cell Biol Toxicol 28:343–368

Lammers T, Hennink WE, Storm G (2008) Tumour-targeted nanomedicines: principles and practice. Br J Cancer 99(3):392–397

Lee TL, Raitano JM, Rennert OM et al (2012) Accessing the genomic effects of naked nanoceria in murine neuronal cells. Nanomedicine 8:599–608

Lettieri T (2006) Recent applications of DNA microarray technology to toxicology and ecotoxicology. Environ Health Perspect 114(1):4–9

Lewinski N, Colvin V, Drezek R (2008) Cytotoxicity of nanoparticles. Small 4(1):26–49

Liao M, Liu H (2012) Gene expression profiling of nephrotoxicity from copper nanoparticles in rats after repeated oral administration. Environ Toxicol Pharmacol 34:67–80

Lipovsek D, Pluckthun A (2004) In-vitro protein evolution by ribosome display and mRNA display. J Immunol Methods 290(1–2):51–67

Mattheakis LC, Bhatt RR, Dower WJ (1994) An in vitro polysome display system for identifying ligands from very large peptide libraries. Proc Natl Acad Sci U S A 91(19):9022–9026

Mazumdar S (2009) Raxibacumab. MAbs 1(6):531–538

Miller AV, Ranatunga SK (2012) Immunotherapies in rheumatologic disorders. Med Clin North Am 96(3):475–496

Niazi JH, Sang BI, Kim YS et al (2011) Global gene response in Saccharomyces cerevisiae exposed to silver nanoparticles. Appl Biochem Biotechnol 164(8):1278–1291

Noble CO, Kirpotin DB, Hayes ME et al (2004) Development of ligand-targeted liposomes for cancer therapy. Expert Opin Ther Targets 8(4):335–353

Odegrip R, Coomber D, Eldridge B et al (2004) CIS display: in vitro selection of peptides from libraries of protein-DNA complexes. Proc Natl Acad Sci U S A 101(9):2806–2810

Peano C, Severgnini M, Cifola I et al (2006) Transcriptome amplification methods in gene expression profiling. Expert Rev Mol Diagn 6(3):465–480

Petrenko V (2008) Evolution of phage display: from bioactive peptides to bioselective nanomaterials. Expert Opin Drug Deliv 5(8):825–836

Porto G, Giordano RJ, Marti LC et al (2011) Identification of novel immunoregulatory molecules in human thymic regulatory CD4+ CD25+ T cells by phage display. PLoS One 6(8):e21702

Poynton HC, Vulpe CD (2009) Ecotoxicogenomics: emerging technologies for emerging contaminants. J Am Water Resour Assoc 45(1):83–96

Poynton HC, Lazorchak JM, Impellitteri CA et al (2011) Differential gene expression in Daphnia magna suggests distinct modes of action and bioavailability for ZnO nanoparticles and Zn ions. Environ Sci Technol 45(2):762–768

Poynton HC, Lazorchak JM, Impellitteri CA et al (2012) Toxicogenomic responses of nanotoxicity in Daphnia magna exposed to silver nitrate and coated silver nanoparticles. Environ Sci Technol 46(11):6288–6296

Prisco A, De Berardinis P (2012) Filamentous bacteriophage fd as an antigen delivery system in vaccination. Int J Mol Sci 13(4):5179–5194

Pritchard L, Corne D, Kell D et al (2005) A general model of error-prone PCR. J Theor Biol 234 (4):497–509

Rothe A, Hosse RJ, Power BE (2006) In vitro display technologies reveal novel biopharmaceutics. FASEB J 20(10):1599–1610

Sergeeva A, Kolonin MG, Molldrem JJ et al (2006) Display technologies: application for the discovery of drug and gene delivery agents. Adv Drug Deliv Rev 58(15):1622–1654

Shim W, Paik MJ, Nguyen DT et al (2012) Analysis of changes in gene expression and metabolic profiles induced by silica-coated magnetic nanoparticles. ACS Nano 6:7665–7680

Shimizu Y, Kanamori T, Ueda T (2005) Protein synthesis by pure translation systems. Methods 36 (3):299–304

Simon DF, Domingos RF, Hauser C et al (2013) RNA-Seq analysis of the effects of metal nanoparticle exposure on the transcriptome of Chlamydomonas reinhardtii. Appl Environ Microbiol 79(16):4774–4785

Singh S, Nalwa HS (2007) Nanotechnology and health safety—toxicity and risk assessments of nanostructured materials on human health. J Nanosci Nanotechnol 7(9):3048–3070

Smith GP (1985) Filamentous fusion phage: novel expression vectors that display cloned antigens on the virion surface. Science 228(4705):1315–1317

Staquicini FI, Ozawa MG, Moya CA et al (2011) Systemic combinatorial peptide selection yields a non-canonical iron-mimicry mechanism for targeting tumors in a mouse model of human glioblastoma. J Clin Invest 121(1):161–173

Tabata N, Sakuma Y, Honda Y et al (2009) Rapid antibody selection by mRNA display on a microfluidic chip. Nucleic Acids Res 37(8):e64

Tonelli RR, Colli W, Alves MJ (2013) Selection of binding targets in parasites using phage-display and aptamer libraries in vivo and in vitro. Front Immunol 3:419. doi:10.3389/fimmu.2012.00419

Torchilin VP (2010) Passive and active drug targeting: drug delivery to tumors as an example. Handb Exp Pharmacol 197:3–53

Tuteja R, Tuteja N (2004) Serial analysis of gene expression (SAGE): application in cancer research. Med Sci Monit 10(6):RA132–RA140

Van Den Hondel CA, Schoenmakers JG (1976) Cleavage maps of the filamentous bacteriophages M13, fd, fl, and ZJ/2. J Virol 18:1024–1039

Wang Z, Gerstein M, Snyder M (2009) RNA-seq: a revolutionary tool for transcriptomics. Nat Rev Genet 10(1):57–63

Wang T, Kulkarni N, Bedi D et al (2011a) In vitro optimization of liposomal nanocarriers prepared from breast tumor cell specific phage fusion protein. J Drug Target 19(8):597–605

Wang T, Kulkarni N, D'Souza GG et al (2011b) On the mechanism of targeting of phage fusion protein-modified nanocarriers: only the binding peptide sequence matters. Mol Pharm 8 (5):1720–1728

Wei Z, Maeda Y, Matsui H (2011) Discovery of catalytic peptides for inorganic nanocrystal synthesis by a combinatorial phage display approach. Angew Chem Int Ed Engl 50 (45):10585–10588

Xu L, Li X, Takemura T et al (2012) Genotoxicity and molecular response of silver nanoparticle (NP)-based hydrogel. J Nanobiotechnol 10:16

Yabe R, Suzuki R, Kuno A et al (2007) Tailoring a novel sialic acid-binding lectin from a ricin-B chain-like galactose-binding protein by natural evolution-mimicry. J Biochem 141(3):389–399

Yamamoto M, Wakatsuki T, Hada A et al (2001) Use of serial analysis of gene expression (SAGE) technology. J Immunol Methods 250(1–2):45–66

Yauk CL, Berndt ML (2007) Review of the literature examining the correlation among DNA microarray technologies. Environ Mol Mutagen 48(5):380–394

Zawada JF (2012) Preparation and testing of E. coli S30 in vitro transcription translation extracts. Methods Mol Biol 805:31–41

Ze Y, Hu R, Wang X et al (2013) Neurotoxicity and gene-expressed profile in brain-injured mice caused by exposure to titanium dioxide nanoparticles. J Biomed Mater Res A. doi:10.1002/jbm.a.34705 [Epub ahead of print]

Zhou C, Jacobsen FW, Cai L et al (2010) Development of a novel mammalian cell surface antibody display platform. MAbs 2(5):508–518

Chapter 8
Genetic Studies on the Effects of Nanomaterials

Renata de Lima and Leonardo Fernandes Fraceto

Abstract The aim of this chapter is to present some of the principal methodologies used to study the effects of toxic substances on DNA, which can be used to analyze aspects of nanomaterial toxicity. It is not the intention here to undertake an in-depth survey of the topic, but rather to highlight the techniques that can be used in nanotoxicity studies. The available methods for evaluating effects on DNA include genotoxicity tests such as (1) the *Allium cepa* chromosome aberration test, (2) the comet analysis, (3) the micronucleus test, and (4) the cytogenetic analysis. In addition, this chapter also described the methodologies for analysis of gene expression that can be applied to the effects of nanomaterials. A description of the characteristics of each method is provided, together with results of selected studies that have evaluated the effects of nanomaterials, and a critical discussion of the main advantages and disadvantages of each technique. The chapter also highlights the challenges and future perspectives for studies of the effects of nanostructured materials on DNA.

8.1 Introduction

The genetic material of all living beings shows great similarity, independent of the specific organism, so that alterations in the genetic material of one type of organism can be used to predict similar effects in others. Its fragility is related to its chemical composition and possible interactions with substances with which it comes into

R. de Lima (✉)
Department of Biotechnology, University of Sorocaba, Rodovia Raposo Tavares S/N, Km 92,5, CEP 18023-000 Sorocaba, São Paulo, Brazil

Universidade Federal de São Carlos, UFSCar, Sorocaba, São Paulo, Brazil
e-mail: renata.lima@prof.uniso.br

L.F. Fraceto
Department of Environmental Engineering, São Paulo State University - UNESP, Av. Três de Março, 511, Alto da Boa Vista, 18087-180 Sorocaba, São Paulo, Brazil

N. Durán et al. (eds.), *Nanotoxicology*, Nanomedicine and Nanotoxicology,
DOI 10.1007/978-1-4614-8993-1_8,

contact. Nucleic acids consist of nucleotide monomers that are connected by phosphodiester bonds to produce the strand of DNA or RNA, with hydrogen bonds formed between the nucleotide bases of one strand with those of another strand to ensure their antiparallel pairing. It is the strength of the phosphodiester bonds, relative to the hydrogen bonds, that serves to maintain the integrity of the genetic material in terms of its sequence, while the weakness of the hydrogen bonds is important for separation of the strands during replication or transcription of the genetic material (Strachan and Read 1999; Wilkins 2013).

The evolutionary process has selected structural arrangements of the genetic material that enable optimum functionality, with precise distributions that ensure correct interaction between different components. The arrangement differs between cell types in terms of number and form of chromosomes; however, in general terms, the DNA strands are wrapped around histone protein cores, forming nucleosomes organized in solenoids, which in turn attach to a protein skeleton to produce the chromatid. The organization of the chromatids depends on the group and species of the organism (Akhmedov et al. 1998; Goetze et al. 2007).

In contrast to biochemical investigations, studies that focus on the genetic material normally pay little attention to the group or cell type of the organism, since it is believed that if a particular substance (including nanomaterials) is capable of causing genetic alterations, then such effects should occur in any type of organism (Amara et al. 2013; Lima et al. 2012a, b). On the other hand, it must be considered the likelihood of the substance being presented to the DNA, which depends on its interaction with the cell, its penetration capacity, and its ability to initiate cellular signaling (which could provoke a range of responses, since different cell types possess characteristic mechanisms and structures). Nevertheless, the occurrence of a positive genetic alteration caused by a specific nanomaterial in a particular cell type virtually guarantees that the same response should be caused in other cells, since the nucleic acids of all living organisms possess similar characteristics.

8.1.1 Maintenance and Function of the DNA

The genetic material has two fundamental functions, namely, transcription for protein synthesis and replication to ensure the transmission of hereditary characteristics between the mother and daughter cells. Various mechanisms act, in one way or another, to protect the cell genome, including those that ensure maintenance of its sequence (preventing genetic mutations), and, equally important, those that maintain its functional characteristics. These include methylations (as in the case of genomic imprinting), as well as the formation of nuclear domains, chromosomal territories, and Cajal bodies in eukaryotic cells (Hubner and Spector 2010; Veschure 2006) (Fig. 8.1).

Alterations of the genetic material, whether in the nucleotide sequences or in the structural arrangements, can lead to undesirable consequences. In somatic cells, an alteration of the genome can lead to the formation of tumors due to loss of control of

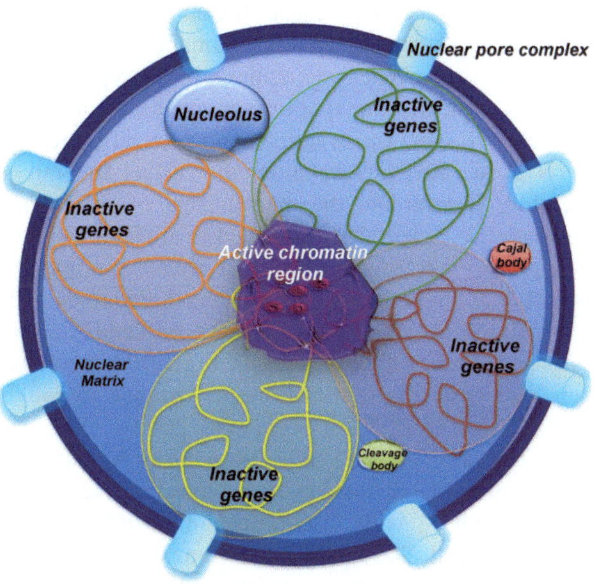

Fig. 8.1 Scheme showing the main nuclear regions

the cell cycle (Harper and Elledge 2007; Foster et al. 2012; Rao and Yamada 2013). As a result, there have been many attempts to evaluate the potential of different substances to cause genetic damage.

Another important point to be considered is the maintenance of the number of copies of functional genes present in each cell, since alterations involving the gain or loss of entire chromosomes or fragments can cause major changes leading to cellular responses that are not compatible with the preexisting programming. This maintenance of chromosomal number and structure is dependent on the arrangement of chromosomes in the cell and the correct induction of mitotic fusion, ensuring the dispersion and separation of the sister chromatids during cell division (Rao and Yamada 2013).

8.1.2 Alterations in the Genetic Material and Formation of Tumors

The control of cell cycle involves proteins such as p53 (protein 53 or tumor protein 53) and pRb (retinoblastoma protein) that are involved in cell maintenance, principally during the process of cell division, and whose role is to ensure that mutations and aneuploidies do not occur (Levine 1997; Jabbur et al. 2000; Han et al. 2002; Giono and Manfredi 2006; Batchelor et al. 2009; Al Hussain and Akhtar 2013). A small fraction of tumors (around 5–10 %) are hereditary in nature, with the remainder being due to environmental interferences that lead to alterations in proteins following unsuccessful cell division.

Damage to the DNA causes the activation of proteins that initiate a cascade of events in an attempt to repair the damage. If successful, this favors the continuation of cell division. When the damage is not repaired, pathways are activated which lead to cell death (the apoptosis cascade), resulting in programmed destruction of the cell to protect the organism and prevent tumor formation (Hoeijmakers 2009; Soria et al. 2012). Substances that affect the integrity of the genetic material therefore interfere in the mechanisms of cell maintenance, leading to cellular responses to repair the damage. The different kinds of DNA lesions depend on the specific type of induction that affects the fidelity of the mechanisms of transcription and replication. The greater the number of breaks, the greater the chance of repair failure, with the cell being steered towards apoptosis, although in some cases, even when the damage is not repaired, the cells continue to divide normally, hence successively transmitting the damage to daughter cells (Zhou and Elledge 2000; Dinant et al. 2008; Soria et al. 2012).

The existing theories concerning tumor formation consider two series of events. One involves the disruption of chromosomes, leading to the loss, gain, or rearrangement of genes, with disequilibrium of alleles and failure in gene expression (Migliore et al. 2011; González-González et al. 2012; Vasconcelos et al. 2013). The other is based on successive mutations of genes linked to control of the cell cycle, causing disturbance of the cycle culminating in formation of a tumor. Some studies have indicated that during tumoral transformation, the cells can generate new spontaneous mutations, with the occurrence of errors in polymerases and mitosis that activate cell division mechanisms (Kadota et al. 2009; Moore et al. 2012; Piazza et al. 2012). This activation of division increases the propensity for generation of new mutations and errors in epigenetically programmed repetitions and expressions.

Chromosome instability arises from the existence of certain common chromosomal regions known as fragile sites, which are liable to difficulties during replication. Studies have shown that alterations in the fragile sites can lead to tumor formation; although the reason for the instability in these regions remains poorly understood, it is believed that they have greater propensity for alterations arising from environmental circumstances (Herzog et al. 2004, 2006; Hudler 2012; Roylance et al. 2011; Stirling et al. 2011). Genetic responses and failures can be investigated in various ways, ranging from observation of the chromosomes to changes in gene expression. Since there are many metabolic control mechanisms that ensure the integrity of the genetic material, there are a wide range of options for observation and evaluation.

Effects on the genetic material can be determined directly using genotoxicity analyses, including tests that permit direct observation of entire or partial chromosomes, such as cytogenetic, micronucleus, and comet tests to evaluate the occurrence of aneuploidies or chromosome breaks and rearrangements (Kumari et al. 2009; Lima et al. 2012a, b). Analysis of mutations at a smaller scale, using molecular genetics techniques including PCR (polymerase chain reaction) and restriction enzyme analysis, can be used to assess the toxicity of nanomaterials in animals, cells, and microorganisms in soil and water, in the latter case enabling consideration of the variability and density of populations (Mo et al. 2012; Abdul Khaliq et al. 2012). It should be pointed out that genotoxicity analyses measure the capacity of the test material to cause changes in the DNA; however, this occurrence does not necessarily imply that the material is mutagenic or carcinogenic. The analysis of gene expression (epigenetic analysis) can also be used to investigate alterations such as the stimulation or suppression of gene transcription due to the presence of a particular substance. In this case, gene expression is not altered due to direct or indirect stimulation or inhibition of synthesis.

8.2 Genotoxicity of Nanomaterials

8.2.1 *Allium cepa* Test

A very wide range of pollutants are found in the environment, some of which include certain nanostructured materials that can be mutagenic. *Allium cepa* was the first plant used as a test organism (in the gene-tox program of the US Environmental Protection Agency) for fast, sensitive, environmental monitoring purposes (Kumari 2009; Lima et al. 2010; Lima et al. 2011, 2012a, b; Lima et al. 2012a). The *Allium cepa* technique, which involves evaluation of chromosomal alterations in root tips, has been validated by the World Health Organization's International Program on Chemical Safety (IPCS), and by the United Nations Environment Program (UNEP), as an efficient test for in situ analysis and monitoring of the genotoxicity of substances present in the environment (Cabrera and Rodrigues 1999).

The test measures chromosomal aberrations caused by chemical agents. In environmental impact assessments, it has been utilized for the determination of genotoxic effects of pollutants in sewage treatment effluents and the monitoring of pesticides in water, amongst other applications (Ma et al. 2005; Leme and Marin-Morales 2009). It essentially consists of exposing onion root tips to solutions containing different concentrations of contaminant, together with a negative control (a medium free of any substance that could cause an effect) and a positive control (using a chemical compound that is known to be genotoxic, such as trifluralin) for comparative purposes. It is recommended to use at least three different concentrations of the contaminant, and to perform the test in triplicate to enable meaningful statistical analysis.

The ploidy of *Allium cepa* is $2n = 16$, which facilitates visualization of the chromosomes in different stages of division (Fig. 8.2). Analysis of the genotoxic effect is somewhat complex, with previous knowledge required of the phases of the cell cycle so that damage can be confidently identified. Differences relative to the normal condition can then be an indication of toxicity (Leme and Marin-Morales 2009).

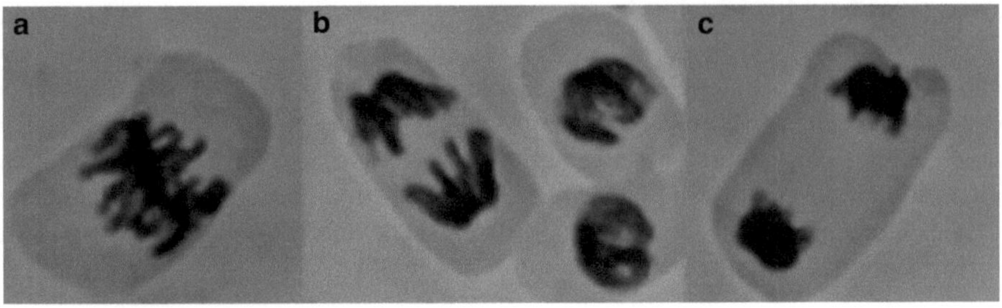

Fig. 8.2 Photographs of onion root cells, with the main phases of cellular division indicated: (**a**) metaphase, (**b**) anaphase, (**c**) telophase

The mitotic index (MI, (8.1)) is calculated by counting the number of cells in division (prophase, metaphase, anaphase, and telophase) and the total number of cells.

$$\mathrm{MI} = \frac{\text{Number of cells in division}}{\text{Total number of cells}} \tag{8.1}$$

In the same test, the numbers of cells showing alterations in the different phases of the cycle are counted, permitting calculation of the proportion of damaged cells (CAS, (8.2)) for each of the applied treatments.

$$\mathrm{CAS} = \frac{\text{Number of damaged cells}}{\text{Number of cells in division}} \tag{8.2}$$

The relative indices are calculated by comparing the mitotic index and alteration index values obtained for each treatment with the values obtained for the negative control. This evaluation of DNA damage using cell analyses can be used to evaluate the genotoxicity of nanostructured materials. Table 8.1 presents some of the published work in which the test has been employed to investigate the toxic effects of such materials.

The advantages of this test are that it is simple and inexpensive, compared to other assays. Importantly, it also allows direct contact between the roots and the material to be tested, avoiding any risk of components of the medium interacting with the nanoparticle surfaces, which could alter the properties of the particles. The genotoxicity evaluation is therefore not compromised by the occurrence of any changes in the physicochemical stability of the nanoparticle system. Disadvantages of the *Allium cepa* test are that analysis of the slides is time-consuming compared to other cell viability tests based on colorimetry and that each treatment requires a large number of cells to be counted.

Table 8.1 Results of published works in which the *Allium cepa* test was used to investigate the toxic effects of nanostructured materials

Material	Principal results of the *Allium cepa* test	Reference
Silver nanoparticles	The effects of silver nanoparticles (<100 nm) were investigated as a function of different concentrations (25, 20, 75, and 100 ppm). The mitotic index values diminished as the concentrations were increased, and the occurrence of chromosomal aberrations indicated that the silver nanoparticles were able to penetrate into the plants and cause breaks in the chromosomes	Kumari et al. (2009)
Titanium dioxide nanoparticles	Titanium dioxide nanoparticles caused chromosomal aberrations that were associated with reduced growth of *Allium cepa* roots. Increased levels of malondialdehyde (MDA) were detected, indicating that lipid peroxidation could be involved in the mechanisms leading to DNA damage	Ghosh et al. (2010)
Chitosan-poly (methacrylic acid) nanoparticles	The effects of differently sized (60, 82, and 111 nm) nanoparticles composed of chitosan (CS) and poly (methacrylic acid) (PMAA) were evaluated according to concentrations (1.8, 18, and 180 mg/L). The results showed that there were no structural or numerical alterations in the DNA, although reduced mitotic index values were observed using 82 and 111 nm particles at higher concentrations, which was indicative of cellular toxicity. No alterations in the mitotic index were observed for 60 nm nanoparticles	Lima et al. (2010)
Zinc oxide nanoparticles	Use of zinc oxide nanoparticles at concentrations of 25, 50, 75, and 100 g/mL showed that the mitotic index values diminished as the concentrations were increased, with increased in the number of chromosomal aberrations	Kumari et al. (2011)

8.2.2 *Comet Assays*

The widely used comet test is highly suitable for studying genotoxicity of nanomaterials. It provides an indication of the ability of the material to cause DNA lesions, although it is essentially a pretest since it measures the damage caused; however, it does not permit observation of any repair processes, since during the test there is no cell division (which is the phase during which most of the repair of the genetic material occurs). Meanwhile, the fact that there is no need for cell division to take place during the test, the procedure is relatively fast and straightforward. It is also inexpensive and can be applied both in vitro and in vivo. The analyses require the use of positive and negative controls to ensure successful completion of all stages of the test. The cells exposed to the nanomaterial (as well as those used as controls) are immobilized on slides using agarose gel, the membranes are lysed, and the genetic material is exposed. This material is submitted to electrophoresis, during which the DNA migrates towards the positive electrode. The migration is proportional to the size of the fragment, with nuclei whose genetic material is fragmented exhibiting a migration in which smaller fragments move faster and larger fragments slowly, creating an appearance that resembles a comet with a tail (Fig. 8.3).

The fragmented nuclei are classified according to the migration of the genetic material and the size of the tail formed, using scores between 0 and 4 (Fig. 8.3). Computer programs are available for the measurement and counting of the "comets" formed. Several studies have shown that the comet test can be applied for the evaluation of nanoparticles, using either cell cultures or different organisms in vivo (Table 8.2).

Advantages of this test are that the investment required is modest, it can be applied for both in vivo and in vitro analyses with equal ease, the counting procedure is simple, and it does not involve the prolonged cell culturing needed for cell division to occur. Disadvantages relate to the sensitivity of the cells, which must be protected in order to avoid possible breaks due to exposure to light, and to the fact that the technique is a pretest, so that caution should be used in interpretation of the results.

8.2.3 Cytogenetic Assays

Cytogenetic analysis has become well established in medical genetics for the detection of aneuploidies, after Lejune in 1959 (Neri and Opitz 2009) discovered altered numbers of chromosomes in patients with Down syndrome. Since then, the techniques used to visualize chromosomes have been improved, enabling greater accuracy in the identification of alterations such as breaks, deletions, and insertions (euploidies). The incorporation of molecular biology in cytogenetic techniques now permits better assessment of damage and the ability to identify points with smaller dimensions. In terms of the genotoxicity of nanomaterials, cytogenetics can be employed directly in organisms (in vivo) or in cultures of cells (in vitro) from different animals including mammals, insects, and mollusks, amongst others, as well as different types of cell from the same organism, such as lymphocytes, fibroblasts, and tumoral cells. For both in vivo and in vitro tests, the technique requires that the cells undergo division, while in the case of in vitro tests the exposure must take place during the culture period (Mateuca et al. 2006).

In order to finalize the test, cell division must be blocked at the metaphase stage, which requires the use of colchicine to block mitotic fusion and prevent karyokinesis. The chromosomes of the cell blocked in metaphase are then visible and can be counted and checked for breaks and other alterations (Fig. 8.4).

Cytogenetic analysis enables the counting of cells that are in division and those that are not, so that as in the *Allium cepa* test it is possible to calculate the mitotic index (MI) by dividing the number of cells in metaphase by the total number of

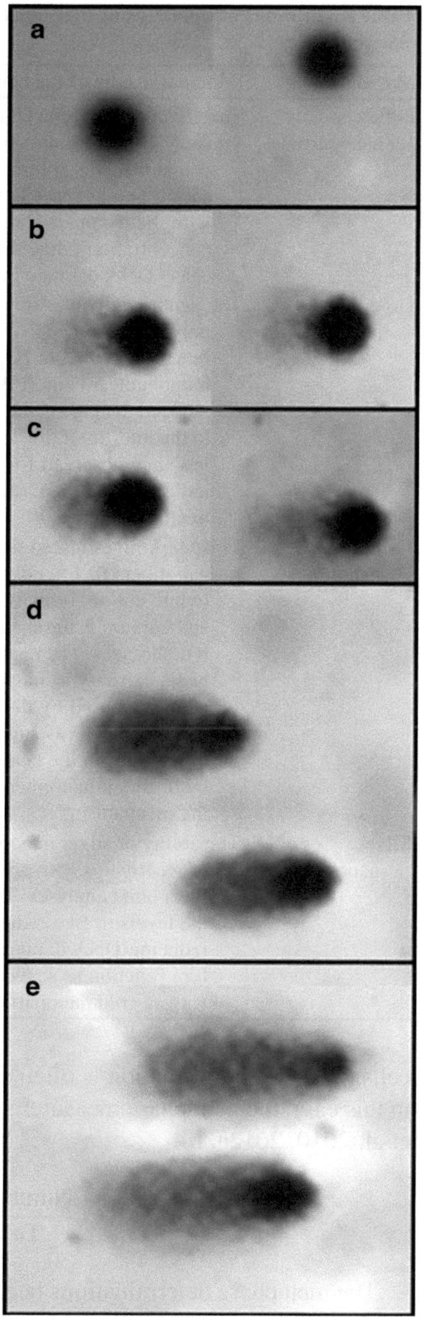

Fig. 8.3 Illustrative scheme showing the levels of tail formation during the comet test: (**a**) level = 0, (**b**) level = 1, (**c**) level = 2, (**d**) level = 3, (**e**) level = 4

Table 8.2 Results of studies using the comet test to investigate the toxic effects of nanostructured materials

Material	Principal results of the comet test	Reference
Fullerene C_{60} nanoparticles	The effect of fullerene C_{60} nanoparticles was studied in rats, using concentrations between 0.5 and 2.5 mg/kg. The comet test was applied to lung cells of the rats, and showed no increase in the % tail DNA for groups treated with the nanoparticles. It was concluded that for this concentration range the particles had no potential to cause DNA damage in the rat lung cells	Ema et al. (2012)
Gold nanoparticles	The genotoxicity of Au nanoparticles stabilized with polyamidoamine (PAMAM) dendrimer and sodium citrate was studied in HepG2 cells and peripheral blood mononuclear cells (PBMC). The Au nanoparticles induced damage in the DNA due to their ease of transport to the interior of the cells. The damage to the PBMC was less than observed for tumoral HepG2 cells, which indicated that the particles could be used in cancer treatment therapy	Paino et al. (2012)
Cellulose nanofibers	This work investigated the genotoxic effects of cellulose nanofibers (0.1 %) obtained from different sources (white cotton, brown cotton, ruby cotton, green cotton, and curaua) using the comet test applied to human lymphocytes. The results indicated that all the nanofibers had a capacity to induce DNA damage, with greatest damage caused by the brown cotton and curaua nanofibers. It was suggested that the observed differences in the comet test results could be explained by different nanofiber transport and intracellular signaling mechanisms	Lima et al. (2012a)
Silver nanoparticles	The ability of silver nanoparticles to suppress the generation of reactive oxygen species was investigated using alkaline comet analyses. The nanoparticles acted to diminish the levels of free radicals produced by H_2O_2, hence reducing DNA damage. The nanoparticles could therefore function as scavengers of free radicals, in a similar way as gold nanoparticles	Flower et al. (2012)

cells (8.3). This information is often seen as an indicator of cytotoxicity; however, in this case, the metaphase measurements provide a measure of genotoxicity (Lima et al. 2010, 2012a, b).

$$\mathrm{MI} = \frac{\text{Number of metaphase cells}}{\text{Total number of cells}} \tag{8.3}$$

The metaphase determinations begin with the counting of chromosome numbers in the cell under analysis. Subsequent analyses are made of breaks (gaps), deletions, chromosome ring formation, inversions, translocations, and possible exchanges between sister chromatids. These analyses are performed using specific colorations. Animal cytogenetic assessment demands the skills of a suitably qualified and

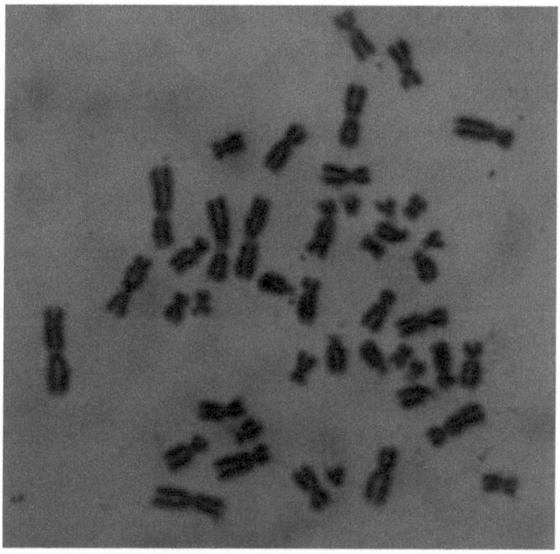

Fig. 8.4 Photograph of a 3T3 cell in metaphase stage

experienced person; hence, there have been few articles published concerning the use of cytogenetic analysis to evaluate genotoxicity. Table 8.3 lists the results obtained in several of the reported studies employing cytogenetic analysis.

A positive aspect of this test is that it enables identification of the location of breaks and the type of damage caused to the DNA, although some of the alterations can only be analyzed using complementary molecular biology procedures such as the FISH (fluorescence in situ hybridization) technique (Skinner et al. 2013). Disadvantages are the difficult analysis, the need to perform cell cultures, even when the analysis is performed in vivo, and the inability to undertake detailed analysis of the metaphase in cultures of cells where the chromosomal number may not remain constant.

8.2.4 Micronucleus Assays

Micronucleus analysis is one of the oldest known techniques for assessment of genotoxicity. The procedure used is similar to the cytogenetic technique, with some alterations, and the analyses can be employed either directly (in vivo) or using cell cultures (in vitro). In some in vivo applications it is possible to use direct smearing of the material, avoiding the need for cell cultures.

The technique involves the detection of the formation of small nuclei resulting from aneuploidies (the loss or gain of chromosomes) or chromosome breaks that could not be repaired during cell division. These small nuclei (micronuclei) contain genetic material that became detached from the principal nucleus, and can occur singly or in groups, depending on the material and the type of damage. The different types of micronuclei can be classified according to the nature of the damage incurred. However, in this test it is impossible to identify the genetic material

Table 8.3 Results of studies using cytogenetic analysis to investigate the toxic effects of nanostructured materials

Material	Principal results	Reference
Poly (lactide-co-glycolide) nanoparticles	Cytogenetic analysis was used to investigate the effects of PLGA nanoparticles (5.4, 54, and 540 μg/mL, polymer mass/volume of solution) on lymphocyte cells. There were no significant changes in mitotic index values, indicating that for the concentration range tested the particles presented no genotoxicity	de Lima et al. (2011)
Chitosan-poly(methacrylic acid) nanoparticles	The genotoxicity of CS/PMAA nanoparticles was evaluated using human lymphocyte cell cultures. There were no numerical or structural alterations in the chromosomes by exposing the cells to different particle sizes (60, 82, and 111 nm) and concentrations (1.8, 18, and 180 mg/L). There were significant alterations in the mitotic index values when the cells were exposed to 82 and 111 nm particles at a concentration of 180 mg/L. The 111 nm nanoparticles caused a diminution of the mitotic index values, while an increase was observed for the 82 nm particles. The differences were attributed to different mechanisms of entry into the cells, as well as interactions with biological macromolecules and cellular signaling processes	Lima et al. (2010)
Carbon nanotubes	This work studied the genotoxic effects of carbon nanotubes (single-walled and multi-walled) on lymphocyte cells. Cytogenetic analysis revealed mitotic inhibition for the treatments using single-walled tubes. In light of the limited number of studies undertaken with nanotubes, it was concluded that the findings were important in terms of risk assessment of the toxicity of these materials	Szendi and Varga (2008)
Alumina (Al_2O_3) and metal alloy (CoCr) particles	The genotoxic effects of alumina ceramic (Al_2O_3) particles were compared with those of cobalt–chromium metal (CoCr alloy) particles, using human fibroblasts. Both types of particle caused numerical changes and structural aberrations in the chromosomes, with extensive lesions caused by the CoCr particles. By comparison, the alumina ceramic particles were weakly genotoxic to the cells	Tsaousi et al. (2010)

contained in the micronuclei, unless specific probex are used to detect predetermined regions (Bouraoui et al. 2013)

This test measures damage that was incurred and could not be repaired. Hence, nanomaterials that lead to the formation of micronuclei cause damage to the genetic material that is effectively irreversible. For this reason, the frequency of appearance of micronuclei is low, so that a large number of cells need to be analyzed in order to detect the few that are damaged. Nonetheless, the analysis is straightforward, compared to cytogenetic techniques.

Tests performed in vitro require the use of cytochalasin B during the cell culture. This compound permits karyokinesis but not cytokinesis, so that two normal nuclei remain in the cell that divided during the culture (Guccini et al. 2012). It is only these cells, with two nuclei that are considered for micronucleus counting, which ensures that the micronuclei identified were formed during exposure to the nanomaterial, and not prior to exposure.

Several studies reported in the literature have used this longstanding test to assess nanomaterial genotoxicity, due to the simplicity of the counting procedure, especially when the investigation is conducted in vivo using a blood smear alone. This is frequently the test of choice in work involving assessment of the cells of mammals. Table 8.4 lists nanostructured materials whose toxicities have been evaluated using micronucleus tests, and describes the main conclusions.

Advantages of the test are that it is relatively simple and can be applied in vivo using only a smear of the material to be analyzed. Disadvantages include the fact that it is difficult to induce micronucleus formation using materials of low toxicity, necessitating the analysis of a large number of cells, and its inability to provide information concerning the identity of the material present in the micronucleus.

8.3 Nanomaterials and Molecular Analysis

8.3.1 DNA Analysis

Molecular analysis of the genetic material can be used to identify the alterations that occur in the DNA during exposure to nanostructured materials, including structural modifications and mutations in the genome. DNA adducts result from exposure to compounds that are either able to interact directly with the genetic material or that during metabolism result in a substance that is able to interact with the DNA. Such interactions induce alterations in the genetic material due to damage that is unable to be repaired; these changes can be detected molecularly (Liao and Liu 2012).

One of the simplest type of molecular assessment involves the detection of breaks in the genetic material using electrophoresis of extracted and purified DNA, enabling detection of alterations in the cells caused by exposure to a nanomaterial that induces disruption of the genetic material and cell apoptosis (Lima et al. 2010). This analysis is relatively simple, although care must be taken to avoid experimental errors during the DNA extraction steps, and consequent generation of false positive results. An appropriate high quality extraction is therefore essential to ensure that the test is valid. Despite these drawbacks, molecular analysis is useful for examination of the genetic material and investigation of

Table 8.4 Studies using the micronucleus test to investigate the toxic effects of nanomaterials

Material	Main results of the micronucleus test	Reference
Cationic nanoparticles	There was a significant increase in micronucleus formation when 16HBE14 cells were exposed to cationic nanoparticles. This was attributed to the generation of reactive oxygen species that caused degradation of the cell membranes, as well as indirect alterations in the mitotic apparatus caused by the particles	Merhia et al. (2012)
Poly(ε-caprolactone)-poly(ethyleneglycol)-poly(-ε-caprolactone) nanomaterials	Nanomaterials composed of poly(-ε-caprolactone)-poly(ethylene glycol)-poly(ε-caprolactone) (PCL-PEG-PCL, PCEC) were prepared and their micronucleus formation indices were determined using the bone marrow cells of mice. No chromosomal aberrations were observed, indicating that the nanomaterials did not present genotoxicity under the experimental conditions employed	Huanga et al. (2010)
Copper oxide nanoparticles	The genotoxic effects of copper oxide nanoparticles were investigated using the Neuro-2A mouse neuroblastoma cell line. Micronucleus analyses showed that after 24 h of treatment, the particles caused an increase in the frequency of both type I micronuclei (smaller than 1/4 the size of the nucleus) and type II micronuclei (between 1/4 and 1/2 the size of the nucleus). Type I micronuclei are considered indicative of clastogenic effects and contain small chromosome fragments, while type II micronuclei reflect aneugenic effects and contain whole chromosomes	Perreault et al. (2012)

the mutagenic effects of nanomaterials in microorganisms, such as bacteria in soil and water. Mutations resulting from exposure lead to alterations in the genome sequences of the organisms, giving rise to new strains, and these changes can be detected without any need for sequencing of the genetic material (Ge et al. 2012).

The choice of technique for use in molecular studies depends on the desired evaluation and on the anticipated genetic alterations. Options include procedures based on the polymerase chain reaction (PCR) and restriction enzymes. Different regions of the genome can be selected depending on the focus of the study, with conserved genomic regions normally being chosen, such as the 16rRNA region responsible for transcription of ribosomal RNA (Nogueira et al. 2012).

A widely used technique for evaluation of conserved genomic regions is PCR-DGGE (denaturing gradient gel electrophoresis), which permits detection of small mutations in the sequence of the region studied. Other techniques that can be used include ARDRA (amplified ribosomal DNA restriction analysis), which is based on conservation of the restriction sites of rDNA, as well as RAPD (random amplified polymorphic DNA), which uses small random primers whose annealing to DNA is affected by the presence or absence of mutations (Gardener and Weller 2001; Schwieger and Tebbe 2000)

Another analytical approach is the study of specific genes. Organisms generally possess genes that are shared with other species and that can be used, for example, to investigate the diversity of microorganisms. However, some genes are specific to individual species, and their presence is therefore indicative of the presence of the microorganisms concerned. This analysis therefore enables detection of the presence or absence of a given organism in a particular environment. For these studies utilize primers that are specific for the target species, and can be either qualitative, with determination of the presence or absence of the organisms of interest, or quantitative, in which case the organisms are quantified using qPCR (real-time PCR).

The molecular analysis of microorganisms is important since they are vital for the sustainable functioning of ecosystems. Analysis of the effects of nanoparticles on microbiota, in terms of the type and degree of genetic alteration, is therefore especially important since such genetic alterations can have wider consequences. Important studies have shown that bacteria respond in different ways to the presence of nanostructured materials in the environment, as illustrated by the examples listed in Table 8.5.

Studies that are more sophisticated use genetic engineering techniques to construct biosensors consisting of plasmids with reporter genes (Shin et al. 2005; Mitchell and Gu 2004; Benton et al. 2007, 2008; Yagi 2007; Yagur-Kroll et al. 2010; Ravikumar et al. 2012). This permits rapid detection of molecular responses of the genetic constructs to specific environmental changes. Drawbacks of molecular genetics methodologies are related to the selection of suitable procedures. Prior knowledge and experience in molecular biology is needed due to the diversity of potential difficulties inherent in these techniques, in order to ensure the quality and validity of the results. Since their application to the effects of nanomaterials and any specific alterations they may cause is relatively new, these techniques should still be used with caution.

8.3.2 Genetic Expression Analysis (Epigenetic Factors)

DNA transcription is an essential step in protein synthesis, and the analysis of mRNA (messenger RNA) provides information concerning the active genes. Studies have shown that nanomaterials can interfere in cellular programming due to alterations in the levels of production of certain key proteins. The activation and

Table 8.5 Summary of important studies that investigated the responses of bacteria exposed to nanomaterials

Material	Principal results using molecular tests	Reference
Metal oxide nanoparticles	The effects of metal oxide nanoparticles (CuO and Fe_3O_4) were studied using two types of soil. Alterations were observed in the composition of the soil bacterial community, which could cause environmental impacts	Ben-Moshe et al. (2013)
Ag, CuO, and ZnO nanoparticles	The effects of Ag, CuO, and ZnO nanoparticles on pathogenic soil bacteria were studied. The work employed a *Pseudomonas putida* KT2440 biosensor construct possessing a plasmid with luxAB reporter genes. The microorganisms were exposed to nanoparticles, with evaluation based on the amount of light emitted. Rapid light reduction was observed using Ag and CuO nanoparticles, indicating cell death, while ZnO nanoparticles exerted a bacteriostatic effect, rather than the bactericidal action of the other nanoparticle types. There was also a different response in terms of nanoparticle aggregation. It was concluded that the nanoparticles showed bacteriostatic and bactericidal activities towards soil bacteria	Gajjar et al. (2009)
Ag nanoparticles	The effects of silver nanoparticles (Ag NPs) on microbial communities in activated sludge were investigated. Sludge samples were exposed to 1 mg/L of the nanoparticles and evaluated in terms of the impact on the structure and the response of the microbial communities. The analyses utilized the PCR-DGGE technique for the 16S RNA region. The results suggested that certain microbial species were highly sensitive to the Ag NPs	Sun et al. (2012)

deactivation of genes occurs due to the binding of proteins known as transcription factors at specific sites (Bleda et al. 2012).

It has been demonstrated that the expression of a variety of genes can be altered due to the presence of a particular type of nanomaterial. However, as expected there is no consistent pattern in the phenomena since each nanomaterial is unique and can affect the genes of different organisms or groups of cells in distinct ways (Table 8.6).

There are various methods of determining gene expression, including those where sets of genes can be analyzed at the same time, such as microarray analysis (DNA chip), and those where knowledge of the behavior of a specific gene is required, such as the RT-PCR (reverse transcription polymerase chain reaction) or qRT-PCR techniques (Husain et al. 2013). Expression analysis is a tool that can be used to improve understanding of the mechanisms of cellular toxicity and the induction of apoptosis. In addition to mRNA, transcription can involve miRNA (microRNA). The latter can interfere in protein translation and inactivate the mRNA, since small changes due to miRNA transcription can lead to substantial

Table 8.6 Results of studies that investigated gene expression as a way for evaluating the toxic effects of nanostructured materials

Material	Principal results	Reference
Cobalt nanoparticles	An evaluation was made of the gene expression of BALB-3T3 A31-1-1 cells, either unexposed or exposed to cobalt nanoparticles (Co-nano) for 72 h. Differential expressions were obtained of 10 sequences whose genes could be used as biomarkers to indicate specific cellular effects resulting from exposure to Co-nano. The results revealed the activation of cellular defense pathways and repair mechanisms. It was shown that molecular techniques could be valuable tools in nanotoxicology research. Although the results did not permit conclusions to be drawn concerning the molecular mechanisms involved, they suggested that treatment with Co-nano in some way activated cell defense and repair pathways	Papis et al. (2007)
Copper nanoparticles	The mechanisms of induction of nephrotoxicity by copper nanoparticles were investigated using in vivo tests of renal gene expression. Male Wistar rats received doses of nanoparticles (50, 100, and 200 mg/kg) for 5 days. The transcriptome of the entire renal genome was evaluated, revealing significant alterations in the expression of many genes, especially those concerning metabolism, cellular processes, and the processing of information about the environment. This included valine, leucine, and isoleucine degradation, complement and coagulation cascades, oxidative phosphorylation, the cell cycle, MAPK (mitogen-activated protein kinase) signaling, and glutathione metabolism, all of which are considered to be related to problems of nephrotoxicity	Liao and Liu (2012)
C_{60} fullerene nanoparticles	Analysis was made of the expression of genes related to the pulmonary system. Animals were exposed, by inhalation, for 4 weeks (6 h per day, 5 days per week) to C_{60} fullerene nanoparticles (0.12 mg/m^3, 4.1×10^4 particles/cm^3, 96 nm diameter) and ultrafine nickel oxide (Uf-NiO) particles (0.2 mg/m^3, 9.2×10^4 particles/cm^3, 59 nm diameter). Following exposure, molecular studies were performed using the pulmonary material, and the gene expression profiles revealed that few of the genes involved in the inflammatory, oxidative stress, apoptosis, and metalloendopeptidase responses were upregulated, while some of the genes associated with the immunological process, including those of the histocompatibility complex (MHC), were upregulated. There were significant differences in induction of expression of the associated genes. It was concluded that C_{60} fullerene should not cause serious pulmonary toxicity when exposure occurs via the inhalation route	Fujita et al. (2009a)

(continued)

Table 8.6 (continued)

Material	Principal results	Reference
TiO_2 nanoparticles	Microarray analyses were performed to determine the expression profile in a culture of human HaCaT keratinocytes. The keratinocytes were exposed for 24 h to three types of TiO_2 nanoparticle (7, 20, and 200 nm), in the absence of light. According to the analyses, the genes involved in the inflammatory response and cell adhesion were upregulated, but not those concerned with oxidative stress and apoptosis. After 24 h, the cells exposed to the 7 nm nanoparticles showed alterations in expression of the genes related to the matrix metalloproteinases (*MPM-9* and *MPM-10*) and cell adhesion (fibronectin *FN-1*, integrin *ITGB-6*, and mucin *MUC-4*). It was concluded that in the absence of light, TiO_2 nanoparticles do not cause any oxidative stress impacts, but can affect cell adhesion in keratinocytes	Fujita et al. (2009b)

cellular alterations (Eastman et al. 2006; Xiu et al. 2010; Wang et al. 2012; Jeon et al. 2012; Gomes et al. 2013; Chew et al. 2012).

It should be remembered that cellular mechanisms often involve a signaling cascade, so that careful selection of the gene to be studied is required, since the fact that the expression of one gene is not altered does not guarantee that other genes in the cascade are not altered, or that the exposure cannot cause tangible alterations in the organism.

8.4 Difficulties and Future Perspectives

Nanoparticle genotoxicity studies are in their infancy, with many difficulties remaining to be resolved, notably in relation to validation of the methodologies employed to investigate the effects of these substances.

It has been shown that at high concentrations, aggregation of nanomaterials can occur, while components present in the environment, such as biological macromolecules, can alter the physicochemical characteristics of the nanomaterial and therefore cause doubt as to whether the observed effect was due to the nanomaterial itself or to the nanomaterial containing the macromolecule. An example is the use of culture media supplemented with substances such as serum albumin and other materials, which could complicate nanotoxicological assessments conducted in vitro and in vivo. Hence, there is a need to consider the physicochemical characteristics of the nanomaterial in the medium to be studied and to try to ensure the same exposure conditions during evaluations performed using different methodologies and different ranges of concentration.

A further point concerns the toxicological evaluation of nanomaterials using the tests already described in the literature for measuring the toxicity of chemical compounds. It is necessary to establish whether the presence of the nanomaterial might interfere in the basic principles of the test. An example is colorimetric analyses, where the nanomaterial might cause light dispersion and lead to erroneous results. Another possible problem is the adhesion of nanostructured materials to the equipment used in the tests (such as slides, tubes, pipette tips, and so on). Further work is needed in order to achieve the ultimate goal of establishing definitive protocols for tests used to evaluate the genotoxicity of nanomaterials. It is hoped that the findings of studies currently in progress will contribute towards this objective.

Acknowledgements Supported by FAPESP, CNPq, Brazilian Network on Nanotoxicology (MCTI/CNPq), INOMAT (CNPq), NanoBioss (CMTI/CNPq).

References

Abdul Khaliq R, Kafafy R, Salleh HM et al (2012) Enhancing the efficiency of polymerase chain reaction using graphene nanoflakes. Nanotechnology 23:455106

Akhmedov AT, Frei C, Tsai-Pflugfelder M et al (1998) Structural maintenance of chromosomes protein C-terminal domains bind preferentially to DNA with secondary structure. J Biol Chem 273:24088–24094

Al Hussain TO, Akhtar M (2013) Molecular basis of urinary bladder cancer. Adv Anat Pathol 20:53–60

Amara S, Slama IB, Omri K, et al (2013) Effects of nanoparticle zinc oxide on emotional behavior and trace element homeostasis in rat brain. Toxicol Ind Health. doi: 10.1177/0748233713491802

Batchelor E, Loewer A, Lahav G (2009) The ups and downs of p53: understanding protein dynamics in single cells. Nat Rev Cancer 9:371–377

Ben-Moshe T, Sammy Frenk S, Dror I et al (2013) Effects of metal oxide nanoparticles on soil properties. Chemosphere 90:640–646

Benton MG, Glasser NR, Palecek SP (2007) The utilization of a Saccharomyces cerevisiae HUG1P-GFP promoter-reporter construct for the selective detection of DNA damage. Mutat Res 633:21–34

Benton MG, Glasser NR, Palecek SP (2008) Deletion of MAG1 and MRE11 enhances the sensitivity of the Saccharomyces cerevisiae HUG1P-GFP promoter-reporter construct to genotoxicity. Biosens Bioelectron 24:736–741

Bleda M, Medina I, Alonso R et al (2012) Inferring the regulatory network behind a gene expression experiment. Nucleic Acids Res 40:168–172

Bouraoui S, Mougou S, Brahem A et al (2013) A combination of micronucleus assay and fluorescence in situ hybridization analysis to evaluate the genotoxicity of formaldehyde. Arch Environ Contam Toxicol 64:337–344

Cabrera GL, Rodriguez DMG (1999) Genotoxicity of soil from farmland irrigated with wastewater using three plant biossays. Mutat Res 426:211–214

Chew WS, Poh KW, Siddiqi NJ et al (2012) Short- and long-term changes in blood miRNA levels after nanogold injection in rats–potential biomarkers of nanoparticle exposure. Biomarkers 17:750–757

Dinant C, Houtsmuller AB, Vermeulen W (2008) Chromatin structure and DNA damage repair. Epigenetics Chromatin 1:9

Eastman PS, Ruan W, Doctolero M et al (2006) Qdot nanobarcodes for multiplexed gene expression analysis. Nano Lett 6:1059–1064

Ema M, Tanaka J, Kobayashi N et al (2012) Genotoxicity evaluation of fullerene C_{60} nanoparticles in a comet assay using lung cells of intratracheally instilled rats. Regul Toxicol Pharmacol 62:419–424

Flower NAL, Brabu B, Revathy M et al (2012) Characterization of synthesized silver nanoparticles and assessment of its genotoxicity potentials using the alkaline comet assay. Mutat Res 742:61–65

Foster SS, De S, Johnson LK, Petrini JH et al (2012) Cell cycle- and DNA repair pathway-specific effects of apoptosis on tumor suppression. Proc Natl Acad Sci U S A 109:9953–9958

Fujita K, Horie M, Kato H et al (2009a) Effects of ultrafine TiO_2 particles on gene expression profile in human keratinocytes without illumination: involvement of extracellular matrix and cell adhesion. Toxicol Lett 191:109–117

Fujita K, Morimoto Y, Ogami A et al (2009b) Gene expression profiles in rat lung after inhalation exposure to C_{60} fullerene particles. Toxicology 258:47–55

Gajjar P, Pettee B, Britt DW et al (2009) Antimicrobial activities of commercial nanoparticles against an environmental soil microbe, *Pseudomonas putida* KT2440. J Biol Eng 3:1–13

Gardener BBM, Weller DM (2001) Changes in populations of rhizosphere bacteria associated with take-all disease of wheat. Appl Environ Microbiol 67:4414–4425

Ge Y, Schimel JP, Holden PA (2012) Identification of soil bacteria susceptible to TiO_2 and ZnO nanoparticles. Appl Environ Microbiol 78:6749–6758

Ghosh M, Bandyopadhyay M, Mukherjee A (2010) Genotoxicity of titanium dioxide (TiO_2) nanoparticles at two trophic levels: Plant and human lymphocytes. Chemosphere 81:1253–1262

Giono LE, Manfredi JJ (2006) The p53 tumor suppressor participates in multiple cell cycle checkpoints. J Cell Physiol 209:13–20

Goetze S, Mateos-Langerak J, van Driel R (2007) Three-dimensional genome organization in interphase and its relation to genome function. Semin Cell Dev Biol 18:707–714

Gomes SI, Soares AM, Scott-Fordsmand JJ et al (2013) Mechanisms of response to silver nanoparticles on *Enchytraeus albidus* (Oligochaeta): survival, reproduction and gene expression profile. J Hazard Mater 254–255:336–344

González-González M, Muñoz-Bellvis L, Mackintosh C et al (2012) Prognostic impact of del(17p) and del(22q) as assessed by interphase FISH in sporadic colorectal carcinomas. PLoS One 7: e42683

Guccini S, Lombardi S, Pisani A et al (2012) Effects of spindle poisons in peripheral human lymphocytes by the in vitro cytokinesis-block micronucleus assay. Mutagenesis 27:749–758

Han Z, Wei W, Dunaway S, Darnowski JW et al (2002) Role of p21 in apoptosis and senescence of human colon cancer cells treated with camptothecin. J Biol Chem 277:17154–17160

Harper JW, Elledge SJ (2007) The DNA damage response: ten years after. Mol Cell 28:739–745

Herzog CR, Bodon N, Pittman B et al (2004) Carcinogen-specific targeting of chromosome 12 for loss of heterozygosity in mouse lung adenocarcinomas: implications for chromosome instability and tumor progression. Oncogene 23:3033–3039

Herzog CR, Desai D, Amin S (2006) Array CGH analysis reveals chromosomal aberrations in mouse lung adenocarcinomas induced by the human lung carcinogen 4-(methylnitrosamino)-1-(3-pyridyl)-1-butanone. Biochem Biophys Res Commun 341:856–863

Hoeijmakers JH (2009) DNA damage, aging, and cancer. N Engl J Med 361:1475–1485

Huanga Y, Gao H, Gou M et al (2010) Acute toxicity and genotoxicity studies on poly(epsilon-caprolactone)-poly(ethylene glycol)-poly(epsilon-caprolactone) nanomaterials. Mutat Res 696:101–106

Hubner MR, Spector DL (2010) Chromatin dynamics. Annu Rev Biophys 39:471–489

Hudler P (2012) Genetic aspects of gastric cancer instability. ScientificWorldJournal 2012:761909

Husain M, Saber AT, Guo C et al (2013) Pulmonary instillation of low doses of titanium dioxide nanoparticles in mice leads to particle retention and gene expression changes in the absence of inflammation. Toxicol Appl Pharmacol 269:250–262

Jabbur JR, Huang P, Zhang W (2000) DNA damage-induced phosphorylation of p53 at serine 20 correlates with p21 and Mdm-2 induction in vivo. Oncogene 19:6203–6208

Jeon SY, Park JS, Yang HN et al (2012) Co-delivery of SOX9 genes and anti-Cbfa-1 siRNA coated onto PLGA nanoparticles for chondrogenesis of human MSCs. Biomaterials 3:4413–4423

Kadota M, Sato M, Duncan D et al (2009) Identification of novel gene amplifications in breast cancer and coexistence of gene amplification with an activating mutation of *PIK3CA*. Cancer Res 69:7357–7365

Kumari M, Mukherjee A, Chandrasekaran N (2009) Genotoxicity of silver nanoparticles in *Allium cepa*. Sci Total Environ 407:5243–5246

Kumari M, Khan SS, Pakrashi S et al (2011) Cytogenetic and genotoxic effects of zinc oxide nanoparticles on root cells of *Allium cepa*. J Hazard Mater 190:613–621

Leme DM, Marin-Morales MA (2009) *Allium cepa* test in environmental monitoring: a review on its application. Mutat Res 682:71–81

Levine AJ (1997) p53, the cellular gatekeeper for growth and division. Cell 88:323–331

Liao M, Liu H (2012) Gene expression profiling of nephrotoxicity from copper nanoparticles in rats after repeated oral administration. Environ Toxicol Pharmacol 34:67–80

Lima R, Feitosa L, Pereira AES et al (2010) Evaluation of the genotoxicity of chitosan nanoparticles for use in food packaging films. J Food Sci 75:N89–N96

Lima R, Pereira AES, Porto RM et al (2011) Evaluation of cyto- and genotoxicity of poly(lactide-co-glycolide) nanoparticles. J Polym Environ 19:196–202

Lima R, Feitosa LO, Maruyama CR et al (2012a) Evaluation of the genotoxicity of cellulose nanofibers. Int J Nanomedicine 7:3555–3565

Lima R, Seabra AB, Durán N (2012b) Silver nanoparticles: a brief review of cytotoxicity and genotoxicity of chemically and biogenically synthesized nanoparticles. J Appl Toxicol 32:867–879

Ma TH, Cabrera GL, Owens E (2005) Genotoxic agents detected by plant bioassays. Rev Environ Health 20:1–13

Mateuca R, Lombaert N, Aka PV et al (2006) Chromosomal changes: induction, detection methods and applicability in human biomonitoring. Biochimie 88:1515–1531

Merhia M, Dombua CY, Brient A et al (2012) Study of serum interaction with a cationic nanoparticle: implications for in vitro endocytosis, cytotoxicity and genotoxicity. Int J Pharm 423:37–44

Migliore L, Migheli F, Spisni R, et al (2011) Genetics, cytogenetics, and epigenetics of colorectal cancer. J Biomed Biotechnol ID 792362:1–19

Mitchell RJ, Gu MB (2004) An Escherichia coli biosensor capable of detecting both genotoxic and oxidative damage. Appl Microbiol Biotechnol 64:46–52

Mo Y, Wan R, Zhang Q (2012) Application of reverse transcription-PCR and real-time PCR in nanotoxicity research. Methods Mol Biol 926:99–112

Moore AS, Faisal A, Castro DG et al (2012) Selective FLT3 inhibition of *FLT3*-ITD + acute myeloid leukaemia resulting in secondary D835Y mutation: a model for emerging clinical resistance patterns. Leukemia 26:1462–1470

Neri G, Opitz JM (2009) Down syndrome: comments and reflections on the 50th anniversary of Lejeune's discovery. Am J Med Genet 149A:2647–2654

Nogueira V, Lopes I, Rocha-Santos T et al (2012) Impact of organic and inorganic nanomaterials in the soil microbial community structure. Sci Total Environ 424:344–350

Paino IMM, Marangoni VS, Oliveira RCS et al (2012) Cyto- and genotoxicity of gold nanoparticles in human hepatocellular carcinoma and peripheral blood mononuclear cells. Toxicol Lett 215:119–125

Papis E, Gornati R, Prati M et al (2007) Gene expression in nanotoxicology research: analysis by differential display in BALB3T3 fibroblasts exposed to cobalt particles and ions. Toxicol Lett 170:185–192

Perreault F, Melegari SP, Costa CH et al (2012) Genotoxic effects of copper oxide nanoparticles in Neuro 2A cell cultures. Sci Total Environ 441:117–124

Piazza R, Valletta S, Winkelmann N et al (2012) Recurrent *SETBP1* mutations in atypical chronic myeloid leukemia. Nat Genet 45:18–24

Rao CV, Yamada HY (2013) Genomic instability and colon carcinogenesis: from the perspective of genes. Front Oncol 3:130. doi:10.3389/fonc.2013.00130

Ravikumar S, Pham VD, Lee SH et al (2012) Modification of CusSR bacterial two-component systems by the introduction of an inducible positive feedback loop. J Ind Microbiol Biotechnol 39:861–868

Roylance R, Endesfelder D, Gorman P et al (2011) Relationship of extreme chromosomal instability with long-term survival in a retrospective analysis of primary breast cancer. Cancer Epidemiol Biomarkers Prev 20:2183–2194

Schwieger F, Tebbe CC (2000) Effect of field inoculation with *Sinorhizobium meliloti* L33 on the composition of bacterial communities in Rhizospheres of a Target Plant (*Medicago sativa*) and a non-target plant (*Chenopodium album*)-linking of 16S rRNA gene-based single-strand conformation polymorphism community profiles to the diversity of cultivated bacteria. Appl Environ Microbiol 66:3556–3565

Shin HJ, Park HH, Lim WK (2005) Freeze-dried recombinant bacteria for on-site detection of phenolic compounds by color change. J Biotechnol 119:36–43

Skinner SO, Sepúlveda LA, Xu H et al (2013) Measuring mRNA copy number in individual Escherichia coli cells using single-molecule fluorescent in situ hybridization. Nat Protoc 8:1100–1113

Soria G, Polo SE, Almouznil (2012) Prime, repair, restore: the active role of chromatin in the DNA damage response. Mol Cell 46(6):722–734

Stirling PC, Bloom MS, Solanki-Patil T et al (2011) The complete spectrum of yeast chromosome instability genes identifies candidate CIN cancer genes and functional roles for ASTRA complex components. PLoS Genet 7:e1002057

Strachan T, Read AP (1999) Human molecular genetics, 2nd edn. Wiley-Liss, New York. http://www.ncbi.nlm.nih.gov/books/NBK7580/

Sun X, Sheng Z, Liu Y (2012) Effects of silver nanoparticles on microbial community structure in activated sludge. Sci Total Environ 443:828–835

Szendi K, Varga C (2008) Lack of genotoxicity of carbon nanotubes in a pilot study. Anticancer Res 28:349–352

Tsaousi A, Jones E, Case CP (2010) The in vitro genotoxicity of orthopaedic ceramic (Al_2O_3) and metal (CoCr alloy) particles. Mutat Res 697:1–9

Vasconcelos DS, da Silva FP, Quintana LG, et al (2013) Numerical aberrations of chromosome 17 and TP53 in brain metastases derived from breast cancer. Genet Mol Res 12. http://iah.iec.pa.gov.br/iah/fulltext/pc/artigos/2013/genetmol res2013prelo.pdf

Veschure PJ (2006) Chromosome organization and gene control: it is difficult to see the picture when you are inside the frame. J Cell Biochem 99:24–35

Wang J, Yi X, Tang H et al (2012) Direct quantification of microRNA at low picomolar level in sera of glioma patients using a competitive hybridization followed by amplified voltammetric detection. Anal Chem 84:6400–6406

Wilkins G (2013) DNA: Twin strands solved the structure. Nature 496:434. doi:10.1038/496434b

Xiu ZM, Gregory KB, Lowry GV et al (2010) Effect of bare and coated nanoscale zerovalent iron on tceA and vcrA gene expression in *Dehalococcoides spp*. Environ Sci Technol 44:7647–7651

Yagi K (2007) Applications of whole-cell bacterial sensors in biotechnology and environmental science. Appl Microbiol Biotechnol 73:1251–1258

Yagur-Kroll S, Bilic B, Belkin S (2010) Strategies for enhancing bioluminescent bacterial sensor performance by promoter region manipulation. Bioeng Bugs 1:151–153

Zhou BB, Elledge SJ (2000) The DNA damage response: putting checkpoints in perspective. Nature 408:433–439

Chapter 9
Cellular Mechanisms in Nanomaterial Internalization, Intracellular Trafficking, and Toxicity

Marcelo Bispo de Jesus and Yvonne L. Kapila

Abstract Nanomaterials are expected to have a significant impact on medicine, although they still need to overcome several challenges before they are widely used. Understanding the molecular interaction of nanomaterials in the context of the cellular environment is crucial for the success of nanomaterials. Therefore, mechanisms responsible for nanomaterial internalization have attracted great attention in the scientific community. These mechanisms greatly impact intracellular trafficking and cellular processing of nanomaterials. Here we discuss the major endocytic pathways by which nanomaterials can be internalized by cells, such as clathrin-mediated endocytosis, caveolae-mediated endocytosis, macropinocytosis, and clathrin- and caveolae-independent endocytosis. In addition, intracellular routing, metabolism of nanomaterials, and undesirable effects of nanotoxicology are discussed. Finally, the role of in vitro studies to evaluate the potential toxic effects of nanomaterials was critically analyzed.

9.1 Introduction

During this last century scientists have revealed many of the most fundamental mysteries about life at the molecular level. This knowledge has given us the power to manipulate and control matter at the nanoscale. This understanding has attracted growing attention in industry and academia, thereby fueling outstanding progress in this field. The design, synthesis, and application of materials at a nanoscale (1–1,000 nm) is called *nanotechnology,* and the nanomaterials created can exhibit very interesting properties (Buzea et al. 2007; McNeil 2005). Given their size,

M.B. de Jesus (✉)
Department of Biochemistry, Institute of Biology, State University of Campinas—UNICAMP, Campinas, Sao Paulo 13083-970, Brazil
e-mail: dejesusmb@gmail.com

N. Durán et al. (eds.), *Nanotoxicology*, Nanomedicine and Nanotoxicology,
DOI 10.1007/978-1-4614-8993-1_9, © Springer Science+Business Media New York 2014

nanomaterials often exhibit different physical (mechanical, electrical, optical, etc.) properties when compared to macroscopic systems. Consequently, these changes in physical properties of nanomaterials are reflected in their chemical properties. The increase in surface area to volume ratio modifies mechanical properties, which in turn make nanomaterials much more reactive than their bulkier material counterparts (McNeil 2005). These novel properties afford nanomaterials' unique applications, enabling them to substantially improve the effectiveness of a number of existing products that could result in a substantial impact on science and technology in the twenty-first century (Yan et al. 2012). Thus, nanotechnology has the potential to introduce dramatic improvements in the human quality of life.

Nanotechnology's ability to manipulate matter provides a vast range of applications, from quantum computers to self-cleaning clothes. Among them, the branch that has been attracting enormous attention from scientists is *nanomedicine*; defined as the medical application of nanotechnology, it allows monitoring, repairing, controlling, and even constructing biological systems at the molecular scale (El-Ansary and Al-Daihan 2009). Nanomedicine is an interdisciplinary field that involves biology, chemistry, physics, medicine, material science, and biomedical engineering and has proven to be a very fertile research field. One of the main reasons for this high productivity is that nanomedicine can offer therapeutic approaches and disease diagnostics that cannot be achieved by conventional strategies (Cabral et al. 2011; El-Ansary and Al-Daihan 2009; Rajendran et al. 2010; Yan et al. 2012). For example, nanomaterials can be used as diagnostic devices identifying diseases at earlier stages, hence increasing the effectiveness of the treatment and decreasing its cost. In addition, some particles have shown therapeutic properties, which opens new possibility for their applications.

Different nanomaterials have been applied in biotechnology, including polymeric nanoparticles, quantum dots, liposomes, polymer-drug conjugates, dendrimers, lipid nanoparticles, silica nanoparticles, carbon fullerenes, nanotubes (single and multi-walled), metal oxides (titanium dioxide, zinc oxide, cerium oxide, iron oxide), and nanoscale metals (silver, gold, copper), among others. Irrespective of their applications, materials at the nanoscale can enter the human body and this may occur via different routes: intravenous, dermal, subcutaneous, inhalation, intraperitoneal, and oral. In this report, we are particularly focused on how this new technology interacts with living things at the cellular level. Due to their size, nanomaterials have greater potential to travel inside living organisms than other materials or larger particles (McNeil 2005). Eventually, these nanomaterials can interact with cells at the plasma membrane and lead to their internalization through the process of *endocytosis*.

Endocytosis is a biological process highly conserved across species and cell types through which cells internalize nutrients, regulate signal transduction, and modulate plasma membrane composition. Endocytosis begins with plasma membrane invagination to bring in the cargo. This is followed by membrane budding, whereby specific proteins (e.g., dynamin) are pinched off so that the cargo can undergo subsequent internalization via the endocytic pathway (Aguilar and Wendland 2005). Inside the cells these vesicles are called endosomes.

Endosomes are decorated with a large variety of proteins on their surface membrane, which has a different composition from the inside surface. Similar to other cargos, nanomaterials can be delivered to different cellular and extra-cellular destinations. Nanomaterials can reach degradative compartments (e.g., late endosomes or lysosomes), be recycled back to the extracellular milieu, be transported across cells (i.e., transcytosis), or reach different organelles (e.g., Golgi apparatus, mitochondria) (Sahay et al. 2010a).

Nevertheless, once inside the cell, the physicochemical characteristics that make nanomaterials so useful can also be the main reason they might be dangerous to cells, and at a higher level to human health. The high reactivity of nanomaterials can lead to toxicity via mechanisms that include induction of oxidative stress, inflammation, organelle dysfunction, and change of cellular morphology. With regard to size, the upper limit of any nanomaterial that can undergo internalization by nonprofessional phagocytic cells was thought to be about 150 nm. However, Gratton and coworkers showed that this upper-size limit needs to be reconsidered, since they demonstrated the internalization of nanoparticles of up to 3 μm by nonprofessional phagocytic cells (Gratton et al. 2008). Therefore, nanotoxicology should not be limited to nanomaterials in the range of 1–100 nm, since microparticles of up to 3 μm can also be internalized via endocytosis (Gratton et al. 2008), and theoretically, once inside the cell, their potential for causing harm is high (Zhao et al. 2011). To avoid these problems, nanomaterials must be engineered from either materials that are biocompatible, nontoxic, and biodegradable or those materials that have minimal toxic effects (Ai et al. 2011).

There has been growing concern with the potential health risks posed by nanoscale materials, and thus, a subdiscipline in this field has emerged, namely, *nanotoxicology* (Donaldson et al. 2004). Nanotoxicology is the study of the toxicity of nanomaterials. Nanotoxicology also deals with understanding the interactions between nanostructures and biological systems, that is, trying to elucidate the relationship between the physicochemical properties (e.g., size, shape, surface chemistry, composition, and aggregation) of nanomaterials and the toxic effects elicited at the cellular level (Oberdörster et al. 2005). The increase in the number of publications related to nanotechnology has drawn the attention of Federal agencies, which underscores the increased importance given to nanotoxicology. Furthermore, some studies have found that publications lack a critical review of the nanotoxicity of new nanomaterials, thereby presenting false-positive results. Thus, agencies have been conducting research that is expected to provide more reliable risk evaluations (Ai et al. 2011; Donaldson et al. 2004; Maynard et al. 2011; Oberdörster et al. 2005).

Although there is concern about the toxicity of new nanomaterials, not all of this can be evaluated in vivo. It has been estimated that a billion dollars and about 30–50 years are needed to conduct traditional in vivo studies on the nanomaterials currently under commercialization (Walker and Bucher 2009). As an alternative, there is a huge campaign in the scientific community to evaluate the potential toxic effects of nanomaterials in vitro. This would contribute not only to a faster development of this field but also to our knowledge about nanomaterial interactions at the molecular and cellular level, which could help us understand their interactions with living things and their environmental impact. Ultimately, the development of nanomaterials and their future commercial applications will be a challenge not only for companies but also for state regulatory agencies that must guarantee their safety for the work force, consumers, and the environment.

9.2 Cellular Endocytosis as a Mechanism of Nanomaterial Internalization

The plasma membrane is selectively permeable, determines the cellular boundary, is not static but dynamic, and has evolved several mechanisms to control the communication between the cytosol and extracellular environment. For example, oxygen, carbon dioxide, and small hydrophobic or nonpolar molecules are able to cross freely across the plasma membrane. Similarly, many small polar molecules, such as ions and amino acids, can get through the plasma membrane via active transport mediated by integral membrane protein pumps or ion channels. In contrast, the plasma membrane is highly impermeable to bigger and polar structures, such as nanomaterials. A mechanism that permits nanomaterial cellular internalization is endocytosis (Doherty and Mcmahon 2009). This mechanism is involved in many normal physiological processes, such as the uptake of extracellular nutrients, regulation of cell surface receptor levels, maintenance of cholesterol homeostasis, and antigen presentation. Moreover, many diseases and pathogenic conditions, including atherosclerosis and diabetes, are the result of abnormalities in endocytic processes. Even pathogens, including viruses, symbiotic microorganisms, and toxins, exploit endocytic pathways to gain entry into the cell. In addition, this ATP-dependent and well-coordinated cellular process, can mediate internalization of several kinds of nanomaterials. It starts with the invagination of the plasma membrane and leads to the internalization of the nanomaterial within an endocytic vesicle in the cytoplasm (Canton and Battaglia 2012; Sahay et al. 2010a).

Endocytosis, performed by professional phagocytes, such as macrophages, neutrophils, monocytes, and dendritic cells, is defined as phagocytosis, whereas pinocytosis is performed by virtually all other eukaryotic cell types (Doherty and Mcmahon 2009; Zhao et al. 2011). Pinocytosis can occur via different morphological and biochemical mechanisms, such as clathrin-mediated endocytosis (CME), caveolae-mediated endocytosis (CvME), macropinocytosis, and clathrin- and caveolae-independent endocytosis. The new pathways that are subclassified as clathrin and caveolae independent include Arf6-dependent, flotillin-dependent, Cdc42-dependent, and RhoA-dependent endocytosis (Canton and Battaglia 2012; Sahay et al. 2010a).

The study of the specific endocytic pathway that is used by nanomaterials is mechanistically important because it determines the intracellular fate of these nanomaterials, and thus is essential for determining their efficiency and possible toxicity. It is well established that this is a cell type-, material composition-, and concentration-dependent process (Hillaireau and Couvreur 2009; Zhao et al. 2011). Here we discuss the major endocytic pathways: CME, caveolin-mediated endocytosis, and macropinocytosis.

9.2.1 Clathrin-Mediated Endocytosis

CME is a key process for carrying out many fundamental physiologic functions in eukaryotic cells. This mechanism is involved in many cellular processes, including the uptake of nutrients, such as transferrin and riboflavin (Bareford et al. 2008; Gao et al. 2005), cholesterol incorporation into cells by low-density lipoprotein (LDL), and downregulation of cell signaling by internalization and degradation of receptors. The involvement of CME in several processes has made this a central topic of study in different fields, and therefore, CME is the best-understood endocytic pathway (González-Gaitán and Stenmark 2003; Hillaireau and Couvreur 2009). Furthermore, viruses seem to take particular advantage of this pathway to gain entry into host cells (Canton and Battaglia 2012; Mudhakir and Harashima 2009).

Clathrin forms a triskelion by combining three clathrin heavy chains and three light chains. This structure is associated with numerous adaptor proteins that help deform the cytoplasmic face of the plasma membrane, forming pits that are pinched off by the protein dynamin (Edeling et al. 2006; Hinrichsen et al. 2006; Marsh and McMahon 1999; Ungewickell and Hinrichsen 2007). Dynamin is a large GTPase responsible for membrane scission, and it is assisted by actin and myosin motor proteins. As the process continues, clathrin-coated vesicles (CCV) are formed in the cytoplasm after the clathrin triskelion helps form a mechanical scaffold on the vesicle surface, then the units are released and recycled back to form new vesicles. This process is very dynamic, and it is estimated that cultured cells take about 1 min to assemble CCV, and hundreds and up to a thousand CCV can be formed every minute (Marsh and McMahon 1999).

CME is particularly important for nanomaterial internalization, since nanomaterials primarily end up in lysosomes. Although some nanocarriers can use this as a mechanism to trigger the release of their contents, most cargo is completely degraded in the lysosomal compartment, a process especially critical for disposal of discarded genetic material. Researchers researchers are trying to outline generalizations about nanomaterial physicochemical properties and the type of endocytic pathway that mediates their processing, and for some structures it has been possible to draw some conclusions. This seems to be the case for lipoplexes, complexes formed by nucleic acids and liposomes. Several studies have suggested that these nanocarriers enter cells via CME (Rejman et al. 2005; Zuhorn et al. 2002). Similarly, different types of nanoparticles seem to be internalized by CME, such as solid lipid nanoparticles (Martins et al. 2012), polysaccharide cationic nanoparticles (Dombu et al. 2010), PLGA-based nanoparticles (Benfer and Kissel 2012), diamond nanoparticles (Faklaris et al. 2009), and silver nanoparticles (Greulich et al. 2011). Furthermore, professional phagocytes, such as macrophages, can use CME to internalize silver nanoparticles (Kim and Choi 2012).

Several inhibitors and inhibitory mechanisms have been used to study and characterize this endocytic process, such as hypertonic sucrose, K^+ depletion, and chlorpromazine (Heuser and Anderson 1989; Madshus et al. 1987; Wang et al. 1993). In addition, the inhibition of actin polymerization with latrunculin A and cytochalasin D, which compromises the internalization via CME, has also been investigated (Sahay et al. 2010a). Furthermore, macromolecules that are normally internalized via CME can be used as endocytic markers to study this process further. The most often used markers are the iron transport protein transferrin and LDL (Sahay et al. 2010a).

9.2.2 Caveolae-Mediated Endocytosis

Caveolae are flask-shaped invaginations of the plasma membrane, which are found in domains enriched with cholesterol and sphingolipids (Bastiani and Parton 2010; Lajoie and Nabi 2007; Nichols 2003). CvME is involved in several biological processes, including transcytosis, cell signaling, lipid regulation, and also many diseases, such as cancer, diabetes, and viral infections (Hayer et al. 2010). CvME is abundant in some cell types, such as muscle cells, endothelial cells, fibroblasts, and adipocytes, and rare in others, such as neurons and leukocytes (Bastiani and Parton 2010; Doherty and Mcmahon 2009).

The conclusive characteristic of caveolae is the presence of caveolin (Cav) proteins. In mammalian cells, the caveolin gene family is comprised of three members: caveolin-1 (Cav-1), caveolin-2 (Cav-2), and caveolin-3 (Cav-3). Cav-1 and Cav-2 are widely found in different body tissues, while Cav-3 seems to be found only in muscles cells. Cav-1 is anchored to the plasma membrane in domains rich in cholesterol, sphingomyelin, saturated fatty acids, and glycosphingolipids (e.g., GM1). These domains can also include glycosylphosphatidylinositol (GPI)-linked proteins (Doherty and Mcmahon 2009; Parton and Simons 2007). Although Cav-1 is necessary for biogenesis of caveolae (Doherty and Mcmahon 2009), its ability to induce membrane curvature is currently under debate. It has been suggested that new cavin proteins form a complex that can stabilize the membrane curvature to produce the classic flask shape of caveolae, therefore regulating caveola structure and function (Hill et al. 2008). In addition, CvME also relies on specific adaptor proteins to mediate this process, and similar to CME, it relies on dynamin to pinch off the budding vesicle. Furthermore, some viruses can also exploit CvME to gain entry into cells, such as Simian virus 40, Shiga toxin, and the cholera toxin B subunit; the latter is used as a marker for CvME. CvME is also responsible for the uptake of nutrients, such as folic acid, albumin, and cholesterol (Hillaireau and Couvreur 2009).

For many years it was thought that the CvME pathway could bypass lysosomal enzymatic degradation. For this reason, this pathway was believed to be beneficial for cellular delivery of drugs sensitive to lysosomal degradation, especially the delivery of peptides, proteins, and nucleic acids among others. Hypothetically, cargo internalized by CvME would lead to caveosomes, special endosomal compartments with neutral pH (Khalil et al. 2006; Pelkmans and Helenius 2002; Pelkmans et al. 2001; Sahay et al. 2010a). Recently, the same authors who described this structure have reported that caveosomes are actually an artifact present in cells overexpressing caveolin (Hayer et al. 2010; Parton and Howes 2010). These findings call for a careful reassessment of the data already published regarding intracellular localization and degradation of nanomaterials. Nonetheless, some researchers found that CvME does avoid or delay lysosomal degradation. For example, highly compacted DNA nanoparticles internalized by CvME exhibited negligible colocalization with LysoTracker, a lysosomal marker, for up to 4-h post-incubation in human bronchial epithelial (BEAS-2B) (Kim et al. 2012). As discussed before, avoiding the lysosomal compartment can be beneficial for the delivery of nucleic acids. Del Pozo-Rodríguez and coworkers observed that solid lipid nanoparticles decorated with proline-rich peptides shifted their internalization pathway from CME to CvME, resulting in higher transfection efficiency in both HEK293 and ARPE-19 cells (del Pozo-Rodríguez et al. 2009). However, recent findings have demonstrated that the CvME pathway can direct nanomaterials to lysosomes (Ekkapongpisit et al. 2012; Sahay et al. 2010b). Given these studies, it is apparent that more work is required to understand nanomaterial trafficking trough CvME and especially the functions associated with caveolin and the fate of nanomaterials in caveolar trafficking (Parton and Howes 2010).

9.2.3 Macropinocytosis

Macropinocytosis was the first endocytic pathway described, although its biological roles have only recently been explored. Macropinocytosis seems to be important for cell motility; therefore, it was studied in the context of cancer progression and metastasis (Lim and Gleeson 2011). It is also important for some specialized cell types, such as professional phagocytes, where it plays significant roles in immune system processes, such as antigen presentation and clearance of apoptotic bodies and necrotic cells. It is a constitutive process in professional phagocytes, but in nonprofessional phagocytes, it has to be induced in response to growth factors, such as epidermal growth factor (Jones 2007; Mercer and Helenius 2009). Interestingly, macropinocytosis can also be induced by bacteria and viruses; thus, these organisms can take advantage of this endocytic pathway to enter host cells. Other pathogens can also enter host cells via macropinocytosis, such as protozoa and prions (Barrias et al. 2012; Kerr and Teasdale 2009; Lim and Gleeson 2011; Mercer and Helenius 2009).

Macropinocytosis can be defined as the bulk uptake of large amounts of extracellular material. It is characterized by the ruffling of the plasma membrane and is induced by global activation of the actin cytoskeleton. In this manner, the entire cell surface starts ruffling or blebbing, and these protrusions fold back and fuse with the plasma membrane. This fusion results in large endocytic vacuoles, called macropinosomes that are heterogeneous in size and morphology (Falcone et al. 2006; Jones 2007). Macropinosomes are easy to distinguish from other vesicles formed during pinocytosis, since they are substantially larger (0.5–10 μm), non-coated structures that shrink during their intracellular maturation. For some time, it was believed that macropinosome formation was a dynamin-independent process, but in some cells this does not seem to be the case (Orth and McNiven 2006). Furthermore, a macropinocytosis-like pathway that is dynamin dependent was recently described for the internalization of quantum dots (Iversen et al. 2012).

Although the definition of macropinocytosis gives the impression of a random process, it is actually a very complex and well-coordinated cellular event (Falcone et al. 2006). External stimuli usually initiate the process through the activation of receptor tyrosine kinases (RTKs), which in turn activate key signaling proteins, such as Ras, PLC, and PI 3-kinase. These proteins coordinate the dynamic changes in actin filaments and generate plasma membrane ruffling (Mercer and Helenius 2009; Swanson and Watts 1995). Additionally, Rho family members and their upstream effectors are also required for the actin polymerization machinery and ruffling (Jones 2007).

Given their nature, macropinosomes can be identified through the use of fluid phase markers, such as dextrans, horseradish peroxidase, and lucifer yellow. Inhibitors can also be used to help examine this process, such as cytochalasin D, which inhibits actin polymerization, and amiloride; however, its underlying mechanism remains unknown (Falcone et al. 2006; Kerr and Teasdale 2009; Lim and Gleeson 2011; Swanson and Watts 1995).

In addition to internalizing viruses, bacteria, and others pathogens, macropinocytosis can also be used to internalize nanomaterials. Macropinocytosis has been suggested as an effective endocytic pathway for drug delivery into cells. Many arginine-rich cell-penetrating peptides such as octo-arginine and human immunodeficiency virus transactivator of transcription (TAT) peptides exploit the process of macropinocytosis to effectively deliver peptides and proteins into cells (Kaplan et al. 2005; Nishimura et al. 2008). Mesoporous silica nanoparticle rods were more effective in the delivery of chemotherapeutic agents in HeLa cells when endocytosed via macropinocytosis (Meng et al. 2011). Particle size is an important determinant for internalization by macropinocytosis, as demonstrated by Vollrath and coworkers, who used using nanoparticles based on poly-(methyl methacrylate) in Hela cells. These authors found that CME is the predominant pathway for internalization of smaller nanoparticles (<200 nm), whereas macropinocytosis is the main pathway for internalization of larger particles (>300 nm) (Vollrath et al. 2012). Nevertheless, small particles can also be internalized by macropinocytosis. Iversen and coworkers showed that even small particles (30 nm), i.e., ricin-coupled quantum dot nanoparticles, are internalized by Hela cells via macropinocytosis (Iversen et al. 2012).

9.2.4 Additional Internalization Pathways for Nanomaterials

Recently, the study of nanomaterial internalization has attracted great attention (Canton and Battaglia 2012; Doherty and Mcmahon 2009; Hansen and Nichols 2009; Hillaireau and Couvreur 2009; Sahay et al. 2010a; Scita and Di Fiore 2010; Verma and Stellacci 2010; Zaki and Tirelli 2010; Zhao et al. 2011). It is important to determine not only the effectiveness of a new nanomaterial but also its cytotoxicity. One must be careful to interpret data before deciding the specific endocytic pathway involved. In addition to CME, CvME, and macropinocytosis, other new endocytic pathways have been described, such as RhoA-, CDC42-, and flotillin-mediated endocytosis. These pathways have already been described for some nanomaterials, and it is expected more pathways will be described (Kasper et al. 2013). Some particles were described as not following the classical pathways (clathrin, caveolin, and macropinocytosis). For example, carboxyl-modified fluorescent polystyrene nanoparticles were internalized by HeLa cells via a nonclassical pathway (Lai et al. 2007). These particles may have taken one of the recently described pathways or a new and yet undescribed pathway. Actually, cells can use more than one endocytic pathway to internalize nanomaterials. For example, dendrimers were internalized via both CvME and macropinocytosis, and in the same paper, the authors described a different formulation of dendrimers that were internalized via CvME, CME, and macropinocytosis (Saovapakhiran et al. 2009). In addition, new mechanisms of nanocarrier-cell surface interactions have been described, such as actin-rich filopodial extensions that are responsible for a majority of the internalization of lipids and polymers carrying DNA into Hela cells (Rehman et al. 2012).

Endocytic markers and inhibitors can be very useful in determining the endocytic internalization pathway of new nanomaterials, however data obtained using these methods should be interpreted with caution. Although the cholera toxin B subunit has been widely used as a marker for CvME, it can also be endocytosed by several pathways, including CME (Torgersen et al. 2001). Furthermore, because the fluorescently labeled form of the toxin can bind to numerous nanomaterials even before being endocytosed, it can lead to false-positive colocalization. Endocytic inhibitors can also present problems, such as toxic effects, and they can be non-specific. For instance, methyl-β-cyclodextrin is commonly used to remove cholesterol from the plasma membrane and disturb CvME, however it can also disturb CME, depending on the concentration used and cell type (Rodal et al. 1999; Subtil et al. 1999). These examples are not intended to exhaustively cover the subject, but to highlight the complexity behind the mechanisms used by cells to internalize nanomaterials. They also suggest that care should be taken in performing experiments and analyzing results to properly determine the specific endocytic pathway used for internalization of nanomaterials by specific cell types. The reader is referred to these articles, which describe specific pitfalls and limitations of techniques for determining the endocytic pathways used by nanomaterials (Vercauteren et al. 2010; Zaki and Tirelli 2010).

New techniques have been developed to study endocytosis and intracellular trafficking of nanomaterials. Recently a novel, highly sensitive, and quantitative technique was developed to elucidate the precise nature of the membrane compartments through which nanomaterials are routed following internalization into cells. The authors used a rapid, multicolor, live-cell, confocal fluorescence microscopy approach that captures images in three dimensions and thereby provides details about the intracellular interactions between nanoparticles and the intracellular components (Sandin et al. 2012). These techniques can add new insight to the rare and fast events by which nanomaterials interact with the intracellular environment.

9.3 Intracellular Trafficking of Nanomaterials

After examining the internalization pathways for nanomaterials, the next areas to be addressed are the intracellular trafficking and localization of nanomaterials. These areas are critical for optimization and characterization of nanomaterials aimed for intracellular-targeted drug delivery. Currently, the impact of nanotechnology on medicine is already significant, since many nanomaterials are approved for clinical use. However, most nanomaterials can only improve the therapeutic index of drugs by reducing their toxicity or enhancing their efficacy. For decades to come, it is expected that nanotechnology will make an enormous impact on medicine and human healthcare; however to achieve this, the next generation of nanomaterials will have to improve the ability to reach specific tissues, cells, and intracellular targets. These advances would improve the nanocarriers efficacy or reduce their toxicity. This need can explain the great attention that has been given to the rational design of nanomaterials and strategies to properly target them to subcellular compartments (Bareford and Swaan 2007; Chou et al. 2010; Murakami et al. 2011; Petros and DeSimone 2010; Prokop and Davidson 2008; Rajendran et al. 2010; Torchilin 2006).

Regardless of the internalization pathway used by cells for taking up nanomaterials, the first compartment that receives the cargo is the early endosome (EE). Currently, EEs are recognized as the main sorting station in the endocytic pathway (Huotari and Helenius 2011). In fact, one of the first steps to confirm that nanomaterials have entered cells through endocytosis is to demonstrate that they are in the EE. The EE is characterized by the presence of the early endosomal antigen-1 (EEA1) protein, the most widely used marker for EE, or Rab5, a protein member of the Ras superfamily of small Rab GTPases (Canton and Battaglia 2012; Scita and Di Fiore 2010). Subsequently, nanomaterials follow the classical intracellular trafficking pathway, meaning that they can be recycled back to the plasma membrane, delivered to the trans-Golgi network or across the cell, or undergo degradation in the lysosomes (Huotari and Helenius 2011; Jovic et al. 2010). Here we will discuss some examples of intracellular trafficking and their consequences. Lysosomal degradation and the recycling of nanomaterials to the plasma membrane are discussed in more detail in the next section.

Among the different fates that nanoparticles can face inside the cell, avoiding lysosomal degradation is the most important for achieving an effective therapeutic effect. Ming and coworkers demonstrated that an antisense oligonucleotide conjugated with a bombesin peptide bypassed lysosomal degradation and accumulated in the trans-Golgi network, representing an efficient system for intracellular delivery of oligonucleotides (Ming et al. 2010). In agreement with these findings, Chang and coworkers showed an accumulation of gold nanoparticles within the endoplasmic reticulum and Golgi apparatus of B16F10 melanoma cells. The authors found that combining gold nanoparticle treatment with radiotherapy resulted in an increase in the apoptotic potential of the therapy, suggesting that gold nanoparticles can improve the clinical outcome of melanoma radiotherapy (Chang et al. 2008). Fichter compared the efficiency of two gene delivery systems, glycofect and linear polyethylenimine polymer, in ER H9c2 rat cardiomyoblasts. They found that glycofect performed better than linear polyethylenimine polymer because the former appeared to bypass lysosomes and was partially taken up in the Golgi apparatus (Fichter et al. 2013). These examples underscore the importance of avoiding lysosomal degradation to enhance the therapeutic efficacy of nanomaterials.

The intracellular trafficking of nanomaterials in polarized cells can lead to a special form of mobilization. In these cells, nanomaterials can end up on the opposite membrane in a process known as transcytosis (Tuma and Hubbard 2003). This route is important for the delivery of substances across capillary endothelial cells, especially across the blood brain barrier (BBB). The BBB is important because it separates the circulating blood from the brain extracellular fluid, posing a major hindrance to the successful delivery of therapeutics to the central nervous system (Agarwal et al. 2009; Pardridge 2007; Tiwari and Amiji 2006). In an attempt to overcome this barrier, Chang and coworkers decorated PLGA nanoparticles with transferrin, which resulted in an effective targeting to CvME and a 20-fold increase accumulation across the BBB (Chang et al. 2009). Harush-Frenkel and coworkers showed that both cationic and anionic polyethylene glycol-polylactide nanoparticles (89.8 ± 4 and 96.4 ± 3 nm, respectively) were internalized via CME into MDCK cells. Interestingly, the cationic nanoparticles efficiently crossed the cells via transcytosis and accumulated at the basolateral membrane, whereas the anionic nanoparticles accumulated in the degradative lysosomal compartments (Harush-Frenkel et al. 2008). This interesting finding was not explored further by these authors, yet merits additional examination. In a biologically inspired example, Georgieva and coworkers decorated nanoparticles with prion peptides to target CvME, and they successfully improved the ability of these nanoparticles to cross the BBB; prion proteins are known to bind to specific receptors and mediate transcytosis from the apical surface of brain endothelial cells (Georgieva et al. 2011). Collectively, these studies point out that in vitro studies can be helpful in understanding and developing nanomaterials capable of solving intricate problems, such as crossing the BBB.

In some cases, the objective goes beyond targeting traditional intracellular sites. In these situations, different strategies must be used to reach these compartments, such as the nucleus. In one such strategy, Chen and coworkers pointed out that the nucleolin protein present at the surface could work as a DNA receptor and shuttle DNA nanoparticles from the membrane into the nucleus (Chen et al. 2008). Later, the same group demonstrated the feasibility of this mechanism in vivo by showing the expression of nucleolin on the apical surface of mouse airway epithelia, suggesting this could be a good target for nonviral gene delivery (Chen et al. 2011). Using a similar approach, Dam and coworkers used a DNA aptamer (AS1411) with high binding affinity to nucleolin to take advantage of the shuttling properties of nucleolin to efficiently target gold nanostars into the nucleus of cancer cells. They found that the nanoconstructs were close to the nuclear membrane and induced changes in nuclear shape. The authors suggested that these changes interfered with nuclear functions, which in turn increased caspase 3 and 7 activity (apoptosis), and also decreased cell viability (Dam et al. 2012). According to Wang and coworkers, titania nanotubes can cross the nuclear membrane, thereby reaching the nucleus of mouse neural stem cells. Given this finding, the authors suggested that these nanotubes can be used for delivery of DNA-targeting drugs (Wang et al. 2010). Although these strategies show promise for the delivery of cargo to the nucleus, care should be taken to evaluate potential toxic side effects associated with the delivery of these nanomaterials to the nucleus.

The cytotoxic effect of nanomaterials is the focus of the next discussion. It is well documented that surface modifications of nanomaterials can affect both their internalization pathway and their intracellular fate. An illustrative example of this behavior was demonstrated using cerium oxide nanoparticle derivatives. Their neutral derivatives were localized mostly in the cytoplasm of cells and hence did not elicit cytotoxicity to cancer cells. Conversely, the negative and positive derivatives that were localized in lysosomes did exhibit cytotoxic effects. The authors explained that the low pH of the lysosomes activated the oxidation of the nanoceria nanoparticles, thereby sensitizing tumors toward radiation therapy (Asati et al. 2010). In another example, several quantum dots were synthesized to extend their intracellular retention time, and the authors found that oxalate-transferrin quantum dots were able to delay their cellular removal both in vitro and in vivo. The authors suggested that these modified quantum dots could have a diverse range of applications, including diagnostic imaging, improved payload release, and decreased nanotoxicity (Wu et al. 2013). Although surface modifications of nanoparticles can be an effective strategy for improving nanomaterial uptake, this approach should be carefully rationalized. After nanoconjugation, the new nanomaterial can be internalized and processed by the cells in a different way than the naked nanomaterial or the free ligand. Indeed, Iversen and coworkers demonstrated that ricin alone was internalized by both dynamin-dependent and dynamin-independent endocytic pathways; however, the nanoconjugated form (ricin quantum dots) did not use these pathways (Iversen et al. 2012).

Since surface modifications of nanomaterials are a common strategy for controlling the endocytic pathway and consequently the intracellular fate of nanomaterials, this topic will be discussed in more detail in this section. Furthermore, recent new methodologies and findings have afforded more details on the intracellular trafficking of nanoparticles. Wang and coworkers have developed an interesting methodology to study intracellular trafficking of nanoparticles. Using dual-color nanoparticle pairs to measure size distribution within caveolae and assembly dynamics in living endothelial cells, they showed that in one caveolae, it is possible to find up to three 20-nm nanoparticles or two 40-nm nanoparticles (Wang et al. 2009). Details about how cells control intracellular trafficking were found using H89, an inhibitor of protein kinase A (PKA). Rehman and coworkers demonstrated that the modulation of PKA activity strongly affected the intracellular trafficking or CME internalization of both poly- and lipoplexes. The authors found that this inhibition channeled the lipoplexes to non-degradative compartments, resulting in a 2–3-fold increase in the transfection efficiency of branched polyethylenimine polymers in Hela cells (Rehman et al. 2011). These findings illustrate new approaches that can be used to help us better understand the intracellular trafficking of nanoparticles to achieve better outcomes. These advances represent the types of approaches that will be necessary to further improve the field of nanotechnology and to achieve its expected impact on medicine.

9.4 Cellular Metabolism of Nanodevices: Biodegradation and Elimination

Ideally, after nanomaterials enter cells and play their biological role, such as delivering drugs or genes, cells should be able to metabolize and eliminate these nanomaterials. Additionally, the metabolites generated should not be toxic. Indeed, nanomaterial metabolism and elimination should be carefully considered, since an optimal balance between lack of toxicity and therapeutic effect can be difficult to achieve. Therefore, studying the intracellular metabolism of nanomaterials is crucial for understanding their overall effect inside cells.

Typically, minutes after internalization, the maturation of the EE starts and the nanomaterials are sorted to different intracellular destinations, eventually ending up in lysosomes. This process begins with the acidification of the EE to a pH of 6.8–6.1, continues to a late endosomes (LE) with a pH of 6.0–4.8, and eventually reaches the lysosomal compartment equipped with degradative enzymes (e.g., proteases, esterases, phosphatases, nucleases, and lipases) and a pH of around 4.5 (Huotari and Helenius 2011). This lysosomal compartment is the limiting step for effective biological-based therapy, such as the delivery of peptides, proteins, or nucleic acids (Varkouhi et al. 2011; Won et al. 2009). For example, del Pozo-Rodríguez and coworkers found that solid lipid nanoparticles (281 $\pm$ 69 nm) used as a gene delivery system were inefficient because CME targets its DNA cargo to lysosomal degradation in ARPE-19 cells (del Pozo-Rodríguez et al. 2008). Although CME is well known for targeting its contents directly to lysosomes, cargo from CvME and macropinocytosis eventually also reach lysosomal compartments (Dharmawardhane et al. 2000; Ekkapongpisit et al. 2012; Sahay et al. 2010a, b). In most cases, after internalization, nanomaterials must escape from endosomes to prevent cargo degradation within lysosomes. Therefore, several strategies have been used to improve endosomal escape (Varkouhi et al. 2011). Caracciolo and coworkers solved this problem by using a protamine/DNA complex coated with a lipid envelope made of cationic 1,2-dioleoyl-3-trimethylammonium propane (DOTAP). This ternary complex was advantageous in terms of endosomal escape and DNA release, resulting in a very efficient gene delivery system for different cell lines (Caracciolo et al. 2011). Another strategy is to take advantage of the drop in pH during endosomal maturation to trigger the release of cargo into the cytoplasm. Wang and coworkers developed gold nanoparticles that can release their content in an acidic environment; these particles efficiently delivered doxorubicin and inhibited the growth of multidrug-resistant MCF-7/ADR cancer cells (Wang et al. 2011a). Similarly, hydroxide nanoparticles successfully delivered methotrexate to cells and thus enhanced drug efficacy due to their anion exchange capacity in acidic pH (Oh et al. 2006). Although some strategies can be used to escape endosomes, lysosomal degradation still remains the limiting step in the efficient delivery of biological cargo.

Another cellular trafficking strategy is known as exocytosis. Exocytosis is used to expel nanomaterials from cells when nanomaterials resist destruction, even after exposure to numerous degradative enzymes, or when cells are exposed to large amounts of nanomaterials and thus need to avoid lysosomal overloading. Exocytosis is a biological process that is highly conserved across species and cell types. It is a process by which intracellular vesicles fuse with the plasma membrane and vesicle contents are released into the extracellular space. It is the major intracellular route for the delivery of proteins and lipids to the plasma membrane, making this process vital for many physiological processes, including membrane expansion during cell division, cell growth and cell migration, establishment of cell polarity, cellular communication, neurotransmission, and the secretion of hormones and cytokines (Ory and Gasman 2011).

Recently, exocytosis was linked to the chronic cytotoxic effects of nanomaterials by regulating their intracellular retention times. Typically, nanomaterials undergo internalization through endocytosis, then the endosomes containing the nanomaterials are transported by dyneins, microtubule-dependent motor proteins. Afterwards, a fraction of internalized nanomaterials can be found in lysosomes and/or transported to the perinuclear region by kinesins, another type of motor proteins. Finally nanomaterials can be sent out of the plasma membrane into the extracellular space via exocytosis. Bae and coworkers have described the same pathway for lanthanide-doped upconverting nanoparticles in HeLa cells, whereby nanoparticles avoid lysosomal degradation and result in high photostability and low cytotoxicity (Bae et al. 2012). Also in HeLa cells, Jiang and coworkers have found a similar pathway for zwitterionic quantum dot nanoparticles. However, these authors found a significant fraction of their nanoparticles in lysosomes, while the remaining fraction was actively transported to the cell periphery and exocytosed with a half-life of 21 min (Jiang et al. 2010). This could indicate that cells try to avoid lysosomal overload, which eventually could result in leaking of lysosomal contents into the cytoplasm, resulting in cytotoxic effects. Even plant cells use exocytosis as a strategy to eliminate single-walled carbon nanotubes to minimize cytotoxicity (Serag et al. 2011). Johnston and coworkers used cellular imaging to show that polystyrene nanoparticles (20 nm) accumulated within distinct areas between adjacent hepatocyte cells, which the authors hypothesized represented bile canaliculi (Johnston et al. 2010). Together these studies point out that exocytosis is a universal process used by eukaryotic cells to minimize cytotoxicity.

As discussed before, surface modifications of nanomaterials can dramatically influence their biological fate inside cells and their targeting to the extracellular space. Gold nanoparticles decorated with different peptides followed different endocytic pathways and thus had a different exocytic profile in human endothelial cells (HUVECs). The authors showed that nanoparticles coupled to the peptide KPRQPSLP were reinternalized by cells after 4 h, whereas particles coupled to the peptide KATWLPPR were progressively exocytosed for a period of 6 h. Interestingly, these gold nanoparticles were internalized, processed by the cells and exocytosed, while still keeping their physicochemical properties (size and zeta potential), showing that these are nonbiodegradable nanoparticles (Bartczak et al. 2012). This finding reinforces the idea that nanoparticle surface modifications alter the way nanoparticles interact with the biological milieu, which subsequently determines their fate.

Another nanomaterial property that influences exocytosis is nanoparticle size. Smaller gold nanoparticles are more quickly exocytosed than larger ones. Dombu and coworkers found that cholesterol depletion in human bronchial epithelial cells increased endocytosis of polysaccharide cationic nanoparticles. Two hypotheses were proposed by the authors to explain the mechanism mediating these effects: (1) cholesterol depletion could lead to caveolae disruption, which in turn would trigger a compensatory effect via CME, or (2) cholesterol depletion could compromise nanoparticle exocytosis. To address this hypothesis, the authors followed nanoparticle exocytosis for up to 240 min in the presence and absence of filipin, which specifically binds to cholesterol. Surprisingly, filipin treatment diminished nanoparticle exocytosis, indicating that exocytosis of nanoparticles in these cells occurred via a cholesterol-dependent pathway (Dombu et al. 2010). Although the findings clearly demonstrate that exocytosis is an important mechanism in nanomaterial processing, further research is needed to determine the underlying mechanism of this process.

Until recently, it was believed that exocytosis of nanoparticles was independent of the permeability glycoprotein (P-gp) pathway (Jiang et al. 2010; Jin et al. 2009; Panyam and Labhasetwar 2003; Wang et al. 2011a). Although many nanoparticles have been developed to try to overcome P-gp-mediated drug efflux, a recent finding showed that nanoparticles can also be substrates for this transporter. Al-Hajaj and coworkers found that functionalized quantum dots (8–10 nm in size) were eliminated from human liver Hep G2 cells and kidney Hek 293 cells via the P-gp transporter (Al-Hajaj et al. 2011). Here it is important to highlight that, the particles' size seems to play an important role in this process; therefore, it is highly unlikely that larger nanomaterials experience the same elimination pathway.

9.5 Mechanisms of Nanomaterial Cytotoxicity

Nanomaterials take advantage of special properties afforded at the nanoscale, such as large surface area, chemical reactivity, physical absorption ability, and quantum size effects, to perform unique functions. Nevertheless, once inside cells, these properties can bring undesirable effects, such as cytotoxicity. These effects can cause local damage that can be repaired by the cells, and they can also get out of control and compromise tissue or organ function, ultimately compromising human life. Thus, the fast development of nanotechnology over the last decades has raised concerns about the effects of these new nanomaterials on human health. Therefore, the evaluation of these effects has culminated into the creation of a new subdiscipline in nanotechnology, namely, nanotoxicology (Donaldson et al. 2004). The establishment of this new field has been marked by the publication "Nanotoxicology: an emerging discipline evolving from studies of ultrafine particles" (Oberdörster et al. 2005) and the introduction of the *Nanotoxicology Journal* in 2007. After this, much attention has been paid to this topic, and several other reviews have since been published (Ai et al. 2011; Alkilany and Murphy 2010; Buzea et al. 2007; El-Ansary and Al-Daihan 2009; Hoet et al. 2004; Jian et al. 2012; Nel et al. 2006). Although studies in this field already existed before it was formally established, it is still not completely equipped with all the necessary scientific tools to properly evaluate all the possible toxic effects nanomaterials can elicit. Therefore, old techniques are being adapted, and new ones have been introduced to help evaluate the hazards and risks associated with nanomaterials (Hillegass et al. 2010; Hussain et al. 2009; Lai 2012; Lai et al. 2012; Love et al. 2012b; Marquis et al. 2009).

Nanomaterial toxicity can emerge from undesirable interactions in the cellular milieu, which can be driven by the physicochemical properties of the nanomaterials, such as retention time inside the cell, surface properties, and toxic metabolites. These undesired interactions can influence a number of cellular events that can lead to cytotoxic effects, such as morphological and structural changes (plasma membrane damage, alterations in cell morphology, and cytoskeleton defects), genotoxicity (gene-expression alterations, DNA damage, micronuclei formation, chromosomal aberrations), and biochemical alterations (oxidative stress, lipid peroxidation, Ca^{2+} release, caspase activation), which in turn can trigger different cellular responses (cell-cycle and proliferation irregularities, inflammatory cytokine production, diminution in mitochondrial function, activation of cell signaling pathways, autophagy, apoptosis, and cell death) (Lai 2012;

Love et al. 2012b; Soenen et al. 2011). Some examples of the most common nanomaterial toxic effects are discussed below.

9.5.1 Morphological and Structural Changes

The plasma membrane represents the first site of interaction for nanomaterials and cells and the first barrier that nanomaterials must overcome to reach their intracellular target. Although the main mechanism by which nanomaterials enter cells is by endocytosis, as already discussed, in some cases nanomaterials can apply a physical stress on the plasma membrane and disrupt it. A disruption in the integrity of the plasma membrane can directly compromise its role as a barrier, leading to intracellular leakage and cell death. For silver nanoparticles, the damage to the cell membrane was reported to be time and concentration dependent (Hussain et al. 2005; Mukherjee et al. 2012). Silver nanoparticles together with other metallic nanoparticles (gold and platinum) were also harmful to bacteria and fungi, basically by disturbing the bacterial wall or fungal membrane (Chwalibog et al. 2010). Silica nanoparticles damaged the plasma membrane and also membranes inside the cells, i.e., the mitochondrial membrane (Sun et al. 2011). This toxic effect appears to be related to particle size; smaller particles seem to have greater potential to cause damage to the plasma membrane. Kasper and coworkers found that small amorphous silica nanoparticles (30 nm) caused greater damage to the plasma membrane of lung epithelial cells (H441) when compared to larger ones (70, 300 nm) (Kasper et al. 2013). Similarly, Liu and coworkers showed that silver nanoparticles damaged plasma membranes at lower concentrations ($\geq$25 μg mL^{-1}) when compared with micro-sized silver particles (100 μg mL^{-1}) (Liu et al. 2011).

Nanomaterials can also disturb the cellular cytoskeleton. This deregulation can compromise chromosome segregation and cytokinesis, inhibiting cell division. Morphological changes in cells exposed to nanoparticles may be due to interferences with the structure and function of the actin cytoskeleton (Asharani et al. 2009). This concept was confirmed through specific staining of the actin cytoskeleton with rhodamine-labeled phalloidin. Gold nanoparticles induced aggregation and formation of dot-like structures of actin filaments in A549 cells (Wang et al. 2011b). Soenen and coworkers found that the actin cytoskeleton was affected by gold nanoparticles at lower levels (50 nM) than that needed to disturb tubulin networks (100 nM), suggesting that actin fibers are more sensitive to nanoparticle-induced deformations (Soenen et al. 2012).

9.5.2 Genotoxicity

Genotoxicity has raised concerns about the safety of nanomaterials. Manufactured nano-/microparticles, such as fullerenes, carbon black, and ceramic fibers, irrespective of their size, induced genotoxic effects in A549 cells (Totsuka et al. 2009). Several reports have also described the genotoxic effect of silver nanoparticles. They appear to cause a wide variety of DNA damage, including DNA double-strand breaks, chromosomal aberrations, chromosomal fusions, and chromosomal fragmentation (AshaRani et al. 2009; Asharani et al. 2009; Kim et al. 2010; Liu et al. 2011). However, there are also particles that display low or no genotoxicity, as demonstrated by Pierscionek and coworkers when they used cerium oxide nanoparticles (5 and 10 μg mL^{-1}) and found that they did not cause any DNA damage or chromosomal changes in cultured eye lens epithelial cells (Pierscionek et al. 2009). Evidence from recent published reports suggests that nanomaterials are involved in mammalian mutagenesis and, possibly, carcinogenesis. Recently, Ng and coauthors published a review covering the latest findings on nanomaterial genotoxicity and the methodologies used in these studies (Ng et al. 2010).

9.5.3 Biochemical Alterations

The generation of reactive oxygen species (ROS) seems to be a central mechanism by which most nanomaterial toxicity is mediated (Markovic et al. 2007; Walker and Bucher 2009). Nanomaterials can generate ROS in different situations and in different cellular environments, either inside cellular compartments or outside in the cytoplasm. After being internalized, the accumulation of nanomaterials in lysosomes can contribute to ROS production. Chen and coworkers showed that iron oxide nanoparticles induced cytotoxicity in U251 cells when entrapped in lysosomes. The iron oxide nanoparticles catalyzed the production of hydroxyl radicals from H_2O_2 via peroxidase-like activity. Surprisingly, the same nanoparticles, when present in the cytosol, decomposed H_2O_2 through catalase-like activity (Chen et al. 2012). Some authors take advantage of the ROS production mediated by nanomaterials to fight cancer cells. Biologically synthesized silver nanoparticles have shown anticancer properties in Hela cells, and increases in intracellular ROS levels seem to play a key role in mediating this effect (Jeyaraj et al. 2013). Similarly, starch-coated silver nanoparticles were toxic to Hela cells by significantly increasing hydrogen peroxide and superoxide production (AshaRani et al. 2009). Sanpui and coworkers reduced silver nanoparticle side effects by impregnating them in chitosan-based nanocarriers. Using a considerably lower concentration of silver nanoparticles (330 ng mL^{-1} at IC_{50}), they efficiently induced apoptosis through ROS generation in human colon cancer cells (HT 29) (Sanpui et al. 2011). Although most publications report that the toxic effects of nanoparticles are mediated by generation of ROS, some particles also show ROS-scavenging properties. For example, cerium oxide nanoparticles can protect biological tissues against radiation-induced damage by protecting them from H_2O_2-induced cell damage (Karakoti et al. 2009). In addition to their free-radical scavenger activity, cerium oxide nanoparticles also increased the expression of superoxide dismutase 2 (Colon et al. 2010). These findings suggest that even nanomaterials with seemingly toxic effects can have useful applications. Therefore, in addition to evaluating the cytotoxic effects of nanomaterials, the controlled production and choice of nanomaterial for performing a specific task are fundamental, since nanomaterial properties and interactions with the cellular environment can dramatically change their behavior.

9.5.4 Cellular Response

The nanomaterial toxic effects discussed here rarely occur alone. In fact a nanomaterial can trigger a multitude of events inside the cell. For example, the generation of ROS triggered by nanomaterials is commonly related to a rise in oxidative stress, leading to cell death, i.e., apoptosis. This toxic profile has been described for silver nanoparticles (Mukherjee et al. 2012) and silica nanoparticles (Sun et al. 2011). In addition, amorphous silica nanoparticles and titanium dioxide nanoparticles can induce cellular inflammatory responses (Kasper et al. 2013; Park et al. 2008). For example, Lunov and coworkers revealed the complex underlying mechanism of toxicity for amino-functionalized polystyrene nanoparticles in human macrophages. Specifically, nanoparticle internalization induced lysosomal rupture and leakage of active cathepsin B to the cytosol, which in turn elicited mitochondrial damage and production of ROS. Subsequently, the accumulation of mitochondrial ROS led to oxidation of the redox-active thioredoxin (TXN) protein that plays an important role in oxidative stress. Meanwhile, thioredoxin-binding protein (TXNIP) was released, which resulted in inflammasome activation and IL-1β production (Lunov et al. 2011). This study illustrates the complexity of cellular responses to nanomaterials and the importance of examining nanomaterial toxicity using different mechanisms.

Autophagy has also emerged as a mechanism to regulate nanomaterial toxicity (Stern et al. 2012). The macroautophagy pathway (herein autophagy) is a highly conserved biological process that involves the sequestration of proteins, lipids, and organelles, followed by their degradation within double-membrane structures called autophagosomes (Kroemer et al. 2010). Li and coworkers showed that gold nanoparticles create an oxidative environment in MRC-5 human lung fibroblasts. The authors suggested that this oxidative environment could affect the regulation of cellular stress response mechanisms and, at the same time, induce the formation of autophagosomes, as a possible attempt to protect the cell from oxidative stress (Li et al. 2010). Ma and coworkers demonstrated that gold nanoparticles compromised lysosomal activity by alkalinization of lysosomal pH in normal rat kidney (NRK) cells. As a result of this, there was an accumulation of autophagosomes caused by the blockade of the autophagic flux (Ma et al. 2011). These studies suggest that autophagy is not only involved in nanomaterial toxicity but also influenced by it. Ultimately, autophagy can be considered a cellular response to foreign bodies (Zabirnyk et al. 2007).

It is worth noting that even without inducing cytotoxic effects, nanomaterials can impact normal biological processes in eukaryotic cells, such as cell signaling and cellular communication. Comfort and coworkers showed that silver nanoparticles reduced Akt and Erk signaling, while gold nanoparticles significantly diminished p-Akt and p-Erk levels and inhibited Akt activity. The authors showed that these alterations to cellular functions occurred both at the protein and genome level (Comfort et al. 2011). Another biological process that can be affected by nanomaterials is cellular communication. Love and coworkers observed that although gold and silver nanoparticles did not alter the viability of murine adrenal medullary chromaffin cells, they did alter their cellular communication. In addition, these authors showed that gold nanoparticles also interfered with cellular adhesion (Love et al. 2012a). Nonporous SiO_2 nanoparticles can also disrupt exocytosis in primary culture mast cells plus cause significant affects on cell viability by inducing hemolysis (Maurer-Jones et al. 2010). These findings support a body of work showing that distinct nanoparticles, such as nonporous SiO_2, gold, or silver nanoparticles, do disrupt the process of exocytosis, typically altering the number of molecules and release kinetics from vesicles (Love et al. 2012a; Maurer-Jones et al. 2010). These alterations in cellular metabolism might be considered subtle and remain unnoticed, since they do not lead to cell death. Nevertheless, they can be very important for some tissues, such as exocytosis in neuronal function and with long time exposure to nanomaterials, as in cell signaling disorders related to chronic diseases. These types of studies can also provide valuable information about nanomaterial toxicity.

9.6 Conclusions and Perspectives

It is tempting to think about a future where nanomaterials can solve diverse types of human problems. Although the first generation of nanoparticles is currently being approved for human use, there is still a long way to go. Over the last several decades, we have made significant progress in the area of cellular internalization and intracellular trafficking of nanomaterials, but broad generalizations still cannot be made. New findings can potentially address this limitation, but still we have to answer some questions: Why do different cells internalize the same nanomaterial in different ways or what makes it a cell type-dependent process? How do cells control intracellular trafficking of nanomaterials? What are the underlying mechanisms that nanomaterials use to escape endosomes? Are there other endocytic pathways? Do endocytic pathways converge on the same compartment?

Current knowledge suggests that broad generalizations cannot be made in this field. Perhaps the variety of cell types and their specific functions do not align with this idea. It may be better to find a balance between understanding the intrinsic characteristics of nanoparticles and knowing the changes that we can make to mediate the changes that we are looking for. The journey toward finding a "magic bullet" has left some things overlooked. For example, the field has focused on metallic nanomaterials, although a better approach might be to consider biodegradable materials, such as polymers and lipids. These materials could be used in new applications because of their intrinsic characteristics and thus give us a wider range of tools for nanomedicine. In addition, it seems that exocytosis plays a more important role in intracellular trafficking of nanomaterials than we initially thought. The field has just started looking in this direction, but we can already formulate new interesting questions from this initial knowledge. For example, does the exocytosed nanomaterial undergo internalization again? Does it take the same route?

As the development of new nanomaterials increases and their human applications become reality, the concerns about the potential hazards of nanomaterials also increase. This led to the establishment of a new field, nanotoxicology. Although this field has just emerged, it has many important issues to address. On one side it faces a great deal of pressure due to the fast growing area of nanotechnology, on the other side, it has the responsibility to examine and guarantee safety for human health and the environment. A potential strategy to address these pressures might be to improve in vitro toxicity assays and studies to evaluate the risk imposed by nanomaterials. Undoubtedly, there are limitations to the in vitro assessment of nanomaterial toxicity. Although, with additional studies these limitations can be addressed and guided by in vivo studies that can lead to better and more specific protocols. In turn, the knowledge gained from in vitro studies can help to predict or anticipate new targets to better define strategies for in vivo evaluations. In addition, distinguishing between normal, transient and real adverse effects of nanomaterials will help. Ultimately, a broader examination is paramount to understanding the impact of nanoparticles on cellular metabolism. These evaluations can help us to determine if nanomaterials will have an affect on chronic or debilitating diseases, such as multiple sclerosis and cancer.

Acknowledgment Support from FAPESP, CNPq, and Brazilian Network on Nanotoxicology (MCTI/CNPq) and NanoBioss (MCTI) are acknowledged.

References

Agarwal A, Lariya N, Saraogi G et al (2009) Nanoparticles as novel carrier for brain delivery: a review. Curr Pharm Des 15:917–925

Aguilar RC, Wendland B (2005) Endocytosis of membrane receptors: two pathways are better than one. Proc Natl Acad Sci U S A 102:2679–2680. doi:10.1073/pnas.0500213102

Ai J, Biazar E, Jafarpour M et al (2011) Nanotoxicology and nanoparticle safety in biomedical designs. Int J Nanomed 6:1117–1127. doi:10.2147/IJN.S16603

Al-Hajaj NA, Moquin A, Neibert KD et al (2011) Short ligands affect modes of QD uptake and elimination in human cells. ACS Nano 5:4909–4918. doi:10.1021/nn201009w

Alkilany AM, Murphy CJ (2010) Toxicity and cellular uptake of gold nanoparticles: what we have learned so far? J Nanopart Res 12:2313–2333. doi:10.1007/s11051-010-9911-8

Asati A, Santra S, Kaittanis C et al (2010) Surface-charge-dependent cell localization and cytotoxicity of cerium oxide nanoparticles. ACS Nano 4:5321–5331. doi:10.1021/nn100816s

AshaRani PV, Low Kah Mun G et al (2009) Cytotoxicity and genotoxicity of silver nanoparticles in human cells. ACS Nano 3:279–290. doi:10.1021/nn800596w
Asharani PVP, Hande MPM, Valiyaveettil SS (2009) Anti-proliferative activity of silver nanoparticles. CORD Conf Proc 10:65–65. doi:10.1186/1471-2121-10-65
Bae YM, Park YI, Nam SH et al (2012) Endocytosis, intracellular transport, and exocytosis of lanthanide-doped upconverting nanoparticles in single living cells. Biomaterials 33:9080–9086. doi:10.1016/j.biomaterials.2012.08.039
Bareford LM, Swaan PW (2007) Endocytic mechanisms for targeted drug delivery. Adv Drug Deliv Rev 59:748–758. doi:10.1016/j.addr.2007.06.008
Bareford LM, Phelps MA, Foraker AB et al (2008) Intracellular processing of riboflavin in human breast cancer cells. Mol Pharm 5:839–848. doi:10.1021/mp800046m
Barrias ES, Reignault LC, De Souza W et al (2012) Trypanosoma cruzi uses macropinocytosis as an additional entry pathway into mammalian host cell. Microbes Infect 14:1340–1351. doi:10.1016/j.micinf.2012.08.003
Bartczak D, Nitti S, Millar TM et al (2012) Exocytosis of peptide functionalized gold nanoparticles in endothelial cells. RSC Adv 4:4470–4472. doi:10.1039/C2NR31064C
Bastiani M, Parton RG (2010) Caveolae at a glance. J Cell Sci 123:3831–3836. doi:10.1242/jcs.070102
Benfer M, Kissel T (2012) Cellular uptake mechanism and knockdown activity of siRNA-loaded biodegradable DEAPA-PVA-g-PLGA nanoparticles. Eur J Pharm Biopharm 80:247–256. doi:10.1016/j.ejpb.2011.10.021
Buzea C, Pacheco II, Robbie K (2007) Nanomaterials and nanoparticles: sources and toxicity. Biointerphases 2:MR17–MR71. doi:10.1116/1.2815690
Cabral H, Nishiyama N, Kataoka K (2011) Supramolecular nanodevices: from design validation to theranostic nanomedicine. Acc Chem Res 44:999–1008. doi:10.1021/ar200094a
Canton II, Battaglia GG (2012) Endocytosis at the nanoscale. Chem Soc Rev 41:2718–2739. doi:10.1039/c2cs15309b
Caracciolo G, Pozzi D, Capriotti AL et al (2011) Factors determining the superior performance of lipid/DNA/protamine nanoparticles over lipoplexes. J Med Chem 54:4160–4171. doi:10.1021/jm200237p
Chang M-Y, Shiau A-L, Chen Y-H et al (2008) Increased apoptotic potential and dose-enhancing effect of gold nanoparticles in combination with single-dose clinical electron beams on tumor-bearing mice. Cancer Sci 99:1479–1484
Chang J, Jallouli Y, Kroubi M et al (2009) Characterization of endocytosis of transferrin-coated PLGA nanoparticles by the blood-brain barrier. Int J Pharm 379:285–292. doi:10.1016/j.ijpharm.2009.04.035
Chen X, Kube DM, Cooper MJ et al (2008) Cell surface nucleolin serves as receptor for DNA nanoparticles composed of pegylated polylysine and DNA. Mol Ther 16:333–342
Chen X, Shank S, Davis PB et al (2011) Nucleolin-mediated cellular trafficking of DNA nanoparticle is lipid raft and microtubule dependent and can be modulated by glucocorticoid. Mol Ther 19:93–102
Chen ZZ, Yin J-JJ, Zhou Y-TY et al (2012) Dual enzyme-like activities of iron oxide nanoparticles and their implication for diminishing cytotoxicity. ACS Nano 6:4001–4012
Chou LYT, Ming K, Chan WCW (2010) Strategies for the intracellular delivery of nanoparticles. Chem Soc Rev 40:233–238
Chwalibog A, Sawosz E, Hotowy A et al (2010) Visualization of interaction between inorganic nanoparticles and bacteria or fungi. Int J Nanomed 5:1085–1094
Colon J, Hsieh N, Ferguson A et al (2010) Cerium oxide nanoparticles protect gastrointestinal epithelium from radiation-induced damage by reduction of reactive oxygen species and upregulation of superoxide dismutase 2. Nanomedicine 6:698–705
Comfort KKK, Maurer EIE, Braydich-Stolle LKL et al (2011) Interference of silver, gold, and iron oxide nanoparticles on epidermal growth factor signal transduction in epithelial cells. ACS Nano 5:10000–10008

Dam DHM, Lee JH, Sisco PN et al (2012) Direct observation of nanoparticle—cancer cell nucleus interactions. ACS Nano 6:3318–3326
del Pozo-Rodríguez A, Delgado D, Solinís MA et al (2008) Solid lipid nanoparticles for retinal gene therapy: transfection and intracellular trafficking in RPE cells. Int J Pharm 360:177–183
del Pozo-Rodríguez A, Pujals S, Delgado D et al (2009) A proline-rich peptide improves cell transfection of solid lipid nanoparticle-based non-viral vectors. J Control Release 133:52–59
Dharmawardhane SS, Schürmann AA, Sells MAM et al (2000) Regulation of macropinocytosis by p21-activated kinase-1. Mol Biol Cell 11:3341–3352
Doherty GJ, Mcmahon HT (2009) Mechanisms of endocytosis. Annu Rev Biochem 78:857–902
Dombu CY, Kroubi M, Zibouche R et al (2010) Characterization of endocytosis and exocytosis of cationic nanoparticles in airway epithelium cells. Nanotechnology 21:355102
Donaldson K, Stone V, Tran CL et al (2004) Nanotoxicology. Occup Environ Med 61:727–728
Edeling MA, Smith C, Owen D (2006) Life of a clathrin coat: insights from clathrin and AP structures. Nat Rev Mol Cell Biol 7:32–44
Ekkapongpisit M, Giovia A, Follo C et al (2012) Biocompatibility, endocytosis, and intracellular trafficking of mesoporous silica and polystyrene nanoparticles in ovarian cancer cells: effects of size and surface charge groups. Int J Nanomed 7:4147–4158
El-Ansary A, Al-Daihan S (2009) On the toxicity of therapeutically used nanoparticles: an overview. J Toxicol 2009:754810
Faklaris O, Joshi V, Irinopoulou T et al (2009) Photoluminescent diamond nanoparticles for cell labeling: study of the uptake mechanism in mammalian cells. ACS Nano 3:3955–3962
Falcone S, Cocucci E, Podini P et al (2006) Macropinocytosis: regulated coordination of endocytic and exocytic membrane traffic events. J Cell Sci 119:4758–4769
Fichter KM, Ingle NP, McLendon PM, Reineke TM (2013) Polymeric nucleic acid vehicles exploit active interorganelle trafficking mechanisms. ACS Nano 7:347–364
Gao H, Shi W, Freund LB (2005) Mechanics of receptor-mediated endocytosis. Proc Natl Acad Sci U S A 102:9469–9474
Georgieva JV, Kalicharan D, Couraud P-O et al (2011) Surface characteristics of nanoparticles determine their intracellular fate in and processing by human blood-brain barrier endothelial cells in vitro. Mol Ther 19:318–325
González-Gaitán M, Stenmark H (2003) Endocytosis and signaling: a relationship under development. Cell 115:513–521
Gratton SEA, Ropp PA, Pohlhaus PD et al (2008) The effect of particle design on cellular internalization pathways. Proc Natl Acad Sci U S A 105:11613–11618
Greulich C, Diendorf J, Simon T et al (2011) Uptake and intracellular distribution of silver nanoparticles in human mesenchymal stem cells. Acta Biomater 7:347–354
Hansen CG, Nichols BJ (2009) Molecular mechanisms of clathrin-independent endocytosis. J Cell Sci 122:1713–1721
Harush-Frenkel OO, Rozentur EE, Benita SS et al (2008) Surface charge of nanoparticles determines their endocytic and transcytotic pathway in polarized MDCK cells. Biomacromolecules 9:435–443
Hayer A, Stoeber M, Ritz D et al (2010) Caveolin-1 is ubiquitinated and targeted to intralumenal vesicles in endolysosomes for degradation. J Cell Biol 191:615–629
Heuser JEJ, Anderson RGR (1989) Hypertonic media inhibit receptor-mediated endocytosis by blocking clathrin-coated pit formation. J Cell Biol 108:389–400
Hill MM, Bastiani M, Luetterforst R et al (2008) PTRF-Cavin, a conserved cytoplasmic protein required for caveola formation and function. Cell 132:113–124
Hillaireau H, Couvreur P (2009) Nanocarriers' entry into the cell: relevance to drug delivery. Cell Mol Life Sci 66:2873–2896
Hillegass JM, Shukla A, Lathrop SA et al (2010) Assessing nanotoxicity in cells in vitro. Wiley Interdiscip Rev Nanomed Nanobiotechnol 2:219–231
Hinrichsen L, Meyerholz A, Groos S et al (2006) Bending a membrane: how clathrin affects budding. Proc Natl Acad Sci U S A 103:8715–8720

Hoet PH, Brüske-Hohlfeld I, Salata OV (2004) Nanoparticles—known and unknown health risks. J Nanobiotechnol 2:12. doi:10.1186/1477-3155-2-12
Huotari J, Helenius A (2011) Endosome maturation. EMBO J 30:3481–3500
Hussain SMS, Hess KLK, Gearhart JMJ et al (2005) In vitro toxicity of nanoparticles in BRL 3A rat liver cells. Toxicol In Vitro 19:975–983
Hussain SM, Braydich Stolle LK, Schrand AM et al (2009) Toxicity evaluation for safe use of nanomaterials: recent achievements and technical challenges. Adv Mater 21:1549–1559
Iversen T, Frerker N, Sandvig K (2012) Uptake of ricin B-quantum dot nanoparticles by a macropinocytosis-like mechanism. J Nanobiotechnol 10:33. doi:10.1371/journal.pone.0005935
Jeyaraj M, Rajesh M, Arun R et al (2013) An investigation on the cytotoxicity and caspase-mediated apoptotic effect of biologically synthesized silver nanoparticles using Podophyllum hexandrum on human cervical carcinoma cells. J Control Release 102:708–717
Jian F, Zhang Y, Wang J et al (2012) Toxicity of biodegradable nanoscale preparations. Curr Drug Metab 13:440–446
Jiang X, Röcker C, Hafner M et al (2010) Endo- and exocytosis of zwitterionic quantum dot nanoparticles by live HeLa cells. ACS Nano 4:6787–6797
Jin H, Heller DA, Sharma R, Strano MS (2009) Size-dependent cellular uptake and expulsion of single-walled carbon nanotubes: single particle tracking and a generic uptake model for nanoparticles. ACS Nano 3:149–158
Johnston HJ, Semmler-Behnke M, Brown DM et al (2010) Evaluating the uptake and intracellular fate of polystyrene nanoparticles by primary and hepatocyte cell lines in vitro. Toxicol Appl Pharmacol 242:66–78
Jones AT (2007) Macropinocytosis: searching for an endocytic identity and role in the uptake of cell penetrating peptides. J Cell Mol Med 11:670–684
Jovic M, Sharma M, Rahajeng J et al (2010) The early endosome: a busy sorting station for proteins at the crossroads. Histol Histopathol 25:99–112
Kaplan IM, Wadia JS, Dowdy SF (2005) Cationic TAT peptide transduction domain enters cells by macropinocytosis. J Control Release 102:247–253
Karakoti AS, Singh S, Kumar A et al (2009) PEGylated nanoceria as radical scavenger with tunable redox chemistry. J Am Chem Soc 131:14144–14145
Kasper J, Hermanns MI, Bantz C et al (2013) Interactions of silica nanoparticles with lung epithelial cells and the association to flotillins. Arch Toxicol 87(6):1053–1065. doi:10.1007/s00204-012-0876-5
Kerr MC, Teasdale RD (2009) Defining macropinocytosis. Traffic 10:364–371
Khalil I, Kogure K, Akita H, Harashima H (2006) Uptake pathways and subsequent intracellular trafficking in nonviral gene delivery. Pharmacol Rev 58:32–45
Kim SS, Choi I-HI (2012) Phagocytosis and endocytosis of silver nanoparticles induce interleukin-8 production in human macrophages. Yonsei Med J 53:654–657
Kim Y-J, Yang SI, Ryu J-C (2010) Cytotoxicity and genotoxicity of nano-silver in mammalian cell lines. Mol Cell Toxicol 6:119–125
Kim AJ, Boylan NJ, Suk JS et al (2012) Non-degradative intracellular trafficking of highly compacted polymeric DNA nanoparticles. J Control Release 158:102–107
Kroemer G, Mariño G, Levine B (2010) Autophagy and the integrated stress response. Mol Cell 40:280–293
Lai DYD (2012) Toward toxicity testing of nanomaterials in the 21st century: a paradigm for moving forward. Wiley Interdiscip Rev RNA 4:1–15. doi:10.1002/wnan.162
Lai SKS, Hida KK, Man STS et al (2007) Privileged delivery of polymer nanoparticles to the perinuclear region of live cells via a non-clathrin, non-degradative pathway. Biomaterials 28:2876–2884
Lai ZW, Yan Y, Caruso F, Nice EC (2012) Emerging techniques in proteomics for probing nano-bio interactions. ACS Nano 6:10438–10448
Lajoie P, Nabi IR (2007) Regulation of raft-dependent endocytosis. J Cell Mol Med 11:644–653

Li JJJ, Hartono DD, Ong C-NC et al (2010) Autophagy and oxidative stress associated with gold nanoparticles. Biomaterials 31:5996–6003
Lim JPJ, Gleeson PAP (2011) Macropinocytosis: an endocytic pathway for internalising large gulps. Immunol Cell Biol 89:836–843
Liu P, Guan R, Ye X et al (2011) Toxicity of nano- and micro-sized silver particles in human hepatocyte cell line L02. J Phys Conf Ser 304:012036
Love SA, Liu Z, Haynes CL (2012a) Examining changes in cellular communication in neuroendocrine cells after noble metal nanoparticle exposure. Analyst 137:3004–3010
Love SA, Maurer-Jones MA, Thompson JW et al (2012b) Assessing nanoparticle toxicity. Annu Rev Anal Chem 5:181–205
Lunov O, Syrovets T, Loos C et al (2011) Amino-functionalized polystyrene nanoparticles activate the NLRP3 inflammasome in human macrophages. ACS Nano 5:9648–9657
Ma XX, Wu YY, Jin SS et al (2011) Gold nanoparticles induce autophagosome accumulation through size-dependent nanoparticle uptake and lysosome impairment. ACS Nano 5:8629–8639
Madshus IHI, Tønnessen TIT, Olsnes SS et al (1987) Effect of potassium depletion of Hep 2 cells on intracellular pH and on chloride uptake by anion antiport. J Cell Physiol 131:6–13
Markovic Z, Todorovic-Markovic B, Kleut D et al (2007) The mechanism of cell-damaging reactive oxygen generation by colloidal fullerenes. Biomaterials 28:5437–5448
Marquis BJ, Love SA, Braun KL et al (2009) Analytical methods to assess nanoparticle toxicity. RSC Adv 134:425–439
Marsh M, McMahon HT (1999) The structural era of endocytosis. Science 285:215–220
Martins SS, Costa-Lima SS, Carneiro TT et al (2012) Solid lipid nanoparticles as intracellular drug transporters: an investigation of the uptake mechanism and pathway. Int J Pharm 430:216–227
Maurer-Jones MA, Lin Y-S, Haynes CL (2010) Functional assessment of metal oxide nanoparticle toxicity in immune cells. ACS Nano 4:3363–3373
Maynard AD, Warheit DB, Philbert MA (2011) The new toxicology of sophisticated materials: nanotoxicology and beyond. Toxicol Sci 120(Suppl 1):S109–S129
McNeil SE (2005) Nanotechnology for the biologist. J Leukoc Biol 78:585–594
Meng H, Yang S, Li Z et al (2011) Aspect ratio determines the quantity of mesoporous silica nanoparticle uptake by a small GTPase-dependent macropinocytosis mechanism. ACS Nano 5:4434–4447
Mercer J, Helenius A (2009) Virus entry by macropinocytosis. Nat Cell Biol 11:510–520
Ming X, Alam MR, Fisher M et al (2010) Intracellular delivery of an antisense oligonucleotide via endocytosis of a G protein-coupled receptor. Nucleic Acids Res 38:6567–6576
Mudhakir D, Harashima H (2009) Learning from the viral journey: how to enter cells and how to overcome intracellular barriers to reach the nucleus. AAPS J 11:65–77
Mukherjee SG, O'Claonadh N, Casey A et al (2012) Comparative in vitro cytotoxicity study of silver nanoparticle on two mammalian cell lines. Toxicol In Vitro 26:238–251
Murakami M, Cabral H, Matsumoto Y et al (2011) Improving drug potency and efficacy by nanocarrier-mediated subcellular targeting. Sci Transl Med 3:64ra2
Nel A, Xia T, Mädler L et al (2006) Toxic potential of materials at the nanolevel. Science 311:622–627
Ng C-T, Li JJ, Bay B-H et al (2010) Current studies into the genotoxic effects of nanomaterials. J Nucleic Acids 2010:1–12
Nichols B (2003) Caveosomes and endocytosis of lipid rafts. J Cell Sci 116:4707–4714
Nishimura SS, Takahashi SS, Kamikatahira HH et al (2008) Combinatorial targeting of the macropinocytotic pathway in leukemia and lymphoma cells. J Biol Chem 283:11752–11762
Oberdörster G, Oberdörster E, Oberdörster J (2005) Nanotoxicology: an emerging discipline evolving from studies of ultrafine particles. Environ Health Perspect 113:823–839
Oh J-M, Choi S-J, Kim S-T et al (2006) Cellular uptake mechanism of an inorganic nanovehicle and its drug conjugates: enhanced efficacy due to clathrin-mediated endocytosis. Bioconjug Chem 17:1411–1417

Orth JDJ, McNiven MAM (2006) Get off my back! Rapid receptor internalization through circular dorsal ruffles. Cancer Res 66:11094–11096
Ory S, Gasman S (2011) Rho GTPases and exocytosis: what are the molecular links? Semin Cell Dev Biol 22:27–32
Panyam J, Labhasetwar V (2003) Dynamics of endocytosis and exocytosis of poly(D,L-lactide-co-glycolide) nanoparticles in vascular smooth muscle cells. Pharm Res 20:212–220
Pardridge WM (2007) Blood-brain barrier delivery. Drug Discov Today 12:54–61. doi:10.1016/j.drudis.2006.10.013
Park E-JE, Yi JJ, Chung K-HK et al (2008) Oxidative stress and apoptosis induced by titanium dioxide nanoparticles in cultured BEAS-2B cells. Toxicol Lett 180:222–229
Parton RG, Howes MT (2010) Revisiting caveolin trafficking: the end of the caveosome. J Cell Biol 191:439–441
Parton RG, Simons K (2007) The multiple faces of caveolae. Nat Rev Mol Cell Biol 8:185–194
Pelkmans L, Helenius A (2002) Endocytosis via caveolae. Traffic 3:311–320
Pelkmans L, Kartenbeck J, Helenius A (2001) Caveolar endocytosis of simian virus 40 reveals a new two-step vesicular-transport pathway to the ER. Nat Cell Biol 3:473–483
Petros RA, DeSimone JM (2010) Strategies in the design of nanoparticles for therapeutic applications. Nat Rev Drug Discov 9:615–627
Pierscionek BK, Li Y, Yasseen AA et al (2009) Nanoceria have no genotoxic effect on human lens epithelial cells. Nanotechnology 21:035102
Prokop A, Davidson JM (2008) Nanovehicular intracellular delivery systems. J Pharm Sci 97:3518–3590. doi:10.1002/jps.21270
Rajendran L, Knölker H-J, Simons K (2010) Subcellular targeting strategies for drug design and delivery. Nat Rev Drug Discov 9:29–42. doi:10.1038/nrd2897
Rehman ZU, Hoekstra D, Zuhorn IS (2011) Protein kinase A inhibition modulates the intracellular routing of gene delivery vehicles in HeLa cells, leading to productive transfection. J Control Release 156:76–84
Rehman ZU, Sjollema KA, Kuipers J et al (2012) Nonviral gene delivery vectors use syndecan-dependent transport mechanisms in filopodia to reach the cell surface. ACS Nano 6:7521–7532
Rejman J, Bragonzi A, Conese M (2005) Role of clathrin- and caveolae-mediated endocytosis in gene transfer mediated by lipo- and polyplexes. Mol Ther 12:468–474
Rodal SK, Skretting G, Garred Ø et al (1999) Extraction of cholesterol with methyl-β-cyclodextrin perturbs formation of clathrin-coated endocytic vesicles. Mol Biol Cell 10:961–974
Sahay G, Alakhova DY, Kabanov AV (2010a) Endocytosis of nanomedicines. J Control Release 145:182–195
Sahay G, Kim JO, Kabanov AV et al (2010b) The exploitation of differential endocytic pathways in normal and tumor cells in the selective targeting of nanoparticulate chemotherapeutic agents. Biomaterials 31:923–933
Sandin P, Fitzpatrick LW, Simpson JC et al (2012) High-speed imaging of rab family small gtpases reveals rare events in nanoparticle trafficking in living cells. ACS Nano 6:1513–1521
Sanpui P, Chattopadhyay A, Ghosh SS (2011) Induction of apoptosis in cancer cells at low silver nanoparticle concentrations using chitosan nanocarrier. ACS Appl Mater Interfaces 3:218–228
Saovapakhiran A, D'Emanuele A, Attwood D et al (2009) Surface modification of PAMAM dendrimers modulates the mechanism of cellular internalization. Bioconjug Chem 20:693–701
Scita G, Di Fiore PP (2010) The endocytic matrix. Nature 463:464–473
Serag MF, Kaji N, Venturelli E et al (2011) Functional platform for controlled subcellular distribution of carbon nanotubes. ACS Nano 5:9264–9270
Soenen SJ, Rivera-Gil P, Montenegro J-M et al (2011) Cellular toxicity of inorganic nanoparticles: common aspects and guidelines for improved nanotoxicity evaluation. Nano Today 6:446–465
Soenen SJS, Manshian BB, Montenegro JMJ et al (2012) Cytotoxic effects of gold nanoparticles: a multiparametric study. ACS Nano 6:5767–5783
Stern ST, Adiseshaiah PP, Crist RM (2012) Autophagy and lysosomal dysfunction as emerging mechanisms of nanomaterial toxicity. Part Fibre Toxicol 9:20. doi:10.1186/1743-8977-9-20

Subtil AA, Gaidarov II, Kobylarz KK et al (1999) Acute cholesterol depletion inhibits clathrin-coated pit budding. Proc Natl Acad Sci U S A 96:6775–6780
Sun L, Li Y, Liu X et al (2011) Cytotoxicity and mitochondrial damage caused by silica nanoparticles. Toxicol In Vitro 25:1619–1629
Swanson JA, Watts C (1995) Macropinocytosis. Trends Cell Biol 5:424–428
Tiwari SB, Amiji MM (2006) A review of nanocarrier-based CNS delivery systems. Curr Drug Deliv 3:219–232
Torchilin VP (2006) Recent approaches to intracellular delivery of drugs and DNA and organelle targeting. Annu Rev Biomed Eng 8:343–375
Torgersen ML, Skretting G, van Deurs B et al (2001) Internalization of cholera toxin by different endocytic mechanisms. J Cell Sci 114:3737–3747
Totsuka Y, Higuchi T, Imai T et al (2009) Genotoxicity of nano/microparticles in in vitro micronuclei, in vivo comet and mutation assay systems. Part Fibre Toxicol 6:23. doi:10.1186/1743-8977-6-23
Tuma PL, Hubbard AL (2003) Transcytosis: crossing cellular barriers. Physiol Rev 83:871–932
Ungewickell EJ, Hinrichsen L (2007) Endocytosis: clathrin-mediated membrane budding. Curr Opin Cell Biol 19:417–425
Varkouhi AK, Scholte M, Storm G et al (2011) Endosomal escape pathways for delivery of biologicals. J Control Release 151:220–228
Vercauteren D, Vandenbroucke RE, Jones AT et al (2010) The use of inhibitors to study endocytic pathways of gene carriers: optimization and pitfalls. Mol Ther 18:561–569
Verma A, Stellacci F (2010) Effect of surface properties on nanoparticle-cell interactions. Small 6:12–21
Vollrath A, Schallon A, Pietsch C et al (2012) A toolbox of differently sized and labeled PMMA nanoparticles for cellular uptake investigations. RSC Adv 9:99
Walker NJ, Bucher JR (2009) A 21st century paradigm for evaluating the health hazards of nanoscale materials? Toxicol Sci 110:251–254
Wang LH, Rothberg KG, Anderson RG (1993) Mis-assembly of clathrin lattices on endosomes reveals a regulatory switch for coated pit formation. J Cell Biol 123:1107–1117
Wang Z, Tiruppathi C, Minshall RD et al (2009) Size and dynamics of caveolae studied using nanoparticles in living endothelial cells. ACS Nano 3:4110–4116
Wang Y, Wang J, Deng X et al (2010) Direct imaging of titania nanotubes located in mouse neural stem cell nuclei. Nano Res 2:543–552
Wang F, Wang Y-C, Dou S et al (2011a) Doxorubicin-tethered responsive gold nanoparticles facilitate intracellular drug delivery for overcoming multidrug resistance in cancer cells. ACS Nano 5:3679–3692
Wang L, Liu Y, Li W et al (2011b) Selective targeting of gold nanorods at the mitochondria of cancer cells: implications for cancer therapy. Nano Lett 11:772–780
Won Y-Y, Sharma R, Konieczny SF (2009) Missing pieces in understanding the intracellular trafficking of polycation/DNA complexes. J Control Release 139:88–93. doi:10.1016/j.jconrel.2009.06.031
Wu L-C, Chu L-W, Lo L-W et al (2013) Programmable cellular retention of nanoparticles by replacing the synergistic anion of transferrin. ACS Nano 7:365–375. doi:10.1021/nn3043397
Yan Y, Such GK, Johnston APR et al (2012) Engineering particles for therapeutic delivery: prospects and challenges. ACS Nano 6:3663–3669. doi:10.1021/nn3016162
Zabirnyk O, Yezhelyev M, Seleverstov O (2007) Nanoparticles as a novel class of autophagy activators. Autophagy 3:278–281
Zaki NM, Tirelli N (2010) Gateways for the intracellular access of nanocarriers: a review of receptor-mediated endocytosis mechanisms and of strategies in receptor targeting. Expert Opin Drug Deliv 7:895–913. doi:10.1517/17425247.2010.501792
Zhao F, Zhao Y, Liu Y et al (2011) Cellular uptake, intracellular trafficking, and cytotoxicity of nanomaterials. Small 7:1322–1337. doi:10.1002/smll.201100001
Zuhorn IS, Kalicharan R, Hoekstra D (2002) Lipoplex-mediated transfection of mammalian cells occurs through the cholesterol-dependent clathrin-mediated pathway of endocytosis. J Biol Chem 277:18021–18028. doi:10.1074/jbc.M111257200

Chapter 10
Cytotoxicity and Genotoxicity of Solid Lipid Nanoparticles

Priscyla D. Marcato and Nelson Durán

Abstract Solid lipid nanoparticles (SLNs) were the last nanoparticles discovered; however, these particles have received great attention as drug delivery system besides the other one. These particles can be produced by several methods, which include methods without use of organic solvents or with only reagents previously approved by Food and Drug Administration (FDA) through the GRAS rules. Furthermore, the easy scale up of production and particles sterilization led to an increase in the industry interest in these nanostructures. Nowadays, there are some products with SLNs on the market and the tendency is that the number of products with this carrier increases significantly in the next years. For this, studies about their toxicity in vitro and in vivo are increasing. SLN exhibits low or no cyto- and genotoxic effect on different cells that were already studied. The cytotoxicity and genotoxicity depend on the SLN composition, more specifically, of the solid lipid and surfactant used in the preparation. Some solid lipid or surfactant can increase the cyto- or genotoxic effect of SLN indicating that SLN composition plays an important role in the cytotoxic and genotoxic effect of these particles.

10.1 Introduction

Liposomes, polymeric nanoparticles, and solid lipid nanoparticles (SLNs) have received particular attention in the encapsulation of drugs or active compounds for topical applications and for other types of administrations (Durán et al. 2011). The development of liposomes was first published in 1965. The preparation of liposomes with entrapped solutes was first demonstrated in 1965 (Bangham et al. 1965) and actually a number of liposome formulations are in clinical use to combat cancer and infectious diseases (Puri et al. 2009).

P.D. Marcato (✉)
Faculty of Pharmaceutical Sciences of Riberão Preto, Universidade de São Paulo, Ribeirão Preto, Sao Paulo, Brazil
e-mail: pmarcato@gmail.com

N. Durán et al. (eds.), *Nanotoxicology*, Nanomedicine and Nanotoxicology,
DOI 10.1007/978-1-4614-8993-1_10, © Springer Science+Business Media New York 2014

Polymeric nanoparticles have been also studied since the end of 1980s whereas SLNs were developed almost a decade later (Fessi et al. 1989). SLN appeared as an alternative carrier system to traditional colloidal carriers such as emulsions, liposomes, and polymeric micro- and nanoparticles. SLNs are nanometric colloidal carriers of around 50–900 nm, composed of physiological lipid, dispersed in water or in aqueous surfactant solution, and they offer special properties such as small size, large surface area, high drug loading, and the interaction of phases at the interface and are useful in enhancement of pharmaceutical effects (Ekambaram et al. 2012). It is quite known that the first SLN systems presented some limitations due to their low drug loading and low stability during storage. These systems were enhanced and it was developed a liquid lipid mixture with solid lipids providing a substitute for SLN and these were named nanostructured lipid carriers (NLCs) (Müller et al. 2002).

Lipid nanoparticles are well tolerated in living systems, since they are made from physiological compounds leading to the metabolic pathways (Rawat et al. 2011a, b; Wissing et al. 2004a). For this purpose, studies focusing on nanotoxicology, that include cytotoxicity and genotoxicity analyses, are of paramount importance (Smijs and Bouwstra 2010).

10.1.1 Solid Lipid Nanoparticles

It was mentioned in the literature that in order to overcome the disadvantages associated with the liquid state of the oil droplets, the liquid lipid was replaced by a solid lipid. A good explanation for the increasing interest in these systems was that lipids enhance oral bioavailability and reduce plasma profile variability. There are facilities for characterization of lipid sources; the SLN production is easily for technology transfer and manufacture scale-up, the lipids show good biocompatibility, low toxicity and the system is physically stable. These systems are an excellent approach to improve the oral bioavailability of the poor water-soluble drugs. SLNs combine all the advantages of polymeric nanoparticles, fat emulsions, and liposomes. As with all the systems SLN has also disadvantages such as particle growth, unpredictable gelation tendency, and unexpected dynamics of polymeric transitions. But it promises the possibility of controlled drug release, increased drug stability, high drug loading, no bio-toxicity, avoidance of organic solvents in preparation methods, and incorporation of lipophilic and hydrophilic drugs is possible (Ekambaram et al. 2012).

Many recent reviews in these structures were published, such as general aspects (Müller et al. 2011; Souto et al. 2011; Khan 2012), in oral applications (Das and Chaudhury 2011; Severino et al. 2012), in pharmaceutical and biomedical applications (Puri et al. 2009; Buse and El-Aneed 2010; Garud et al. 2012; Waghmare et al. 2012), and in cancer (Mathur et al. 2010; Ekambaram et al. 2012).

10.1.2 Nanostructured Lipid Carriers

NLCs were developed to overcome the limitations associated with the SLN. SLN components are solid lipids, while NLCs are produced by a mixture of specially blended solid lipid (long chain) with liquid lipid (short chain). The final lipid matrix exhibited a melting point depression compared to the original solid lipid, but they are maintained as solid at body temperature (Pardeike et al. 2009). As mentioned above the disadvantages of SLN include limited drug-loading capacity, drug expulsion during storage, and relatively high water content in the dispersions (70–99.9 %) (Westesen et al. 1997; Mehnert and Mader 2001).

NLCs have a higher drug-loading capacity for many active compounds, minimizing the potential expulsion of active compounds for longer periods of storage (Mehnert and Mader 2001; Müller et al. 2007). NLCs possess numerous features that are advantageous for the topical route of application, since these carriers are composed of physiological and biodegradable lipid, exhibiting low systemic toxicity and low cytotoxicity (Müller et al. 1997; Joshi and Patravale 2008) as it will be showed later.

An important aspect of SLN is also the stability and this must be considered from the particle size distribution and the lipid crystalline state view. The size is a critical safety factor for parental administration and affects the biodistribution and RES clearance mechanisms. The crystalline state is correlated with active incorporation, active release, and the particle geometry (Pragati et al. 2009). Probably this factor could affect significantly the cytotoxicity and genotoxicity of these nanoparticle types.

10.2 Solid Lipid Nanoparticles

10.2.1 Preparation

The main components in an SLN are a solid lipid (at room and body temperature), in the presence of a surfactant (to prevent agglomerations) and water. Some common lipids used are triglycerides (e.g., myristyl myristate (MM)), partial glycerides (e.g., glyceryl monostearate), fatty acids (e.g., palmitic acid), steroids (e.g., cholesterol), and waxes (e.g., cetyl palmitate (CP)). The choice of the best surfactant is a very important factor. In Table 10.1 is shown the most common ingredients used in the preparation of SLN (Marcato 2008).

The most common methods used for the preparation of SLN are hot microemulsion, emulsification and solvent evaporation, solvent diffusion, and high pressure homogenization (HPH), and the latter being the most used since the scale-up is easy and led to sterilize the sample after preparation directly from the homogenizer (Durán et al. 2011).

Table 10.1 Lipids and surfactants used in the preparation of SLN

	Structure
Lipid	
Stearic acid	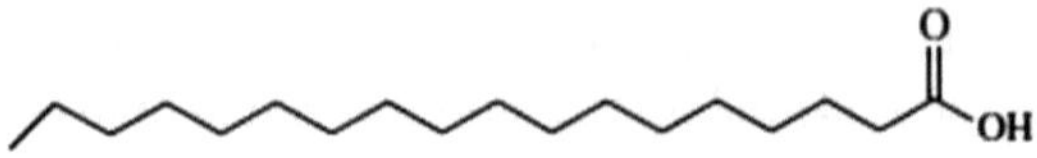
Glyceryl monostearate	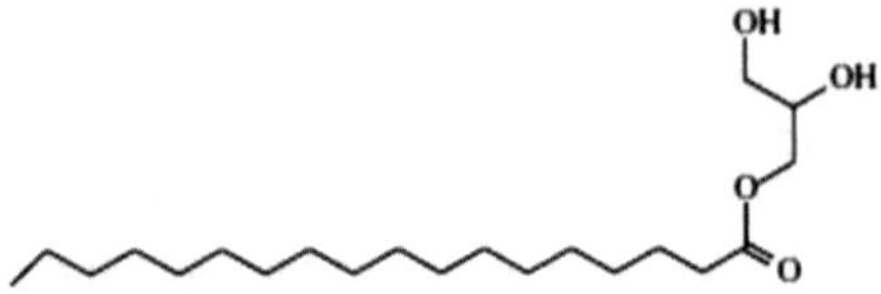
Tristearin	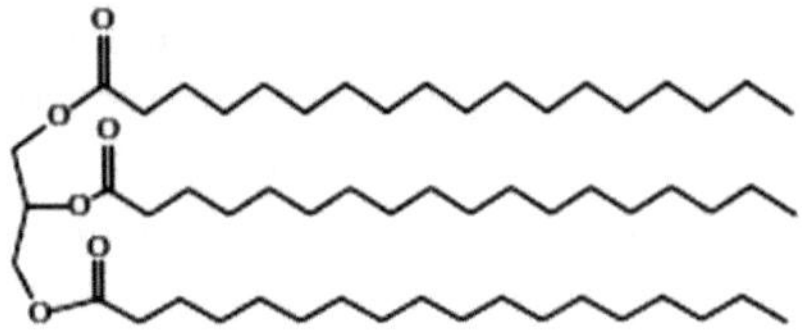
Surfactant	
Poloxamer F68	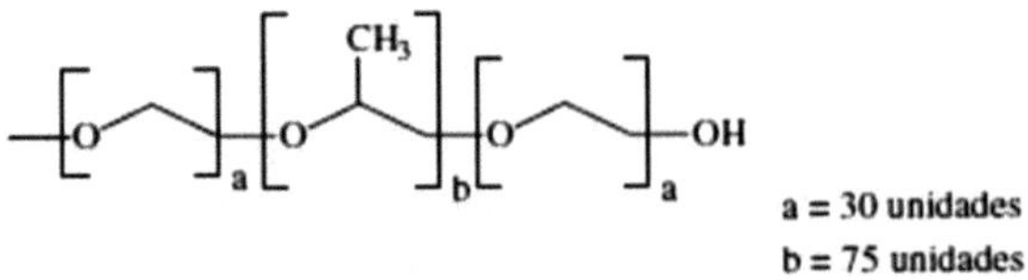
Polysorbate 80	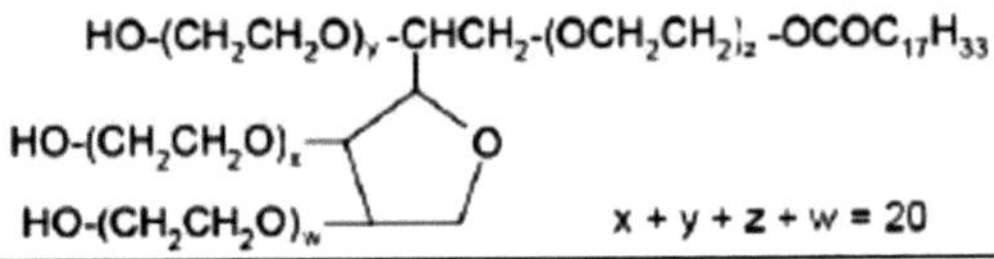

10.2.1.1 Heated Microemulsion

The microemulsion method was developed by Gasco (2007) and modified by other authors. The microemulsion is prepared with 10 % of melted lipid (e.g., stearic acid), 15 % of surfactant (e.g., polysorbate 20 or 60), more than 10 % of co-surfactant (e.g., Poloxamer), and water. The procedure starts with the lipid melted (5–10 °C above of fusion temperature of lipid), which is added into a hot surfactant and co-surfactant solution, under agitation, forming oil/water microemulsion. Under agitation is cold at 2–3 °C, leading to solidification of the particles (Mehnert and Mader 2001; Pedersen et al. 2006). This method has a disadvantage due to the excess water which is necessary to be removed at a high surfactant and co-surfactant concentration. This excess of water is possible to be removed by ultracentrifugation, lyophilization, or dialysis.

SLN dispersion can be used as granulation fluid for transferring in to solid product (tablets, pellets) by granulation process. High-temperature gradients facilitate rapid lipid crystallization and prevent aggregation (Ekambaram et al. 2012).

10.2.1.2 Emulsion and Solvent Evaporation

The preparation of SLN by the emulsion and solvent evaporation method led to the encapsulation of thermosensitive molecules and it consists in a mixture of lipid and water-immiscible organic solvent (e.g., chloroform) added under agitation to a surfactant solution, forming an oil/water emulsion and then the organic solvent evaporated under low pressure. The disadvantages of this method are the residual solvent in the final product and the concomitant production of microparticles (Wissing et al. 2004a, b; Üner 2006).

10.2.1.3 Solvent Diffusion

The solvent diffusion method was previously used for preparation of polymeric nanoparticles. This method is simple and does not require special equipment, but it is associated with the disadvantages of requiring organic solvents and difficulties related to scale-up. In this case, the lipid is mixed with a water-miscible solvent (e.g., acetone) and is added to a surfactant solution under agitation and then the solvent evaporated under lower pressure (Hou et al. 2003; Hu et al. 2006).

10.2.1.4 High Pressure Homogenization

In the HPH method the dispersion is homogenized at high pressure (500–2,000 bar) through a narrow cavity (few micrometers) and accelerated over a short distance at high velocity (~100 km/h). High shear stress and cavitation forces produce submicron particles (Mehnert and Mader 2001).

HPH can be carried out by two different methods: hot homogenization and cold homogenization. In hot HPH, the whole process is carried out at temperatures above the melting point (T_m) of the lipid (5–10 °C) together with the active compound (in general a hydrophilic one) which is added to a hot surfactant solution under agitation forming a pre-emulsion and then homogenized at high pressure (500–2,000 bar). After the emulsion homogenization, the dispersion is cooled to form the SLNs (Fig. 10.1). The size of the particles depends also on the number of cycles used, but in general, 1–3 cycles are sufficient for nanoparticles to form. This method is applied mostly in the case of lipophilic drugs (Mehnert and Mader 2001; Marcato 2008). The hot homogenization method produces, in general, nanoparticles below 500 nm.

In the cold homogenization method the lipid is melted (5–10 °C above of T_m of lipid) together with the active compound (it is possible to encapsulate thermosensitive substances due to a short time of high temperature) and then solidified with liquid nitrogen or dry ice and grinded in a power mill obtaining microparticles (50–100 μm). Then this is added to a surfactant solution at room temperature under agitation and, after that, homogenized at high pressure

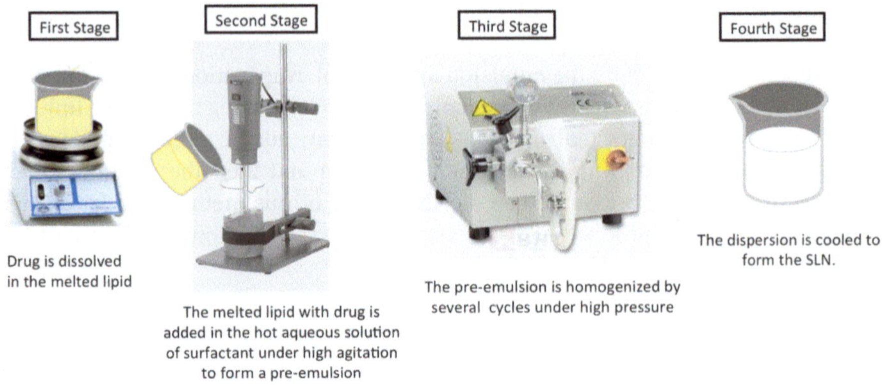

Fig. 10.1 Schematic representation of the methodology used in the preparation of SLN and NLC, using high pressure homogenization

(500–2,000 bar). The solid state of the matrix decreases the migration of the hydrophilic active compound to the aqueous phase, increasing the encapsulation efficiency. Cold homogenization produces microparticles with larger sizes than hot homogenization. However, a size reduction can be obtained through increase of the pressure and using 3–7 cycles in the homogenization (Üner 2006; Üner et al. 2007).

The advantages of the HPH method (hot or cold) are the easy scale-up and the production of SLN without the use of organic solvents. However, the high pressure can produce the coalescence of the particles and the high temperature used in the process can increase the degradation rate of the drug as well as of the particle (Hu et al. 2006).

10.2.2 Characterization

Many techniques have been used for the characterization of SLN such as X-ray diffraction (XRD), differential scanning calorimetry (DSC), proton nuclear magnetic resonance (NMR), and microscopy techniques (Durán et al. 2011).

10.2.2.1 X-Ray Diffraction

Polymorphism is observed in SLN due to the organization of alkane chains into different packing patterns: α (hexagonal), β′ (orthorhombic), and β (triclinic). SLN exhibited several degrees of orders. For example, hexagonal packing is the most disordered form, orthorhombic packing is less disordered, and triclinic packing is the most organized (Allais et al. 2003; Bunjes et al. 2007). These observations implicate that for the industrial application of SLN it is necessary to control clearly its polymorphism. This careful control is important since the polymorphic structure has a direct influence on the encapsulation efficiency and release of the drug during storage. The technique most commonly used to characterize the polymorphic structure of SLN is XRD. This technique determines the length of long and short spacing of the lipid lattice, allowing amorphous and crystalline materials to be differentiated. Also this technique serves to evaluate the influence of constituent oil in long spacing of the nanocrystals of NLC. A good combination of this technique with DSC allows amorphous solids and liquids to be differentiated (Üner 2006; Müller et al. 2000; Ruktanonchai et al 2008).

10.2.2.2 Differential Scanning Calorimetry

Information of the physical state and crystallinity of SLN is reached from DSC technique. The crystallinity of the particles is an important parameter due to the influence on the encapsulation efficiency, the release rate in its application, as well as on the release of the drug during storage. DSC analysis led to determine the fusion temperature and enthalpy of the particles. It is known that a high value for the fusion enthalpy indicates a high level of organization in the crystal lattice. This is due to the fusion of a highly organized crystal (perfect crystal) requiring more energy to overcome the forces of cohesion in the crystal lattice. This data could show the crystallization behavior of particles that could influence the release of the drug. During storage, the lipid can be converted from the α to the β' form under certain conditions of temperature, light exposure, and water loss from the SLN dispersion (Müller et al. 2000; Teeranachaideekul et al. 2008b).

10.2.2.3 Proton Nuclear Magnetic Resonance

NMR technique characterizes the liquid lipid domains within SLNs, e.g., mobility and arrangement of the molecules of oil, through the technique of ^{1}H NMR spectroscopy. NMR data can verify whether a lipid is in the liquid or semisolid/solid state through the difference in the relaxation time of protons in the liquid state and the semisolid/solid state, since the protons in the liquid state have a narrower signal of higher amplitude than those in the semisolid/solid state (Mehnert and Mader 2001; Teeranachaideekul et al. 2008a).

10.2.2.4 Microscopy Techniques

One of the most appropriated microscopic techniques is cryogenic transmission. Electron microscopy (Cryo-TEM) is used in order to obtain direct images of SLN. High-resolution-TEM allowed to determine the size (diameter) distribution and determination of the crystal structure (Hou et al. 2003). Atomic force microscopy (AFM) has also been used in the characterization of SLN (Chen et al. 2006).

10.3 Toxic Effects of Solid Lipid Nanoparticles and Nanostructured Lipid Carriers Administrated In Vitro and In Vivo

10.3.1 Culture Cells

Many lipid nanoparticles, in which is included SLN, exhibited non-cytotoxicity (Liu et al. 2008; Joshi and Müller 2009; Yuan et al. 2010). Due to the diversity of compositions of the formulations, the biocompatibility is difficult to evaluate. From few years, it has appeared in the literature the necessity to evaluate the toxicity of SLN first in vitro, then ex vivo, and finally in vivo for comparison with other nanostructures (Nassimi et al. 2009, 2010). Some examples will be described and discussed in view of this lack of toxicity data in SLN and NLC reviewed in the literature.

One of the first studies that was done in this area was by researching the viability of human granulocytes incubated with magnetite-loaded SLN and it was shown to have given an ED_{50} over 10 % (Müller et al. 1996a). In another study with human granulocytes, SLNs were found to be ten times less cytotoxic than PLA nanoparticles and 100-fold less cytotoxic than butylcyanoacrylate nanoparticles (Müller et al. 1996b). Schöler et al. (2000) investigated the action of SLN effects in peritoneal macrophages which showed neither direct nor indirect cytotoxic effects. There were no immunomodulatory effects since no changes in the secretion of interleukin-6, pro-inflammatory cytokines, interleukin-12, and TNF-alpha were observed. They found that the cytotoxicity was dependent on the lipid used as matrix material; for example, stearic acid showed pronounced cytotoxicity, whereas triglycerides showed no major cytotoxic effects (Schöler et al. 2002). Stabilizers such as Tyloxapol and Lutrol exerted a different tolerability/cytotoxicity effect in cell cultures and internalized in the cells and localized in the cytoplasm. Tyloxapol-stabilized SLN led to a cytostatic effect and caused moderate delayed cytotoxicity differently than Lutrol (Kristl et al. 2008).

The NLCs were prepared with monostearin dissolved in warm ethanol. The resultant organic solution was dispersed into distilled water under mechanical stirring. The obtained pre-emulsion (melted lipid droplet) was then cooled to room temperature till NLC dispersion was obtained (134.4 ± 21.0 nm, zeta potential -45.5 ± 3.0 mV). The cytotoxicity by MTT assay was carried out on MCF-7 (human breast cancer cells) (IC_{50} of 455.5 ± 27.7 μg/mL) and MCF-7/ADR (multidrug-resistant variant) (IC_{50} of 469.7 ± 30.0 μg/mL), SKOV3 (human ovarian cancer cells) (IC_{50} of 487.9 ± 31.9 μg/mL), and SKOV3-TR30 (multidrug-resistant variant) (IC_{50} 498.9 ± 35.9 μg/mL). The results showed that NLC showed very high value of IC_{50} in the two kinds of cell lines above and their multidrug-resistant variants, suggesting these lipid materials were safe for the use of drug carriers (Zhang et al. 2008).

SLNs using cetyl palmitate (CP) (189.0 ± 1.8 nm and zeta potential −34.7 ± 4.1 mV), myristyl myristate (MM) (185.4 ± 6.3 nm and zeta potential −31.9 ± 2.1 mV), and cetyl esters (CE) (197.5 ± 3.6 nm and zeta potential −30.5 ± 1.5 mV) were produced by hot HPH. The solid lipid was heated to around 10 °C above its melting point. Afterwards, the mixture was added to a hot aqueous solution of Pluronic F68 under high agitation in an Ultra-turrax to form a pre-emulsion. The pre-emulsion was homogenized using a high pressure homogenizer, applying three homogenization cycles at 600 bar and cooled to form the SLNs. The results obtained in cytotoxicity by MTT assays on BALB/c 3T3, and HaCaT cell lines showed that the cytotoxic effect was influenced by the lipid matrix. The lipid that showed smaller reduction in cell viability in the tested concentration range was the CE. Although in some concentrations of SLN cell viability was reduced, until the highest concentration tested (500 μg/mL), the IC_{50} was not reached for the three lipids in both cells showing no cytotoxic potential of the tested SLN dispersions (Marcato et al. 2011; Ridolfi et al. 2011).

NLC prepared by melt emulsification followed by ultrasonication and coated with polysorbate 80, producing particles of 90.7 ± 4.28 nm size, showed in vitro no significant cytotoxic effects on leukemic EL-4 cells (Sharma et al. 2011).

In another work, the cytotoxicity of three different formulations of SLN was studied. The first one was tristearin SLN (F4: Tween 80, 128 ± 4.1 nm, zeta potential −18.0 mV; F5: SDS, 110 ± 4.5 nm, zeta potential −26.0 mV), the second was solid white vaseline USP SLN (F1: Tween 80, 116.0 ± 3.6 nm, and zeta potential −39.0 mV), and the third one was tristearin SLN (F5: SDS, 116 ± 3.5 nm, and zeta potential −39 mV). For preparation of both SLNs, the lipid was solubilized in a solution of chloroform/methanol. Organic solvents were removed and the lipid layer was melted by heating to 5 °C above the lipid melting point. After that, an aqueous phase was prepared by dissolving Tween 80 or SDS in ultra-pure water to produce the preparation and heated to the molten lipid phase temperature. The hot aqueous phase was added to the molten lipid phase and ultrasonicated and then the hot nanoemulsion is allowed to cool at room temperature. The cytotoxicity of SLN was evaluated by MTT assay using monkey kidney fibroblasts (Vero) and dog kidney fibroblasts (MDCK). IC_{50} for Vero cells were 489 ± 4.9 μg/mL for F1, 682 ± 6.7 μg/mL for F4, and 247 ± 7.2 μm/mL for F5. IC_{50} for MDCK cells were >1,000 μm/mL for F1, >1,000 μg/mL for F4, and 603 ± 2.7 μg/mL for F5. These results showed that nanoparticles F1 and F5 were more harmful to Vero cell lineage than F4. However, F5 was the only SLN showing cytotoxicity to MDCK cell lineage. Although these values are showing low toxicity, they suggest that these nanoparticles interfere in mitochondrial metabolism of cells (Silva et al. 2012). This test evaluates the mitochondrial function as a measurement of cell viability, which allows the detection of dead cells before they lose their integrity and shape (Meunier et al. 1995).

10.3.2 In Vivo Assays

The tolerability in vivo was carried out by multiple high dose bolus injections. The administered dose of SLN in mice corresponded to a sixfold bolus injection of 100 g solid lipid to man (75 kg body weight). The histology of the hepatic and splenic tissues revealed that, in general, all administered particles are well tolerated. If alterations were observed, they were reversible within 6 weeks after injection (Müller et al. 1995; Weyghers et al. 2006).

NLC-based topical gel for the treatment of in vivo inflammation and allied conditions was studied. The NLC was produced dissolving phospholipon 90G in methanol and mixed with an acetone solution containing a blend of stearic acid and oleic acid. The mixture was then added dropwise to pluronic F68 solution at 70 °C, under high agitation in Ultra-turrax homogenizer, obtaining a pre-emulsion. This pre-emulsion was ultrasonicated (20 W) for 15 min to prevent the crystallization of lipids. The o/w emulsion obtained was subsequently cooled down to room temperature with continuous stirring, and the lipid was recrystallized to form NLC with particle sizes between 233 and 286 nm and zeta potential of −14.8 mV. The NLC-unloaded gel did not exhibit reaction found on the skin. Therefore, it can be assured that the gel formulation can be used for topical application. The anti-inflammatory activity of the optimized formulation was evaluated by the carrageenan-induced hind paw inflammation method on Wistar rats. Values for the percentage edema rate for NLC-gel were compared with the saline control and the differences were considered insignificant (Patel et al. 2012).

SLN particles (100 nm), with a potential application as carriers for pulmonary administration, were prepared by rotor-stator homogenization, using Compritol SLM and different surfactant (Poloxamer 188) concentrations and emulsification times. The particles showed spherical shape and smooth surface. In vivo assessment was carried out in rats by intratracheal instillation of either placebo or SLN dispersion, and by bronchoalveolar lavage for cytological analysis. Total cell counts showed no significant differences between placebo and SLN groups. Regarding cytology, percentage of polymorphonuclear neutrophils and macrophages did not significantly differ between groups. These results suggest that a single intratracheal administration of the SLN does not induce a significant inflammatory airway response in rats and that the SLN might be a potential carrier for encapsulated drug by the pulmonary route (Sanna et al. 2003).

NLCs with 600 nm diameter were prepared by microemulsion in order to determine the genotoxicity through *Drosophila melanogaster* (Benford et al. 2000). Although it was not possible to establish a dose/toxic effect of NLC and the descendent of *D. melanogaster* generated, the preliminary results showed that NLC is not genotoxic.

10.4 Toxicity of Solid Lipid Nanoparticles and Nanostructured Lipid Carriers

SLN composites with lipid cholesteryl-butyrate (Chol-but) (12 %), Epikuron 200 (containing about 95 % of soy phosphatidylcholine), and sodium taurocholate (3 %) were prepared by microemulsion method. The average diameter of SLNs was around 100–150 nm with zeta potential of −29 mV and their shape was spherical. Chol-but SLNs were tested in vitro and proved to be an effective and suitable pro-drug of butyrate in cancer treatment. Cytotoxic activity analysis (MTT test at 72 h) performed on four human glioma cell cultures (U87, U373, Lipari, DF) showed that Chol-but SLNs exhibited an IC_{50} of around 0.18–0.25 mM for all cell cultures, indicating that Chol-but SLNs are able to induce a great cell growth inhibition (Brioschi et al. 2008).

Cationic SLNs were prepared by a modified solvent-emulsification method using cholesteryl oleate, glyceryl trioleate, L-α-dioleoyl phosphatidylethanolamine (DOPE), and 3-β-[*N*-(*N′*,*N′*-dimethylaminoethane)-carbamoyl] cholesterol hydrochloride (DC-cholesterol). The cytotoxicity of SLN (117 nm size and zeta potential of +41.8 mV) was performed using human breast adenocarcinoma cells (MDAMB435 cells) by Cell Counting Kit-8 assay. The concentration of SLN was varied from 3 to 72 μg/mL, the same range for transfection experiments. SLN did not induce any damage to cells up to 48 μg/mL and only around 20 % at the concentration of 72 μg/mL. The cytotoxicity study clearly revealed that SLN was not toxic, suggesting that SLN could be used as no cytotoxic core to transfection of molecules (Kim et al. 2008). Similar study was recently published (Doktorovova et al. 2012).

Cationic SLNs were prepared by phase inversion temperature (PIT) method using three cationic lipids (tetradecyltrimethyl ammonium bromide (CTAB), dimethyldioctadecylammonium bromide (DDAB), and *N*-[1-(2,3-dioleoyloxy)propyl]-*N*,*N*,*N*-trimethylammonium chloride [DOTAP]). The SLN with DOTAP exhibited a higher zeta potential and smaller particle size (462.9 nm) than the other particles (SLN-CTAB, SLN-DDAB). Thus, the cytotoxicity of these particles on two models of cell cultures (human prostate cancer androgen-non-responsive DU-145 cells and primary cultures of rat astrocytes) was evaluated. DU-145 cells resulted to be more sensible, exhibiting an IC_{50} of 125 μg/mL, whereas to primary cultures of rat astrocytes, the IC_{50} was 500 μg/mL. Probably this difference was due to different proliferative capacity of the two cellular types as well as to their different cellular membrane composition (Carbone et al. 2012).

SLNs capped with chitosan (SLN-chitosan) were produced by hot HPH using a solid lipid (myristyl myristate) and Pluronic® F68—as stabilizer. The solid lipid was heated at 65 °C and added to a hot aqueous solution containing Pluronic® F68 and chitosan (pH 4.3) under high agitations in an Ultra-turrax to form a pre-emulsion. The pre-emulsion was homogenized by three homogenization cycles at 600 bar and cooled in an ice bath to form the nanoparticles. SLN-chitosan showed larger size (284.8 nm) than SLN without chitosan (SLN) (162.7 nm) and positive surface charge (zeta potential +55.9 mV), whereas SLN showed negative zeta potential (−31.9 mV). Cytotoxicity was carried out by the MTT assay with HaCaT cell lines, and both particles (SLN-chitosan and SLN) were not toxic to these cell lines indicating that these particles are an interesting carrier system (Ridolfi et al. 2012).

Cytotoxicity of SLN and NLC was studied. SLNs were prepared with myristyl myristate, as solid lipid, and NLC was composited by the same solid lipid and Liponate® GC (a mixture of triglycerides of caprylic and capric acids) as oily phase. Both particles were produced by HPH method. The diameters of SLN and NLC were 188.02 ± 7.07 (zeta potential of −26.91 mV) and 167 ± 14 nm (zeta potential of −28.83 mV), respectively. The cytotoxicity of the nanoparticles by MTT assay on two distinct cell lines (mouse 3T3 fibroblasts and human keratinocytes (HaCaT) cultured cells) was tested. Under the experimental conditions (0.5–4.2 mmol/L lipid concentration), SLN and NLC can be considered safe, since they did not affect the survival of either cell lines (~100 % cell viability) (Barbosa et al. 2013).

Tripterine-loaded NLCs (T-NLCs) were prepared by solvent evaporation using Precirol ATO-5 (solid lipid), Labrafil M 1944CS (liquid lipid), soybean lecithin (stabilizer), D-α-tocopherol polyethylene glycol succinate (TPGS), and Ste-R_6L_2 (as cell-penetrating peptide [CPP]). The cytotoxicity was performed by MTT assay to investigate in vitro intestinal toxicity using Caco-2 cells for this assay. The cells were divided into three treatment groups: those treated with (a) CPPs-coated T-NLCs (CT-NLCs −115.9 nm, zeta potential +13.6 mV), (b) tripterine-loaded NLCs (T-NLCs −102.4 nm, zeta potential −26.2 mV), and (c) tripterine solution (dimethyl sulfoxide, 0.1 %). At the same dose, the viability of cells treated with CT-NLCs was lower than that of the cells treated with T-NLCs, but was markedly higher than the viability of cells treated with tripterine solution. A substantial decrease was noted in the viability of cells treated with tripterine especially at a high concentration compared with those treated with the CT-NLCs and T-NLCs. The IC_{50} values of the tripterine solution, T-NLCs, and CT-NLCs were 0.316, 1.171, and 0.885 μg/mL, respectively (Chen et al. 2012). The T-NLCs' higher toxicity than that of the CT-NLCs could be related to the negative charge on the surface of the nanoparticles.

10.5 Size and Chemical Structure Effect of Solid Lipid Nanoparticles on Cytotoxicity and Genotoxicity

SLN consisting of stearic acid or different kinds of adeps solidus (hard fat is a mixture of mono-, di-, and triglyceride esters of higher saturated fatty acids) was produced when these components were formulated with lecithin, sodium taurocholate, polysorbate 80, and stearylamine using melt-emulsification method. The cytotoxicity and physical properties of SLNs for dermal applications showed a significant dependence from the formulation process. Droplet size and zeta potential of submicron emulsions depended on the composition of the co-surfactant blend used. The viability of J774 macrophages, mouse 3T3 fibroblasts, and HaCaT keratinocytes was significantly reduced in the presence of stearylamine. Survival of macrophages was highly affected by stearic acid and stearylamine. In general a viability of more than 90 % was observed when semisynthetic glycerides or hard fat was employed to formulate nanoparticles (Weyenberg et al. 2007).

SLN was prepared by HPH technique using as lipid phase hydrogenated palm oil (S154) and hydrogenated soybean lecithin (Lipoid S100) and as aqueous phase water, oleyl alcohol, thimerosal, and Sorbitol. Both phases were mixed and homogenized at 1,000 bar by 20 cycles, obtaining particles with 145.00 ± 3.39 nm size and zeta potential of -19.50 ± 1.80 mV. The MTT assay was conducted for the cytotoxicity on human breast cancer cell lines, MCF-7 and MDA-MB231. The IC_{50} of SLN was time-dependent and different between the tested breast cell lines (287.5 ± 17.7 μg/mL for MCF-7 and 290.2 μg/mL for MDA-MB231). The IC_{50} values indicated that SLN had low cytotoxicity to these cells. The low cytotoxicity of the SLN can be attributed to lecithin and components of the aqueous phase used, especially the nonionic emulsifier Abbasalipurkabir et al. (2011a, b).

10.6 Comments and Future Perspectives

SLN and NLS are attractive carriers of actives used as cosmetic and pharmaceutical products. An advantage of NLS is its easy production in large scale and sterilization that led to increase the industry interest in this carrier. Different methods are used for their preparations, with the absence of organic solvents or only reagents previously approved by Food and Drug Administration (FDA) through the GRAS rules. A disadvantage is sometimes the low loading efficiency; however, this can be possible to circumvent, building different kind of SLN or NLC that exhibits different capacities for oral, nasal, parenteral, dermic, and ophthalmic administrations. This chapter showed that SLN or NLC, in general, is nontoxic to cell cultures or in vivo assays. However, it depends on many factors, such as size, chemical components, capped surface, or surface charge. In order to maintain the safety condition it must be studied case by case. It is not possible to generalize their behavior in cells or in vivo studies. Curiously, very few studies related to their genotoxicity were found in the actual literature. In the next year a larger emphasis must be done in this direction.

Acknowledgment The authors acknowledge the FAPESP, INOMAT (MCT/CNPq), NanoBioss (MCTI), and Brazilian Network of Nanotoxicology (MCT/CNPq).

References

Abbasalipurkabir R, Salehzadeh A, Abdullah R (2011a) Delivering tamoxifen within solid lipid nanoparticles. Pharm Technol 36:74–79

Abbasalipurkabir R, Salehzadeh A, Abdullah R (2011b) Cytotoxicity effect of solid lipid nanoparticles on human breast câncer cell lines. Biotechnology 10:528–633

Allais C, Keller G, Lesieur P et al (2003) X-ray diffraction/calorimetry coupling a tool for polymorphism control. J Therm Anal Calorim 74:723–728

Bangham AD, Standish MM, Watkins JC (1965) Diffusion of univalent ions across the lamellae of swollen phospholipids. J Mol Biol 13:238–252

Barbosa RM, da Silva CMG, Bella TS et al (2013) Cytotoxicity of solid lipid nanoparticles and nanostructured lipid carriers containing the local anesthetic dibucaine designed for topical application. J Phys Conf 429:012035

Benford D, Hanley B, Bottrill K (2000) Biomarkers as predictive tools in toxicity-testing. Altern Lab Anim 28:119–131

Brioschi A, Zara GP, Calderoni S et al (2008) Cholesterylbutyrate solid lipid nanoparticles as a butyric acid prodrug. Molecules 13:230–254

Bunjes H, Steiniger F, Richter W (2007) Visualizing the structure of triglyceride nanoparticles in different crystal modifications. Langmuir 23:4005–4011

Buse J, El-Aneed A (2010) Properties, engineering and applications of lipid-based nanoparticle drug-delivery systems: current research and advances. Nanomedicine (Lond) 5:1237–1260

Carbone C, Tomasello B, Ruozi B et al (2012) Preparation and optimization of PIT solid lipid nanoparticles via statistical factorial design. Eur J Med Chem 49:110–117

Chen H, Chang X, Du D et al (2006) Podophyllotoxin-loaded solid lipid nanoparticles for epidermal targeting. J Control Release 110:296–306

Chen Y, Yuan L, Zhou L et al (2012) Effect of cell-penetrating peptide-coated nanostructured lipid carriers on the oral absorption of tripterine. Int J Nanomed 7:4581–4591

Das S, Chaudhury A (2011) Recent advances in lipid nanoparticle formulations with solid matrix for oral drug delivery. AAPS PharmSciTech 12:62–76

Doktorovova S, Silva AM, Gaivão I et al (2012) Genotoxicity evaluation of cationic solid lipid nanoparticles in Caco-2 cells. In: Proceedings of the ninth central European symposium on pharmaceutical technology, Dubrovnik

Durán N, Teixeira Z, Marcato PD (2011) Topical application of nanostructures: solid lipid, polymeric and metallic nanoparticles. In: Guterres SS, Beck R, Polhmann A (eds) Nanocosmetics and nanomedicines: new approaches for skin care. Part II, Chap. 4. Springer, Berlin, pp 69–100

Ekambaram P, Sathali AAH, Priyanka K (2012) Solid lipid nanoparticles: a review. Sci Rev Chem Commun 2:80–102

Fessi H, Puisieux F, Devissaguet JP et al (1989) Nanocapsule formation by interfacial polymer deposition following solvent displacement. Int J Pharm 55:R1–R4

Garud A, Singh D, Garud N (2012) Solid lipid nanoparticles (SLN): method, characterization and applications. Int Curr Pharm J 1:384–393

Gasco MR (2007) Lipid nanoparticles: perspectives and challenges. Adv Drug Deliv Rev 59:377–378

Hou D, Xie C, Huang K, Zhu C (2003) The production and characteristics of solid lipid nanoparticles (SLNs). Biomaterials 24:1781–1785

Hu F-Q, Jiang S-P, Du Y-Z et al (2006) Preparation and characteristics of monostearin nanostructured lipid carriers. Int J Pharm 314:83–89

Joshi MD, Müller RH (2009) Lipid nanoparticles for parenteral delivery of actives. Eur J Pharm Biopharm 71:161–172

Joshi M, Patravale V (2008) Nanostructure lipid carrier (NLC) based gel of celecoxib. Int J Pharm 346:124–132

Khan S (2012) Solid lipid nanoparticles: a review. World J Pharm Sci 1:96–115

Kim HR, Kim IK, Bae KH et al (2008) Cationic solid lipid nanoparticles reconstituted from low density lipoprotein components for delivery of siRNA. Mol Pharm 5:622–631

Kristl J, Teskac K, Milek M et al (2008) Surface active stabilizer tyloxapol in colloidal dispersions exerts cytostatic effects and apoptotic dismissal of cells. Toxicol Appl Pharmacol 232:18–25

Liu J, Gong T, Fu H et al (2008) Solid lipid nanoparticles for pulmonary delivery of insulin. Int J Pharm 356:333–334

Marcato PD (2008) Preparation, characterization and application in drugs and cosmetics of solid lipid nanoparticles. Rev Eletron Fárm 6:1–37

Marcato PD, Caverzan J, Rossi-Bergmann B et al (2011) Nanostructured polymer and lipid carriers for sunscreen. Biological effects and skin permeation. J Nanosci Nanotechnol 11:1880–1886

Mathur V, Satrawala Y, Rajput MS et al (2010) Solid lipid nanoparticles in cancer therapy. Int J Drug Deliv 2:192–199

Mehnert W, Mader K (2001) Solid lipid nanoparticles: production, characterization and applications. Adv Drug Deliv Rev 47:165–196

Meunier V, Bourrie M, Berger Y et al (1995) The human intestinal epithelial cell line Caco-2: pharmacological and pharmacokinetic applications. Cell Biol Toxicol 11:187–194

Müller RH, Mehnert W, Lucks JS et al (1995) Solid lipid nanoparticles (SLN): an alternative colloidal carrier system for controlled drug delivery. Eur J Pharm Biopharm 41:62–69

Müller RH, Maassen S, Weyhers H et al (1996a) Cytotoxicity of magnetite-loaded polylactide, polylactide/glycolide particles and solid lipid nanoparticles. Int J Pharm 138:85–94

Müller RH, Maassen S, Weyhers H et al (1996b) Phagocytic uptake and cytotoxicity of solid lipid nanoparticles (SLN) sterically stabilized with poloxamine 908 and poloxamer 407. J Drug Target 4:161–170

Müller RH, Rühl D, Runge S et al (1997) Cytotoxicity of solid lipid nanoparticles as a function of the lipid matrix and the surfactant. Pharm Res 14:458–462

Müller RH, Maëder K, Gohla S (2000) Solid lipid nanoparticles (SLN) for controlled drug delivery: a review of the state of the art. Eur J Pharm Biopharm 50:161–177

Müller RH, Radtke M, Wissing SA (2002) Solid lipid nanoparticles (SLN) and nanostructured lipid carriers (NLC) in cosmetic and dermatological preparations. Adv Drug Deliv Rev 54: S131–S155

Müller RH, Petersen RD, Hommoss A et al (2007) Nanostructured lipid carriers (NLC) in cosmetic dermal products. Adv Drug Deliv Rev 59:522–530

Müller RH, Shegokar R, Cornelia M et al (2011) 20 years of lipid nanoparticles (SLN and NLC): present state of development and industrial applications. Curr Drug Discov Technol 8:207–227

Nassimi M, Schleh C, Lauenstein HD et al (2009) Low cytotoxicity of solid lipid nanoparticles in in vitro and ex vivo lung models. Inhal Toxicol 21(suppl 1):104–109

Nassimi M, Schleh C, Lauenstein HD et al (2010) A toxicological evaluation of inhaled solid lipid nanoparticles used as a potential drug delivery system for the lung. Eur J Pharm Biopharm 75:107–116

Pardeike J, Hommoss A, Müller RH (2009) Lipid nanoparticles (SLN, NLC) in cosmetic and pharmaceutical dermal products. Int J Pharm 366:170–184

Patel D, Dasgupta S, Dey S et al (2012) Nanostructured lipid carriers (NLC)-based gel for the topical delivery of aceclofenac: preparation, characterization, and in vivo evaluation. Sci Pharm 80:749–764

Pedersen N, Hansen S, Heydenreich AV et al (2006) Solid lipid nanoparticles can effectively bind DNA, streptavidin and biotinylated ligands. Eur J Pharm Biopharm 62:155–162

Pragati S, Kuldeep S, Ashok S et al (2009) Solid lipoid nanoparticles: a promising drug delivery technology. Int J Pharm Nanotechnol 2:509–516

Puri A, Loomis K, Smith B et al (2009) Lipid-based nanoparticles as pharmaceutical drug carriers: from concepts to clinic. Crit Rev Ther Drug Carrier Syst 26:523–580

Rawat MK, Jain A, Singh S (2011a) In vivo and cytotoxicity evaluation of repaglinide-loaded binary solid lipid nanoparticles after oral administration to rats. J Pharm Sci 100:2406–2417

Rawat MK, Jain A, Singh S (2011b) Studies on binary lipid matrix based solid lipid nanoparticles of repaglinide: in vitro and in vivo evaluation. J Pharm Sci 100:2366–2378
Ridolfi DM, Marcato PD, Machado D et al (2011) In vitro cytotoxicity assays of solid lipid nanoparticles in epithelial and dermal cells. J Phys Conf 304:012032
Ridolfi M, Marcato PD, Justo GZ et al (2012) Solid lipid nanoparticles coated with chitosan as carrier for topical delivery of tretinoin. Colloids Surf B Biointerfaces 93:36–40
Ruktanonchai U, Limpakdee S, Meejoo S et al (2008) The effect of cetyl palmitate crystallinity on physical properties of gamma-oryzanol encapsulated in solid lipid nanoparticles. Nanotechnology 19:095701
Sanna V, Kirschvink N, Gustin P et al (2003) Preparation and in vivo toxicity study of solid lipid microparticles as carrier for pulmonary administration. AAPS PharmSciTech 5:1–7
Schöler N, Zimmermann E, Katzfey U et al (2000) Preserved solid lipid nanoparticles (SLN) at low concentrations do cause neither direct nor indirect cytotoxic effects in peritoneal macrophages. Int J Pharm 196:235–239
Schöler N, Hahn H, Muller RH et al (2002) Effect of lipid matrix and size of solid lipid nanoparticles (SLN) on the viability and cytokine production of macrophages. Int J Pharm 231:167–176
Severino P, Andreani T, Macedo AS et al (2012) Current state-of-art and new trends on lipid nanoparticles (SLN and NLC) for oral drug delivery. J Drug Deliv 2012:750891
Sharma P, Dube B, Sawant K (2011) Development and evaluation of nanostructured lipid carriers of cytarabine for treatment of meningeal leukemia. J Nanosci Nanotechnol 11:6676–6682
Silva AH, Filippin-Monteiro FB, Mattei B et al (2012) In vitro biocompatibility of solid lipid nanoparticles. Sci Total Environ 432:382–388
Smijs TG, Bouwstra JA (2010) Focus on skin as a possible port of entry for solid nanoparticles and the toxicological impact. J Biomed Nanotechnol 6:469–484
Souto EB, Severino P, Santana MHA et al (2011) Solid lipid nanoparticles: classical methods of lab production. Quim Nova 34:1762–1769
Teeranachaideekul V, Boonme P, Souto EB et al (2008a) Influence of oil content on physicochemical properties and skin distribution of Nile red-loaded NLC. J Control Release 128:134–141
Teeranachaideekul V, Souto EB, Müller RH et al (2008b) Physicochemical characterization and in vitro release studies of ascorbyl palmitate-loaded semi-solid nanostructured lipid carriers (NLC gels). J Microencapsul 25:111–120
Üner M (2006) Preparation, characterization and physico-chemical properties of solid lipid nanoparticles (SLN) and nanostructured lipid carriers (NLC): their benefits as colloidal drug carrier systems. Pharmazie 61:375–386
Üner M, Radtke M, Wissing AS (2007) Solid lipid nanoparticles and nanostructured lipid carriers. Encyclop Nanosci Nanotechnol 10:43–56
Waghmare AS, Grampurohit ND, Gadhave MV et al (2012) Solid lipid nanoparticles: a promising drug delivery system. Int Res J Pharm 3:100–107
Westesen K, Bunjes H, Koch MHJ (1997) Physicochemical characterization of lipid nanoparticles and evaluation of their drug loading capacity and sustained release potential. J Control Release 48:223–236
Weyenberg W, Filev P, Van den Plas D et al (2007) Cytotoxicity of submicron emulsions and solid lipid nanoparticles for dermal application. Int J Pharm 337:291–298
Weyghers H, Ehlers S, Hahn H et al (2006) Solid lipid nanoparticles (SLN)—effects of lipid composition on in vitro degradation and in vivo toxicity. Pharmazie 61:539–544
Wissing SA, Müller RH, Manthel L et al (2004a) Structural characterization of Q10-loaded solid lipid nanoparticles by NMR spectroscopy. Pharm Res 21:400–405
Wissing SA, Kayser O, Müller RH (2004b) Solid lipid nanoparticles for parenteral drug delivery. Adv Drug Deliv Rev 56:1257–1272
Yuan H, Zhang W, Du YZ et al (2010) Ternary nanoparticles of anionic lipid nanoparticles/protamine/DNA for gene delivery. Int J Pharm 392:224–231
Zhang XG, Miao J, Dai YQ et al (2008) Reversal activity of nanostructured lipid carriers loading cytotoxic drug in multi-drug resistant cancer cells. Int J Pharm 361:239–244

Chapter 11
Cytotoxicity and Genotoxicity of Biogenically Synthesized Silver Nanoparticles

Nelson Durán, Amedea B. Seabra, and Renata de Lima

Abstract In recent years, the development of nanotechnology has been focused on the development of protocols to synthesize important technological and medical metallic nanoparticles, such as silver nanoparticles, based on clean, nontoxic, biocompatible, and environmentally friendly approaches. "Green" synthesis of nanoparticles can be successfully performed extracellularly or intracellularly by organisms such as bacteria, yeast, fungi, algae, and plant extracts. Only in the recent past, biogenic syntheses of metal nanoparticles have gained significant attention. Silver nanoparticles (AgNPs) are considered one of the most important and commonly used metallic nanoparticles, in particular in medical applications, due to their known antimicrobial activities. In this scenario, this chapter discusses the recent developments on the biogenic synthesis of AgNPs by bacteria, yeast, fungi, algae and plants, highlighting the advantages and drawbacks of biogenic syntheses methods. Moreover, in order to propose any biological applications of AgNPs, it is mandatory to detailed investigate the toxicity of this nanomaterial. In this context, this chapter also discusses recent progress on the in vitro and in vivo cytotoxicity and genotoxicity of biogenic and chemically synthesized AgNPs. Although important progresses have been reached in this domain, there is still a necessity of more and detailed studies on the toxicity of AgNPs, in particular on biogenic AgNPs. Therefore, this chapter hopes to be a source of inspiration for more studies on the biogenic syntheses of AgNPs and the fully characterization of their toxic effects on humans and on the environment.

N. Durán (✉)
Biological Chemistry Laboratory, Institute of Chemistry, Universidade Estadual de Campinas, CEP 13083-970, Campinas, São Paulo, Brazil

Center of Natural and Human Sciences, Universidade Federal do ABC – UFABC, Santo André, São Paulo, Brazil
e-mail: duran@iqm.unicamp.br

N. Durán et al. (eds.), *Nanotoxicology*, Nanomedicine and Nanotoxicology,
DOI 10.1007/978-1-4614-8993-1_11,

11.1 Introduction

In recent years, the accelerate expansion of nanotechnology has led to the production of various commercially available nano-sized materials, such as silver nanoparticles (AgNPs) (Ahmed et al. 2008), which received considerable attention due to their potent antimicrobial activities (Durán et al. 2010). AgNPs can be synthesized by traditional chemical methods, which are considered aggressive to the environment and to human health. Biogenic syntheses of AgNPs have been emerging as an attractive "green" alternative to synthesize nanomaterials for diverse applications (Bansal et al. 2012).

AgNPs have been used in many commercial products (Hood 2004; Wijnhoven et al. 2009) with different chemical and physical properties (Wise et al. 2010; AshaRani et al. 2008). The increased uses of AgNPs in several applications led to a concern regarding the toxicity of these nanoparticles and the safe use of these nanomaterials, as discussed in diverse subareas of nanotechnology, such as nanobiotechnology and nanomedicine (Brayner 2008; Panda et al. 2011). In fact, nanoparticles may have higher toxicity than bulk materials (Donaldson et al. 1999; Xiong et al. 2011). Nanoparticle toxicity direct impacts human health and the environment, and more investigations are needed in this topic (Nel et al. 2006; Lewinski et al. 2008; Ju-Nam and Lead 2008).

Important studies have demonstrated that size and the chemical nature of nanoparticles coating can lead to different toxic effects at the cellular, subcellular, and biomolecular levels, such as genes and proteins (Gurr et al. 2005; Chi et al. 2009). Most of the in vitro assays to characterize the toxicity of AgNPs have been carried out in different cellular models, such as human lung fibroblasts (AshaRani et al. 2009). The majority of the studies reported an increase in the oxidative stress and severe lipid peroxidation, as observed in the case of fish brain tissue upon exposure to nanomaterials (Oberdörster 2004). In general, the proposed mechanisms for these effects consider the induction of reactive oxygen species (ROS) formation by AgNPs, which is associated with DNA damage, apoptosis, and necrosis (Arora et al. 2008; Kim et al. 2009a, b; Foldbjerg et al. 2011). In vitro exposure to AgNPs showed a reduction in glutathione levels and the observation of lipid peroxidation (Arora et al. 2008; Kim et al. 2009a, b). In this scenario, complete elucidation of the toxicity mechanisms of AgNPs is of paramount importance, and more studies based on the cytotoxicity and genotoxicity of AgNPs are still necessary. In this regard, many important reviews described the methodologies currently available for in vitro and in vivo genotoxicological studies of nanomaterials, including AgNPs (Ng et al. 2010; Johnston et al. 2010; Rico et al. 2011; Gonzalez et al. 2011; de Lima et al. 2012; Doak et al. 2012). Therefore, the aim of this chapter is to present and discuss recent progress on the synthesis of AgNPs by using environmentally friendly approaches, which minimizes negative impacts of nanoparticles on human health, and also the evaluation of the toxicity of AgNPs, highlighting the necessity of more studies in this exciting area.

11.2 Biogenic Silver Nanoparticles: Preparation and Characterization

Traditional chemical methods of synthesis of metallic nanoparticles, in particular AgNPs, employ toxic reagents, release harmful byproducts to the environment, consume a lot of energy, and use expensive chemicals. Moreover, chemical syntheses result in the absorbance of toxic chemicals on the surface of nanoparticles (Bansal et al. 2012). In contrast, microorganisms and plants extracts are considered interesting nanofactories in the fabrication of metallic nanoparticles, in a novel concept of environmentally friendly synthesis of nanomaterials (Antony et al. 2011; Bai et al. 2011; Kumar et al. 2012a). In this context, the use of plant extracts and microorganisms in the synthesis of NPs has been emerged as a biocompatible, "green," and exciting approach. In fact, biogenic synthesis of NPs has been gained a tremendous attention in recent years, since it is cost-effective and environmentally friendly. Either uni- or multicellular organisms have been shown immense potential for the synthesis of NPs, since they are able to synthesize AgNPs either intracellulary or extracellulary (Birla et al. 2009). Biogenic synthesis of inorganic NPs employs mild experimental conditions, such as pH, pressure, and temperature, leading to the formation of NPs coated with capping layer formed by proteins and/or lipids that confer physiological solubility and stability for the nanoparticles (Banu et al. 2011; Fayaz et al. 2011; Li et al. 2011). In this scenario, this section highlights recent developments on the biogenic syntheses of AgNPs by bacteria, yeast, fungi, algae, and plant extracts. The most common techniques used to characterize the obtained biogenic AgNPs are UV-visible spectroscopy, transmission electron microscopy (TEM), scanning electron microscopy (SEM), X-ray diffraction, Fourier transform infrared spectroscopy (FTIR), and Raman spectroscopy (Kora et al. 2012).

11.2.1 Bacterial Synthesis

Recently, many bacterial species have been successfully used to extracellularly synthesize AgNPs (Bai et al. 2011; Bansal et al. 2012). Biogenic synthesis of AgNPs by bacteria is carried out by the incubation of silver ions with cell filtrates of bacteria at room temperature for minutes–hours.

Incubation of bacterial culture isolated of *Pseudomonas stutzeri* AG259 with silver ions led to the formation of AgNPs with controlled size and distinct morphology within the periplasmic space of the bacteria (Klaus et al. 1999; Joerger et al. 2000; Klaus-Joerger et al. 2001). Recently, marine actinomycetes strain of *Streptomyces albidoflavus* was used to synthesize either extracellular or intracellularly AgNPs (Prakasham et al. 2012). The obtained AgNPs showed high stability in aqueous solution. This stabilization could be attributed to the secretion of proteins by the bacterium in the reaction mixture. Moreover, biogenic synthesized AgNPs revealed antimicrobial activities against both Gram-negative and Gram-positive bacterial strains (Prakasham et al. 2012). Fayaz et al. (2011) reported the preparation of stable AgNPs through the exposition of cell-free extract of the bacterium *Geobacillus stearothermophilus* to $AgNO_3$ solution. The stability of AgNPs was attributed to the presence of capping proteins on nanoparticle surface, as evidenced by FTIR and sodium dodecyl sulfate polyacrylamide gel electrophoresis (SDS PAGE). The formation of AgNPs was confirmed by the detection of the characteristic optical absorption band in the UV-visible region (plasmon band at ca. 420 nm). TEM analysis of the AgNPs revealed that the particles have spherical shape and polydispersed pattern, with an average size of dimensions of 5–35 nm. Moreover, X-ray diffraction pattern confirmed that NPs are face-centered cubic in shape and FTIR indicated the presence of reducing enzymes and capping proteins on the NPs surface (Fayaz et al. 2011).

Nowadays, several different bacteria have been extensively used to synthesize AgNPs. Table 11.1 summarizes some important and recent studies based on the preparation of AgNPs by different organisms, including bacteria.

11.2.2 Yeasts Synthesis

Compared with other microorganisms, such as bacteria and fungi, there are a few reports describing the use of yeasts to produce AgNPs. In fact, yeasts can be successfully employed to synthesize metallic NPs, and this approach should be more investigated. Extracellular synthesis of AgNPs was observed in silver-tolerant yeast strains MKY3 when challenged with 1 mM silver ions, in the log phase of yeast growth (Kowshik et al. 2003). Recently, biosynthesis of AgNPs by using yeast biomass was reported (Mourato et al. 2011). Extremophilic yeast strain was isolated from acid mine drainage in Portugal. Exposition of washed yeast cells with Ag ions produced AgNPs with diameter size smaller than 20 nm (Mourato et al. 2011).

11.2.3 Fungal Synthesis

Mycosynthesis is the synthesis of nanoparticles by fungi. The term "mycosynthesis" was used for the first time by Ingle et al. (2008) describing the synthesis of nanoparticles by *Fusarium acuminatum*. Overall, fungi possess some important advantages as a potent microorganism for the synthesis of AgNPs. It is relatively easy to culture fungi, the NPs synthesis is mostly extracellular, and the synthesized NPs have a good polydispersivity, size, and stability (Gade et al. 2010a; Birla et al. 2009).

Rai et al. (2009) proposed the term "Myconanotechnology" to include research carried out on nanoparticles synthesized by fungi. Many fungal species have been

Table 11.1 List of important organisms that synthesize silver nanoparticles (AgNPs)

Microorganism	Size (nm)	Reference
Bacteria		
Rhodobacter sphaeroides	9.5 ± 0.3	Bai et al. (2011)
Geobacillus stearothermophilus	5–35	Fayaz et al. (2011)
Brevibacterium casei	10–50	Kalishwaralal et al. (2010)
Pseudomonas aeruginosa	13 ± 3.7	Kumar and Mamidyala (2011)
Morganella spp.	10–50	Parikh et al. (2011)
Bacillus sp.	5–15	Pugazhenthiran et al. (2009)
Escherichia coli	15–50	Zaki et al. (2011)
Bacillus megaterium	15–50	Zaki et al. (2011)
Acinetobacter sp.	15–50	Zaki et al. (2011)
Stenotrophomonas maltophilia	15–50	Zaki et al. (2011)
Continuation-yeasts		
Extremophilic yeast strain	<20	Mourato et al. (2011)
Yeast strain MKY3	2–5	Kowshik et al. (2003)
Fungi		
Rhizopus stolonifer	3–20	Banu et al. (2011)
Phoma glomerata	60–80	Birla et al. (2009)
Geotricum sp.	30–50	Jebali et al. (2011)
Puccinia graminis	30–120	Kirthi et al. (2012)
Aspergillus tamarii	25–50	Kumar et al. (2012a)
Penicillium purpurogenum	8–10	Nayak et al. (2011)
Cochliobolus lunatus	3–21	Salunkhe et al. (2011)
Chrysosporium keratinophilum	24–51	Soni and Prakash (2012)
Verticillium lecanii	20–50	Soni and Prakash (2012)
Fusarium oxysporum f.sp.	20–40	Soni and Prakash (2012)
Phaenerochaete chrysosporium	5–200	Vigneshwaran et al. (2006)
Cladosporium cladosporioides	10–100	Balaji et al. (2009)
Algae		
Chlamydomona reinhardtii	5–35	Barwal et al. (2011)
Gracilaria dura	4–8	Shukla et al. (2012)
Spirulina platensis	6–10	Govindaraju et al. (2008)
Plant extracts		
Rhizophora apiculata	19–42	Antony et al. (2011)
Iresine herbstii	44–64	Dipankar and Murugan (2012)
Tribulus terrestris	16–28	Gopinath et al. (2012)
Boswellia serrata	7.5 ± 3.8	Kora et al. (2012)
Terminalia Chebula	≤100	Kumar et al. (2012b)
Prosopis juliflora	11–19	Raja et al. (2012)
Tridax procumbens	≤20	Rajasekharreddy et al. (2010)
Jatropha curcas	≤20	Rajasekharreddy et al. (2010)
Calotropis gigantea	≤20	Rajasekharreddy et al. (2010)
Solanum melongena	20–40	Rajasekharreddy et al. (2010)
Datura metel	20–40	Rajasekharreddy et al. (2010)
Carica papaya	20–40	Rajasekharreddy et al. (2010)
Citrus aurantium	20–40	Rajasekharreddy et al. (2010)
Ocimum sanctum	6–110	Rao et al. (2013)

(continued)

Table 11.1 (continued)

Microorganism	Size (nm)	Reference
Cocos nucifera coir	23 ± 2	Roopan et al. (2013)
Cissus quadrangularis	50–1,000	Valli and Vaseeharan (2012)
Artemisia nilagirica	70–90	Vijayakumar et al. (2013)
Annona squamosa	20–100	Vivek et al. (2012)
Stevia rebaudiana	2–50	Yilmaz et al. (2011)

explored for the production of different metal nanoparticles of different shapes and sizes. Chen et al. (2003) reported the extracellular formation of AgNPs using *Phoma* species. Some soil-borne fungi like *Aspergillus fumigatus* are able to synthesize AgNPs when the cell extract was challenged with aqueous silver ions (Bhainsa and D'souza 2006). Among different fungal genera used for the synthesis of NPs, the genus *Fusarium* was extensively used. Ahmad et al. (2003) used for the first time the *Fusarium oxysporum* for the synthesis of AgNPs and opened a new avenue in the field of nanobiotechnology. Durán et al. (2005) studied the extracellular production of metal NPs by several strains of the fungus *F. oxysporum*. Similarly, other *Fusarium* species like *F. oxysporum* strain 5115 (Mohammadian et al. 2007), *F. semitectum* (Basavaraja et al. 2008), *F. acuminatum* (Ingle et al. 2008), *F. solani* (Ingle et al. 2009), and *F. culmorum* (Bawaskar et al. 2010) have been successfully used for the synthesis of AgNPs.

Recently, Kumar et al. (2012a) reported the mycogenic synthesis of AgNPs by *Aspergillus tamari*. Ag ions were reduced by the fungal extracellular filtrate leading to the formation of AgNPs, after 30 min of incubation. SEM images revealed that the nanoparticles were spherical with size ranging from 25 to 50 nm. UV-visible spectrophotometry confirmed the presence of the characteristic plasmon band at 420 nm; X-ray diffraction confirmed the crystalline face-centered cubic of AgNPs. Moreover, FTIR indicated that AgNPs were coated with proteins secreted by the fungus, allowing stabilization of AgNPs in aqueous solution (Kumar et al. 2012a). Narayanan and Sakthivel (2010) discussed in detail the biogenic syntheses of metallic nanoparticles, including AgNPs by microbes, in particular by fungi. Table 11.1 summarizes recent important papers based on the mycosynthesis of AgNPs.

11.2.4 Algae Synthesis

The unicellular algae *Chlamydomonas reinhardtii* was used as a model system to investigate the role of cellular proteins in the synthesis of AgNPs (Barwal et al. 2011). Cell-free extract (in vitro) of *C. reinhardtii* and in vivo cells produced AgNPs of size range 5 ± 1 to 15 ± 2 nm and 5 ± 1 to 35 ± 5 nm, respectively. The authors have identified several cellular proteins related to the synthesis of AgNPs by matrix-assisted laser desorption/ionization-mass spectrometry (MALDI-MS), such as ATP synthase, superoxide dismutase, carbonic anhydrase, and ferredoxin-NADP$^+$ reductase. This elegant paper provides an evidence for the involvement of oxidoreductive proteins in biosynthesis and stabilization of AgNPs (Barwal et al. 2011).

Shukla et al. (2012) reported the formation of AgNPs by the reduction of silver ions, at room temperature, with the addition of agar extracted from the red algae *Gracilaria dura*. AgNPs were found to be spherical in shape, with a good polydispersivity with average size of 6.0 ± 2 nm, as reveled by TEM analysis. Similarly, Govindaraju et al. (2008) reported the formation of several metallic nanoparticles, including AgNPs from *Spirulina platensis*.

11.2.5 Plants Synthesis

Rai et al. (2008) stated that plants have emerged as a simple, cost-effective and eco-friendly system to rapidly synthesize NPs. In this scenario, Persimmon (*Diospyros kaki*) leaf extract led to the formation of AgNPs of 15–90 nm size (Song and Kim 2008). Krishnaraj et al. (2010) reported the rapidly biosynthesis of AgNPs (within 30 min) using leaf extract of *Acalypha indica* and their activity on water-borne bacterial pathogens. Antibacterial activity of biogenically synthesized AgNPs showed effective inhibitory activity against water-borne pathogens, viz., *Escherichia coli* and *Vibrio cholera* (Krishnaraj et al. 2010).

Gade et al. (2010b) reported the green synthesis of AgNPs by *Opuntia ficus-indica*. They evaluate antibacterial activity of synthesized AgNPs against *E. coli* and *S. aureus*, in combination with commercially available antibiotics, such as ampicillin, gentamicin, kanamycin, streptomycin, and vancomycin. Antibacterial activities of the antibiotics increased due to combination with AgNPs, since AgNPs are known to possess antimicrobial properties. The authors also proposed a mechanism for the biogenic synthesis of AgNPs, which involved the biomolecule quercetin (Gade et al. 2010b). Recently, Kumar et al. (2012b) reported the reduction of silver ions to AgNPs by aqueous extract of *Terminalia chebula*, within only 20 min. The formation of AgNPs was confirmed by surface plasmon resonance at 452 nm using UV-visible spectrophotometer. In addition, high-resolution TEM (HR-TEM) revealed the formation of anisotropic nanostructures of pentagons, spherical and triangular-shaped AgNPs, with diameter less than 100 nm. Selected area electron diffraction pattern (SAED) confirmed the crystalline nature of AgNPs. Atomic force microscopy (AFM) images indicated the presence of coating comprised by oxidized polyphenols on the surface of AgNPs. Indeed, in biogenic synthesis of AgNPs by plant extracts, polyphenols act as reducing agent and also as a capping material leading to stabilization of AgNPs in aqueous solution (Kumar et al. 2012b). Narayanan and Sakthivel (2011) published detailed and updated review article based on the biogenic synthesis of metallic nanoparticles, including AgNPs by plant extracts and algae. Table 11.1 summarizes some relevant and recent studies based on green synthesis of AgNPs by plant extracts.

11.3 Toxic Effects of Silver Nanoparticles Administrated In Vitro and In Vivo

In recent years, nanotechnology has been rapidly expanding leading to important impacts on different areas such as health, environment, and economy. In fact, the commercialization of products comprised by nanomaterials is now a reality. Among commercially available nanocompounds, AgNPs are the most used (Ahmed et al. 2008), mainly due to their potent microbicidal activity (Durán et al. 2010). The rapid development of nanotechnology has been leading to an increasing concern related to possible toxic effects of nanomaterials, including human health and environmental impacts (Brayner 2008; Panda et al. 2011). In this scenario, the detailed investigation of nanoparticles toxicity has been emerging as an important area of research. The great interest of the scientific community by toxicological evaluations of nanomaterials is relatively new, and it has been increasing in recent years. However, detailed studies on nanoparticles toxicity, including AgNPs, are still limited and relatively unexplored (de Lima et al. 2012). Therefore, the great appeal of nanotechnology can be considered the better evaluation of the cytotoxicity and genotoxicity of nanoparticles, in particular, biogenic synthesized metallic nanoparticles. In this scenario, this section summarizes a survey of recent evaluations of cytotoxicity and genotoxicity of AgNPs, in special biogenic synthesized AgNPs, and highlights the necessity of more studies in the field.

11.3.1 Culture Cells

Vivek et al. (2012) reported the in vitro cytotoxicity of biogenic AgNPs synthesized by plant extract of *Annona squamosa*. The AgNPs synthesized by plant leaf extract showed to be spherical in shape with average diameter ranging from 20 to 100 nm. The in vitro cytotoxicity of these biogenic AgNPs were evaluated against normal epithelial cells (HBL-100) and human breast cancer cell (MCF-7). The authors reported a dose-dependent cytotoxicity effect. The inhibitory concentrations (IC50), in which 50 % of cells die, were found to be 50 μg/mL, 30 μg/mL and 80 μg/mL, 60 μg/mL for AgNPs against MCF-7 and normal HBL-100 cells at 24- and 48-h incubation, respectively (Vivek et al. 2012). Moreover, it was also observed an induction of apoptosis. Overall, the results revealed cytotoxic effects of biosynthesized AgNPs against breast cancer cell line, in comparison with normal breast cell line, indicating that these nanoparticles might be used in cancer treatments.

A similar work describes the preparation of biogenic AgNPs by plant extract of *Iresine herbstii* (Dipankar and Murugan 2012). The obtained AgNPs were found to be face-centered cubic in shape, with size ranging from 44 to 64 nm, and capped with plant components (Dipankar and Murugan 2012). The in vitro cytotoxicity of plant-mediated AgNPs was evaluated towards HeLa cancer cells with the trypan blue assay. AgNPs exhibited a potent toxic effect (CL_{50} of 51 μg/mL) on cancer cells with 88 % death of HeLa cells upon treatment with 300 μg/mL of AgNPs (Dipankar and Murugan 2012).

However, the toxicity of AgNPs towards culture cells is still under study. Several studies did not observe toxicity on human culture cells treated with capped AgNPs (diameters ranging from 6 to 80 nm) below 10 mg/mL (Hussain et al. 2005; AshaRani et al. 2009; Lu et al. 2010; Foldbjerg et al. 2011). An interesting paper investigated the toxicity of biogenic AgNPs (size of 25–45 nm) synthesized by *Alternaria alternata* towards human lymphocytes using comet assay (Sarkar et al. 2011). The authors reported that up to 50 mg/mL of biogenic AgNPs, no DNA damage was observed. DNA damage was only reported upon incubation of human lymphocytes with over 300 mg/mL of AgNPs.

On contrary, some papers described different degrees of toxicity upon incubation of cell cultures with AgNPs, at concentrations up to 50 mg/mL (de Lima et al. 2012). For example, culture of human mesenchymal stem cells incubated with 0.1 mg/mL of albumin-capped AgNPs (average size 46 nm) (Hackenberg et al. 2011) or human glioblastoma cells starch incubated with capped AgNPs (sizes 6–20 nm) (AshaRani et al. 2008) showed genotoxicity up to 50 mg/mL of AgNPs. Another interesting work showed that albumin-coated AgNPs (average size 70 nm) were found to be more genotoxic on mouse peritoneal macrophage cell line (genotoxicity at around 2 mg/mL) (Park et al. 2010) in comparison with AgNPs capped with polysaccharides (average size 25 nm), on mouse embryonic stem and fibroblasts cells, which exhibited genotoxicity at 50 mg/mL (Ahmed et al. 2008). In general and with a few exceptions, human culture cells were observed to be less sensitive compared to mouse culture cells, upon treatment with capped AgNPs, independently of the chemical nature of the capping (de Lima et al. 2012). Obviously, this tendency should be further investigated.

11.3.2 Calf Thymus DNA

It was observed that the genotoxicity of AgNPs alone (sizes 20–50 nm) was weak on calf thymus DNA (ctDNA); however, in the presence of detergent (cetylpyridinium bromide), the NPs showed significant genotoxicity, mainly owing to the presence of the surfactant (Chi et al. 2009). This result is important from an environmental point of view, since considerable amounts of AgNPs are eliminated to water (rivers, oceans) and probably readily interacting with surfactant presented in the water, leading to a strong genotoxicity owing to the formation of AgNPs–detergent interactions (Chi et al. 2009).

11.3.3 In Vivo Assays

Many papers have reported in vivo toxicity studies of AgNPs by different routes of administration (Tian et al. 2007; Vlachou et al. 2007; Trop et al. 2006; Samberg et al. 2010). For example, Rahman et al. (2009) reported that intraperitoneal administration of AgNPs may lead to alterations of gene expression, suggesting the neurotoxic effects of these nanoparticles. Furthermore, several reports describe oral administration of AgNPs (Cha et al. 2008; Kim et al. 2009a, b). After oral exposure, AgNPs and/or silver ions are assumed to translocate from the gut into the blood, systemically inducing liver damage. However, it is necessary to carefully investigate these statements, since it is necessary to distinguish whether the systemic distribution of silver is due to the presence of AgNPs or silver ions in the liver, and this topic deserves more studies, as stated by Johnston et al. (2010). Concerning to the possible oxidation of AgNPs to silver ions, an in vivo study based on intravenous administration of AgNPs on mouse supported the antiplatelet properties of these nanoparticles (AgNPs are considered to have antiplatelet properties) (Shrivastava et al. 2009). It was suggested that there is a low possibility of oxidation of AgNPs into silver ions inside the platelets due to the high silver ionization potential (735 kJ mol^{-1}), since this high ionization potential is difficult to reach in an intracellular environment. In view of this fact the observed antiplatelet property effects are more likely to be due to AgNPs rather than to silver ions. It is noticed that this same effect was observed in human platelets (patients with type 2 diabetes mellitus) (Shrivastava et al. 2009).

In a recent study, the mycosynthesis of AgNPs by filamentous fungus *Cochliobolus lunatus*, their characterization, and larvicidal effects were reported (Salunkhe et al. 2011). Toxic effects of biogenic AgNPs towards *Aedes aegypti* and *Anopheles stephensi*, which are responsible for diseases of public health importance, were investigated. Potent mortality effects on second, third, and fourth instar larvae of *A. aegypti* and *A. stephensi* were found to correlate with concentrations of AgNPs. The observed larvicidal activity of mycogenic AgNPs was related to the penetration of nanoparticles through larvae membrane. Moreover, toxicity studies were also carried out against nontarget fish species *Poecilia reticulata*. Common organisms found in this kind of habits are *A. aegypti* and *A. stephensi*. Interestingly, mycogenic AgNPs did not exhibit noticeable effects on *P. reticulata* after either 24 or 48 h of exposure at the level of LC50 and LC90 values against fourth instar larvae of *A. aegypti* and *A. stephensi*. These results indicate that biogenic AgNPs might be used as a potent larvicidal agent, with no toxic effects to the environment (Salunkhe et al. 2011).

Interestingly, recent studies suggested that in vivo toxicity of AgNPs is dependent on the organism. In this scenario, in vivo studies demonstrated that mice are more sensitive to capped AgNPs in comparison with fish. Indeed, some papers showed that AgNPs capped with starch (8–15 nm), or with bovine serum albumin (10–20 nm) (AshaRani et al. 2008), or with polyvinyl alcohol (5–35 nm) (AshaRani et al. 2011) were no genotoxic towards zebra fish embryos up to nanoparticles concentration of 25 mg/mL. Genotoxicity was observed by increasing the Nanoparticles concentration to values over 100 mg/mL, in special in the case of albumin-capped AgNPs, which lead to embryos apoptosis and also abnormalities (AshaRani et al. 2008). However, at low nanoparticle dose, such as 1 mg mL^{-1}, this effect was found to be very discrete (Ordzhonikidze et al. 2009).

Very few studies investigated the pulmonary toxicity of AgNPs. As stated by Johnston et al. (2010), there is limited information available related to this topic. Studies based on the investigation of pulmonary toxicity of AgNPs are important since proinflammatory and oxidative potentials of AgNPs within the lung are known to drive nanoparticles toxicity (Sung et al. 2009; Hyun et al. 2008; Ji et al. 2007). Aspects such as exposure time to nanoparticles, their concentrations, routes of applications, and particle sizes are extremely important to define nanoparticle toxicity. AgNP translocation and their potential accumulation within secondary targets, such as liver, spleen, and brain, following pulmonary exposure, were reported. However, Johnston et al. (2010) stated that, until now, it is not possible to confirm that the silver content of any cell distribution is due to AgNPs or to silver ions, since the methods used are not sensible enough to discriminate between them. It is known that in biological systems both forms of silver can exist owing to degradation and/or metabolism of the nanoparticles, indicating that further studies are required in this field.

Moreover, several applications of AgNPs are based on dermatological uses, such as the production of wound dressing containing AgNPs. In this regards, it is important to further investigate possible toxic effects associated with administration of AgNPs on either wounded skin or normal skin. It is important to clearly define the hazards associated with dermal exposure to AgNPs, since liver has been suggested as a secondary target for nanoparticles toxicity (de Lima et al. 2012 and reference therein).

11.4 Aspects That Determine Nanoparticles Cytotoxicity and Genotoxicity

Several aspects including size, surface area, surface reactivity, composition, surface charge, and capping molecules are known to influence cell toxicity. In general, reactivity of nanomaterials in biological medium is directly related to the observed toxic effects of nanoparticles, and particle size controls the reactivity of these nanomaterials (Brown et al. 2001; Duffin et al. 2007). Indeed, small nanoparticles can cause several side effects in lung, due to their higher ability for cell invasion (Oberdörster et al. 2005). Hence, smaller sized particles were found to be more toxic than bigger sized particles (Panda et al. 2011; Gaiser et al. 2012). However, some exception was reported, as in the case of nonmetallic nanoparticles, such as polymeric nanoparticles. Toxicity of chitosan nanoparticles indicated that smaller particles might be used more safely than larger particles, at higher concentrations, and probably others factors could be involved in nanoparticle toxicity (Lima et al. 2010). Overall, in the case of metallic nanoparticles such as AgNPs, it can be observed a tendency that small particles have higher reactivity and thus produce higher toxicity (Ordzhonikidze et al. 2009; Kim et al. 2011). However, it must be noted that particle size is not the only factor that determines nanoparticle toxicity,

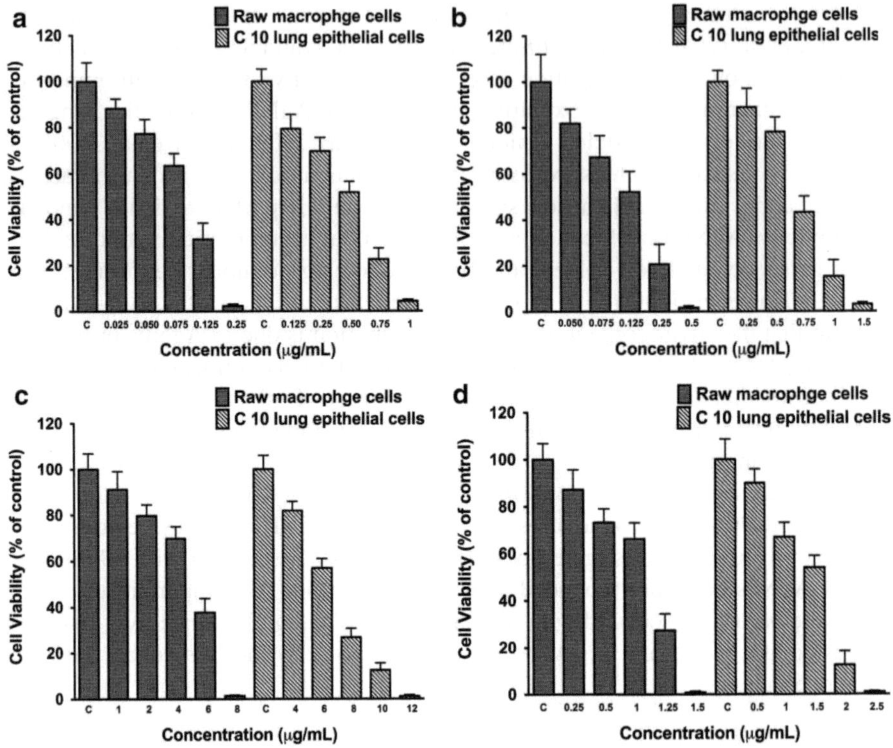

Fig. 11.1 Results of MTT assay of the mouse macrophage and lung epithelial cells exposed to the four different types of silver nanoparticles at various concentrations: (**a**) poly(diallyldimethylammonium) chloride-AgNPs, (**b**) biogenic AgNPs, (**c**) uncoated-AgNps, and (**d**) oleate-Ag nanoparticles NPs. Reproduced from Suresh et al. 2012 by permission of American Chemical Society

since important parameters such as synthesis procedures, presence and nature of capping agents, surface charge, and aggregation are also important (Panda et al. 2011; Suresh et al. 2012). In fact, Suresh et al. (2012) investigated the cytotoxicity of effects of AgNPs with different surface coatings on two different cell lines. Figure 11.1 shows the cytotoxicity of mouse macrophage and lung epithelial cells treated with four different types of AgNPs: (a) poly (diallyldimethylammonium) chloride-AgNPs; (b) biogenic AgNPs, synthesized by bacteria *Shewanella oneidensis*; (c) uncoated-AgNPs (colloidal-AgNPs); and (d) oleate-AgNPs. The 3-(4,5-dimethylthiazol-2-yl)-2,5-diphenyltetrazolium bromide (MTT) assay measured mitochondrial integrity and oxidative metabolism and is an indirect measure of cell viability. All different AgNPs tested showed a dose-dependent cytotoxicity in both cell lines. However, depending on the type of AgNPs different cytotoxic parameters were obtained. For example, in the case of mouse macrophage cells, the inhibitory concentration values were found to be 0.1, 0.125, 1.1, and 4.9 μg/mL for poly(diallyldimethylammonium) chloride-AgNPs, biogenic AgNPs, oleate-AgNPs, and uncoated-AgNPs, respectively. For lung epithelial cells, the inhibitory concentration values were found to be 0.45, 0.7, 1.6, and 6.3 μg/mL for poly(diallyldimethylammonium) chloride-AgNPs, biogenic AgNPs, oleate-AgNPs, and uncoated-AgNPs, respectively. It can be concluded that uncoated-AgNPs were the less toxic to both cell lines, followed by olate-AgNPs, biogenic AgNPs, and poly(diallyldimethylammonium) chloride-AgNPs (Suresh et al. 2012). These results reveal the impact of synthesis procedure and surface coating in determining nanoparticle toxicity.

11.5 Comments and Future Perspectives

It can be concluded that biogenic synthesis of metallic NPs, in particular AgNPs, has been emerging as an important and fruitful domain of nanotechnology. Despite the fact that several papers have successfully described different routes to biogenically synthesize AgNPs, as stated in this chapter, the mechanisms involved in the biogenic production of AgNPs have not yet been elucidated. Therefore, elucidation of the mechanistic aspects of some biological systems in the synthesis of nanoparticles needs more detailed studies (Durán et al. 2011).

Moreover, by reviewing the actual literature based on AgNPs and toxicity, it can be concluded that there are few studies based on cytotoxicity and genotoxicity of biogenic AgNPs, in comparison with chemically synthesized AgNPs. The lack of more studies based on toxicity of biogenic AgNPs can be understood by considering that biogenic synthesis of metallic nanoparticles is a relatively new methodology (de Lima et al. 2012). By analyzing and comparing published papers on toxic effects of AgNPs, the determination of AgNPs toxicity (both cytotoxicity and genotoxicity) is a complex issue, since many different parameters play important roles. In fact, controversial papers showing different effects of toxicity of AgNPs can be understood by considering several parameters, such as different methodologies to synthesize the nanoparticles, nanoparticles size, shape, presence and nature of capping agents, surface area and charge, and finally, the diverse kinds of toxic evaluation tests employed (de Lima et al. 2012). The uses of different organisms and/or culture cells to investigate AgNPs toxicity make the comparison between different studies a difficult and complex task, leading to inconclusive results in some cases. In this context, it is necessary to elaborate standard protocols to carefully analyze nanoparticles toxicity in order to decrease the possible discrepancies related to final conclusions and comparison between different works.

Therefore, with the present information, it is premature to conclude whether biogenic nanoparticles are less genotoxic/cytotoxic compared with chemically synthesized nanoparticles. However, some considerations and tendencies can be postulated with the actual literature, such as the lower toxicity of AgNPs compared with silver ions (Ordzhonikidze et al. 2009; Kim et al. 2011; Griffitt et al. 2008; Park and Choi 2010; Panda et al. 2011; Gaiser et al. 2012). Panda et al. (2011) investigated the induction of cellular death of cultured cells of *Allium cepa* incubated with silver ions (Ag^+), silver complexes (AgCl), capped biogenic AgNPs (AgNP-biogenic), and commercially uncapped AgNPs from Sigma (AgNP-Sigma), and they found that the induction of cell death followed the order: Ag^+ ions > colloidal AgCl > AgNP-commercial from Sigma > AgNP-biogenic, showing that the less toxic form of silver was biogenic synthesized capped AgNPs. Similar results were reported for biogenic AgNPs for different applications, such as wound healing antileishmaniases, antifungal, and antibacterial effects (Melo et al. 2011; Marcato and Durán 2011; Rossi-Bergmann et al. 2012; Marcato et al. 2012a, b).

Taken together, it can be assumed that biogenic synthesis of AgNPs has been emerging as an attractive and important route to prepare nanoparticles for several applications. In general, biogenic AgNPs are less toxic in comparisons with chemically synthesized nanoparticles. This chapter highlights the necessity of more mechanist studies on the biogenic synthesis of AgNPs, in order to optimize the green synthesis process, and more studies based on the toxicity of AgNPs, in particular, genotoxicity studies of biogenic AgNPs.

Acknowledgement Support from FAPESP, CNPq and Brazilian Network on Nanotoxicology (MCTI/CNPq), and NanoBioss (MCTI) is acknowledged.

References

Ahmad A, Mukherjee P, Senapati S et al (2003) Extracellular biosynthesis of silver nanoparticles using the fungus *Fusarium oxysporum*. Colloids Surf B Biointerfaces 28:313–318

Ahmed M, Karns M, Goodson M et al (2008) DNA damage response to different surface chemistry of silver nanoparticles in mammalian cells. Toxicol Appl Pharmacol 233:404–410

Antony JJ, Sivalingam P, Siva D et al (2011) Comparative evaluation of antibacterial activity of silver nanoparticles synthesized using *Rhizophora apiculata* and glucose. Colloids Surf B Biointerfaces 88:134–140

Arora S, Jain J, Rajwade JM et al (2008) Cellular responses induced by silver nanoparticles: *in vitro* studies. Toxicol Lett 79:93–100

AshaRani PV, Wu YL, Gong Z et al (2008) Toxicity of silver nanoparticles in zebrafish models. Nanotechnology 19:255102

AshaRani PV, Mun GLK, Hande MP et al (2009) Cytotoxicity and genotoxicity of silver nanoparticles in human cells. ACS Nano 3:279–290

AshaRani PV, Lianwu Y, Gong Z et al (2011) Comparison of the toxicity of silver, gold and platinum nanoparticles in developing zebrafish embryos. Nanotoxicology 5:43–54

Bai H-J, Yang B-S, Chai C-J et al (2011) Green synthesis of silver nanoparticles using *Rhodobacter sphaeroides*. World J Microbiol Biotechnol 27:2723–2728

Balaji DS, Basavaraja S, Deshpande R et al (2009) Extracellular biosynthesis of functionalized silver nanoparticles by strains of *Cladosporium cladosporioides* fungus. Colloids Surf B Biointerfaces 68:88–92

Bansal V, Bharde A, Ramanathan R et al (2012) Inorganic materials using 'unusual' microorganisms. Adv Colloid Interface Sci 179:150–168

Banu A, Rathod V, Ranganath E (2011) Silver nanoparticle production by *Rhizopus stolonifer* and its antibacterial activity against extended spectrum b-lactamase producing (ESBL) strains of Enterobacteriaceae. Mater Res Bull 46:1417–1423

Barwal I, Ranjan P, Kateriya S et al (2011) Cellular oxido-reductive proteins of *Chlamydomonas reinhardtii* control the biosynthesis of silver nanoparticles. J Nanobiotechnology 9:56. doi:10.1186/1477-3155-9-56

Basavaraja S, Balaji SD, Legashetty A et al (2008) Extracellular biosynthesis of silver nanoparticles using the fungus *Fusarium semitectum*. Mater Res Bull 43:1164–1170

Bawaskar M, Gaikwad S, Ingle A et al (2010) A new report on mycosynthesis of silver nanoparticles by *Fusarium culmorum*. Curr Nanosci 6:376–380

Bhainsa KC, D'souza SK (2006) Extracellular biosynthesis of silver nanoparticles using the fungus *Aspergillus fumigatus*. Colloids Surf B Biointerfaces 47:160–164

Birla SS, Tiwari VV, Gade AK et al (2009) Fabrication of silver nanoparticles by *Phoma glomerata* and its combined effect against *Escherichia coli*, *Pseudomonas aeruginosa* and *Staphylococcus aureus*. Lett Appl Microbiol 48:173–179

Brayner R (2008) The toxicological impact of nanoparticles. Nano Today 3:48–55

Brown DM, Wilson MR, MacNee W et al (2001) Size-dependent proinflammatory effects of ultrafine polystyrene particles: a role for surface area and oxidative stress in the enhanced activity of ultrafines. Toxicol Appl Pharmacol 175:191–199

Cha K, Hong HW, Choi YG et al (2008) Comparison of acute responses of mice livers to short-term exposure to nano-sized or micro-sized silver particles. Biotechnol Lett 30:1893–1899

Chen JC, Lin ZH, Ma XX (2003) Evidence of the production of silver nanoparticles via pretreatment of *Phoma* sp.3.2883 with silver nitrate. Lett Appl Microbiol 37:105–108

Chi Z, Li R, Zhao L et al (2009) A new strategy to probe the genotoxicity of silver nanoparticles combined with cetylpyridine bromide. Spectrochim Acta A Mol Biomol Spectrosc 72:577–581

de Lima R, Seabra AB, Durán N (2012) Silver nanoparticles: a brief review of cytotoxicity and genotoxicity of chemically and biogenically synthesized nanoparticles. J Appl Toxicol 32:867–879

Dipankar C, Murugan S (2012) The green synthesis, characterization and evaluation of the biological activities of silver nanoparticles synthesized from *Iresine herbstii* leaf aqueous extracts. Colloids Surf B Biointerfaces 98:112–119

Doak SH, Manshian B, Jemkins GJ et al (2012) In vitro genotoxicity testing strategy for nanomaterials and the adaptation of current OECD guidelines. Mutat Res 745:104–111

Donaldson K, Stone V, MacNee W (1999) The toxicology of ultrafine particles. In: Maynard LA, Howards CA (eds) Particulate matter properties and effects upon health. Bios Scientific, Oxford, pp 115–127

Duffin R, Tran L, Brown D et al (2007) Proinflammogenic effects of low-toxicity and metal nanoparticles *in vivo* and *in vitro*: highlighting the role of particle surface area and surface reactivity. Inhal Toxicol 19:849–856

Durán N, Marcato PD, Alves OL et al (2005) Mechanistic aspects of biosynthesis of silver nanoparticles by several *Fusarium oxysporum* strains. J Nanobiotechnol 3:8–14

Durán N, Marcato PD, de Conti R et al (2010) Potential use of silver nanoparticles on pathogenic bacteria, their toxicity and possible mechanism of action. J Braz Chem Soc 21:949–959

Durán N, Marcato PD, Durán M et al (2011) Mechanistic aspects in the biogenic synthesis of extracellular metal nanoparticles by peptides, bacteria, fungi, and plants. Appl Microbiol Biotechnol 90:1609–1624

Fayaz AM, Girilal M, Rahman M et al (2011) Biosynthesis of silver and gold nanoparticles using thermophilic bacterium *Geobacillus stearothermophilus*. Process Biochem 46:1958–1962

Foldbjerg R, Dang DA, Autrup H (2011) Cytotoxicity and genotoxicity of silver nanoparticles in the human lung cancer cell line, A549. Arch Toxicol 85:743–750

Gade A, Ingle A, Whiteley C et al (2010a) Mycogenic metal nanoparticles: progress and applications. Biotechnol Lett 32:593–600

Gade A, Gaikwad S, Tiwari V et al (2010b) Biofabrication of silver nanoparticles by *Opuntia ficus-indica*: *in vitro* antibacterial activity and study of the mechanism involved in the synthesis. Curr Nanosci 6:370–375

Gaiser BK, Fernandes TF, Jepson MA et al (2012) Interspecies comparisons on the uptake and toxicity of silver and cerium dioxide nanoparticles. Environ Toxicol Chem 31:144–154

Gonzalez L, Sanderson BJ, Kirsh-Volders M (2011) Adaptations of the in vitro MN assay for the genotoxicity assessment of nanomaterials. Mutagenesis 26:185–191

Gopinath V, MubarakAli D, Priyadarshini NM et al (2012) Biosynthesis of silver nanoparticles from *Tribulus terrestris* and its antimicrobial activity: a novel biological approach. Colloids Surf B Biointerfaces 96:69–74

Govindaraju K, Basha SK, Kumar G et al (2008) Silver, gold and bimetallic nanoparticles production using single-cell protein (*Spirulina platensis*) Geitler. J Mater Sci 43:5115–5122

Griffitt RJ, Luo J, Gao J et al (2008) Effects of particle composition and species on toxicity of metallic nanomaterials in aquatic organisms. Environ Toxicol Chem 27:1972–1978

Gurr JR, Wang AS, Chen CH et al (2005) Ultrafine titanium dioxide particles in the absence of photoactivation can induce oxidative damage to human bronchial epithelial cells. Toxicology 213:66–73

Hackenberg S, Scherzed A, Kessler M et al (2011) Silver nanoparticles: evaluation of DNA damage, toxicity and functional impairment in human mesenchymal stem cells. Toxicol Lett 201:27–33

Hood E (2004) Nanotechnology, diving into the unknown. Environ Health Perspect 112: A747–A749

Hussain SM, Hess KL, Gearhart JM et al (2005) *In vitro* toxicity of nanoparticles in BRL 3A rat liver cells. Toxicol In Vitro 19:975–983

Hyun JS, Lee BS, Ryu HY et al (2008) Effects of repeated silver nanoparticles exposure on the histological structure and mucins of nasal respiratory mucosa in rats. Toxicol Lett 182:24–28

Ingle A, Gade A, Pierrat S et al (2008) Mycosynthesis of silver nanoparticles using the fungus *Fusarium acuminatum* and its activity against some human pathogenic bacteria. Curr Nanosci 4:141–144

Ingle A, Gade A, Bawaskar M et al (2009) *Fusarium solani*: a novel biological agent for the extracellular synthesis of silver nanoparticles. J Nanopart Res 11:2079–2085

Jebali A, Ramezani F, Kazemi B (2011) Biosynthesis of silver nanoparticles by *Geotrichum* sp. J Cluster Sci 22:225–232

Ji JH, Jung JH, Kim SS et al (2007) Twenty-eight-day inhalation toxicity study of silver nanoparticles in Sprague–Dawley rats. Inhal Toxicol 19:857–871

Joerger R, Klaus T, Granqvist CG (2000) Biologically produced silver-carbon composite materials for optically functional thin film coatings. Adv Mater 12:1407–1409

Johnston HJ, Hutchison G, Christensen FM et al (2010) A review of the *in vivo* and *in vitro* toxicity of silver and gold particulates: particle attributes and biological mechanisms responsible for the observed toxicity. Crit Rev Toxicol 40:328–346

Ju-Nam Y, Lead JR (2008) Manufactured nanoparticles: an overview of their chemistry, interactions and potential environmental implications. Sci Total Environ 400:396–414

Kalishwaralal K, Deepaka V, Pandiana SR et al (2010) Biosynthesis of silver and gold nanoparticles using *Brevibacterium casei*. Colloids Surf B Biointerfaces 77:257–262

Kim S, Choi JE, Cho J et al (2009a) Oxidative stress-dependent toxicity of silver nanoparticles in human hepatoma cells. Toxicol In Vitro 23:1076–1084

Kim Y, Suh HS, Cha HJ et al (2009b) A case of generalized argyria after ingestion of colloidal silver solution. Am J Ind Med 52:246–250

Kim JS, Sung JH, Ji JH et al (2011) *In vivo* genotoxicity of silver nanoparticles after 90-day silver nanoparticle inhalation exposure. Saf Health Work 2:34–38

Kirthi AV, Rahuman AA, Jayaseelan C et al (2012) Novel approach to synthesis silver nanoparticles using plant pathogenic fungi, *Puccinia graminis*. Mater Lett 81:69–72

Klaus T, Granqvist CG, Joerger R et al (1999) Silver-based crystalline nanoparticles, microbially fabricated. Proc Natl Acad Sci U S A 96:13611–13614

Klaus-Joerger T, Joerge R, Olsson E et al (2001) Bacteria as workers in the living factory: metal accumulating bacteria and their potential for material sciences a review. Trends Biotechnol 19:15–20

Kora AJ, Sashidhar RB, Arunachalama J (2012) Aqueous extract of gum olibanum (*Boswellia serrata*): a reductant and stabilizer for the biosynthesis of antibacterial silver nanoparticles. Process Biochem 47:1516–1520

Kowshik M, Ashtaputre S, Kharrazi S et al (2003) Extracellular synthesis of silver nanoparticles by a silver-tolerant yeast strain MKY3. Nanotechnology 14:95–100

Krishnaraj C, Jagan EG, Rajasekar S et al (2010) Synthesis of silver nanoparticles using *Acalypha indica* leaf extracts and its antibacterial activity against water borne pathogens. Colloids Surf B Biointerfaces 76:50–56

Kumar CG, Mamidyala SK (2011) Extracellular synthesis of silver nanoparticles using culture supernatant of *Pseudomonas aeruginosa*. Colloids Surf B Biointerfaces 84:462–466

Kumar RR, Priyadharsani KP, Thamaraiselvi K (2012a) Mycogenic synthesis of silver nanoparticles by the Japanese environmental isolate *Aspergillus tamari*. J Nanopart Res 14:860. doi:10.1007/s11051-012-0860-2

Kumar KM, Sinha M, Mandal BK et al (2012b) Green synthesis of silver nanoparticles using *Terminalia chebula* extract at room temperature and their antimicrobial studies. Spectrochim Acta A Mol Biomol Spectrosc 91:228–233

Lewinski N, Colvin V, Drezedk R (2008) Cytotoxicity of nanoparticles. Small 4:26–49

Li X, Xu H, Chen Z-S et al (2011) Biosynthesis of nanoparticles by microorganisms and their applications. J Nanomater Article ID 270974. doi:10.1155/2011/270974

Lima R, Feitosa L, Pereira AES et al (2010) Evaluation of the genotoxicity of chitosan nanoparticles for use in food packaging films. J Food Sci 75:89–96

Lu W, Senapati D, Wang S et al (2010) Effect of surface coating on the toxicity of silver nanomaterials on human skin keratinocytes. Chem Phys Lett 487:92–96

Marcato PD, Durán N (2011) Biogenic silver nanoparticles: applications in medicines and textiles and their health implications. In: Rai M, Durán N (eds) Metal nanoparticles in microbiology, Springer, Germany, Chap. 11, p 249–267

Marcato PD, Durán M, Huber SC et al (2012a) Biogenic silver nanoparticles and its antifungal activity as a new topical transungual drug. J Nano Res 20:99–107

Marcato PD, Nakasato G, Brocchi M et al (2012b) Biogenic silver nanoparticles: antibacterial and cytotoxicity applied to textile fabrics. J Nano Res 20:69–76

Melo PS, Marcato PD, Huber SC, Ferreira IR, de Paula LB, Almeida ABA, Durán N, Torloni S, Seabra AB, Alves OL (2011) Nanoparticles in treatment of thermal injured rats: is it safe? J Phys Conf Ser 304:012027

Mohammadian A, Shojaosadati SA, Habibi-Rezaee M (2007) *Fusarium oxysporum* mediates photogeneration of silver nanoparticles. Scientia Iranica 14:323–326

Mourato A, Gadanho M, Lino AR et al (2011) Biosynthesis of crystalline silver and gold nanoparticles by extremophilic yeasts. Bioinorg Chem Appl Article ID 546074. doi:10.1155/2011/546074

Narayanan KB, Sakthivel N (2010) Biological synthesis of metal nanoparticles by microbes. Adv Colloid Interface Sci 156:1–13

Narayanan KB, Sakthivel N (2011) Green synthesis of biogenic metal nanoparticles by terrestrial and aquatic phototrophic and heterotrophic eukaryotes and biocompatible agents. Adv Colloid Interface Sci 169:59–79

Nayak RR, Pradhan N, Behera D et al (2011) Green synthesis of silver nanoparticle by *Penicillium purpurogenum* NPMF: the process and optimization. J Nanopart Res 13:3129–3137

Nel A, Xia T, Madler L et al (2006) Toxic potential of materials at the nanolevel. Science 311:622–627

Ng CT, Li JJ, Bay BH et al (2010) Current studies into the genotoxic effects of nanomaterials. J Nucleic Acid Article ID 947859. doi: 10.4061/2010/947859

Oberdörster E (2004) Manufactured nanomaterials (fullerenes, C60) induce oxidative stress in the brain of juvenile largemouth bass. Environ Health Perspect 112:1058–1062
Oberdörster G, Oberdörster E, Oberdörster J (2005) Nanotoxicology: an emerging discipline evolving from studies of ultrafine particles. Environ Health Perspect 113:823–839
Ordzhonikidze CG, Ramaiyya LK, Egorova EM et al (2009) Genotoxic effects of silver nanoparticles on mice *in vivo*. Acta Naturae (Russia) 3:99–101
Panda KK, Achary VMM, Krishnaveni R et al (2011) *In vitro* biosynthesis and genotoxicity bioassay of silver nanoparticles using plants. Toxicol In Vitro 25:1097–1105
Parikh RY, Ramanathan R, Coloe PJ et al (2011) Genus-wide physicochemical evidence of extracellular crystalline silver nanoparticles biosynthesis by *Morganella* spp. PLoS ONE 6: e21401
Park BSY, Choi J (2010) Geno- and ecotoxicity evaluation of silver nanoparticles in freshwater crustacean Daphnia magna. Environ Eng Res 15:23–27
Park AEJ, Yi J, Kim Y et al (2010) Silver nanoparticles induce cytotoxicity by a Trojan-horse type mechanism. Toxicol In Vitro 24:872–878
Prakasham RS, Kumar BS, Kumar YS et al (2012) Characterization of silver nanoparticles synthesized by using marine isolate *Streptomyces albidoflavus*. J Microbiol Biotechnol 22:614–621
Pugazhenthiran N, Anandan S, Kathiravan G et al (2009) Microbial synthesis of silver nanoparticles by *Bacillus* sp. J Nanopart Res 11:1811–1815
Rahman MF, Wang J, Patterson TA et al (2009) Expression of genes related to oxidative stress in the mouse brain after exposure to silver-25 nanoparticles. Toxicol Lett 187:15–21
Rai M, Yadav A, Gade A (2008) CRC 675—currents trends in phytosynthesis of metal nanoparticles. Crit Rev Biotechnol 28:277–284
Rai M, Yadav A, Bridge P et al (2009) Myconanotechnology: a new and emerging science. Appl Mycol 258–267
Raja K, Saravanakumar A, Vijayakumar R (2012) Efficient synthesis of silver nanoparticles from *Prosopis juliflora* leaf extract and its antimicrobial activity using sewage. Spectrochim Acta A Mol Biomol Spectrosc 97:490–494
Rajasekharreddy P, Rani PU, Sreedhar B (2010) Qualitative assessment of silver and gold nanoparticle synthesis in various plants: a photobiological approach. J Nanopart Res 12:1711–1721
Rao YS, Kotakadi VS, Prasad TNVKV et al (2013) Green synthesis and spectral characterization of silver nanoparticles from Lakshmi tulasi (*Ocimum sanctum*) leaf extract. Spectrochim Acta A Mol Biomol Spectrosc 103:156–159
Rico CM, Majumdar S, Duarte-Gardea M et al (2011) Interaction of nanoparticles with edible plants and their possible implications in the food chain. J Agric Food Chem 59:3485–3498
Roopan SM, Madhumitha RG, Rahuman AA et al (2013) Low-cost and eco-friendly phyto-synthesis of silver nanoparticles using *Cocos nucifera* coir extract and its larvicidal activity. Ind Crop Prod 43:631–635
Rossi-Bergmann B, Pacienza-Lima W, Marcato PD et al (2012) *In vivo* Leishmaniasis Study: biogenic and chemical silver nanoparticles. J Nano Res 20:89–97
Salunkhe RB, Patil SV, Patil CD et al (2011) Larvicidal potential of silver nanoparticles synthesized using fungus *Cochliobolus lunatus* against *Aedes aegypti* (Linnaeus, 1762) and *Anopheles stephensi* Liston (Diptera; Culicidae). Parasitol Res 109:823–831
Samberg ME, Oldenburg SJ, Monteiro-Riviere NA (2010) Evaluation of silver nanoparticle toxicity in skin *in vivo* and keratinocytes *in vitro*. Environ Health Perspect 118:407–413
Sarkar J, Chattopadhyay D, Patra S et al (2011) *Alternaria alternata* mediated synthesis of protein capped silver nanoparticles and their genotoxic activity. Dig J Nanomater Biostruct 6:563–573
Shrivastava S, Bera T, Singh SK et al (2009) Characterization of antiplatelet properties of silver nanoparticles. ACS Nano 3:1357–1364

Shukla MK, Singh RP, Reddy CRK et al (2012) Synthesis and characterization of agar-based silver nanoparticles and nanocomposite film with antibacterial applications. Bioresour Technol 107:295–300
Song JY, Kim BS (2008) Biological synthesis of bimetallic Au/Ag nanoparticles using Persimmon (*Diopyros kaki*) leaf extract. Korean J Chem Eng 25:808–811
Soni N, Prakash S (2012) Fungal-mediated nano silver: an effective adulticide against mosquito. Parasitol Res 111:2091–2098
Sung JH, Ji JH, Park JD et al (2009) Subchronic inhalation toxicity of silver nanoparticles. Toxicol Sci 108:452–461
Suresh AK, Pelletier DA, Wang W et al (2012) Cytotoxicity induced by engineered silver nanocrystallites is dependent on surface coatings and cell types. Langmuir 28:2727–2735
Tian J, Wong KK, Ho CM et al (2007) Topical delivery of silver nanoparticles promotes wound healing. ChemMedChem 2:129–136
Trop M, Novak M, Rodl S et al (2006) Silver coated dressing acticoat caused raised liver enzymes and argyria-like symptoms in burn patient. J Trauma 60:648–665
Valli JS, Vaseeharan B (2012) Biosynthesis of silver nanoparticles by *Cissus quadrangularis* extracts. Mater Lett 82:71–173
Vigneshwaran N, Kathe AA, Varadarajan PV et al (2006) Biomimetics of silver nanoparticles by white rot fungus, *Phaenerochaete chrysosporium*. Colloids Surf B Biointerfaces 53:55–59
Vijayakumar M, Priya K, Nancy FT et al (2013) Biosynthesis, characterisation and anti-bacterial effect of plant-mediated silver nanoparticles using *Artemisia nilagirica*. Ind Crop Prod 41:235–240
Vivek R, Thangam R, Muthuchelian K et al (2012) Green biosynthesis of silver nanoparticles from *Annona squamosa* leaf extract and its *in vitro* cytotoxic effect on MCF-7 cells. Process Biochem 47:2405–2410
Vlachou E, Chipp E, Shale E et al (2007) The safety of nanocrystalline silver dressings on burns: a study of systemic silver absorption. Burns 33:979–985
Wijnhoven SWP, Peijnenburg WJGM, Herberts CA et al (2009) Nano-silver—a review of available data and knowledge gaps in human and environmental risk assessment. Nanotoxicology 3:109–138
Wise JP Sr, Goodale BC, Wise SS et al (2010) Silver nanospheres are cytotoxic and genotoxic to fish cells. Aquat Toxicol 97:34–41
Xiong D, Fang T, Yu L et al (2011) Effects of nano-scale TiO_2, ZnO and their bulk counterparts on zebraûsh: acute toxicity, oxidative stress and oxidative damage. Sci Total Environ 409:1444–1452
Yilmaz M, Turkdemir H, Kilic AM et al (2011) Biosynthesis of silver nanoparticles using leaves of *Stevia rebaudiana*. Mater Chem Phys 130:1195–1202
Zaki S, Kady MFE, Abd-El-Haleem D (2011) Biosynthesis and structural characterization of silver nanoparticles from bacterial isolates. Mater Res Bull 46:1571–1576

Chapter 12
Cytotoxicity and Genotoxicity of Iron Oxides Nanoparticles

Amedea B. Seabra and Paula S. Haddad

Abstract The interest in the development of nanoparticles for diverse applications, mainly biomedical and technological purposes, has been greatly increasing in recent years. Among the nanostructured materials, metallic nanoparticles, in particular, iron oxide magnetic nanoparticles have been the focus of intensive research. Recently, the biomedical applications of iron oxide magnetic nanoparticles, such as magnetite (Fe_3O_4) and maghemite (γ-Fe_2O_3), have been increasing. Due to their special properties, such as small sizes and superparamagnetic behavior at room temperature, these nanoparticles find important pharmacological applications, such as drug delivery and contrast agents in magnetic resonance imaging. Iron oxide nanoparticles, with different sizes and coating surfaces, can be synthesized by well-established physical, chemical, and, more recent, biogenic techniques. Biogenic synthesis of iron oxide nanoparticles has emerged as a new and environment-friendly approach to obtain biocompatible nanomaterials. It must be noted that, for biomedical or technological applications of iron oxide nanoparticles, it is of paramount importance to fully characterize the in vitro and in vivo toxicity of these nanoparticles. In recent years, important studies have been characterized the cyto- and genotoxicity, as well as the biological consequences due to in vivo administration of iron oxide nanoparticles. In despite of these advances in toxicological evaluations of these nanoparticles, there are still some important questions to be answered. For biomedical or technological applications, it is mandatory to characterize in details the toxicity of this nanomaterial, as well as its fate upon in vivo administration. In this regard, this chapter summarizes the recent progress in the synthesis of iron oxide nanoparticles and the in vitro and in vivo characterization of nanoparticle toxicities.

A.B. Seabra (✉)
Exact and Earth Sciences Department, Universidade Federal de São Paulo, Rua São Nicolau, 210, 09913-030 Diadema, Sao Paulo, Brazil
e-mail: amedea.seabra@unifesp.br

N. Durán et al. (eds.), *Nanotoxicology*, Nanomedicine and Nanotoxicology, 265
DOI 10.1007/978-1-4614-8993-1_12,

The present chapter highlights the drawbacks and challenges that still need to be overcome regarding the toxicity of iron oxide nanoparticles, in order to propose their safe use.

12.1 Introduction

Magnetic iron oxide nanoparticles comprised by magnetite (Fe_3O_4) or maghemite (γ-Fe_2O_3) are the most important nanoparticles (NPs) employed for several applications, such as cancer treatment (hyperthermia), drug delivery, contrast agents in magnetic resonance image (MRI), cell separation, ferrofluids, electronic devices, and catalysis (Gupta and Gupta 2005; Haddad and Seabra 2012). Magnetic iron oxide NPs with size smaller than 100 nm are usually found in a superparamagnetic state at room temperature (Batlle and Labarta 2002; Kucheryavy et al. 2013). In the past decade, there have been several reports that describe the synthesis of superparamagnetic iron oxide nanoparticles (SPIONs), with appropriate surface chemistry, which can be used for numerous biomedical applications (Gupta and Gupta 2005; Laurent et al. 2008; Wahajuddin 2012).

Generally, there are three methods to synthesize magnetic iron oxide NPs: physical, chemical, and biogenic. Traditional physical and chemical methods are able to produce NPs with highly controlled sized, good polydispersity and desired surface chemical composition. However, these traditional physical and chemical methods usually employ hazardous chemical reagents, which are cost-effective, and produce toxic sub-products (Marchiol 2012). Biogenic methods to synthesize iron oxide NPs have been emerging as an attractive alternative to obtain biocompatible NPs, in a "green" and environment-friendly approach. Several microorganisms and plant extracts have been reported to synthesize iron oxide NPs for biomedical applications (Bharde et al. 2006; Durán and Seabra 2012). The present chapter discusses recent progress on the synthesis of iron oxide NPs based on traditional physical and chemical methods, as well as new strategies based on biogenic synthesis of these NPs. In all cases, there are several techniques that have been applied to characterize the produced NPs, such as transmission electron microscopy (TEM), scanning electron microscopy (SEM), X-ray diffraction (XRD), and Fourier transform infrared spectroscopy (FTIR).

However, besides the detailed characterization of iron oxide NPs, toxicity is of paramount importance for biomedical and technological applications. For biomedical applications, iron oxide NPs should have a narrow size distribution, while the surface of the NP should be coated with biocompatible hydrophilic materials, so that the coated NPs can form a stable aqueous dispersion at physiological pH (Rai and Durán 2011; Haddad and Seabra 2012). In this context, this chapter presents and discusses recent progress on toxicological evaluations of cyto- and genotoxicities of iron oxide NPs, as well as the biological impacts of in vivo administration of these NPs in several animal modes. Although the number of studies based on toxicity of iron oxide NPs has been increasing in the last few years, there are still several drawbacks and challenges to be overcome in the near future, in order to propose a safe use of iron oxide NPs to human beings and to the environment. In this scenario, this chapter hopes to be a source of inspiration to scientists to investigate the in vitro and in vivo toxicity and fates of iron oxide NPs, since there are still several key questions to be answered. Detailed investigation of the toxicity of metallic NPs is a promising and exciting area of the expertise to be explored.

12.2 Synthesis of Iron Oxide Nanoparticles

The synthesis and characterization of iron oxide NPs have been the focus of exhaustive investigations in recent years (Asuhan et al. 2012; Tsai et al. 2012). These NPs must be biocompatible, nontoxic, and non-immunogenic in order to be used in biomedical applications. Upon in vivo administration, the particle size should be small enough to stay in the circulation after the internalization to pass through the capillary systems of skins, avoiding vessel embolism (Ranade and Hollinger 2004). Moreover, for biomedical or technological applications, it is desirable to produce NPs with low polydispersity and heterogeneity and biocompatible coating (Haddad et al. 2009; Zhou et al. 2012).

There are many procedures to obtain biomedically applicable iron oxide NPs, based on physical, chemical, and biogenic techniques (Haddad et al. 2004; Jiang et al. 2004; Laurent et al. 2008; Shopska et al. 2013). Overall, the most common approaches to produce magnetic NPs are based on physical vapor deposition, mechanical attrition, chemical routes in aqueous solution, and, more recently, biogenic syntheses. This section summarizes the synthesis of NPs by different routes.

12.2.1 Physical Synthesis

Important physical methods to synthesize iron oxide NPs have been recently reported (Singh et al. 2012). These physical techniques are based on mechanical milling (Chakka et al. 2006), co-sputtering (Chien 1991), mechanic subdivision of metallic aggregates, melting spinning (Singh et al. 2010), and evaporation of a metal in a vacuum by resistive heating or laser ablation (Jun et al. 1960). In the co-sputtering method the elements of solid granules are simultaneously deposited on a substrate by sputtering technique; meanwhile in the melting spinning technique, two metallic elements are heated together, and then the melted solution is ejected on the surface of a rotating metal drum, resulting in high cooling rates (Chien 1991).

Physical methods based on vacuum and gas-phase systems present important advantages over other methods, such as the production of high-quality NPs, free of contaminants (Wang et al. 2012). The laser synthesis is another important technique, in which noncrystalline NPs in a fluid carrier gas are combined with very sharp temperature gradients induced by high-energy laser source, leading to NPs development in wall-less environment (Wang et al. 2012). Spray and laser pyrolyses have been shown to be excellent techniques for direct and continuous production of well-defined magnetic NPs, under controlled experimental conditions. These techniques have a great potential to physically synthesize NPs for useful applications in vitro and in vivo, mainly due to the high production rates of NPs. It must be noted that there are some differences between spray and laser pyrolysis methods to synthesize NPs. In spray pyrolysis, the ultrafine particles are usually aggregated into larger particles, while in laser pyrolysis the ultrafine particles are less aggregated due to the shorter reaction time (Singh et al. 2010).

12.2.2 Chemical Synthesis

Chemical methods in aqueous solutions are still considered the best techniques to produce iron oxide NPs with controllable size and dispersivity (Willard et al. 2004; Molina et al. 2013). Indeed, chemical synthesis permits a better control of particle size and its distribution, morphology, and agglomerate size through the manipulation of experimental parameters that determine nucleation, growth, and coalescence (Haddad et al. 2009; Haddad and Seabra 2012). Moreover, chemical synthesis techniques allow surface modification of the particles during synthesis, providing additional functionality to the NPs. The general strategy for preparing monodisperse NPs in a solution medium is to separate the nucleation and growth phases.

In general, synthesis of particles in aqueous solution occurs by chemical reactions forming stable nuclei with subsequent particle growth (LaMer and Dinegar 1950). A rupture nucleation event first occurs when the monomer concentration rapidly grows over critical supersaturation, without additional formation of nuclei (Souza et al. 2009; Zhu et al. 2012). Upon the addition of precipitating agents and oxidizing reagents to the solution, which contains the reactants, reactions occur, and the solution becomes supersaturated. The thermodynamically equilibrated state is restored by condensation of nuclei of the reaction product and controlled by the kinetics of the nucleation and growth. The produced nuclei then grow at the same rate, giving monodisperse particles (Haddad et al. 2009). Once formed, these NPs have a high surface area and agglomerate easily to minimize their surface energy. An appropriate covering agent has to be used to stabilize these NPs and thus avoiding agglomeration. Figure 12.1 represents this mechanism.

Recently, Haddad and Seabra (2012) presented an overview focusing on the chemical synthesis of various magnetic NPs. Indeed, there are several chemical routes to produce iron oxide NPs, including coprecipitation (Molina et al. 2013), sol–gel synthesis (Haddad et al. 2004; Becuwe et al. 2012), microemulsion synthesis (Laurent et al. 2008; Maleki et al. 2012), sonochemical reaction (Nguyen

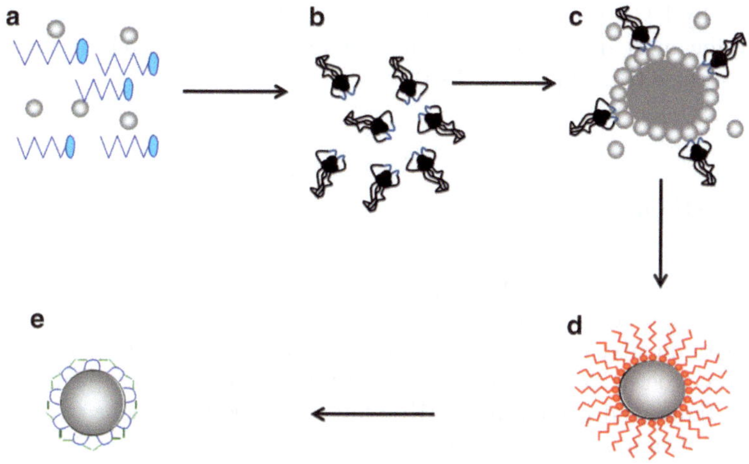

Fig. 12.1 Schematic representation of chemical formation of iron oxide NPs. (**a**) Metal and ligands, (**b**) reduction and nucleation, (**c**) adsorption, (**d**) reduction on surface and growth, (**e**) stabilization of nanoparticle

et al. 2012), hydrothermal reaction (Chen and Xu 1998; Khollam et al. 2002), and thermal decomposition (Haddad et al. 2009; Park et al. 2011), among others.

Nanostructured materials including magnetite (Fe_3O_4) and maghemite (γ-Fe_2O_3) have been successfully synthesized in aqueous solution by chemical methods. However, it is difficult to distinguish between these two phases (magnetite and maghemite) by diffraction X-rays techniques, since both reflections have similar positions. Few studies reported the difficulties to distinguish between magnetite and maghemite phases of iron oxide NPs. A technique typically employed to differentiate these phases is Mossbauer spectroscopy, since the obtained spectra can clearly discriminate these two phases (Lima et al. 2010).

A series of experimental parameters, such as pH, reaction temperature, precursor, and ligand, have been studied to control NP morphology, size, and quantity. As nanoparticle sizes affect their properties, a narrow size distribution is essential (Haddad and Seabra 2012). It is known that almost all biochemical reactions are conducted in an aqueous environment. However, most of chemical syntheses are carried out in organic solvents, leading to NPs coated with hydrophobic ligands. Therefore, to overcome this barrier, a variety of phase-transfer methods have been developed to transfer NPs from organic to aqueous solution (Haddad and Seabra 2012). Common examples of this phase transfers include the coating of NPs with mercaptoundecanoic acid and other mono- and dimercapto alkane carboxylic acids (Abad et al. 2005; Lin et al. 2008). The advantage of this system is that the carboxylic acid terminal group not only confers water solubility but also represents a site for further chemical functionalization. Replacement of hydrophobic ligand as oleic acid by hydrophilic ligand as dimercaptosuccinic acid (DMSA) led to nanoparticles containing sulfhydryl groups on the surface; these groups can be identified by FTIR spectroscopy (Molina et al. 2013). This technique is useful to identify the most important stretching vibrations of the ligand attached on the particle surface (Rutledge et al. 2010).

12.2.3 Biogenic Synthesis

The expansion of eco-friendly approaches to synthesize iron oxide NPs has increased in the last years (Krumov et al. 2009; Li et al. 2011; Durán and Seabra 2012). Specifically, iron oxide NPs can be obtained by a variety of organisms. Nowadays, iron oxide NPs can be successfully synthesized by organisms such as bacteria, fungi, and plant extracts (Krumov et al. 2009; Gudadhe et al. 2011; Li et al. 2011; Durán and Seabra 2012).

Magnetite NPs with narrow size distribution in the range of 10–50 nm were successfully synthesized by the bacteria *Geobacter sulfurreducens* (Bryne et al. 2011). The bacterium was grown in anaerobic environment without light at ambient temperature. A similar work reported the biogenic synthesis of iron oxide NPs comprised of Fe_3O_4 with average size of 10 nm by dissimilatory iron-reducing bacterium, *Shewanella* sp. (Jung et al. 2008).

Several papers describe the use of fungi to synthesize an iron oxide NP, such as *Verticillium* sp. and *Fusarium oxysporum*, which are acidophilic fungi (Bharde et al. 2006). The great advantage of using fungi to mediated synthesis of metallic NPs is the extracellular production of the NPs. Treating fungi with iron salts led the extracellular formation of magnetite (Bharde et al. 2006).

Finally, green synthesis of iron oxide NPs can also be performed by plant extracts (Herrera-Becerra et al. 2010). Tannins are natural and nontoxic polyphenolic compound present in plants, which act as reducing agent leading to the formation of iron oxide NPs (Andjelkovic et al. 2006). Large quantities of tannins accumulate in different parts of plants, to encourage plant defense. Tannins can be oxidized to quinones, while iron salts are reduced to iron NPs. Green synthesis of iron oxide NPs led to the formation of particles coated with biomolecules, such as tannins leading to NP stabilization and avoiding oxidative reactions (Herrera-Becerra et al. 2010).

Biogenic syntheses of NPs have been emerging as a novel and eco-friendly approach. However, it should be noted that there are still some drawbacks to be overcome in the near future. Indeed, complete control of NP size and monodispersity are still major challenges to be overcome. Furthermore, for biomedical applications of NPs, it is mandatory to have a detailed morphological structural and magnetic characterization of the synthesized NPs. The commonly employed techniques are TEM, small-angle X-ray scattering (SAXS), X-ray powder diffraction (XRD), and superconducting quantum interference device (SQUID).

12.3 Toxic Effects of Iron Oxides Administrated In Vitro and In Vivo

Iron oxide NPs, such as magnetite (Fe_3O_4) and hematite (Fe_2O_3), have many important biomedical and industrial applications; moreover, SPIONs have already been clinically commercialized (Doak et al. 2009; Hong et al. 2011). In this context, nanotoxicology has been becoming an increasingly contentious issue in recent years (Doak et al. 2009). Although important progress on the elucidation of in vitro and in vivo toxicity of iron oxide NPs has been achieved, there are still controversial reports regarding the toxicity and fate of these nanomaterials. Therefore, detailed studies on the nanotoxicity of magnetic iron oxide NPs, including their in vivo fate, are required. This section highlights some recent important reports.

12.3.1 Culture Cells

The majority of the studies on the toxicity of iron oxide NPs are based on in vitro assays. Auffan et al. (2006) investigated the cytotoxicity and genotoxicity of nano-γ-Fe_2O_3 particles coated with *meso*-2,3-dimercaptosuccinic acid (DMSA) on normal human fibroblast culture cell. NPs were chemically synthesized by the coprecipitation of Fe^{2+} and Fe^{3+} method, leading to roughly spherical particles with a mean diameter of 6 nm. Figure 12.2 shows the fibroblasts viability from 2 to 48 h period assessed by WST-1 assay, which measures the decrease in the metabolically active cells (Auffan et al. 2006). It can be observed that cell viability significantly decreased from NP concentrations ranging from 10^{-6} to 10^{-3} g/L. However, metabolic mitochondrial activity of cells statistically increased upon treatment with10^{-1} g/L of DMSA-NPs, following 24 and 48 h exposure duration (Fig. 12.2).

The genotoxicity effects of DMSA–iron oxide NPs on cultured cells were evaluated at different NP concentrations (from 10^{-6} to 10^{-1} g/L) by comet assay (Auffan et al. 2006). This assay is based on single cell gel electrophoresis, which shows breakage of DNA strand as a consequence of either direct DNA damage caused by incubation with NPs or indirect effects due to DNA repair process. The authors demonstrated that DMSA-NPs did not cause genotoxic effects on normal human fibroblast culture cell in all tested NP concentrations. Although the absence of DNA breaks, as revealed by the genotoxicity test, cannot be considered as a proof of absence of carcinogenic effect, DMSA-coated NPs were responsible to increase the biocompatibility of the iron oxide NPs. Indeed, the toxicity of iron oxide NPs is greatly influenced by the chemical nature of the coating, rather than the nanoparticle itself (Auffan et al. 2006).

Besides NP coating, the size of particles is also known to play a key role in the toxicity of the material. Recently, Guichard et al. (2012) investigated the in vitro cytotoxicity and genotoxicity of commercially available Fe_3O_4 (magnetite) and

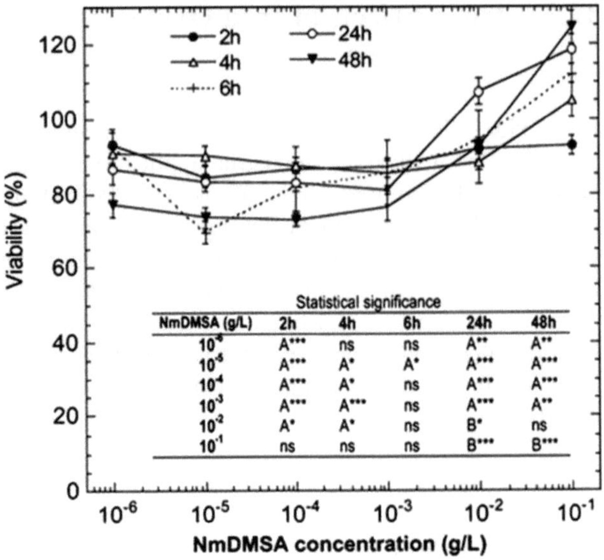

NmDMSA (g/L)	2h	4h	6h	24h	48h
10^{-6}	A***	ns	ns	A**	A**
10^{-5}	A***	A*	A*	A***	A***
10^{-4}	A***	A*	ns	A***	A***
10^{-3}	A***	A***	ns	A***	A**
10^{-2}	A*	A*	ns	B*	ns
10^{-1}	ns	ns	ns	B***	B***

Fig. 12.2 Fibroblast viability over 48 h period incubation with DMSA-NPs at different concentrations and statistically significance. A: Decreased mitochondrial activity. B: Increased mitochondrial activity. Statistical significance: ns, no significance; *$P < 0.05$; **$P < 0.01$; ***$P < 0.001$. Reproduced from Auffan et al. (2006) by permission of American Chemical Society

Fe_2O_3 (hematite) particles (purchased from Sigma-Aldrich) on Syrian hamster embryo (SHE) cells. The authors incubated SHE cells with nanosized iron oxide particles (size in the range of 14–35 nm, as revealed by TEM) and particles in the sub-microscale (size ranging from 147 to 530 nm). After 72 h of treatment, higher cytotoxic effects, including the higher production of reactive oxygen species (ROS), were observed for SHE cells exposure to nanosized Fe_2O_3 particles in comparison with sub-microscale Fe_2O_3 particles. In contrast, nano- and sub-micro-sized Fe_3O_4 particles did not significantly produce intracellular ROS. In fact, both nano- and sub-micro-sized Fe_3O_4 did induce cytotoxicity on SHE cells, at all tested concentrations (Guichard et al. 2012). Regarding the genotoxicity of nano- and sub-micro-sized Fe_2O_3 and Fe_3O_4 particles, no significant increase in DNA damage was detected whatever the concentration tested, after 24 h of exposure. In this case, genotoxicity results indicate the absence of particle size on SHE cells (Guichard et al. 2012).

Several publications demonstrated different toxicity results concerning to the effects of coating composition of iron oxide NPs. An interesting study compares the cytotoxicity and genotoxicity of chemically synthesized SPIONs coated with various functional groups (Hong et al. 2011). The effects of SPIONs with different sizes (average diameter of 10 nm and ranging from 100 to 150 nm), coated with different surface functional groups, were evaluated on L-929 fibroblast cells. SPIONs were coated with the molecules—tetraethyl orthosilicate (TEOS), (3-aminoprophyl) trimethoxysilane (APTMS), and TEOS-APTMS or citrate—leading to the SPIONs coated with the following function groups: hydroxyl (OH), carboxylic (COOH), and amine (NH_2), respectively. Test based on the activity of mitochondrial dehydrogenase was assayed to evaluate the effects of SPIONs coated with different molecules on F-929 cells. The results showed a dose-dependent reduction of cell viability after 24 h of exposure for SPIONs coated with different ligands, suggesting that SPION concentration is more critical than surface modification. Similarly, the authors observed that all surface modified SPIONs increased ROS generation in a dose-dependent manner, with the exception of SPIONs coated with citrate. Moreover, all surface modified SPIONs caused LDH leakage from cells treated, indicating loss of membrane integrity upon exposure of L-929 cells for 24 h with SPIONs coated with different functional groups (Hong et al. 2011).

Concerning to the genotoxicity L-929 cell exposure to SPIONs coated with different ligands, comet assay showed a concentration-dependent increase in the DNA damage, with exception of cell treated with TEOS-coated SPIONs. It can be concluded that positively charged particles might enter into the nucleus of the cell, interacting with DNA, which is negatively charged owing to its phosphate groups. Thus, genotoxicity effects can be dependent on the surface charge of the NP. This assumption was reinforced by observing that positively charged SPIONs, such as those coated with APTMS, showed to be firmly attached to the cell surface, in comparison with negatively charged NPs, such as those coated with TEOS, as revealed by SEM and TEM images. Overall, SPIONs coated with different ligands showed similar cytotoxic results on L-929 cells; however, noticeable responses in the genotoxicity were observed, possibly due to variations of SPIONs surface charges (Hong et al. 2011).

An interesting work characterized the cytotoxicity of polycationic iron oxide NPs, comprised by Fe_3O_4, coated with poly(ethylenimine) (PEI) and with poly (ethylene glycol) (PEG), leading to the formation of particles with varying surface charges (Hoskins et al. 2012). Results revealed that PEI-NPs had average particle size of 30 nm and surface charge of -16.9 mV, while PEG-PEI-NPs exhibit particle size of 40 nm and surface charge of +31.2 mV. Both nanoparticles were incubated with three different human cell lines: neuroblastoma (SH-SYSY), breast cancer (MCF-7), and macrophage-like differentiated V937 cells. Negatively surface charged PEI-NPs showed an increased cytotoxic effect and also ROS generation on all three cell lines, in comparison to positively surface charged PEG-PEI-NPs. Indeed, pegylation of the PEI-NPs reduced the cytotoxicity of the NPs.

In this scenario, a similar work investigated the cytotoxicity effects of iron oxide NPs coated with different molecules (Schweiger et al. 2011). Cytotoxic effects of poly(ethylenimine)-coated γ-Fe_2O_3 NPs(PEI-NPs) and poly-(ethylenimine)-graft-poly(ethylene glycol)-coated γ-Fe_2O_3 NPs (PEI-g-PEG-NPs) were evaluated in adenocarcinomic epithelial cells. The authors found IC_{50} values of 6.8 and 160 μg/mL [Fe] for PEI-NPs and PEI-g-PEG-NPs, respectively. The higher cytotoxicity of PEI-NPs was assigned to the strong cationic PEI character, which can induce defects into cell lipid bilayers. The introduction of hydrophilic PEG moieties into the polymer backbone reduced NP cytotoxicity, indicating that this coating might find important biomedical applications (Schweiger et al. 2011).

Overall, size and coating are the most important parameters that dictate NP cytotoxicity. Ying and Hwang (2010) evaluated the cytotoxicity of iron oxide NPs with different coatings and sizes in A3 human T lymphocytes. Cells were incubated with iron oxide NPs with two different sizes (10 and 50 nm) and two different surface coating (amine and carboxyl groups). Smaller size iron oxide NPs were found to be more cytotoxic than those of larger size. Moreover, carboxyl-coated iron oxide NPs caused a higher cytotoxicity in comparison with amine-coated NPs Ying and Hwang (2010).

A biocompatible magnetic iron oxide NP was obtained by grafting thermosensitive poly(*N*-isopropylacrylamide) (PNIPAM) on the surface of silica (SiO_2)-coated Fe_3O_4 NPs with particle size of ca. 19 nm (Lien et al. 2011). Cytotoxicity studies were performed by treating Chinese hamster ovary (CHO-K1) cells with different concentrations of PNIPAM-grafted NPs for 24 h. The results showed that PNIPAM-grafted NPs were cytocompatible at concentrations up to 0.5 mg/mL, and these NPs did not induce morphological changes upon cell exposure for 24 and 108 h. Thus, this biocompatible magnetic NP can find important biomedical applications, such as drug delivery system (Lien et al. 2011).

Concerning the biomedical applications of magnetic NPs, SPIONs were successfully used as drug carrier vehicle (Munnier et al. 2011). The anticancer drug doxorubicin (DOX) was bound to the surface of SPIONs through a preformed DOX-Fe^{2+} complex. Free DOX and DOX-SPIONs were incubated with MCF-7 cancer cells at concentrations of 8.6 μmol/L for 4 h. DOX-SPIONs induced higher cancer cell mortality (40 %), in comparison with free DOX (10 %), whereas SPIONs alone were not cytotoxic. These results highlight the importance of SPIONs in cancer treatments and indicate that more studies are still necessary in this area (Munnier et al. 2011).

Incorporation of NPs by cells is an important aspect related to the toxicity of iron oxide NPs that deserves special attention. In this context, the uptake of commercial iron oxide magnetite NPs in human alveolar epithelial-like type II cells (A549) by endocytosis was observed by TEM analysis (Konczol et al. 2011). TEM images showed that the particles were found as agglomerates in cytoplasm-bound vesicle, however, none of them in the nucleus. Phagocytosed iron oxide NPs caused slight cytotoxic effects, increased generation of ROS, and a concentration-dependent DNA damage. Moreover, large particles caused less genotoxicity in comparison with smaller particles, indicating size-dependent genotoxicity (Konczol et al. 2011).

12.3.2 *In Vivo Assays*

Toxicity of iron oxide NPs based on in vivo studies can give further insights into the complete biological effects of NPs, such as their fate, accumulation site, pharmacokinetics, biodistribution, interaction with biomolecules, excretion, and degradation (Li and Chen 2011). Compared with in vitro studies, fewer reports investigated the in vivo impact of iron oxide NPs, and more studies in this area are still necessary. Overall, in vivo studies based on administration of iron oxide NPs indicate that toxic effects are highly dependent on NP size and chemical nature of coating (Mahmoudi et al. 2012).

Iron oxide NPs can enter living organisms through active or passive pathways, and several reports described iron accumulation in organs and tissues upon different exposure routes for a certain period. For example, inhalation of polyvinylpyrrolidone (PVP)–iron oxide NPs by mice led to iron accumulation in various organs, such as brain, testes, spleen, lung, and liver (Kwon et al. 2008). Similarly, femoral artery administration of iron oxide NPs at a dose of 12 mg Fe/Kg in rats resulted in accumulation of iron oxide NPs in liver and spleen (Ma et al. 2008). Several papers described degradation of iron oxide NPs in animal tissues after a couple of weeks following administration of these NPs (Jain et al. 2008).

Concerning to in vivo toxicity, intratracheal instillation of Fe_2O_3 NPs (sizes of 22 and 280 nm) induced oxidative stress in the lung (Zhu et al. 2008). However, the presence of a coating on NP surface can significantly enhance the biocompatibility of the NP. For instance, femoral artery administration of alginate-coated iron oxide NPs was found to be biocompatible in hemolysis aspects (Ma et al. 2008).

Iron oxide NPs are metabolized in vivo leading to the formation of iron ions, which in turn are incorporated by erythrocytes as hemoglobin, acting as iron pool (Weissleder et al. 1989). In this regard, infusion of ferrofluids in humans was not found to be toxic to the patients (Lübbe et al. 1996a). Similarly, intravenous administration of magnetic iron oxide NPs with average size of 100 nm did not cause toxic effects in mice (Lübbe et al. 1996b). It must be noted that the metabolism of iron oxide NPs processed after their entrance to the intercellular medium, leading to a temporary increase in iron amounts in the serum (Weissleder et al. 1989). These free iron ions can induce oxidative stress, and thus high doses of SPIONs should be avoided for in vivo applications (Mahmoudi et al. 2012).

Although there are several reports regarding the importance of in vivo studies based on administration of iron oxide NPs by different routes in several animal modes, there is still a necessity of detailed investigation of complete toxicological evaluations of in vivo administration of iron oxide NPs, in order to propose a safe future use of this interesting nanomaterial.

12.4 Comments and Future Perspectives

There are no doubts that iron oxide NPs are considered promising nanomaterials with several important biomedical and technological applications in diverse fields of expertise. Nowadays, iron oxide NPs with controlled size and dispersity can be successfully synthesized by different physical, chemical, and biogenic methods. Biogenic synthesis of iron oxide NPs can be considered the most promising method to prepare metallic NPs, since it is an environment-friendly approach. However, there is still a pressing necessary need for more studies to optimize the biogenic synthesis of these NPs. Most of publications related to SPIONs are still based on the synthesis and characterization of these nanomaterials. Although the interest of the scientific community for the toxicity of magnetic iron oxide NPs has been increasing in the last few years, there is still a lack of detailed information regarding the safe uses of these NPs. In this context, it is evident that nanomaterials cannot be considered as the same manner as bulk materials with regard to their safe uses. Thus, it is of vital importance to evaluate the in vitro and in vivo toxicity of iron oxide NPs.

It can be concluded from the discussion that assessment of toxicity of iron oxide NPs is a complex issue that needs to be further characterized. Several aspects dictate NPs toxicity, such as NP size, size distribution, surface morphology, surface charge density, chemical nature of the coating material, dosage, cell types, and exposure routes. Hence, toxicological results from different studies are difficult to compare, since these parameters may vary between the studies. In this context, more investigations are still needed to better understand the impact of iron oxide NPs on cells, tissues, organs, and organisms in order to extrapolate the potential effects on human health and propose a completely safe use of these promising and interesting nanomaterials.

Acknowledgements FAPESP, CNPq, CAPES, INOMAT (MCT/CNPq), and Brazilian Network in Nanotoxicology (MCT-CNPq).

References

Abad JM, Mertens SFL, Pita M et al (2005) Functionalization of thioctic acid-capped gold nanoparticles for specific immobilization of histidine-tagged proteins. J Am Chem Soc 127:5689–5694

Andjelkovic M, Van Camp J, de Meulenaer B et al (2006) Iron-chelation properties of phenolic acids bearing catechol and galloyl groups. Food Chem 98:23–31

Asuhan S, Wan HL, Zhao S et al (2012) Water-soluble, mesoporous Fe_3O_4: synthesis, characterization, and properties. Ceram Int 38:6579–6584

Auffan M, Decome L, Rose J et al (2006) *In vitro* interactions between DMSA-coated maghemite nanoparticles and human fibroblasts: a physicochemical and cyto-genotoxical study. Environ Sci Technol 40:4367–4373

Batlle X, Labarta A (2002) Finite-size effects in fine particles: magnetic and transport properties. J Phys D Appl Phys 35:R15–R42

Becuwe M, Rouge P, Gervais C et al (2012) A new sensitive organic/inorganic hybrid material based on titanium oxide for the potentiometric detection of iron(III). J Colloid Interface Sci 88:130–136

Bharde A, Rautaray D, Bansal V et al (2006) Extracellular biosynthesis of magnetite using fungi. Small 2:135–141

Bryne JM, Telling ND, Coker VS et al (2011) Control of nanoparticle size, reactivity and magnetic properties during the bioproduction of magnetite by *Geobacter sulfurreducens*. Nanotechnology 22(45):455709. doi:10.1088/0957-4484/22/45/455709

Chakka VM, Altuncevahir B, Jin ZQ et al (2006) Magnetic nanoparticles produced by surfactant-assisted ball milling. J Appl Phys 99:08E912–08E915

Chen D, Xu R (1998) Hydrothermal synthesis and characterization of nanocrystalline Fe_3O_4 powders. Mater Res Bull 33:1015–1021

Chien CL (1991) Granular magnetic solids. J Appl Phys 69:5267–5272

Doak SH, Griffiths SM, Manshian B (2009) Confounding experimental considerations in nanogenotoxicology. Mutagenesis 24:285–293

Durán N, Seabra AB (2012) Metallic oxide nanoparticles: state of the art in biogenic syntheses and their mechanisms. Appl Microbiol Biotechnol 95:275–288

Gudadhe J, Bonde SR, Gaikwad SC et al (2011) *Phoma glomerata*: a novel agent for fabrication of iron oxide nanoparticles. J Bionanosci 5:138–142

Guichard Y, Schmit J, Darne C (2012) Cytotoxicity and genotoxicity of nanosized and microsized titanium dioxide and iron oxide particles in Syrian hamster embryo cells. Ann Occup Hyg 56:631–644

Gupta AK, Gupta M (2005) Synthesis and surface engineering of iron oxide nanoparticles for biomedical applications. Biomaterials 26:3995–4021

Haddad PS, Seabra AB (2012) Biomedical applications of magnetic nanoparticles. In: Martinez AI (ed) Iron oxides: structure, properties and applications. Nova, Nova York, pp 165–188

Haddad PS, Duarte EL, Baptista MS et al (2004) Synthesis and characterization of silica-coated magnetic nanoparticles. Prog Colloid Polym Sci 128:232–238

Haddad PS, Rocha TR, Souza EA et al (2009) Interplay between crystallization and particle growth during the isothermal annealing of colloidal iron oxide nanoparticles. J Colloid Interface Sci 339:344–350

Herrera-Becerra R, Rius JL, Zorrilla C (2010) Tannin biosynthesis of iron oxide nanoparticles. Appl Phys A 100:453–459

Hong SC, Lee JH, Lee J et al (2011) Subtle cytotoxicity and genotoxicity differences in superparamagnetic iron oxide nanoparticles coated with various functional groups. Int J Nanomedicine 6:3219–3231

Hoskins C, Cuschieri A, Wang L (2012) Hybrid gold-iron oxide nanoparticles as a multifunctional platform for biomedical application. J Nanobiotechnology 10:15. http://www.jnanobiotechnology.com/content/10/1/15

Jain TK, Reddy MK, Morales MA (2008) Biodistribution, clearance, and biocompatibility of iron oxide magnetic nanoparticles in rats. Mol Pharm 5:316–327

Jiang WQ, Yang HC, Yang SY et al (2004) Preparation and properties of superparamagnetic nanoparticles with narrow size distribution and biocompatible. J Magn Magn Mater 283:210–214

Jun WK, Willens RH, Duwez P (1960) Non-crystalline structure in solidified gold–silicon alloys. Nature 187:869–870

Jung H, Kim JW, Choi H et al (2008) Synthesis of nanosized biogenic magnetite and comparison of its catalytic activity in ozonation. Appl Catal B: Environ 83:208–213

Khollam YB, Dhage SR, Potdar HS et al (2002) Microwave hydrothermal preparation of submicron-sized spherical magnetite (Fe_3O_4) powders. Mater Lett 56:571–577

Konczol M, Ebeing S, Goldenberg E et al (2011) Cytotoxicity and genotoxicity of size-fractionated iron oxide (magnetite) in A549 human lung epithelial cells: role of ROS, JNK, and NF-kappa B. Chem Res Toxicol 24:1460–1475

Krumov N, Perner-Nochta I, Oder S et al (2009) Production of inorganic nanoparticles by microorganisms. Chem Eng Technol 32:1026–1035

Kucheryavy P, He J, John VT et al (2013) Superparamagnetic iron oxide nanoparticles with variable size and an iron oxidation state as prospective imaging agents. Langmuir 29:710–716

Kwon JT, Hwang SK, Jin H et al (2008) Body distribution of inhaled fluorescent magnetic nanoparticles in the mice. J Occup Health 50:1–6

LaMer VK, Dinegar RH (1950) Theory, production and mechanism of formation of monodispersed hydrosols. J Am Chem Soc 72:4847–4854

Laurent S, Forge D, Port M et al (2008) Magnetic iron oxide nanoparticles: synthesis, stabilization, vectorization, physicochemical characterizations, and biological applications. Chem Rev 108:2064–2110

Li Y-F, Chen C (2011) Fate and toxicity of metallic and metal-containing nanoparticles for biomedical applications. Small 7:2965–2980

Li X, Xu H, Chen ZS et al (2011) Biosynthesis of nanoparticles by microorganisms and their applications. J Nanomater Art. ID 270974. doi:10.1155/2011/270974

Lien Y-H, Wu T-M, Wu J-H et al (2011) Cytotoxicity and drug release behavior of PNIPAM grafted on silica-coated drug release behavior grafted on silica-coated iron oxide nanoparticles. J Nanopart Res 13:5065–5075
Lima JE, de Biasi E, Vasquez MM et al (2010) Surface effects in the magnetic properties of crystalline 3 nm ferrite nanoparticles chemically synthesized. J Appl Phys 108:103919–103929
Lin SY, Chen NT, Sum SP et al (2008) Ligand exchanged photoluminescent gold quantum dots functionalized with leading peptides for nuclear targeting and intracellular imaging. Chem Commun 39:4762–4764
Lübbe AS, Bergemann C, Riess H et al (1996a) Preclinical experiences with magnetic drug targeting: tolerance and efficacy. Cancer Res 56:4694–4701
Lübbe AS, Bergemann C, Riess H et al (1996b) Clinical experiences with magnetic drug targeting: a phase I study with 4′-epidoxorubicin in 14 patients with advanced solid tumors. Cancer Res 56:4686–4693
Ma H-L, Qi X-R, Ding W-X (2008) Magnetic targeting after femoral artery administration and biocompatibility assessment of superparamagnetic iron oxide nanoparticles. J Biomed Mater Res 84A:598–606
Mahmoudi M, Hofmann H, Rothen-Rutishauser B et al (2012) Assessing the in vitro and in vivo toxicity of superparamagnetic iron oxide nanoparticles. Chem Rev 112:2323–2338
Maleki H, Simchi A, Imani M et al (2012) Size-controlled synthesis of superparamagnetic iron oxide nanoparticles and their surface coating by gold for biomedical applications. J Magn Magn Mater 324:3997–4005
Marchiol L (2012) Synthesis of metal nanoparticles in living. Italian J Agron 7:274–282
Molina MM, Seabra AB, de Oliveira MG et al (2013) Nitric oxide donor superparamagnetic iron oxide nanoparticles. Mat Sci Eng C 33:746–751
Munnier E, Cohen-Jonathan S, Herve K et al (2011) Doxorubicin delivered to MCF-7 cancer cells by superparamagnetic iron oxide nanoparticles: effects on subcellular distribution and cytotoxicity. J Nanopart Res 13:959–971
Nguyen DP, Trinh XS, Hoang TC et al (2012) Amorphous iron-chromium oxide nanoparticles prepared by sonochemistry. J Non-Cryst Solids 358:537–543
Park S, Lee H, Goo S et al (2011) Composition behavior of carbon nanotube reinforced thermotropic liquid crystalline polymers. J Appl Polym Sci 122:2060–2070
Rai M, Durán N (eds) (2011) Metal nanoparticles in microbiology, 1st edn. Springer, Germany
Ranade VV, Hollinger MA (eds) (2004) Drug delivery systems. Pharmacology and toxicology: basic and clinical aspects, 2nd edn. New York, CRC Press
Rutledge RD, Warner CL, Pittman JW et al (2010) Thiol-ene induced diphosphonic acid functionalization of superparamagnetic iron oxide nanoparticles. Langmuir 26:12285–12292
Schweiger C, Pietzonka C, Heverhagen J (2011) Novel magnetic iron oxide nanoparticles coated with poly(ethylene imine)-g-poly(ethylene glycol) for potential biomedical application: synthesis, stability, cytotoxicity and MR imaging. Int J Pharm 408:130–137
Shopska M, Cherkezova-Zheleva ZP, Paneva DG et al (2013) Biogenic iron compounds: XRD, Mossbauer and FTIR study. Cent Eur J Chem 11:215–227
Singh SP, Chakradhar RPS, Rao JL et al (2010) EPR, FTIR, optical absorption and photoluminescence studies of Fe_2O_3 and CeO_2 doped ZnO-Bi_2O_3-B_2O_3 glasses. J Alloys Compd 493:56–262
Singh AP, Mettenborger A, Golus P et al (2012) Photoelectrochemical properties of hematite films grown by plasma enhanced chemical vapor deposition. Int J Hydrogen Energ 37:13983–13988
Souza EA, Winnischofer H, Haddad P et al (2009) Effects of particles size and synthesis temperature on the structural properties of the Ni nanoparticles: insights about the formation of the fcc-Ni structure. AIP Conf Proc 1092:135–137
Tsai ZT, Tsai FY, Yang WC et al (2012) Preparation and characterization of ferrofluid stabilized with biocompatible chitosan and dextran sulfate hybrid biopolymer as a potential magnetic resonance imaging (MRI) T2 contrast agent. Mar Drugs 10:2403–2414

Wahajuddin AS (2012) Superparamagnetic iron oxide nanoparticles: magnetic nanoplatforms as drug carriers. Int J Nanomedicine 7:3445–3471

Wang Z, Li X, Gao M et al (2012) One-step preparation of amorphous iron nanoparticles by laser ablation. Powder Technol 215:147–150

Weissleder R, Stark DD, Engelstad BL et al (1989) Superparamagnetic iron oxide: pharmacokinetics and toxicity. Am J Roentgenol 152:167–173

Willard MA, Kurihara LK, Carpenter EE et al (2004) Chemically prepared magnetic nanoparticles. Int Mat Rev 49:125–170

Ying E, Hwang H-M (2010) *In vitro* evaluation of the cytotoxicity of iron oxide nanoparticles with different coatings and different sizes in A3 human T lymphocytes. Sci Total Environ 408:4475–4481

Zhou Z, Huang D, Bao J et al (2012) A synergistically enhanced T1–T2 dual-modal contrast agent. Adv Mater 24:6223–6228

Zhu M-T, Feng W-Y, Wang B et al (2008) Comparative study of pulmonary responses to nano- and submicron-sized ferric oxide in rats. Toxicology 247:102–111

Zhu J, Wei S, Gu H et al (2012) One-pot synthesis of magnetic graphene nanocomposites decorated with core@double-shell nanoparticles for fast chromium removal. Environ Sci Technol 46:977–985

Chapter 13
Poloxamers as Drug-Delivery Systems: Physicochemical, Pharmaceutical, and Toxicological Aspects

Daniele R. de Araújo, Alisson Oshiro, Deyse Cardoso da Silva, Alessandra Cristina Santos Akkari, Joyce Cristine de Mello, and Tiago Rodrigues

Abstract Poloxamers (PL) are copolymers A-B-A type consisting of ethylene oxide (EO) and propylene oxide (PO) units in a triblock EO_x–PO_y–EO_x arrangement. These copolymers are interesting due to their ability for temperature-dependent gel formation, as a result of their self-assembling in micelles. Several studies have demonstrated the application of the thermoreversible copolymers as drug-delivery systems in order to prolong the drug release, to sustain the effectiveness, and also to reduce local and/or systemic toxicity, connecting the expertise of different research fields such as Biochemistry, Nanotechnology, Biopharmaceutics, Pharmacology, and Toxicology. Then, the purpose of this chapter involves a discussion about PL copolymers in the light of those research fields.

The development of one-type or binary PL carriers systems is a function of composition (type of copolymer, differences on EO/PO units number, molecular weight), physicochemical properties (hydrophilic-lipophilic balance, cloud point, critical micellar concentration), and structural parameters such as micellar size, temperature for micelles, and hydrogels assembling. Besides, for focusing on how PL can be useful to achieve sustained drug release it is necessary to consider their pharmacological properties (such as the ability to inhibit the P-glycoprotein), in vitro (cytotoxicity or cytoprotection and their mechanisms) and in vivo toxicological evaluation (biocompatibility and regulatory aspects).

D.R. de Araújo (✉)
Centro de Ciências Naturais e Humanas, Universidade Federal do ABC – UFABC, Av. dos Estados 5001, Bangú, CEP 09210-580 Santo André, Sao Paulo, Brazil
e-mail: daniele.araujo@ufabc.edu.br

N. Durán et al. (eds.), *Nanotoxicology*, Nanomedicine and Nanotoxicology, DOI 10.1007/978-1-4614-8993-1_13,

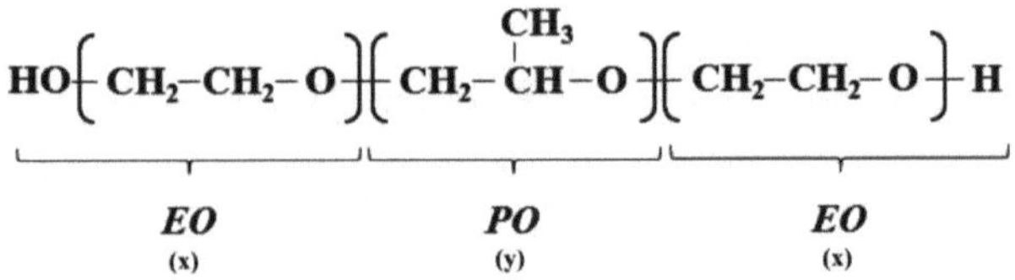

Fig. 13.1 Basic chemical structure of poloxamers copolymers

13.1 Physicochemical Aspects: Chemical Structure, Micellization, and Gelling Phenomena

Poloxamers (PL) are a family of more than 30 different surface active agents, consisting of triblock copolymers (Fig. 13.1) with two hydrophilic poly(ethylene oxide) (PEO) blocks and a hydrophobic poly(propylene oxide) (PPO) block arranged in a basic EO_a–PO_b–EO_a structure. The number of ethylene oxide (EO) and propylene oxide (PO) units varies according to each copolymer (Schmolka 1972; Kabanov et al. 2002a, b; Quadir 2005). These copolymers are commercially known with the trade names of Pluronic®, Lutrol®, and Synperonic® and are available in solid, liquid, and paste state depending on the percentage of EO for each poloxamer copolymer structure (Dumortier et al. 2006; Yapar and Inal 2012).

PL are synthesized by anionic polymerization. The variation of their molecular characteristics, which can be adjusted during the synthesis process, could confer to them different physical and chemical properties, mainly related to EO:PO ratio, i.e., hydrophilic/hydrophobic parts, block architecture, and molecular weight (MW) (Vadnere et al. 1984; Alexandridis and Hatton 1995; Su et al. 2002; Artzner et al. 2007). Considering this, the capacity of altering the composition of the copolymers during the synthesis allows the obtention of compounds with different properties for many applications including their use for emulsification, detergency, wetting, foaming, lubrication, bioprocessing, including cosmetic and pharmaceutical applications, especially for drug release (Mitchard et al. 1992; Alexandridis 1994; Su et al. 2002; Fusco et al. 2006).

The hydrophilic-lipophilic balance (HLB) of PL, which value cover a range from 1 to 30, is characterizaed by their amphiphilic structure as a dependence of EO and PO units (Kozlov et al. 2000; Fusco et al. 2006). A high HLB value refers to a more hydrophilic PL that represents a higher percentage of hydrophilic EO units when compared to the number of the hydrophobic PO units. On the other hand, low HLB value indicates a more lipophilic copolymer, which has lower content of EO than PO units (Kozlov et al. 2000). Due to their amphiphilic features, PL also presents a surface activity, consisting of a nonionic surfactant self-assembled in micelles (Schmolka 1972; Kabanov et al. 2002a, b; Escobar-Chávez et al. 2006).

13.1.1 Micelles and Poloxamer Thermoreversible Hydrogels: Self-Association and Structural Studies

Micellization process has an enormous importance for drug-delivery, since it can be formed when the concentration of the copolymer is higher than its critical micelle concentration (CMC) (Alexandridis and Hatton 1995; Mata et al. 2005). The CMC values for PL decrease according to the high number of PO units (considering a constant number of EO units), indicating that the micellization process is mainly a function of the hydrophobic block. In addition, for copolymers with high molecular weight the formation of micelles is favored due to their ability of self-association (Alexandridis and Hatton 1995).

Another important physicochemical property is the cloud point, i.e., the temperature at which the PL phase separates from water and a new system is formed, where the aqueous solution phase is in equilibrium with the copolymer phase (Campese et al. 2003). For PL with high percentage of EO the cloud point is elevated due to the strongest interactions between the hydrophilic EO block and water molecules, which is a good solvent for EO units in a range from 0 to 100 °C. In contrast, for PL with high percentage of hydrophobic PO content the cloud point is low (Alexandridis 1994). Table 13.1 shows some physicochemical properties of PL.

Poloxamer micellization process is deeply temperature-dependent due to the difference in the solvation of EO and PO blocks (Zhang and Lam 2006). For this reason, at low temperatures PL exist as unimers form, when the temperature increases the formation of the micelles (considering the copolymer concentration above the CMC) are favored due to the dehydration of the PO block (Hvidt et al. 2002; Artzner et al. 2007). The structure of the micelles formed by PL can be spherical, rod-like, or lamellar depending on the hydrophobic/hydrophilic ratio, number of the EO or PO blocks, and temperature (Nagarajan 1999; Kabanov et al. 2002a, b; Nambam and Philip 2012). Nevertheless, any kind of micelles has a hydrophobic core consisting mainly of PO and a hydrophilic shell formed by hydrated EO segments, probably interacting by hydrogen bonds and/or Van der Waals forces (Vadnere et al. 1984; Hvidt et al. 2002; Oh et al. 2004; Kadam et al. 2010). The dehydrated core of the micelles allows the non-covalent incorporation of hydrophobic drugs improving the solubility, the metabolic stability, and the circulation time of the drug (Oh et al. 2004).

The micellization is followed by the gelation process and the liquid crystalline gel phases in water can be formed in cubic, hexagonal, and lamellar phases (Ivanova et al. 2002). Above the sol-gel transition temperature (T_{gel}), the aqueous solutions of PL form hydrogels, which is totally reversible, i.e., at higher temperatures copolymers solutions liquefy again and characterize the thermoreversible gelation of PL. This mechanism refers to micelar desolvation and swelling, forming cross-link between the polymer micelles, which confers an increase in the viscosity and consistency of the polymer solution, leading to a gel state upon heating (Guzmán et al. 1992).

Table 13.1 Physicochemical properties of some PL copolymers

Poloxamer	MW	Number of EO units	Number of PO units	Cloud point (°C)*	HLB	CMC (M)
L35	1,900	21.59	16.38	73	19	5.3×10^{-3}
L43	1,850	12.61	22.33	42	12	2.2×10^{-3}
L44	2,200	20.00	22.76	65	16	3.6×10^{-3}
L61	2,000	4.55	31.03	24	3	1.1×10^{-4}
L62	2,500	11.36	34.48	32	7	4.0×10^{-4}
L64	2,900	26.36	30.00	58	15	4.8×10^{-4}
L81	2,750	6.25	42.67	20	2	2.3×10^{-5}
L92	3,650	16.59	50.34	26	6	8.8×10^{-5}
L101	3,800	8.64	58.97	15	1	2.1×10^{-6}
L121	4,400	10.00	68.28	14	1	1.0×10^{-6}
P84	4,200	38.18	43.45	74	14	7.1×10^{-5}
P85	4,600	52.27	39.66	85	16	6.5×10^{-5}
P103	4,950	33.75	59.74	86	9	6.1×10^{-6}
P104	5,900	53.64	61.03	81	13	3.4×10^{-6}
P105	6,500	73.86	56.03	91	15	6.2×10^{-6}
P123	5,750	39.20	69.40	90	8	4.4×10^{-6}
F68	8,400	152.73	28.97	>100	29	4.8×10^{-4}
F87	7,700	122.50	39.83	>100	24	9.1×10^{-5}
F88	11,400	207.27	39.31	>100	28	2.5×10^{-4}
F98	13,000	236.36	44.83	>100	28	7.7×10^{-5}
F108	14,600	265.45	50.34	>100	27	2.2×10^{-5}
F127	12,600	200.45	65.17	>100	22	2.8×10^{-6}

Note: * refers to 1 % PL in water; L (liquid), P (paste), and F (flakes); MW and HLB values were provided by manufacturer monographs (BASF®, USA). CMC values were determined using pyrene probe (Kozlov et al. 2000; Alexandridis and Hatton 1995; Kabanov et al. 2002a, b; Fusco et al. 2006)

The thermoreversible nature of these copolymers is extremely important in drug delivery because, handling the PL concentration, it allows their administration as liquid at room temperature (20–25 °C) and then form a gel in situ in contact with body fluids (35–37 °C), providing controlled drug release (Mitchard et al. 1992; Escobar-Chávez et al. 2006; Kidowaki et al. 2006; Artzner et al. 2007; Overstreet et al. 2013). Besides the temperature, it is worth noting that the presence of different salts and solvents influences the micellization and gelation processes. For instance, the addition of salts strongly alters the cloud point and the CMC of PL, and these parameters can be reduced or increased for different salts. Furthermore, it was noticed that the presence of NaCl reduced the influence of temperature on micelar size and hydration of PL-F127, an FDA approved PL largely used for drug-delivery studies (Alexandridis and Hatton 1995; Pandit and Wang 1998). In regard to gelation phenomena, the presence of salt decreases remarkably the sol-gel temperature and can modify the rheological properties of the hydrogel (Jiang et al. 2008).

The process of PL self-association also depends on the solvent due to the characteristic of PL which self-assembles in selective solvents with great affinity for hydrophilic EO block and less for hydrophobic PPO block. Therefore, variation on type of solvents modifies the copolymer behavior, since the interaction PL-solvent is different for each solvent. Previous studies (Liu and Chu 2000; Hvidt et al. 2002) describe the use of different solvents and additional steps to promote sample purification using polar solvents, especially to remove possible manufacturing impurities that might be present in bulk material. The solvents also can influence the stability of liquid crystalline gel phase (Ivanova et al. 2002).

The physicochemical characterization of the PL solutions is an important step to describe different systems to each application. Some of the methods for micelles sizes determination and hydrogels structure studies and also other characteristics are described below. The following variables can influence the final drug-delivery system, such as PL type and its molecular weight, EO and PO chains size, type of solvent, salt effect, binary composition, total copolymer concentration and temperature.

PL-micelles size can be determined by Laser Light Scattering (LLS), also known as Photon Correlation Spectroscopy (PCS). It is considered a rapid method for routine measures for determining size, particles distribution, and polydispersity Index (PI) (Gaumet et al. 2008). The usual method for size measurements by dynamic light scattering uses a He–Ne laser operating at a wavelength of 633 nm and may use the noninvasive backscatter technology of 173° or more commonly described 90°. This technique calculates the Brownian motion of the suspended particles to measure its size distribution. The Brownian motion of the micelles is dependent on the diffusivity of the micelles and the hydrodynamic radius (RH) of the solute particles can be calculated according to the (13.1) and Stokes–Einstein (13.2) (Nambam and Philip 2012), where Γ is the relaxation or correlation constant, D is translational diffusion coefficient, and q is the scattering vector, k_B is the Boltzmann constant, and η is the viscosity of the solvent.

$$\Gamma = D.q.2 \tag{13.1}$$

$$\mathrm{RH} = k_B T / 6\pi\eta D \tag{13.2}$$

In spite of being a largely described technique, the literature points that the particle size measurement should be combined with other methods such as Scanning Electron Microscopy (SEM), if applicable. The LLS method presents some limitations, such as for measuring the particle size of high polydisperse populations, and can be influenced by aggregates or large particles (Gaumet et al. 2008; Mora-Huertas et al. 2010; Parmar et al. 2011; Nambam and Philip 2012).

For that reason, the particle size description for PL micelles has a wide range of variability in the literature. Analysis by PCS method has described micelles sizes with hydrodynamic radius ranging from 2 to 30 nm. For Pluronic® P85 and Pluronic® P123 the micelar radius was 8.1 ± 0.3 and 9.1 ± 0.5 nm, respectively. On the other hand, the 50:50 binary composition of Pluronic® P85 and Pluronic® P123 presented a micellar radius of 10.6 ± 0.6 nm in water at 25 °C (Newby et al. 2009). Micelle size of Pluronic® F127 has been reported to 13 nm in the same conditions of previous study (Sharma et al. 2008). Besides, other study with Pluronic® F127 showed a hydrodynamic radius value of 10 nm in three different angles (45°, 90°, and 133°), indicating the formation of spherical micelles aggregates for this PL (Zhang et al. 2005).

Other physicochemical technique usually applied to characterize PL systems is the differential scanning calorimetry (DSC). Calorimetric analyses are used to determine the $T_{onset,}$ which corresponds to the CMT (Critical Micellization Temperature) of the PL solution and the final temperature for the micellization process (T_{endset}). The peak area can be calculated to determine the micellization enthalpy (Artzner et al. 2007).

DSC analysis also gives information on copolymer thermoreversible properties when testing the sample on heating/cooling/ heating ramps. Data analysis has shown the major dependence on PO content in thermodynamic parameters, since its increasing molecular weight promotes an enthalpy variation. Previous studies reported that micellization is an endothermic process with positive enthalpy. Besides, there is a tendency of higher molecular weight PL to present more negative free energies during the micellization process (Alexandridis and Hatton 1995; Mitchard et al. 1992). Additionally, it has been described that the heat capacity of PL makes less positive when the molecular mass of the PO content is above 750 and the polymer starts to become insoluble in water (Mitchard et al. 1992; Artzner et al. 2007).

The presence of salts can significantly decrease the micelles formation temperature enhancing the desolvation of PO by displacing the water hydrogen bounded to other oxygen of PO (Jiang et al. 2008). Desolvation in the presence of salts promotes exposure of hydroxyl groups and formation of cross-links of weak interactions among PL unimers, mainly hydrogen and van der Waals interactions (Vadnere et al. 1984). High copolymer concentrations reduces the intermicellar distances and the micelar swelling, contributing to the copolymers chains contact and therefore reducing the micellization temperature (Guzmán et al. 1992). However, thermodynamic studies do not provide information about hydrogels structures, which can be elucidated by rheological, small angle X-ray scattering (SAXS), and Small Angle Neutron Scattering (SANS) studies.

Most of PL copolymers form hydrogels and their thermoreversible properties provide both Newtonian and non-Newtonian behavior as a dependence of the concentration and T_{gel} (Dumortier et al. 2006).

PL in solutions are presented as monomers and according to the temperature (at sufficient PL concentration) they exhibit organization in micelles aggregates, evoking an increase on the viscosity parameters (Artzner et al. 2007).

PL rheological studies are mainly focused on shear stress versus shear rate (or flow curve studies) and oscillatory studies for determining the G′ storage modulus and G″ loss modulus. In solution, PL presents lower G′ values than G″ and at high temperatures the elastic modulus (G′) prevails over the viscous modulus (G″) (Dumortier et al. 2006). The development of drug-controlled release formulations by studying rheological characteristics of the hydrogels can provide information about the rate of dissolution and the rate of drug release, since oscillatory studies (G′, G″ versus frequency) for semisolid state (gel) allow to obtain a relationship with the gel consistency.

Regarding supramolecular organization, studies using SAXS/SANS have been performed to describe different types of PL. A phase transition is observed when increasing copolymer concentration or after changes in composition of binary systems, being described different geometric phases like lamellar, face-centered cubic (FCC), body-centered cubic (BCC), and hexagonal (Wu et al. 1997; Jiang et al. 2008; Mortensen et al. 2008; Newby et al. 2009).

At low temperatures (about 15 °C) and concentrations above 15 % (wt/wt), PL are soluble in water and the solution presents a lamellar structure, as observed for Pluronic® F127. If the temperature and/or concentration are increased, phase transitions from hexagonal to FCC or BCC phases are observed (Fig. 13.2). It is also important to notice that the commercially available PL may contain up to 20 % of diblock copolymer (EO-PO) or even isolated EO chains impurities, which might lead to different BCC or FCC phases depending on the purification method employed for the sample. These impurities, in some levels, can induce toxicity depending on the concentration and the PL biomedical application (Singh-Joy and McLain 2008; Mortensen et al. 2008).

13.2 Pharmaceutical and Toxicological Aspects: In Vitro and In Vivo Evaluation

The fundamental properties for pharmaceutical application of PL as drug-delivery systems are: (1) high molecular weight; (2) covalent and non-covalent interactions with drugs, solving problems related to the solubility, stability, and permeability, (3) biological functions, such as the application of some types of PL as P-glycoprotein inhibitors, and (4) stimulus-sensitive features (Kabanov et al. 2002a, b; Dumortier et al. 2006).

The responsivity to local stimuli is an essential factor in the formation of hydrogels composed of PL and its applications as delivery systems for drugs or in situ delivery systems. One of the most studied PL, the PL 407 (or Pluronic® F127), as well as other copolymers such as PL 188 and PL 403, has been considered of great interest as carrier system for drugs and gene therapy (considering the slow diffusion of the molecule incorporated across intermicellar spaces) (Sharma et al. 2008), cellular and tissue engineering (as substrates for hematopoietic cells, osteoblasts, and healing process for burn tissues) (Schmolka 1972, Weinand et al. 2006; Higuchi et al. 2006), as well as vehicles for many types of pharmaceutical forms (solutions for intravenous, subcutaneous, intramuscular, oral, ophthalmic, and topical formulations) (Klouda and Mikos 2008).

An important point is that the hydrogel composition is critical for the formulation development, since reports in the literature showed that the association of different types of PL (as binary systems, for example) is an alternative for

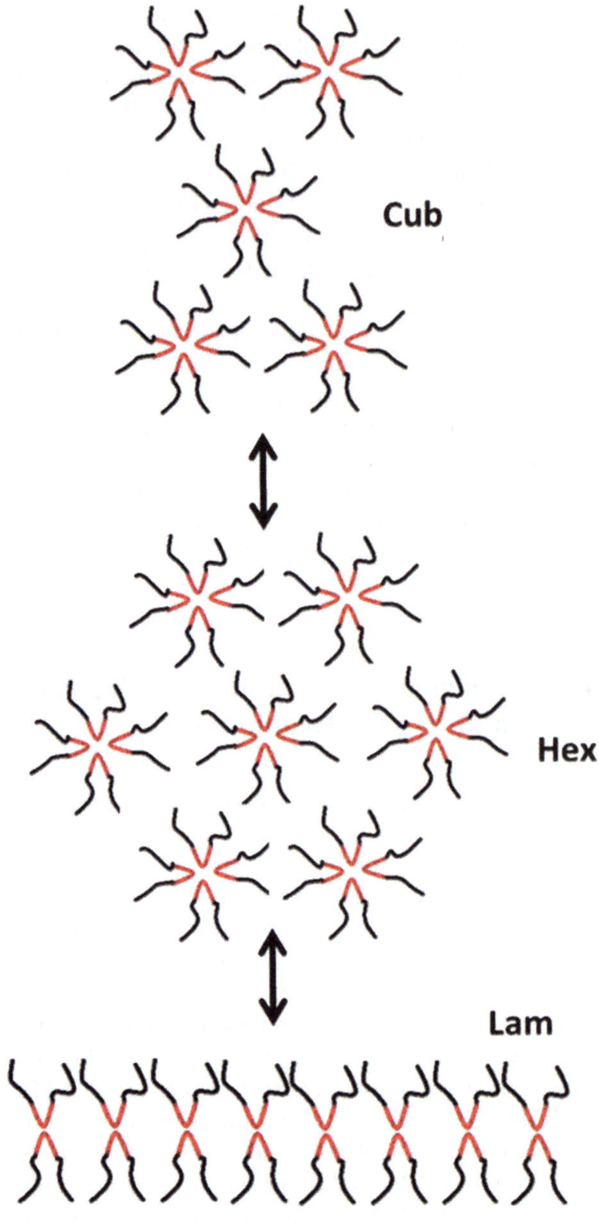

Fig. 13.2 Supramolecular arrangements and phase transitions observed for poloxamers hydrogels. Lam (lamellar phase), Hex (hexagonal phase), and Cub (cubic phase)

modulating the biopharmaceutical profile, micelles structure, the hydrogel rheological properties, formulation stability, solubility and release rate of the incorporated drug (Harrison et al. 2005; Chaibundit et al. 2007; Newby et al. 2009).

13.2.1 In Vitro Toxicology: Focus on Cytotoxicity and Cytoprotection

Several drug-delivery systems have been used as therapeutic strategies in the treatment of many diseases. Among them are polymeric micellar delivery systems, such as PL that exhibit great versatility by acting only as carriers of drugs facilitating its solubility or also contributing to the expected pharmacological effects. Thus, the study of the interactions of these compounds with cellular models is an important initial step to define their therapeutic applications (Alvarez-Lorenzo et al. 2011).

Initially, the poloxamers were considered biologically inert and used to enhance the efficiency in delivering drugs, to decrease the degradation time, or to increase the tissue exposure to the drug. Several studies have demonstrated that poloxamers present a relatively low toxicity in cultured cells or even cytoprotection; however, some compounds exhibited cytotoxicity. Although the molecular mechanisms of PL-induced cytoprotection/cytotoxicity remain unclear, some advances have been reached in recent studies, since PL can exhibit cytotoxicity in tumor cells or enhance the effect of antitumor drugs leading some authors to suggest their use in the antitumor chemotherapy (Gong et al. 2012; Batrakova et al. 2010; Alakhova et al. 2010; Kabanov et al. 2002a, b).

Among the mechanisms of cytotoxicity are the inhibition of the multidrug resistance P-glycoprotein (Pgp), the modulation of the membrane fluidity, and the impairment of ATP synthesis (Batrakova and Kabanov 2008). The cytoprotection seems to be promoted by membrane sealing effects (Cheng et al. 2012), as discussed above. Some in vitro effects of PL in cultured cells are summarized on Table 13.2.

13.2.1.1 Membrane Interaction: Relationship Between Cytotoxicity and Cytoprotection

Different interactions between cell membranes and PL related to their biological effects have been described (Cheng et al. 2012). The HLB is a parameter correlated to the changes promoted by the PL in membranes, since higher HLB values indicate higher hydrophilicity of the compound and its possible insertion in cell membrane. Pluronic® F127, which is widely used in formulations for drug delivery, has HLB 29 and causes slight changes in membranes as compared with the Pluronic® L61 with HLB 3 (Croy and Kwon 2006; Demina et al. 2005). The L61 is able to accelerate the flip-flop movement in bilayers and its effect on the permeability of the lipid bilayer depends on the membrane composition (Zhirnov et al. 2005).

Hydrophobic PL as PL L181 (Pluronic® L61) may insert into the lipid bilayer and permeabilize membranes, facilitating the permeation of small molecules and affecting the membrane structure and function. Moreover, relatively hydrophilic PL as PL 188 (Pluronic® F68) tend to remain on the surface of the membrane acting as

Table 13.2 In vitro studies with different types of poloxamers or Pluronics®

Poloxamer	In vitro model	Results	References
L43 or P123	Associated with F68 for delivery of curcumin in MCF-7/ADR cells	Increasing the cytotoxicity	Zhao et al. (2012)
L61 or P181	In mixed micelles with mPEG-PLA for delivery of paclitaxel in MCF-7/ADR cells	Increasing the cytotoxicity	Li et al. (2010)
L35 or P105	In association with of polyethylene glycol and polycaprolactone in doxorubicin-treated K562/ADR cells	Increasing the cytotoxicity	Han et al. (2011)
F68 or P188	Protection from mechanical injury in primary neurons cells	Neuroprotection	Luo et al. (2013)
	Protection of skeletal muscle cells after exposure to irradiation	Greater survival in treated cells	Greenebaum et al. (2004)
F127 or P407	toxicity in MCF-7 cells associated with graphene nanosheet and doxorubicin	facilitates the entry of doxorubicin into cells	Hu et al. (2012)
P85 or P235	Toxicity in A549 cell of co-delivery system, P85-PEI/TPGS/PTX/shSur to overcome paclitaxel resistance	Increasing the cytotoxicity	Shen et al. (2012)
	Effects in mitochondria isolated from MCF-7/ADR cells	ATP depletion	Alakhova et al. (2010)

a sealant or dampening fluctuations in the membrane (Cheng et al. 2012). Such characteristics were observed in another study that compared F127 and L61 and showed that the more hydrophobic L61 was able to disrupt vesicles and destroy bilayers (Chieng and Chen 2009). The authors suggest that this feature is determinant for the cytotoxicity of L61 but not for F127.

Different mechanisms were proposed to explain the cytoprotective ability of PL 188. The prevention of the traumatic neuronal damage (Luo et al. 2013) and the irradiation-induced necrosis in skeletal muscle cells (Greenebaum et al. 2004) were attributed to the maintenance of the integrity of damaged membranes by this copolymer. Also, the successful use of PL 188 in the treatment of burns was attributed to its ability to insert in cell membranes, to decrease the capillary blood flow stasis, and to recovery the lysozyme inactivated by heat (Yuhua et al. 2012).

Two mechanisms of interaction of PL and cell membranes have been described for poloxamers: peripheral interaction for hydrophilic poloxamers and adsorption followed by insertion for hydrophobic poloxamers. Insertion refers to the interaction that promotes membrane perturbation, while only adsorption promotes repair or delay of the loss of integrity (Fig. 13.3). For an efficient drug-delivery system development, an appropriate balance among those interactions must be considered (Cheng et al. 2012; Wang et al. 2010).

Particularly, for the antitumor chemotherapy, formulations have showed the combination of PL in order to promote drug-delivery by micellar systems, which reduce the leakage of drug in normal tissues and increase the retention time of the drug into the tumor cells. Furthermore, PL promotes changes in membrane

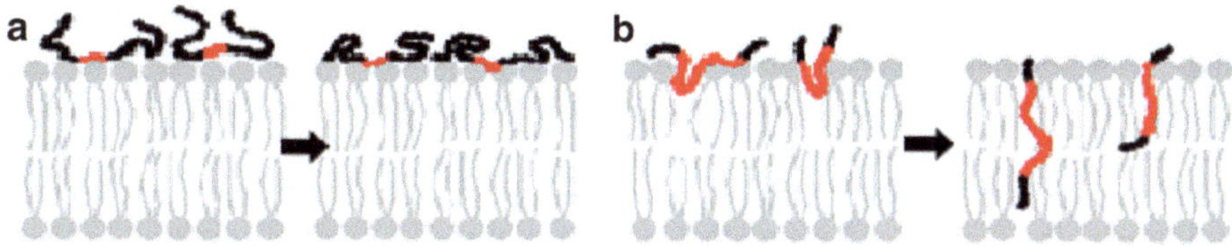

Fig. 13.3 Schematic representation of the interactions of poloxamers with lipid bilayers. Hydrophilic poloxamers interact with membrane surface and remain peripherally (**a**), and the adsorption followed by insertion is proposed for the hydrophobic poloxamers (**b**)

properties to potentiate the desirable pharmacological effect. For example, L81, L61, and P85 are known to inhibit the P-glycoprotein, a protein involved in multidrug resistance in antitumor therapy, and one of the mechanisms involved seems to be the perturbation in the membrane stability (Batrakova and Kabanov 2008).

13.2.1.2 Mechanisms of Cytotoxicity: Inhibition of P-Glycoprotein and Mitochondrial Disfunctions

The Pgp is an ATP-binding cassette transporter found on tumor cells resistant and described as a protein that decreases the drug accumulation inside the cells. Physiological activities such as extrusion of xenobiotics and endogenous molecules have been attributed to Pgp. These proteins have four transmembrane and two cytoplasmic domains, which are involved in the formation of the binding site and ATP interaction site, respectively (Fu and Arias 2012; Montesinos et al. 2012). The expression and overexpression of Pgp is common in tumors that exhibit multiple drug resistance (MDR) including leukemia, breast cancer, hepatocellular carcinoma, and various sarcomas (Ieiri 2012; McDevit and Callaghan 2007; Arora et al. 2005). In these cells, Pgp function is to promote drug efflux and decrease its intracellular concentration, increasing the tumor resistance to chemotherapy agents with affinity for Pgp such as doxorubicin, imatinib, methotrexate, paclitaxel, vinblastine, and vincristine (McDevit and Callaghan 2007; Balayssac et al. 2005).

As mentioned above, the incorporation of PL in lipid bilayers induces changes on its properties resulting in the inhibition of the activity of Pgp. One of these changes is the decrease of the microviscosity which can alter the conformational structure of Pgp and, thus, change the function of the protein (Batrakova and Kabanov 2008). These effects are not dependent on the copolymer concentration and they occur only at low concentrations, close to CMC. It was shown that high concentrations of PL may restore the activity of Pgp (Montesinos et al. 2012), but the selectivity of PL to Pgp inhibition is a preferential action in "lipids rafts," where it is believed that Pgp is located (Batrakova and Kabanov 2008).

Despite the inhibitory activity of Pgp by PL, few cytotoxic effects are reported by these copolymers in the absence of an antitumor drug. However, when PL are associated with an antitumor agent a potentiation effect of the drug was observed, mainly in MDR-positive tumors. Thus, these polymeric systems are useful to improve the solubility of poor water-soluble antitumor drugs, increasing their bioavailability and, consequently, enhancing the antitumor effect. Also, due to the preferential action of MDR (Multidrug Resistance) cells, PL may increase the antitumor effectiveness of drugs that will be extruded by MDR proteins in these tumors. In this context, Zhao et al. (2012) demonstrated that the PL systems containing curcumin exhibit higher cytotoxicity in MCF-7/ADR cells than in MCF-7 cells. The same effect was also observed in systems containing paclitaxel in MCF-7 and A549 cells (Shen et al. 2012; Li et al. 2010), showing that polymeric delivery systems composed of PL and doxorubicin also showed higher cytotoxicity than the isolated drug by improvement of drug uptake and/or enhancement effect in cells resistant (Hu et al. 2012; Han et al. 2011).

The potentiation of antitumor effect of classical drugs by PL has been directly attributed to the inhibition of Pgp, increasing the concentration of the drug inside the tumor cells. However, other unknown/unexploited mechanisms may also take place. Further studies suggest that inhibition of this protein can restore normal apoptosis cascade, commonly inhibited in resistant cells (Minko et al. 2005; Kabanov et al. 2002a, b). It was shown that the ATP depletion promoted by PL may result from their interaction with mitochondria, which would contribute to the induction of cell death (Montesinos et al. 2012; Batrakova et al. 2010).

PL cross plasma membranes and may reach the mitochondrial membranes. It was reported as the inhibition of mitochondrial respiration by poloxamers by impairing the electron transport in complexes I and IV, promoting the dissipation of the mitochondrial transmembrane potential and ATP depletion. Also it may increase in free radicals generation and mitochondrial permeabilization with cytochrome *c* release, both phenomena involved in the cell death signaling (Batrakova and Kabanov 2008).

The effects described above were observed for Alakhova et al. (2010) using low concentrations of Pluronic® P85 in mitochondria isolated from MCF/ADR and such effects on mitochondria were more pronounced in cells with MDR phenotype, suggesting a high selectivity of PL to cells overexpressing Pgp. Furthermore, there is a positive correlation among the decrease of ATP intracellular levels and cytotoxicity in MDR cells. PL such as Pluronic® L61 and Pluronic® P85 are more potent than Pluronic® F127, a PL highly efficient delivery of drug but with less ability to promote ATP depletion (Kabanov et al. 2003).

13.2.2 *In Vivo Toxicology*

Regulatory agencies such as Food and Drug Administration (FDA) ensure that new drugs or technologies have acceptable results in terms of safety for the risk-benefit ratio. These agencies are responsible for assessing quality, efficacy, and safety of new products using a rigorous post-market surveillance (a process known as pharmacovigilance) to certify that the drug or technology is safe when used and also that possible problems will be quickly identified and treated (Gaspar and Ducan 2009).

Recently, PL have shown to be efficient drug-delivery systems (Singh-Joy and McLain 2008; Kabanov et al. 2002a, b), since drugs are incorporated into the core of the micelles, directed by their physicochemical properties, resulting in increased solubility, metabolic stability, and circulation time for the drug (Kabanov et al. 2002a, b). For this reason, reports in the literature have showed in vitro and in vivo studies evaluating the safety of PL.

PL block copolymers have been introduced in the late 1950. These copolymers are approved by FDA and have been proposed for diverse pharmaceutical applications, as well as are used in cosmetic products (Singh-Joy and McLain 2008), listed in the United States and European Pharmacopoeia as the trademarks Pluronic®, Synperonic®, or Tetronic® (Dumortier et al. 2006). PL are also widely used for manufacturing food and pharmaceutical products, as surfactants, emulsifying agents, cleansing agents, solubilizing agents and currently are used in 141 cosmetic products at concentrations from 0.005 % to 20 % (Singh-Joy and McLain 2008). For pharmaceutical products, commercially available formulations present in their composition Pluronic® F68 and/or Pluronic® F127 alone or in combination, for parenteral prilocaine/lidocaine and meloxicam, tegaserode, and gabapentin in tablets and also acyclovir as a cream form for topical use.

The main reason for the wide use of PL, especially PL 407, is their biocompatibility and approval by the Food and Drug Administration (FDA) for clinical use (Kabanov et al. 2002a, b; Singh-Joy and McLain 2008), despite some reports have shown that PL 407 induces alterations in lipid metabolism (significantly increase on serum triglyceride and cholesterol). However, it is necessary to emphasize that these complications occurred after use of relatively high doses (0.5 to 1 g/kg, intraperitoneal route) (Li et al. 1996; Blonder et al. 1999; Johnston et al. 2000). Besides, the use of PL 407 can alter the renal filtration capacity (reduced glomerular filtration) after intramuscular injection, using an inulin/PL 407 gel as formulation model (Johnston and Miller 1985, 1989; Abe et al. 1990; Li et al. 1996). In contradiction, another study pointed that PL 407 is relatively safe, since there were no observed differences on renal elimination parameters after intramuscular or subcutaneous administration (Pec et al. 1992).

13.3 Comments and Future Perspectives

Although synthesized in the 50's, poloxamers (PL) copolymers have been considered of great interest as carrier system for drugs and gene therapy (since the diffusion of the molecule incorporated in the spaces inside and intermicelares is relatively slow), cellular and tissue engineering (as substrates for hematopoietic cells, osteoblasts, and restoration of burn tissues), as well as vehicles for a variety of pharmaceutical formulations looking forward to their clinical applications for intravenous, subcutaneous, intramuscular, oral, ophthalmic, and topical routes.

One of the reasons for the various clinical applications of PL is due to their thermoreversible properties, high biocompatibility, and some pharmacological/toxicological properties such as immunomodulation, induction of cellular growth, healing after surgical procedures, inhibition of the multidrug resistance P-glycoprotein, the modulation of membrane fluidity, and also the impairment of ATP synthesis.

The sensibility to local stimuli is an essential factor in the formation of hydrogels composed of PL and their applications as delivery systems. Due to their thermoreversible capacity and water retention, these copolymers are fluids prior to administration (and low temperatures), but present sol-gel transition under physiological conditions without interference of chemical agents (as organic solvents), reducing their toxic potential when used for parenteral routes. Even some reports have described alterations in hepatic lipid metabolism and reduction on renal filtration capacity, in general. PL are relatively safe and widely used for manufacturing food and pharmaceutical products, as surfactants, emulsifying agents, cleansing agents, solubilizing agents, and currently cosmetic products.

Finally, the use of PL with different physicochemical characteristics (such as different hydrophylic lipophylic balance) could be an interesting alternative to investigate the properties of binary systems for sustained release of drugs, since changes on composition and proportion of these copolymers can modulate the drug solubility, the micellar structure, the rheological properties, and the release rate, making them effective bases for emulsions, nanoparticles, and also lipid vesicles.

References

Abe T, Sasaki M, Nakajima H et al (1990) Evaluation of Pluronic F127 as a base for gradual release of anticancer drug. Gan To Kagaku Ryoho 17:1546–1550

Alakhova DY, Rapoport NY, Batrakova EV et al (2010) Differential metabolic responses to pluronic in MDR and non-MDR cells: a novel pathway for chemosensitization of drug resistant cancers. J Control Release 142:89–100

Alexandridis P (1994) Surface activity of poly(ethylene oxide)-block-poly(propylene oxide)-block-poly(ethylene oxide) copolymers. Langmuir 10:2604–2612

Alexandridis P, Hatton AT (1995) Poly(ethylene oxide)-poly(propylene oxide)-poly(ethylene oxide) block copolymer surfactants in aqueous solutions and at interfaces: thermodynamics, structure, dynamics, and modeling. Colloid Surf A: Physicochem Eng Aspects 96:1–46

Alvarez-Lorenzo C, Sosnik A, Concheiro A (2011) PEO-PPO block copolymers for passive micellar targeting and overcoming multidrug resistance in cancer therapy. Curr Drug Targets 12:1112–1130

Arora A, Seth K, Kalra N et al (2005) Modulation of P-glycoprotein-mediated multidrug resistance in K562 leukemic cells by indole-3-carbinol. Toxicol Appl Pharmacol 202:237–243

Artzner F, Geiger S, Olivier A et al (2007) Interactions between poloxamers in aqueous solutions: micellization and gelation studied by differential scanning calorimetry, small angle X-ray scattering, and rheology. Langmuir 23:5085–5092

Balayssac D, Authier N, Cayre A et al (2005) Does inhibition of P-glycoprotein lead to drug-drug interactions? Toxicol Lett 156:319–329
Batrakova EV, Kabanov AV (2008) Pluronic block copolymers: evolution of drug delivery concept from inert nanocarriers to biological response modifiers. J Control Release 130:98–106
Batrakova EV, Li S, Brynskikh AM et al (2010) Effects of pluronic and doxorubicin on drug uptake, cellular metabolism, apoptosis and tumor inhibition in animal models of MDR cancers. J Control Release 143:290–301
Blonder JM, Baird L, Fulfs JC, Rosenthal GJ (1999) Dose-dependent hyperlipidemia in rabbits following administration of poloxamer 407 gel. Life Sci 65:261–266
Campese GM, Rodrigues EMG, Tambourgi EB et al (2003) Determination of cloud-point temperatures for different copolymers. Braz J Chem Eng 20:335–337
Chaibundit C, Ricardo NMPS, Costa FLL et al (2007) Effect of ethanol on the micellization and gelation of pluronic P123. Langmuir 23:9229–9236
Cheng C, Wang J, Kausik R et al (2012) Nature of interactions between PEO-PPO-PEO triblock copolymers and lipid membranes: (II) role of hydration dynamics revealed by dynamic nuclear polarization. Biomacromolecules 13:2624–2633
Chieng YY, Chen SB (2009) Interaction and complexation of phospholipid vesicles and triblock copolymers. J Phys Chem 113:14934–14942
Croy SR, Kwon GS (2006) Polymeric micelles for drug delivery. Curr Pharm Des 12:4669–4684
Demina T, Grozdova I, Krylova O et al (2005) Relationship between the structure of amphiphilic copolymers and their ability to disturb lipid bilayers. Biochemistry 44:4042–4054
Dumortier G, Grossiord JL, Agnely F et al (2006) A review of poloxamer 407 pharmaceutical and pharmacological characteristics. Pharm Res 23:2709–2728
Escobar-Chávez JJ, López-Cervantes M, Naïk A et al (2006) Applications of thermo-reversible pluronic F-127 gels in pharmaceutical formulations. J Pharm Pharm Sci 9:339–358
Fu D, Arias IM (2012) Intracellular trafficking of P-glycoprotein. Int J Biochem Cell Biol 44:461–464
Fusco S, Borzacchiello A, Netti PA (2006) Perspectives on: PEO-PPO-PEO triblock copolymers and their biomedical applications. J Bioact Compat Polym 21:149–164
Gaspar R, Ducan R (2009) Polymeric carriers: preclinical safety and the regulatory implications for design and development of polymer therapeutics. Adv Drug Deliv Rev 61:1220–1231
Gaumet M, Vargas A, Gurny R et al (2008) Nanoparticles for drug delivery: the need for precision in reporting particle size parameters. Eur J Pharm Biopharm 69:1–9
Gong J, Chen M, Zheng Y et al (2012) Polymeric micelles drug delivery system in oncology. J Control Release 159:312–323
Greenebaum B, Blossfield K, Hannig J et al (2004) Poloxamer 188 prevents acute necrosis of adult skeletal muscle cells following high-dose irradiation. Burns 30:539–547
Guzmán M, García FF, Molpeceres J et al (1992) Polyoxyethylene-polyoxypropylene block copolymer gels as sustained release vehicles for subcutaneous drug administration. Int J Pharm 80:119–127
Han M, Diao Y, Jiang H et al (2011) Molecular mechanism study of chemosensitization of doxorubicin-resistant human myelogenous leukemia cells induced by a composite polymer micelle. Int J Pharm 420:404–411
Harrison WJ, Aboulgassem GJ, Elathrem FAI et al (2005) Regression of poloxamer 407-induced atherosclerotic lesions in C57BL/6 mice using atorvastatin. Langmuir 21:6170–6178
Higuchi A, Aoki N, Yamamoto T et al (2006) Bioinert surface of pluronic-immobilized flask for preservation of hematopoietic stem cells. Biomacromolecules 7:1083–1089
Hu H, Yu J, Li Y et al (2012) Engineering of a novel pluronic F127/graphene nanohybrid for pH responsive drug delivery. J Biomed Mater Res A 100:141–188
Hvidt S, Trandum C, Batsberg W (2002) Effects of poloxamer polydispersity on micellization in water. J Colloid Interface Sci 250:243–250

Ieiri I (2012) Functional significance of genetic polymorphisms in P-glycoprotein (MDR1, ABCB1) and breast cancer resistance protein (BCRP, ABCG2) (2012). Drug Metab Pharmacokinet 27:85–105

Ivanova R, Lindman B, Alexandridis P (2002) Effect of pharmaceutically acceptable glycols on the stability of the liquid crystalline gels formed by poloxamer 407 in water. J Colloid Interface Sci 252:226–235

Jiang J, Li C, Lombardi C et al (2008) The effect of physiologically relevant additives on the rheological properties of concentrated Pluronic copolymer gels. Polymer 49:3561–3567

Johnston TP, Miller SC (1985) Toxicological evaluation of poloxamer vehicles for intra-muscular use. J Parenter Sci Technol 39:83–88

Johnston TP, Miller SC (1989) Inulin disposition following intramuscular administration of an inulin/poloxamer gel matrix. J Parenter Sci Technol 43:279–286

Johnston TP, Baker JC, Hall D et al (2000) Regression of poloxamer 407-induced atherosclerosis lesions in C57BL/6 mice using atorvastatin. Atherosclerosis 149:303–314

Kabanov AV, Batrakova EV, Alakhov VY (2002a) Pluronic block copolymers for overcoming drug resistance in cancer. Adv Drug Deliv Rev 54:759–779

Kabanov A, Batrakova E, Alakhov VY (2002b) Pluronic® copolymers as novel polymer therapeutics for drug and gene delivery. J Control Release 82:189–212

Kabanov AV, Batrakova EV, Alakhov VY (2003) An essential relationship between ATP depletion and chemosensitizing activity of Pluronic block copolymers. J Control Release 91:75–83

Kadam Y, Yerramilli U, Bahadur A et al (2010) Micelles from PEO-PPO-PEO block copolymers as nanocontainers for solubilization of a poorly water soluble drug hydrochlorothiazide. Colloids Surf B Biointerfaces 83:49–57

Kidowaki M, Zhao C, Kataoka T et al (2006) Thermoreversible sol–gel transition of an aqueous solution of polyrotaxane composed of highly methylated α-cyclodextrin and polyethylene glycol. Chem Commun (Camb) 39:4102–4103

Klouda L, Mikos AG (2008) Pluronic block copolymers as novel polymer therapeutics for drug and gene delivery. Eur J Pharm Biopharm 68:34–45

Kozlov MY, Melik-Nubarov NS, Batrakova E et al (2000) Relationship between pluronic block copolymer structure, critical micellization concentration and partitioning coefficients of low molecular mass solutes. Macromolecules 33:3305–3313

Li C, Palmer WK, Johnston TP (1996) Disposition of poloxamer 407 in rats following a single intraperitoneal injection assessed using a simplified colorimetric assay. J Pharm Biomed Anal 14:659–666

Li X, Li P, Zhang Y et al (2010) Novel mixed polymeric micelles for enhancing delivery of anticancer drug and overcoming multidrug resistance in tumor cell lines simultaneously. Pharm Res 27:1498–1511

Liu T, Chu B (2000) Formation of homogeneous gel-like phases by mixed triblock copolymer micelles in aqueous solution: FCC to BCC phase transition. J Appl Cryst 33:727–730

Luo C, Chen X, Li Q et al (2013) Poloxamer 188 attenuates in vitro traumatic brain injury-induced mitochondrial and lysosomal membrane permeabilization damage in cultured primary neurons. J Neurotrauma 30(7):597–607. doi:10.1089/neu.2012.2425

Mata JP, Majhi PR, Guo C et al (2005) Concentration, temperature, and salt-induced micellization of a triblock copolymer Pluronic L64 in aqueous media. J Colloid Interface Sci 292:548–556

McDevit CA, Callaghan R (2007) How can we best use structural information on P-glycoprotein to design inhibitors? Pharmacol Ther 113:429–441

Minko T, Batrakova EV, Li S et al (2005) Pluronic block copolymers alter apoptotic signal transduction of doxorubicin in drug-resistant cancer cells. J Control Release 105:269–278

Mitchard NM, Beezer AE, Mitchell JC et al (1992) Thermodynamic analysis of scanning calorimetric transitions observed for dilute aqueous solutions of ABA block copolymers. J Phys Chem 96:9507–9512

Montesinos RN, Béduneau A, Pellequer Y et al (2012) Delivery of P-glycoprotein substrates using chemosensitizers and nanotechnology for selective and efficient therapeutic outcomes. J Control Release 161:50–61
Mora-Huertas C, Fessi H, Elaissari A (2010) Polymer-based nanocapsules for drug delivery. Int J Pharm 385:113–142
Mortensen K, Batsberg W, Hvidt S (2008) Effects of PEO-PPO diblock impurities on the cubic structure of aqueous PEO-PPO-PEO pluronics micelles: fcc and bcc ordered structures in F127. Macromolecules 41:1720–1727
Nagarajan R (1999) Solubilization of hydrocarbons and resulting aggregate shape transitions in aqueous solutions of Pluronic® (PEO–PPO–PEO) block copolymers. Colloids Surf B Biointerfaces 16:55–72
Nambam J, Philip J (2012) Effects of interaction of ionic and nonionic surfactants on self-assembly of PEO–PPO–PEO triblock copolymer in aqueous solution. J Phys Chem B 116:1499–1507
Newby GE, Hamley IW, King SM et al (2009) Structure, rheology and shear alignment of Pluronic block copolymer mixtures. J Colloid Interface Sci 329:54–61
Oh KT, Bronich TK, Kabanov AV (2004) Micellar formulations for drug delivery based on mixtures of hydrophobic and hydrophilic Pluronic® block copolymers. J Control Release 94:411–422
Overstreet DJ, Huynh R, Jarbo K et al (2013) In situ forming, resorbable graft copolymer hydrogels providing controlled drug release. J Biomed Mater Res A 101(5):1437–1446. doi:10.1002/jbm.a.34443
Pandit NK, Wang D (1998) Salt effects on the diffusion and release rate of propranolol from poloxamer 407 gels. Int J Pharm 167:183–189
Parmar A, Singh K, Bahadur A et al (2011) Interaction and solubilization of some phenolic antioxidants in Pluronic® micelles. Colloids Surf B Biointerfaces 86:319–326
Pec EA, Wout ZG, Johnston TP (1992) Biological activity of urease formulated in poloxamer 407 after intraperitoneal injection in the rat. J Pharm Sci 81:626–630
Quadir A (2005) Characterization of newly developed micronized poloxamers for poorly soluble drugs. Basf Pharma Solutions, Miami
Schmolka IR (1972) Artificial skin I. Preparation and properties of pluronic F-127 gels for treatment of burns. J Biomed Mater Res 6:571–582
Sharma PK, Reilly MJ, Bhatia SK et al (2008) Effect of pharmaceuticals on thermoreversible gelation of PEO-PPO-PEO copolymers. Colloids Surf B Biointerfaces 63:229–235
Shen J, Yin Q, Chen L et al (2012) Co-delivery of paclitaxel and survivin shRNA by pluronic P85-PEI/TPGS complex nanoparticles to overcome drug resistance in lung cancer. Biomaterials 33:8613–8624
Singh-Joy SD, McLain VC (2008) Safety assessment of poloxamers 101, 105, 108, 122, 123, 124, 181, 182, 183, 184, 185, 188, 212, 215, 217, 231, 234, 235, 237, 238, 282, 284, 288, 331, 333, 334, 335, 338, 401, 402, 403, and 407, poloxamer 105 benzoate, and poloxamer 182 dibenzoate as used in cosmetics. Int J Toxicol 27:93–128
Su Y, Wang J, Liu H (2002) Melt, hydration, and micellization of the PEO–PPO–PEO block copolymer studied by FTIR spectroscopy. J Colloid Interface Sci 251:417–423
Vadnere M, Amidon G, Lindebaum S et al (1984) Thermodynamic studies on the gel-sol transition of some pluronic polyols. Int J Pharm 22:207–218
Wang J, Chin J, Marks JD et al (2010) Effects of PEO-PPO-PEO triblock copolymers on phospholipid membrane integrity under osmotic stress. Langmuir 26:12953–12961
Weinand C, Pomerantseva I, Neville C et al (2006) Hydrogel-beta-TCP scaffolds and stem cells for tissue engineering bone. Bone 38:555–563
Wu C, Liu T, Chu B (1997) Characterization of the PEO-PPO-PEO triblock copolymer and its application as a separation medium in capillary electrophoresis. Macromolecules 30:4574–4583

Yapar EA, Inal O (2012) Poly(ethylene oxide)–poly(propylene oxide)-based copolymers for transdermal drug delivery: an overview. Trop J Pharmaceut Res 11:855–866

Yuhua S, Ligen L, Jiake C et al (2012) Effect of Poloxamer 188 on deepening of deep second-degree burn wounds in the early stage. Burns 38:95–101

Zhang Y, Lam YM (2006) Study of mixed micelles and interaction parameters for polymeric nonionic and normal surfactants. J Nanosci Nanotechnol 6:3877–3881

Zhang Y, Lam Y, Tan W (2005) Poly(ethylene oxide)–poly(propylene oxide)–poly(ethylene oxide)-g-poly (vinylpyrrolidone): association behavior in aqueous solution and interaction with anionic surfactants. J Colloid Interface Sci 285:74–79

Zhao L, Du J, Duan Y et al (2012) Curcumin loaded mixed micelles composed of Pluronic P123 and F68: preparation, optimization and in vitro characterization. Colloids Surf B Biointerfaces 97:101–108

Zhirnov AE, Demina TV, Krylova OO et al (2005) Lipid composition determines interaction of liposome membranes with Pluronic L61. Biochim Biophys Acta 1720:73–83

Chapter 14
Polymeric Nanoparticles: In Vivo Toxicological Evaluation, Cardiotoxicity, and Hepatotoxicity

Solange C. Garcia, Silvia S. Guterres, Guilherme B. Bubols, Rachel P. Bulcão, Mariele F. Charão, and Adriana R. Pohlmann

Abstract The applications of nanoparticles (NPs) in therapeutics have motivated the increasing development of studies in the field of nanotoxicology. This chapter focused to provide a critical analysis of the available literature about the toxicity and safety of biodegradable polymeric nanoparticles, particularly in terms of cardiotoxicity and hepatotoxicity. The most commonly applied methods in NP toxicity studies are also discussed, and their limitations concerning the specific properties of NPs, once a key point to obtain accurate and reliable in vitro and in vivo toxicological evaluations, are to guarantee an appropriate physicochemical characterization of the nanoparticles. Large surface area, high absorption capacity, the aggregation state, and surface coating of nanoparticles are also intrinsic properties of NPs that can interfere with the results. For further studies, a challenge to be overcome is the standardization of experiments, especially regarding the consensus in the way to express the administered dose of nanoparticles. Finally, a diagram for a nanotoxicological evaluation using in vitro and in vivo models is proposed.

14.1 Introduction

Nanotechnology in Health Sciences is a promising tool to transfer scientific knowledge from bench to clinic. Up to now, different nanoparticles have been studied as new medicines and diagnosis devices, but few among them reached the market either as an intermediate or a final product. Besides the effectiveness, a new medicine must be safe to be approved before industrialization. Polymeric nanoparticles are highlighted in the field of drug delivery because of their

S.C. Garcia (✉)
Laboratório de Toxicologia (LATOX), Departamento de Análises, Faculdade de Farmácia, Universidade Federal do Rio Grande do Sul, Avenida Ipiranga 2752, Porto Alegre, Rio Grande do Sul 90610-000, Brazil
e-mail: Solange.garcia@ufrgs.br

N. Durán et al. (eds.), *Nanotoxicology*, Nanomedicine and Nanotoxicology, 299
DOI 10.1007/978-1-4614-8993-1_14, © Springer Science+Business Media New York 2014

biocompatibility and, in some cases, their biodegradability. The term "polymeric nanoparticles" include a diversity of macromolecular and supramolecular structures. The most investigated are the polymeric nanospheres, composed of a matrix of polymer stabilized by surfactants or emulsifiers, which applications are oral delivery, intramuscular and subcutaneous injection, and targeted deliver (Brannon-Peppas 1995). Secondly, the polymeric nanocapsules are useful drug carriers based on a vesicular supramolecular structure. The macromolecular component is restrained at the oil–water interface acting as a barrier to coalescence and to drug diffusion (Mora-Huertas et al. 2010). A variant of nanocapsule, the lipid-core nanocapsule, has been recently proposed as a new carrier able to more effectively control the drug release, as well as the cell penetration and the drug targeting (Pohlmann et al. 2013). Another type of relevant nanoparticle are dendrimers. These macromolecules are highly branched, reactive, and three-dimensional with all bonds coming up from a central core (Liu and Fréchet 1999). A variety of applications of dendrimers have been proposed in the last years in pharmaceutical and biomedical fields. Particularly, nanoparticles in the form of dendrimers have been suggested as valuable tools in therapeutics, because their inner cavity can be used for the drug entrapment, allowing a controlled release of the active compound.

According to a search done in the ISI database (Web of Science), the number of articles relating to the use of dendrimers and polymeric nanoparticles in the pharmaceutical area is growing exponentially (Fig. 14.1). In parallel, in recent years, there is also interest in evaluating aspects of toxicity of these structures.

Toxicological effects are closely related to the type of nanoparticle (NP) under investigation. Therefore, for a multiple comparison of nanotoxicity studies, NP characterization is mandatory. Studies commonly report that NPs may trigger oxidative damages or inflammation, among other effects, but these studies rarely investigate biodegradable polymeric NPs, and when they do, nanoparticles are loaded with drugs rather than drug-unloaded NPs (the nanocarrier itself). In this sense, knowledge from systemic toxicological studies and studies about specific tissues or organs, such as liver or cardiovascular system, are still necessary. There is a certain scientific tendency to suppose that polymeric nanoparticles are more biocompatible than carbon nanotubes or metallic NPs. However, it should be kept in mind that only after careful nanotoxicological studies are conducted, these assumptions could be properly confirmed to guarantee the safety of polymeric NPs towards human health.

The promising therapeutical applications of NPs in drug delivery have motivated increasing studies in nanotoxicology, which has become an important field to assure the safety of these potential new formulations. A critical analysis of the available literature, though, indicates that much knowledge is still necessary about the toxicity of polymeric nanoparticles. In particular, there is a lack of data about the impact of the nanostructure itself on the safety of biodegradable nanocarriers. Considering that although it is expected that biodegradable NPs are biocompatible, careful acute, subchronic, and chronic toxicity studies targeting hepatotoxicity and

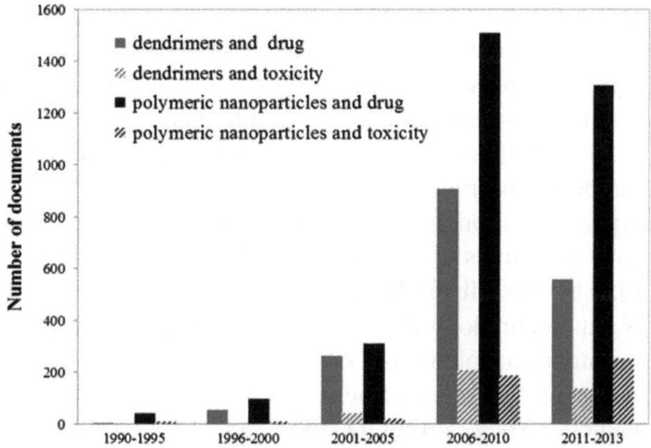

Fig. 14.1 Histogram showing the number of documents relating to dendrimers and polymeric nanoparticles in therapeutics and toxicity

cardiotoxicity are still necessary to guarantee that the drug carriers do not exhibit toxicological problems before they are adopted in therapeutics as drug nanocarriers.

14.2 In Vivo Models: Advantages and Disadvantages

Although numerous toxicological studies have been performed with nanoparticles (NPs), to date these studies have not resulted in the creation of a set of rules applicable to many of the new biodegradable NPs used for biomedical applications, like polymeric nanoparticles and dendrimers (Laurent et al. 2012).

The understanding of the relationship between the physical and chemical properties of the nanostructure and their in vivo behavior is complex. The interactions of nanostructures with biological components (proteins, lipids, and cells), toxicokinetic behavior (biodistribution, metabolism, and clearance) and toxicodynamics (e.g., effects on the immune response) play an important role in the studies on the safety of nanomaterials (NM). The toxicity of NPs may be directly related to the surface chemistry, size, shape, composition, and aggregation; hence the general properties of NPs are very important in such studies. In this context, in vivo studies need to be performed and, for that, the above-mentioned aspects should be carefully evaluated.

In vivo studies enable the identification of the organs that could potentially be injured from the animal exposure to nanostructures and provide a molecular basis of the tissue stress (Fischer and Chan 2007). Once the target tissue affected by NP administration is found, the investigation of the probable toxicity mechanism should be performed, in particular by means of in vitro studies.

Despite the tremendous potential benefit of polymeric nanoparticles in the field of biomedical applications, knowledge about their tissue bioavailability and toxicity in animals is still limited. Different animal models such as mice, rats, rabbits, and swine must be used to study the fate and toxicity of polymeric NPs in vivo. In addition, there are numerous advantages of animal models; they are living organisms presenting the complexity of hundreds of tissues and all physiological reactions and interactions are maintained.

In relation to in vivo experimental studies, over the last 20 years a rapid progress has been observed in the development and use of alternative methods to animal experimentation in toxicology (Kandarova and Letasiova 2011). Moreover, the three principles known as the "3Rs" (reduce, refine, and replace) were defined in 1959 by Russell and Burch (1959). In parallel, in the past 3 decades a rapid progress of nanotechnology and its challenges emerged. In fact, the alternative models have an important role to assess toxicological effects, some of which have already been validated (ESAC statements on alternative methods) and accepted by regulatory agencies, such as OECD, FDA, and EPA. Taking into account the relevance of studying hazard, exposure, and risk assessment of NMs, the regulatory agencies are working to provide a high quality, science-based, and internationally harmonized program to evaluate safety and toxicity. For instance, OECD is addressing NMs which require risk assessments for human health and the environment. In this case, those nanomaterials are mostly based on inorganic components, such as fullerenes, carbon nanotubes, silver nanoparticles, iron nanoparticles, nanoclays, and gold, among others. Dendrimers are the only organic-based NM listed.

Hartung (2011) reported that several American agencies announced a coalition to facilitate the implantation of alternative methods. They proposed a shift from primarily in vivo animal studies to in vitro assays, in vivo assays with lower organisms, and computational modeling for toxicity assessments. The development of novel approaches is needed, since traditional approaches might have even greater limitations for NPs than for other industrial chemicals and may not offer the throughput and velocity to cope with the dynamic developments in nanotechnologies (Hartung 2011).

Besides, in vivo studies when compared to in vitro cytotoxicity assays offer further insights into the integrated biological effect of NPs, such as the identification of the accumulation site and toxicological profiles within a specific organ (Li and Chen 2011). Also, the lack of correlation between in vivo and in vitro effects has recently been demonstrated (Sayes et al. 2007), indicating that in vitro experiments must have in vivo validation in order to be useful. Intercellular effects such as macrophage and dendritic cell recruitment require that the cells and their proper signaling occur in its natural state and preserved cytoarchitecture (Sayes et al. 2007). Some differences also warrant the development of novel methods to define equivalent doses between in vitro and in vivo exposures in order to improve correlations between the two testing systems. One difference is the high concentrations/doses used in most traditional in vitro studies. An extremely high dose rate (dose administered per unit of time) is another issue of in vitro studies (Han et al. 2012).

Additionally, the characteristics of nanostructures appear to influence a possible methodological interference when in vitro tests are adopted. In order to circumvent these issues, some steps should be carefully followed. The first step is a complete characterization of the NM. Secondly, literature should be searched for reports on interferences such as the pH influence and presence of metals. Third, studies should use preferably more than one method to assess the same objective, for example, two different methods should be chosen for cell viability among those available, e.g. MTT, lactate dehydrogenase (LDH), and neutral red. As a matter of fact, Kroll et al. (2011) have been demonstrating clear in vitro interferences depending on the NM analyzed, which may result in false interpretations for cell viability, oxidative stress and inflammation, thus reflecting in inaccurate data about the NP toxicity (Kroll et al. 2009, 2011). Different authors have also reported that the high adsorption capacities, different optical properties, and increased catalytic activities of NPs can influence the results of many in vitro toxicity assays, leading to the misinterpretation of results (Han et al. 2012; Kroll et al. 2009, 2011; Worle-Knirsch et al. 2006; Casey et al. 2007; Belyanskaya et al. 2007). Moreover, through their large specific surface area, NPs could also adsorb essential nutrients in cell culture medium, making it difficult to interpret some cytotoxicity results (Guo et al. 2008). Importantly, the absence of standardized methodologies and guidelines also makes it difficult to compare the toxicity assessments from different research groups. This delays the advances in nanotoxicology and results in much apprehension regarding the possible adverse health and environmental implications of NMs (Dhawan and Sharma 2010).

It is also important to consider that methodological interferences can be a problem of in vitro and also in vivo experiments. Laboratory tests using analytical techniques such as optical diffraction, electrical impedance, scatter laser, and others, performed after the exposure of the NP in biological systems (either in vitro or in vivo), may produce false results. Bulcão et al. (2012) have detected that the hematological parameters obtained from rats treated with lipid-core nanocapsules (LNC), a kind of vesicular biodegradable poly(ε-caprolactone) nanoparticle, were misinterpreted due to an interference on blood counting analysis using cell counter equipment, probably for interference of electrical impedance (Bulcão et al. 2012).

Therefore, many advantages and disadvantages are involved in toxicological evaluation of NMs in both in vivo and in vitro studies (Table 14.1). However, in vivo research still poses a beneficial model for gaining immediate and direct knowledge of human exposure to such xenobiotics (Clift et al. 2011).

Another important point is the surface of NMs which is routinely modified with the intention of controlling NM–protein interactions. For example, the modification of the nanoparticle surface with polyoxygenated polymers or surfactants is a commonly used strategy within the area of drug delivery which increases the circulation time of drugs by preventing/delaying recognition by the reticuloendothelial system through the introduction of "stealth" properties and reducing protein adsorption. This suggests that the inclusion of physiologically relevant dispersants influences the toxicity of NMs, which may be mediated by the interactions that occur between proteins contained in the different biological media and the NMs.

According to Johnston et al. (2012), the formation of NM–protein complexes has a variety of implications on the biological and toxicological response following

Table 14.1 Comparison of the major characteristics between in vivo and in vitro models to evaluate the toxicity of NPs in biological systems

In vivo models	In vitro models
Time-consuming	Faster
Expensive	Less expensive
Ethical issues	Devoid of any ethical issues
Laborious	More convenient
Provide data on mechanism	Provide detail on mechanism
Toxicokinetic study is possible	Toxicokinetic study is not possible
Cytoarchitecture preserved	No preservation of cytoarchitecture
Evaluation of hormonal effects	No evaluation of hormonal effects
Study of chronic effects	No study of chronic effects
Study of transport mechanisms (blood, lymph, bile)	No study of transport mechanisms (blood, lymph, bile)

exposure. The attached proteins may block the NM surface, and thereby reduce NM reactivity and toxicity; may result in the bolus delivery of active protein to the cell surface; also, these proteins may undergo changes in their structural conformation and be denatured preventing active protein activity. The toxicity of NMs may be underestimated due to their interference with the assays used to evaluate their toxicity, as cited previously, and the adsorption of proteins onto the NM surface can improve the dispersion of NM suspensions, preventing against NM agglomeration and aggregation (Johnston et al. 2012).

Regarding polymeric nanocapsules, few studies concerning their toxic effects have been conducted so far. Bulcão et al. (2012) performed a systematic toxicological screening on rats after LNC intraperitoneous (i.p.) administration, in which the acute toxicity was studied after a single dose and an observation period of 14 days, and the subchronic toxicity after daily i.p. administration of repeated doses over 30 days. Wistar rats were monitored for routine behavioral abnormalities and changes in body weight. Upon sacrifice, tissues were observed for macroscopic abnormalities. In this work, the accumulation of nanocapsules in the peritoneal cavity (aggregation) was demonstrated. Thus, i.p. administration could be considered inadequate to evaluate acute and subchronic toxicities at high doses. Tissue sections of liver spleen, heart, kidney, and brain were fixed and stained with hematoxylin and eosin to observe microscopic abnormalities. One of the challenges was choosing the dose, considering absence of previous in vivo studies for i.p. administration of LNC for reference, so the maximum volume permitted for these routes has been used. As a result, after exposure to the higher dose volume, rats which received LNC by i.p. route presented a granulomatous foreign body reaction (Bulcão et al. 2012).

Similar studies have been reported in the literature. Huang et al. (2010) performed a study on acute and genotoxic effects of copolymers and nanomaterials based on poly(ε-caprolactone) (PCL) intravenously administered to rats which did not show any adverse effects (Huang et al. 2010). In a previous work, Fang et al. (2009) mainly focused on the acute organ toxicity of Balb/c mice by subcutaneous (s.c.) injection of triblock copolymer based on PCL (Fang et al. 2009). The formulation containing the triblock copolymer was thought to be nontoxic. Furthermore, another study based on in vitro and in vivo toxicological investigations conducted in mice indicated that nanospheres of PCL-*block*-methoxypoly(ethylene glycol) (MePEG-PCL) did not show signs of toxicity by i.p. administration (Kim et al. 2003). The authors suggested that the diblock copolymer nanospheres were a biocompatible drug delivery system. Despite the above-mentioned studies, no extensive research has been conducted to analyze the toxicity of nanocarriers per se.

There are a few studies that used oral administration of NPs, which is the route preferably selected for studying toxicity. The main routes of administration (injection methods) are parenteral, such as intraperitoneal (Bernardi et al. 2008, 2009; Frozza et al. 2010; Kim et al. 2003), intravenous (i.v.) (acute treatment performed by Huang et al. 2010), and subcutaneous (Fang et al. 2009).

Dendrimers are also important polymeric nanoparticles. Preliminary toxicity and immunological evaluation were also performed for amine-terminated poly (amidoamine) (PAMAM) dendrimers in male swiss-webster mice by i.v. administration (Roberts et al. 1996). Animals were monitored for routine behavioral abnormalities and changes in body weight; after sacrifice, tissues were observed for macroscopic abnormalities. Also, serum samples for the immunogenicity testing were obtained from sets of two New Zealand white rabbits to serve as nonimmune controls. The animals were then injected s.c. and two subsequent booster shots, at the same dose as the first, were given at 3-week intervals. Blood samples were collected 10 days after. As described by Sadekar and Ghandehari (2012) this study was one of the first reports of the in vivo evaluation of dendrimers. It was a preliminary in vivo evaluation of toxicity of cationic PAMAM dendrimers at a fixed dose and showed that PAMAM toxicity increased with the increase in the generation and surface charge density of amine groups (Sadekar and Ghandehari 2012).

Recent in vivo studies have focused on establishing the maximum tolerated doses for different generations and surface charges of PAMAM dendrimers administered orally and intravenously to CD-1 mice (Greish et al. 2012). The findings from Greish et al. (2012) were in agreement with the previous study from Roberts et al. (1996). Overall, these studies revealed that PAMAM dendrimers showed similar toxicity trends when administered orally and i.v., with the higher generation cationic dendrimers being more toxic than their lower generation counterparts and the anionic dendrimers being less toxic than the cationic ones. Blood analysis showed decreased levels of fibrinogen, platelets, and high levels of fibrin degradation products (FDP) in groups treated with amine-terminated dendrimers, suggesting the occurrence of both pro- and anti-coagulation processes similar to disseminated intravascular coagulation (DIC). Intravascular coagulation and hemorrhage were also observed in these animals (Greish et al. 2012; Roberts et al. 1996).

Likewise, as previously mentioned, correlation of in vitro and in vivo toxicity and bioavailability data also remains a challenge. There are no established correlations to extrapolate nontoxic concentrations obtained from in vitro data to be used as doses for in vivo experiments. Hence, especially the in vivo oral toxicity profiles need to be established for the different generations and surface modifications of PAMAM dendrimers (Sadekar and Ghandehari 2012).

Mukherjee et al. (2010) demonstrated that an interaction of the dendrimers with the components of the cell culture medium could elicit a secondary or indirect toxic response (Mukherjee et al. 2010). The interaction appeared to be a selective and dynamic physical adsorption of medium constituents (Lynch and Dawson 2008), and as such, the response may be dependent on the medium used. According to the authors, the UV/visible absorption spectroscopic analysis indeed indicates that the additional hydrocortisone may influence the interaction of the dendrimer nanoparticles with the medium and, so far, it is unknown how such an effect would translate from in vitro to in vivo situations. In the same study, spectroscopic evidence suggests that there was significant protein adsorption by the particles in the cell growth media and that this effect increases with dendrimer generation. Subsequently, the toxic response was best understood in terms of an indirect effect as a result of medium depletion. In spite of the best correlation of the physicochemical properties with the toxicity found in in vitro mammalian cytotoxicity tests, data indicate an indirect response, a direct interaction and possibly the internalization by the cells cannot be ruled out and studies are ongoing to further elucidate the contributions of such interactions (Mukherjee et al. 2010).

An alternative strategy was studying the toxicity of PAMAM in zebrafish, as reported by Heiden et al. (2007). Despite the obvious differences in physiology between fish and humans, the zebrafish offers an ideal platform for an in vivo, whole-animal, medium to high throughput screen in the field of drug discovery, as well as a cost-effective complement to mouse models of human disease (Kari et al. 2007; Lieschke and Currie 2007). However, as zebrafish are evolutionarily more distant from humans, findings from such experiments will need to be repeated in other animal models before they can be directly correlated to humans (Guyon et al. 2007). Nevertheless, the zebrafish model is ideal for screening NMs for use as therapeutics, as well as obtaining reliable information regarding the toxicity of different NMs at the whole-animal level which can then be used to extrapolate adverse effects of NM exposure to humans and other vertebrates (Heiden et al. 2007).

The previously discussed studies on the in vivo nanotoxicity of drug-unloaded polymeric nanoparticles are compiled in Table 14.2.

Further, technological development as an analytical tool allows a systematic study of the biological effects including cellular uptake, trafficking, distribution, excretion, interaction with genes, proteins, metabolites, and other biomolecules and has been proposed as “omics.” The omics technologies are particularly well suited to evaluate toxicity in both in vitro and in vivo systems. Metabolomics, specifically, is an important tool to rapidly screen for biomarkers related to predefined pathways or processes in biofluids and tissues. Oxidative stress, for instance, has been implicated as a potential mechanism of toxicity for NPs and it is generally difficult

Table 14.2 In vivo studies based on the acute and chronic toxicity evaluations of drug-unloaded biodegradable polymeric nanoparticles on different animal models

Nanoparticle	Concentration	Model	Route	Test/treatment	Toxicity	References
PCL-PEG	100 mg/mL 1 g/kg/b.w.—MTD 200 μL (vol. admin.)	Rat	i.v.	Acute toxicity and genotoxic	Absence of toxicity	Wang et al. (2009)
PEG-PCL-PEG hydrogel	200 mg/kg 500 mg/kg 500 mg/kg Max: 10–25 g/kg	Rat	Intrapleural i.p. s.c.	Acute toxicity	PECE hydrogel was non-toxic after i.pl., i.p., or s.c. administration	Gong et al. (2009)
PCL-PEG-PCL (PCEC)	2.4 g/kg	Rat	i.v.	Acute toxicity and genotoxic	Did not cause any acute toxicity and genotoxicity in their experimental conditions	Huang et al. (2010)
Triblock copolymer PCL (PCL-PEG-PCL)	100 mg PCL–PEG–PCL 5 g/kg b.w.	BALB/c mice	s.c.	Acute toxicity	No toxic response or histopathological change was observed.	Fang et al. (2009)
MPEG-PCL micelles	200 mg/kg each	Mice	i.v.	The experiment was terminated on day 25. Weight loss, ruffled fur, appetite, diarrhea, cachexia, skin ulceration, or toxic deaths were observed	RBW fluctuated slightly but no significant difference was observed. The pattern of the morphological changes of the internal organs exhibited significant difference. The heart tissue in mice exhibited typical damage of focal destruction of myofibrils	Zheng et al. (2011)

(continued)

Table 14.2 (continued)

Nanoparticle	Concentration	Model	Route	Test/treatment	Toxicity	References
PCL lipid-core nanocapsules (LNC)	Acute: 18.03, 36.06, 72.12 × 10^{12} LNC/kg Subchronic: 6.01, 12.02, 18.03 × 10^{12} LNC/kg	Wistar rats	i.p.	Acute and subchronic toxicities	Did not cause acute and subchronic toxicity in their experimental conditions. A granulomatous body reaction was observed on histopathological analyses in high-dose-treated groups (acute and subchronic)	Bulcão et al. (2012)
Chitosan and PEG PLGA nanoparticles	0.2 mL of 20 mg/mL	Balb/C mouse	Oral and i.p.	Uptake and acute immune response	Absence of toxicity	Semete et al. (2010)
Chitosan nanoparticles	10, 20, and 40 μg/mL, with particle sizes of 200 or 340 nm	Zebrafish embryos		The hatching rate, mortality, and malformations of zebrafish embryos at 96 h post-fertilization were observed under stereomicroscopes	200 nm caused malformations, including a bent spine, pericardial edema, and an opaque yolk in zebrafish embryos	Hu et al. (2011)

i.v. intravenous, *i.p.* intraperitoneal, *s.c.* subcutaneous, *PEG* poly(ethylene glycol), *PCL* poly(ε-caprolactone), *PLGA* poly(D,L-lactic-acid-*co*-glycolic acid) polymer, *RBW* relative body weight

LNC lipid-core nanocapsules containing (PCL) poly(ε-caprolactone) as polymer

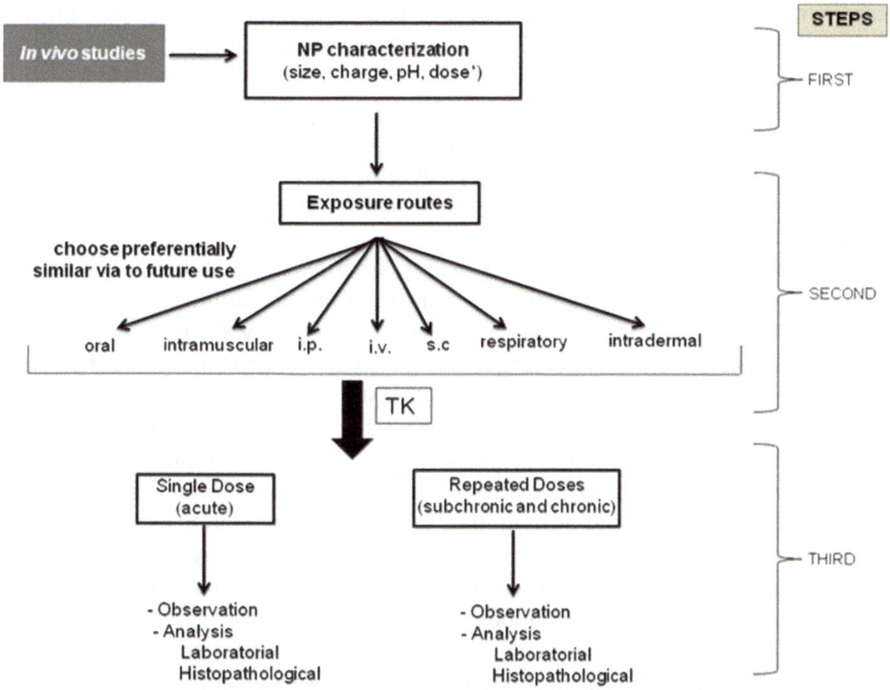

Fig. 14.2 Representative diagram of key elements required for in vivo toxicological evaluations of nanoparticles (*asterisk*, number of particles per volume and/or surface area). *TK* toxicokinetic, *i. p.* intraperitoneal, *i.v.* intravenous, *s.c.* subcutaneous

to be measured by conventional methods. In this context, metabolomics can provide mechanistic insights into nanotoxicity (Schnackenberg et al. 2012).

In summary, a sequence of key steps required for in vivo studies applied to the toxicological evaluation of nanoparticles is presented in Fig. 14.2, according to the previous discussions in this section.

14.3 Cardiotoxicity of Polymeric Nanoparticles

14.3.1 Background

The heart, in association to the vascular system, is vital to maintaining body homeostasis and distributing nutrients and oxygen in the bloodstream to several organs and tissues. In this context, potential cardiotoxic agents may trigger systemic toxicological effects as a consequence of the severe cardiac effects. MacDonald and Robertson (2009) reported that among very many drugs withdrawn from marketplace in the period of 1998 to 2008, the major cause was the cardiovascular toxicological effects such as fatal arrhythmias, sudden death due to QT_C prolongation, heart attack/stroke, and others (MacDonald and Robertson 2009). Recently, Qureshi et al. (2011) have also shown that cardiovascular system problems were responsible for the withdrawal of new drugs from 1980 to 2009 (Qureshi et al. 2011). Therefore, cardiotoxicity stands out as one of the most undesirable drug-induced effects and, for this reason, a careful evaluation of potential negative effects in the cardiovascular system elicited by new drug formulations is necessary.

According to ICH protocols for Safety Pharmacology Studies for Human Pharmaceuticals, a preclinical toxicity evaluation on the cardiovascular system is recommended prior to clinical trials, as well as the toxicity on the respiratory and central nervous systems. Regarding the cardiotoxicity, the heart rate (HR), blood pressure (BP), and electrocardiogram (ECG) should be included in the analysis (ICH 2001). These assessments enable the prediction of some effects such as QT interval prolongation, impairment in the conduction, and cardiac contractibility. The ECG, for instance, may indicate events like myocardial infarction, ventricular overload (right or left), and arrhythmias. Further, histopathological analyses must be conducted, in addition to other in vivo, in vitro, and/or ex-vivo tests (ICH 2001, 2005). In order to follow these recommendations, the tests should be conducted in both acute toxicological evaluations and in repeated doses protocols, because otherwise (for instance in repeated doses with limited points) some authors agree that these evaluations lack the broadness necessary to detect a real cardiotoxic potential.

On the other hand, preclinical studies present some limitations. Yang and Papoian (2012) described some gaps when testing tyrosine kinase inhibitors (TKI) as drugs indicated for the treatment of advanced cancer. These drugs may induce cardiotoxicity in animals despite the lack of histopathological alterations. In this case, the preclinical cardiotoxicity tests may not detect the potential cardiotoxic effects observed in humans, and these authors draw attention to the limitations in preclinical cardiotoxicity assessments and suggest the need to evaluate cardiotoxic effects in patients with more detailed tests (Yang and Papoian 2012). According to the International Conference on Harmonisation (ICH) Guidance S9 “Nonclinical Evaluation for Anticancer Pharmaceuticals” (ICH S9 2009), however, detailed tests of prospective drugs to be used in patients with advanced or end stage cancer are not recommended. Therefore, a thorough and careful review of the cardiovascular system should be performed in preclinical tests, utilizing additional tests whenever possible, but in some cases clinical studies may provide relevant responses to questions that preclinical studies fail to answer.

Additionally, ex-vivo assessments could be used in cardiotoxicity tests, for example the isolated perfused heart methodology with Langendorff apparatus and perfusion with Krebs-Henseleit buffer (Blasi et al. 2012), which allows the evaluation of cardiac output, ventricular contractibility, vascular resistance, edema, and others (Yang and Papoian 2012).

In relation to in vitro tests, cardiac safety analysis has been performed mostly through cell culture assays utilizing neonatal rat cardiac myocytes with analysis of

LDH liberation, indicating myocyte damage (Yang and Papoian 2012). In this same line, primary fetal human cardiac myocytes or immortalized cardiac cell lines can also be adopted for the evaluation of cardiac cell damage. These lineages, however, could be contaminated especially with nonhuman fibroblasts and another limitation is that the electrophysical characteristics are not determined in these cultured lineages, thus lacking data about the real cardiac function, which could be properly obtained by cultivating pluripotent stem cell-derived cardiac myocytes (Shell et al. 2011).

Considering the relevance of cardiotoxic effects, the future pharmacological applications of drugs incorporated in biodegradable nanoparticles will demand toxicological studies, especially to evaluate their potential cardiotoxicity.

14.3.2 Cardiotoxicity Studies

It is known that chemical composition, size and route of exposure, among others factors, play an important role in the development or not of toxicological effects, thus these factors may directly affect cardiotoxicity. There are reports indicating that some nanoparticles are able to increase oxidative damage and inflammation, or trigger myocardial infarction, hypertension, arrhythmias, and thrombosis (Mann et al. 2012). However, the first step to compare these studies is to characterize the nanoparticles, which will facilitate the analysis of these studies or the discussion of NP effects. For this matter, this section will focus on studies evaluating the cardiotoxicity of biodegradable nanoparticles.

Although some studies addressing the general toxicity or cardiotoxicity have been published, there is a lack of studies evaluating the toxic potential of drug-unloaded nanoparticles in high doses administrated in repeated doses (subchronic or chronic treatments). The studied nanoparticles are predominantly loaded with drugs which are normally well-known and the preclinical evaluations are performed with therapeutic doses from the nanoencapsuled drugs.

Another important consideration should be made about the routes of administration. Especially in relation to biodegradable nanoparticles, it has not clearly been demonstrated that they could be orally absorbed. On the other hand, studies utilizing intraperitoneal (i.p.) or intravenous (i.v.) administration for prolonged periods (repeated doses—chronic treatments) become unviable by several reasons, including the stress for the animals. These limitations contribute to form gaps in the evaluations performed in nanotoxicology, making it difficult to know if in fact the nanoparticle presents toxicological effects and therefore health risks. In this scenario, there is still lack of in vivo studies that evaluate the systemic or more specifically cardiac tissue toxicological effects of biodegradable nanoparticle.

The studies available on the possible effects of NP on the cardiovascular system have not evaluated the cardiotoxic effects of drug-unloaded NPs, but when chemotherapy drugs were nanoencapsulated and this could reduce cardiotoxicity. It is also noted that the routes of administration, NP concentrations, or the treatment times are different in these studies, making it difficult to perform proper inter-study comparisons.

In a 2007 study, it has been demonstrated that halofantrine, a drug used to treat malaria, when entrapped in poly(ε-caprolactone) nanocapsules and intravenously administered to male Wister rats at a single and especially high dose of halofantrine, the halofantrine-loaded nanocapsules were able to increase the LD50 of halofantrine to nearly the double of concentration compared to free halofantrine, as well as the mean time to death, being of 90 vs. 30 min, respectively (Leite et al. 2007). In relation to cardiotoxicity, electrocardiography (ECG) was performed in anesthetized rats. The animals that received nanoencapsuled halofantrine demonstrated decrease of prolongation QT interval compared to unencapsulated halofantrine, while the group treated with vehicle nanocapsules showed no alterations in the ECG. The main mechanism responsible for the cardiotoxic effect of halofantrine is the inhibition of potassium channels, corroborating the prolongation of QT interval (Wesche et al. 2000). The authors demonstrated that halofantrine-loaded nanocapsules were efficient to decrease the cardiotoxicity frequently observed in the treatment with halofantrine, considering intravenous administration, concentration of 100 and 150 mg/kg, at 2 mL/min, and a single dose treatment (Leite et al. 2007).

Another chemotherapy drug widely studied is the doxorubicin which is related to induce cardiac congestion and cardiomyopathy. Authors demonstrated that doxorubicin nanoencapsulated with polymeric nanoparticles (Cai et al. 2010; Jain et al. 2011a; Pramanik et al. 2012; Zheng et al. 2011) and dendrimers (Kaminskas et al. 2012; Lee et al. 2011; Xiao et al. 2011) reduced the drug cardiotoxicity with an increase of its antitumoral effect. The cardiovascular evaluations were performed by histopathological assessment, ECG, serum enzymes as LDH and CK-MB, cardiac protein as Troponin I. The administration route (oral, subcutaneous, intravenous), treatment time, and the concentrations of NPs were variable.

Thus, studies have not explored the possible cardiotoxic effects of drug-unloaded NPs in high concentrations as expected for toxicological studies, and such studies are extremely necessary considering that cardiovascular disease stands for the first reason for the withdraw of drugs from market.

14.4 Hepatotoxicity of Polymeric Nanoparticles

14.4.1 Background

Liver is a key organ in the detoxification of drugs and xenobiotics, being responsible for the drug biotransformation by the action of different liver enzymes, especially the microsomal cytochrome P450 enzyme group. In order to determine hepatic damage, important biomarkers are evaluated. Transaminases, such as alanine transaminase (ALT) and aspartate transaminase (AST), are intracellular liver enzymes released from cells as a consequence of hepatocellular damage. Also, the enzyme LDH serves as an additional biomarker to estimate liver cell damage, once it is also released after cell damage. Another test frequently adopted in vitro assay is the cell viability by the reduction of 3-[4,5-dimethylthiazol-2-yl]-2,5-diphenyl tetrazolium bromide (MTT), which indicates the mitochondrial integrity and activity, thus applicable to estimate the hepatotoxicity of the exposure of hepatic cells to NPs.

Hepatotoxicity, following MacDonald and Robertson (2009), was the second cause of withdraw of medicaments between 1998 and 2008, representing the second reason why drugs were removed from commercialization. In addition to drug testing, NPs when used as drug carrier devices should also be tested for hepatotoxicity, and recent works aimed to investigate liver damage in both situations when drugs are incorporated to nanoparticles or drug-unloaded NPs themselves. While drug-carrying nanoparticles are especially formulated as an attempt to reduce the previously known hepatotoxicity of a certain drug, drug-unloaded NPs could be easily identified as ideal targets for toxicological evaluation due to the lack of complete toxicity studies with drug-unloaded NPs, thus requiring a much broader investigation or toxicological screening. Therefore, biodegradable nanoparticles shall not be considered as presumably absent of toxicity to biological systems for their application in drug delivery, and in this context the next sections will approach the hepatotoxicity of biodegradable polymeric nanoparticles.

14.4.2 In Vitro Studies

Cell culture experiments for the assessment of hepatotoxicity are based on cytotoxicity tests, such as MTT and LDH release, in most reports in the literature. Moreover, polymeric nanoparticles have been tested incorporated to drugs, especially antitumoral agents, in human hepatoma HepG2 cell lineages and others. Few in vitro studies focused on the toxicity evaluation of drug-unloaded nanoparticles. Table 14.3 depicts the studies discussed in this section.

Considering the in vitro hepatotoxicity of drug-loaded nanoparticles, the studies on HepG2 cells focus on the evaluation of antitumoral activity by analyzing the cytotoxicity enhancement of nanoencapsulated drugs, thus the toxicity to liver cells is desirable, although nontumoral cells are not usually evaluated in these studies. Comparatively, in vivo studies enable the hepatotoxicity evaluation of nanoparticles with antitumoral drugs, when tested against animals bearing several cancer types other than hepatic tumors (as discussed in Sect. 14.4.3). Drug-unloaded NPs, however, when tested by in vitro models could indicate their hepatotoxicity.

Liu et al. (2012) evaluated the in vitro toxicity of docetaxel-loaded PEG-PCL nanoparticles (DOC-NPs) against murine hepatic carcinoma cell line H22. The authors report that drug-unloaded NPs did not elicit cytotoxicity to H22 cells in concentrations up to 20 μg/mL. However, MTT analyses showed that DOC-NPs

Table 14.3 Summary of in vitro and in vivo studies on the evaluation of hepatotoxicity of biodegradable polymeric nanoparticles

Nanoparticle	Dose/admin.	Study design	*n*	Toxicity	References
In vitro studies					
DOC-PEG-PCL-NPs	0.02–200 ng/mL, 37 °C, 48 h	H22 cells (8 × 10^3/well), 24 h adhesion	3	DOC-NPs had similar cytotoxicity to free DOC in MTT test	Liu et al. (2012)
Lectin-A-chain-[P (MDS-*co*-CES)] (Lectin-A-NPs)	0.01–10 ppm, 37 °C, 4 h	HepG2 cells (1 × 10^4/well), 24 h adhesion	8	Viability >80 % for vehicle NPs in MTT; Lectin-A-NPs were cytotoxic	Lee et al. (2008)
Gal-PGA-PLA-NPs	0–100 μg/mL, 37 °C, 48 h	HepG2 cells (5 × 10^4/well), 16 h adhesion	n.d.	Vehicle NPs showed no cytotoxicity in MTT test	Liang et al. (2005)
PLA-PEG-NPs	0.001–0.10 mg/mL, 37 °C, 24 h	HepG2 cells (1 × 10^5/well), 16 h adhesion	n.d.	Vehicle NPs showed no cytotoxicity in MTT test	Zhang et al. (2007)
PDS-BOC and PDS-NH_2	0.1–0.7 mg/mL, 37 °C, 24 h (MTT) and 0.1–0.5 mg/mL (LDH), 37 °C, 24 h	L-02 cells (1 × 10^4/well), 24 h adhesion	3	Both vehicle NPs were cytotoxic in MTT and LDH tests	Qiu et al. (2013)
In vivo studies					
TRAIL-PEG-PLL-NPs	100 μg of TRAIL equivalents/mouse every 4 days for 24 days, i.v.	HCT-116 tumor-bearing male Balb/c mice; subchronic	5	Reduction in TRAIL-induced hepatotoxicity (H&E)	Lim et al. (2011)
DOX-loaded mannosylated SLNPs	5 mg/kg b.w., single dose, i.v.	A549 tumor-bearing male Balb/c mice; acute	6	Decrease in DOX-induced damage to liver cells by ALT, AST, and ALP enzymes[a]	Jain et al. (2010)
PEO-PCL-NPs (PTX, TAM and PTX + TAM-loaded NPs)	(a) PTX: 20 mg/kg (100 μL), (b) TAM: 70 mg/kg (100 μL), and PTX + TAM: premixed (a + b) (200 μL). Two doses: day 1 and 24, for 24 days, i.v.	SKOV3/SKOV3$_{TR}$ tumor-bearing female Charles River *nu*/*nu* mice; subchronic	7	No hepatotoxicity by ALT, AST enzymes, and H&E liver histology[b]	Devalapally et al. (2008)
USN-PLGA-NPs	(i) 15 mg/kg/day; daily injections for 15 days, i.p. and (ii) 15 mg/kg/day; daily injections for 7 days, i.p.	(i) Healthy male Swiss mice; subchronic, and (ii) Sarcoma 180-bearing male Swiss mice	6[c]	(i) Reduction in ALT and AST activities[a]; (ii) H&E results showed lower hepatotoxicity of USN after encapsulation[a]	Santos et al. (2006)

TAM-PLGA-NPs	TAM 3 mg/kg b.w., repeated dose (every 3 days) for 30 days, oral admin.	Female Sprague–Dawley rats with DMBA-induced breast cancer, subchronic	6	Hepatotoxicity by H&E, ALT, AST decreased with TAM-PLGA-NPs[a]	Jain et al. (2011b)
SYN-loaded Eudragit® RS100-NPs	SYN-NPs i.p. 125 mg/kg for 7 days prior to APAP or i.p. SYN-NPs 125 mg/kg 1 h after APAP	Swiss albino mice, 12 h-fasted and 300 mg/kg APAP single dose, i.p.	12	SYN-NPs (pre- and post-treat.) decreased ALT, AST, ALP[a], and regenerated hepatocytes in H&E	Das et al. (2011)
PCL-NPs (LNC)	Acute: 18.03×10^{12}, 36.06×10^{12} and 72.12×10^{12} LNC/kg, i.p. Subchronic: 6.01×10^{12}, 12.02×10^{12} and 18.03×10^{12} LNC/kg, 28 days, i.p.	Healthy male Wistar rats in acute (i) and subchronic (ii) tests	(i) 6–9; (ii) 6	No hepatotoxicity by ALT, AST, ALP, LDH, ALB, TP[b], and highest LNC doses induced changes in H&E	Bulcão et al. (2012)

i.v. intravenous, *i.p.* intraperitoneal, *PEG* poly(ethylene glycol), *PLL* poly-L-lysine, *PLA* poly-(L-lactide), *SLNPs* solid lipid nanoparticles, *P(MDS-co-CES)* copolymer poly{(*N*-methyldietheneamine sebacate)-*co*-[(cholesteryl oxocarbonylamido ethyl) methyl bis(ethylene) ammonium bromide] sebacate}, *PEO* poly(ethylene oxide), *PCL* poly(ε-caprolactone), *PDS-BOC* PEGylated polyphosphazene-modified *tert*-butyloxycarbonyl-L-serine methyl ester-NPs, *PDS-NH₂* PEGylated polyphosphazene-modified-L-serine methyl ester-NPs, *PLGA* 50/50 poly(D,L-lactic-acid-*co*-glycolic acid) polymer, *DOC* docetaxel, *TRAIL* tumor necrosis factor (TNF)-related apoptosis-inducing ligand, *DOX* doxorubicin, *PTX* paclitaxel, *TAM* tamoxifen, *USN* usnic acid, *SYN* silymarin, *APAP* *N*-acetyl-*p*-aminophenol (paracetamol), *DMBA* 7,12-dimethylbenz[*a*]anthracene, *MTT* 3-[4,5-dimethylthiazol-2-yl]-2,5-diphenyl tetrazolium bromide, *H&E* haematoxylin and eosin staining, *ALT* alanine aminotransferase, *AST* aspartate aminotransferase, *ALP* alkaline phosphatase, *LDH* lactate dehydrogenase, *TP* total protein, *n.d.* not declared

[a]Significant differences in drug-NP treatment versus free drug

[b]No significant differences observed

[c]$n = 6$ for both (i) and (ii) protocols

had similar cytotoxic activity compared to the free DOC treatment. Furthermore, an in vivo experiment with H22 tumor-bearing mice has been conducted, which detected a higher cytotoxicity of DOC-NPs versus free DOC in the reduction of tumor volume (Liu et al. 2012).

Lectin A-chain, an anticancer glycoprotein, was incorporated to [P(MDS-*co*-CES)] nanoparticles by Lee et al. (2008). Cytotoxicity of the drug-unloaded nanoparticles was found to increase with increasing polymer concentrations, so the authors optimized this concentration and fixed it at 40 ppm for HepG2 cells, in which cell viability was greater than 80 %. In these conditions, Lectin A-chain incorporated to P(MDS-*co*-CES)-NPs showed higher cytotoxicity to hematoma cells even in serum-containing medium, indicating that the glycoprotein uptake by HepG2 cells occurred resulting in antitumoral activity (Lee et al. 2008).

Liang et al. (2005) developed NPs based on poly-(L-lactide)-*block*-poly(-γ-glutamic acid) (PGA-PLA-NPs) which were conjugated with galactosamine (Gal) in order to specifically target HepG2 cells. The authors reported that Gal-PGA-PLA-NPs were not hepatotoxic in the MTT viability test and that conjugation with Gal enabled the nanoparticles to be internalized by tumoral hepatocytes, but not by Hs68 fibroblasts, due to the fact that galactosamine-coated nanoparticles presented a specific interaction with HepG2 cells by ligand-receptor binding (Liang et al. 2005).

Zhang et al. (2007) reported that poly(ethylene glycol) (PEG)-coated poly-(L-lactide) nanoparticles (PLA-PEG-NPs) were not cytotoxic to HepG2 cells in the tested concentrations in the MTT test (Zhang et al. 2007). In a very interesting study, Qiu et al. (2013) have recently proposed the use of normal hepatocytes (L-02 cells) in the nanotoxicological evaluation and tested polyphosphazene nanoparticles modified by PEGylation and by serine methyl ester (PDS-NH_2-NPs), thus containing the active chemical group –NH_2. Once amino groups may enhance the toxicity of these nanoparticles, another nanoparticle was developed with *tert*-butyloxycarbonyl groups (PDS-BOC-NPs) in order to protect the amino group. In this study, both PDS-NH_2-NPs and PDS-BOC-NPs were cytotoxic to normal hepatocytes, in the MTT and LDH release tests, and induced apoptosis in a dose-dependent manner, while necrosis was found only in PDS-NH_2-NPs in annexin V/FITC-PI, caspase-3, and caspase-8 analyses. In addition, PDS-NH_2-NPs elicited a higher cytotoxic activity via MTT and LDH than PDS-BOC-NPs. These specific nanoparticles showed not to be suitable for application in drug delivery after the in vitro nanotoxicological evaluation, but these results indicate the need to carry out in vitro toxicity tests with additional promising polymeric nanocarriers prior to animal tests (Qiu et al. 2013).

Taking into account the diversity of in vitro studies performed in the general evaluation of toxicity of nanoparticles, e.g., studies with different cell lineages, some points are recognized as key elements in these studies, as shown in Fig. 14.3, and the presence of these steps in nanotoxicity evaluations is recommended in order to guarantee reproducible results.

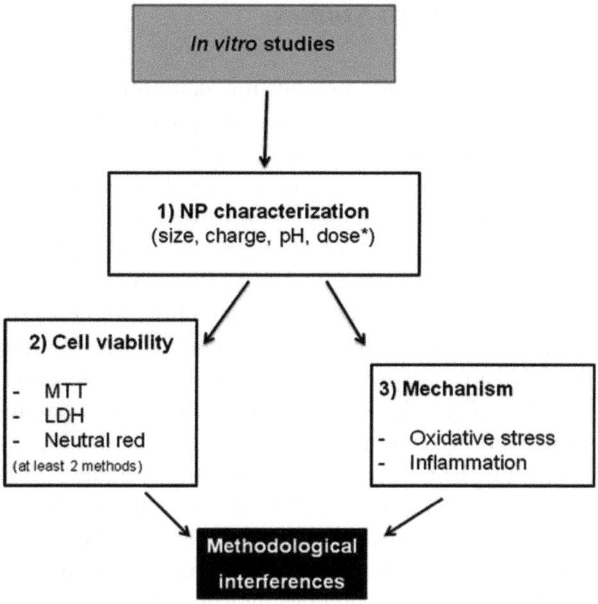

Fig. 14.3 Representative diagram of key elements required for in vitro toxicological evaluations of nanoparticles (*asterisk*, number of particles per volume and/or surface area). Methodological interferences need to be excluded

A critical point of in vitro studies is to have the caution of testing for methodological interferences in the analyses, such as spectroscopy or quenching of fluorescence dye by NPs (Liu et al. 2012; Liu and Chang 2008). Nanoparticles have large surface area, therefore a high absorption capacity and other interferences may be due to the aggregation state and coating of NPs with different proteins, leading to the formation of artifacts.

In relation to the NP characterization, size and dispersion agents may influence the results. Importantly, a standardization of the tested doses as number of particles/ volume and surface area could contribute to a better comparison among studies. Further, in relation to cytotoxicity tests, it is required to perform at least two techniques with different physicochemical principles to guarantee that the results are reliable, for example by choosing two among the LDH assay, the MTT test for mitochondrial activity and the neutral red for lysosome incorporation.

After the characterization and cell viability or cytotoxicity evaluation, a mechanistic approach such as the assessment of oxidative stress and inflammation could be performed. Oxidative stress can be determined by flow cytometry by measuring intracellular reactive oxygen species (ROS) in the dichlorofluorescein (DCF) assay, which results in the formation of a fluorescent product. Inflammatory biomarkers, e.g., cytokines, are widely determined by enzymatic immunoassays, but there are reports on the adsorption of cytokines by nanoparticles such as carbon nanoparticles and metal oxides (Kroll et al. 2009) and that non-sterile nanoparticles could be contaminated with bacteria and thus induce inflammatory responses.

14.4.3 In Vivo Studies

Animal models are crucial for determining the toxicity of new drugs and formulations, as already discussed in Sect. 14.2. In relation to nanocarrier devices, in particular, most studies are carried out with drug-loaded nanoparticles, in which the focus is to evaluate markers related to the most prominent toxicity of the drug incorporated to the nanoparticles, for instance the hepatic toxicity. The in vivo hepatotoxicity studies of biodegradable polymeric nanoparticles described in this section are also summarized in Table 14.3.

Cancer research is constantly searching for new drug candidates or the improvement of the existing therapeutic agents. Therefore, intense efforts are being made to diminish adverse or toxic effects of drug options currently available in chemotherapy, particularly by incorporating these drugs to polymeric nanoparticles. In a study with HCT-116 tumor-bearing mice, Lim et al. (2011) administered tumor necrosis factor (TNF)-related apoptosis-inducing ligand (TRAIL) incorporated to NPs prepared with PEG and poly(L-lysine) intravenously in Balb/c mice, once TRAIL alone presents hepatotoxicity. The TRAIL-PEG-NP treatment exerted an effective tumor suppressing activity, showed better pharmacokinetic profile than TRAIL, and notably, histological findings indicated that the formulation was also able to reduce the TRAIL-induced hepatotoxicity, which is promising for the application of TRAIL as an anticancer agent (Lim et al. 2011).

Similarly, further studies have also investigated the potential reduction of hepatotoxicity obtained after encapsulation of anti-carcinogenic compounds, such as tamoxifen and paclitaxel, in female mice bearing ovarian adenocarcinoma (Devalapally et al. 2008) and the plant-isolated compound usnic acid (Santos et al. 2006). In the first study, paclitaxel-loaded and tamoxifen-loaded nanocapsules, as well as the combination of paclitaxel and tamoxifen in poly (oxyethylene)-poly(ε-caprolactone) nanoparticles, were neither able to elicit any significant alteration in hepatocellular damage by ALT and AST intracellular enzymes nor able to induce histological changes in liver tissues from treated animals (Devalapally et al. 2008). According to the other work, published by Santos et al. (2006), the encapsulation of usnic acid into poly(lactide-*co*-glycolide) nanocapsules reduced usnic acid-induced hepatotoxicity in normal rats, as observed by the diminished ALT and AST transaminase activities, and also in tumor-bearing rats, by the observation of mild histological damages in comparison to unencapsulated USN-treated animals (Santos et al. 2006).

Jain et al. (2011b) have studied the influence of orally administered tamoxifen-loaded poly(lactide-*co*-glycolide) nanoparticles in the liver toxicity of tamoxifen to female rats bearing chemically induced breast cancer. The transaminases and histological findings indicated that unencapsulated tamoxifen enhanced liver damage and that tamoxifen-loaded poly(lactide-*co*-glycolide) nanoparticles could preserve cellular integrity and prevent hepatotoxicity. Interestingly, the authors suggested that the decreased hepatotoxicity after encapsulation of tamoxifen could be attributed to an escape from first-pass metabolism (Jain et al. 2011b).

Silymarin (SYN) is a polyphenolic compound with antioxidant and hepatoprotective effects. Das et al. (2011) prepared and tested silymarin-loaded Eudragit® RS100 polymeric nanoparticles against an animal model of hepatocellular damage induced by paracetamol. SYN-NPs have been administered to rats before and after paracetamol injury, and enzymatic activities of ALT, AST, and ALP indicated that SYN-NPs were able to decrease cell damages in both pre- and posttreatments compared to unencapsulated SYN. Also, liver histological sections indicated that SYN-NPs induced significant regeneration of paracetamol-induced necrotic injury (Das et al. 2011).

In comparison to most of the previous studies, current literature is extremely deficient in studies focusing on the toxicological evaluation of biodegradable NPs themselves (drug-unloaded nanoparticles). According to a recent study from Bulcão et al. (2012), the i.p. administration of drug-unloaded lipid-core PCL nanoparticles in rats induced no alterations in several liver damage biomarkers after both acute and subchronic toxicity protocols. Histological sections, however, indicated the formation of a granulomatous foreign body reaction in liver and spleen only after highest LNC doses, for acute and subchronic evaluations (Bulcão et al. 2012).

14.5 Final Remarks

In a nutshell, current literature indicates that few studies are available based on nanotoxicity of polymeric nanoparticles, although interest in this field is enormous as observed by the increasing number of recent publications. For a proper evaluation and comparison among the studies available and to be performed in the future, some standardization of protocols is recommended, and some important points require much attention, such as the testing for methodological interferences and including specific particularities for nanoparticle toxicity evaluations.

Despite the advantages and disadvantages inherent to in vitro or in vivo models, in relation to nanoparticles the adoption of in vivo toxicological tests is very important for allowing evaluations of toxicokinetics, routes of absorption, interaction with biomolecules, and for detecting any damaged target tissues after general toxicological effects. It should be noted that for in vitro and in vivo studies, it is necessary to investigate possible methodological interferences in the analyses, once NPs could interfere in spectroscopy or quenching of fluorescence dye. Also, some properties (large surface area, high absorption capacity, or the aggregation state and coating of NPs with different proteins) could lead to the formation of artifacts in the assays and thus interfere with the results. The relevance of utilizing at least two different in vitro assays to investigate the same parameters has also been pointed out in order to assure the reliability of the results.

Finally, as suggested in Fig. 14.4, some steps should be followed in the toxicity evaluation of nanoparticles based on in vitro and subsequently in vivo models. Nanoparticles should be completely characterized previously to any nanotoxicity

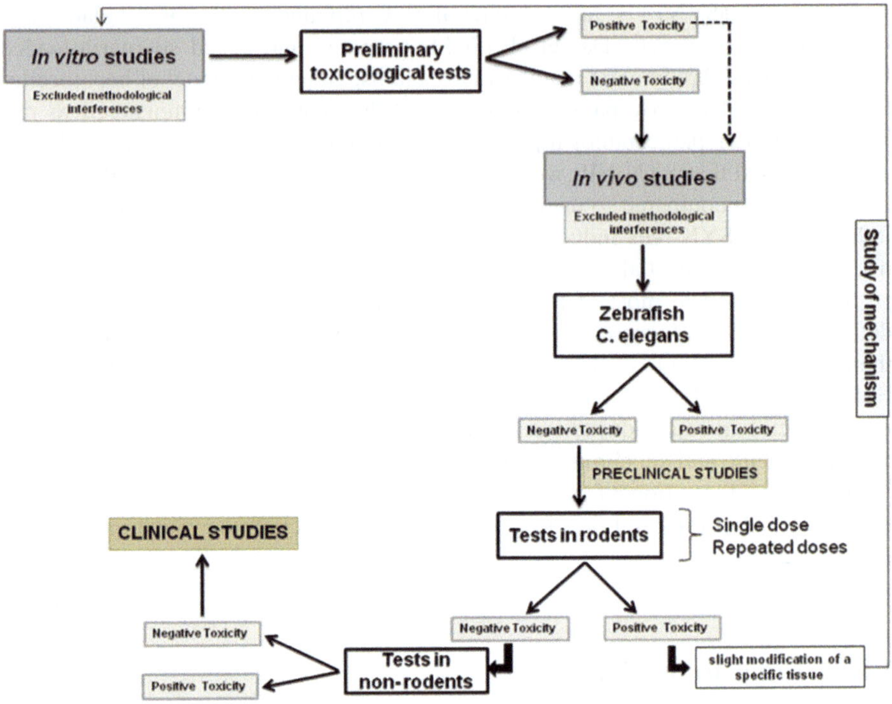

Fig. 14.4 Proposed diagram for a nanotoxicological evaluation using in vitro and in vivo models. This scheme proposes that once a positive toxicity is found in vitro (with cell culture and/or ex-vivo model, for an example) for a NP, alternative in vivo models (e.g., *C. elegans* and/or zebrafish) could also be adopted, while NPs with negative in vitro toxicity should proceed with further in vivo nanotoxicological evaluation. It should be noted that in vitro tests may elucidate possible mechanisms involved in slight toxicological modifications of specific tissues (doses expressed as number of particles per volume and/or surface area)

studies, considering that size and presence of dispersion agents may influence the results, by leading to methodological interferences or to a true toxicological response. Moreover, it is difficult to compare studies that use different units of nanoparticle concentration, so for a better comparison of studies, a consensus should be achieved on how to establish the tested doses, such as number of particles/volume or surface area and then a standardization in this regard is essential to compare inter-study findings.

References

Belyanskaya L, Manser P, Spohn P et al (2007) The reliability and limits of the MTT reduction assay for carbon nanotubes–cell interaction. Carbon 45:2643–2648

Bernardi A, Frozza RL, Jager E et al (2008) Selective cytotoxicity of indomethacin and indomethacin ethyl ester-loaded nanocapsules against glioma cell lines: an in vitro study. Eur J Pharmacol 586:24–34

Bernardi A, Braganhol E, Jager E et al (2009) Indomethacin-loaded nanocapsules treatment reduces in vivo glioblastoma growth in a rat glioma model. Cancer Lett 281:53–63
Blasi E, Heyen J, Patyna S et al (2012) Sunitinib, a receptor tyrosine kinase inhibitor, increases blood pressure in rats without associated changes in cardiac structure and function. Cardiovasc Ther 30:287–294
Brannon-Peppas L (1995) Recent advances on the use of biodegradable microparticles and nanoparticles in controlled drug delivery. Int J Pharm 16(1):1–9
Bulcão RP, Freitas FA, Venturini CG et al (2012) Acute and subchronic toxicity evaluation of poly (epsilon-caprolactone) lipid-core nanocapsules in rats. Toxicol Sci 132:162–176
Cai S, Thati S, Bagby TR et al (2010) Localized doxorubicin chemotherapy with a biopolymeric nanocarrier improves survival and reduces toxicity in xenografts of human breast cancer. J Control Release 146:212–218
Casey A, Herzog E, Davoren M et al (2007) Spectroscopic analysis confirms the interactions between single walled carbon nanotubes and various dyes commonly used to assess cytotoxicity. Carbon 45:1425–1432
Clift MJ, Gehr P, Rothen-Rutishauser B (2011) Nanotoxicology: a perspective and discussion of whether or not in vitro testing is a valid alternative. Arch Toxicol 85:723–731
Das S, Roy P, Auddy RG et al (2011) Silymarin nanoparticle prevents paracetamol-induced hepatotoxicity. Int J Nanomedicine 6:1291–1301
Devalapally H, Duan Z, Seiden MV et al (2008) Modulation of drug resistance in ovarian adenocarcinoma by enhancing intracellular ceramide using tamoxifen-loaded biodegradable polymeric nanoparticles. Clin Cancer Res 14:3193–3203
Dhawan A, Sharma V (2010) Toxicity assessment of nanomaterials: methods and challenges. Anal Bioanal Chem 398:589–605
Fang F, Gong CY, Dong PW et al (2009) Acute toxicity evaluation of in situ gel-forming controlled drug delivery system based on biodegradable poly(epsilon-caprolactone)-poly(ethylene glycol)-poly(epsilon-caprolactone) copolymer. Biomed Mater 4:025002
Fischer HC, Chan WC (2007) Nanotoxicity: the growing need for in vivo study. Curr Opin Biotechnol 18:565–571
Frozza RL, Bernardi A, Paese K et al (2010) Characterization of trans-resveratrol-loaded lipid-core nanocapsules and tissue distribution studies in rats. J Biomed Nanotechnol 6:694–703
Gong C, Shi S, Dong P et al (2009) Synthesis and characterization of PEG-PCL-PEG thermosensitive hydrogel. Int J Pharm 365:89–99
Greish K, Thiagarajan G, Herd H et al (2012) Size and surface charge significantly influence the toxicity of silica and dendritic nanoparticles. Nanotoxicology 6:713–723
Guo L, Von Dem Bussche A, Buechner M et al (2008) Adsorption of essential micronutrients by carbon nanotubes and the implications for nanotoxicity testing. Small 4:721–727
Guyon JR, Steffen LS, Howell MH et al (2007) Modeling human muscle disease in zebrafish. Biochim Biophys Acta 1772:205–215
Han X, Corson N, Wade-Mercer P et al (2012) Assessing the relevance of in vitro studies in nanotoxicology by examining correlations between in vitro and in vivo data. Toxicology 297:1–9
Hartung T (2011) From alternative methods to a new toxicology. Eur J Pharm Biopharm 77:338–349
Heiden TC, Dengler E, Kao WJ et al (2007) Developmental toxicity of low generation PAMAM dendrimers in zebrafish. Toxicol Appl Pharmacol 225:70–79
Hu YL, Qi W, Han F et al (2011) Toxicity evaluation of biodegradable chitosan nanoparticles using a zebrafish embryo model. Int J Nanomedicine 6:3351–3359
Huang Y, Gao H, Gou M et al (2010) Acute toxicity and genotoxicity studies on poly-(epsilon-caprolactone)- poly(ethyleneglycol)-poly(epsilon-caprolactone) nanomaterials. Mutat Res 696:101–106

International Conference on Harmonisation of Technical Requirements for Registration of Pharmaceuticals for Human Use (ICH) (2001) In: S7A: safety pharmacology studies for human pharmaceuticals, Geneva
International Conference on Harmonisation of Technical Requirements for Registration of Pharmaceuticals for Human Use (ICH) (2005) In: S7B: the nonclinical evaluation of the potential for delayed ventricular repolarization (QT interval prolongation) by human pharmaceuticals, Geneva
International Conference on Harmonisation of Technical Requirements for Registration of Pharmaceuticals for Human Use (ICH) (2009) In: S9: nonclinical evaluation for anticancer pharmaceuticals, Geneva
Jain A, Agarwal A, Majumder S et al (2010) Mannosylated solid lipid nanoparticles as vectors for site-specific delivery of an anti-cancer drug. J Control Release 148:359–367
Jain AK, Swarnakar NK, Das M et al (2011a) Augmented anticancer efficacy of doxorubicin-loaded polymeric nanoparticles after oral administration in a breast cancer induced animal model. Mol Pharm 8:1140–1151
Jain AK, Swarnakar NK, Godugu C et al (2011b) The effect of the oral administration of polymeric nanoparticles on the efficacy and toxicity of tamoxifen. Biomaterials 32:503–515
Johnston H, Brown D, Kermanizadeh A et al (2012) Investigating the relationship between nanomaterial hazard and physicochemical properties: informing the exploitation of nanomaterials within therapeutic and diagnostic applications. J Control Release 164:307–313
Kaminskas LM, McLeod VM, Kelly BD et al (2012) A comparison of changes to doxorubicin pharmacokinetics, antitumor activity, and toxicity mediated by PEGylated dendrimer and PEGylated liposome drug delivery systems. Nanomedicine 8:103–111
Kandarova H, Letasiova S (2011) Alternative methods in toxicology: pre-validated and validated methods. Interdiscip Toxicol 4:107–113
Kari G, Rodeck U, Dicker AP (2007) Zebrafish: an emerging model system for human disease and drug discovery. Clin Pharmacol Ther 82:70–80
Kim SY, Lee YM, Baik DJ et al (2003) Toxic characteristics of methoxy poly(ethylene glycol)/poly(epsilon-caprolactone) nanospheres; in vitro and in vivo studies in the normal mice. Biomaterials 24:55–63
Kroll A, Pillukat MH, Hahn D et al (2009) Current in vitro methods in nanoparticle risk assessment: limitations and challenges. Eur J Pharm Biopharm 72:370–377
Kroll A, Dierker C, Rommel C et al (2011) Cytotoxicity screening of 23 engineered nanomaterials using a test matrix of ten cell lines and three different assays. Part Fibre Toxicol 8:9
Laurent S, Burtea C, Thirifays C et al (2012) Crucial ignored parameters on nanotoxicology: the importance of toxicity assay modifications and "cell vision". PLoS One 7:e29997
Lee AL, Wang Y, Ye WH et al (2008) Efficient intracellular delivery of functional proteins using cationic polymer core/shell nanoparticles. Biomaterials 29:1224–1232
Lee IH, Yu MK, Kim IH et al (2011) A duplex oligodeoxynucleotide-dendrimer bioconjugate as a novel delivery vehicle for doxorubicin in in vivo cancer therapy. J Control Release 155:88–95
Leite EA, Grabe-Guimaraes A, Guimaraes HN et al (2007) Cardiotoxicity reduction induced by halofantrine entrapped in nanocapsule devices. Life Sci 80:1327–1334
Li YF, Chen C (2011) Fate and toxicity of metallic and metal-containing nanoparticles for biomedical applications. Small 7:2965–2980
Liang HF, Yang TF, Huang CT et al (2005) Preparation of nanoparticles composed of poly (gamma-glutamic acid)-poly(lactide) block copolymers and evaluation of their uptake by HepG2 cells. J Control Release 105:213–225
Lieschke GJ, Currie PD (2007) Animal models of human disease: zebrafish swim into view. Nat Rev Genet 8:353–367
Lim SM, Kim TH, Jiang HH et al (2011) Improved biological half-life and anti-tumor activity of TNF-related apoptosis-inducing ligand (TRAIL) using PEG-exposed nanoparticles. Biomaterials 32:3538–3546

Liu ZC, Chang TM (2008) Long-term effects on the histology and function of livers and spleens in rats after 33% toploading of PEG-PLA-nano artificial red blood cells. Artif Cells Blood Substit Immobil Biotechnol 36:513–524

Liu M, Fréchet JMJ (1999) Designing dendrimers for drug delivery. Pharm Sci Technol Today 2 (10):393–401

Liu Q, Li R, Zhu Z et al (2012) Enhanced antitumor efficacy, biodistribution and penetration of docetaxel-loaded biodegradable nanoparticles. Int J Pharm 430:350–358

Lynch I, Dawson K (2008) Protein–nanoparticle interactions. Nanotoday 3:40–47

MacDonald JS, Robertson RT (2009) Toxicity testing in the 21st century: a view from the pharmaceutical industry. Toxicol Sci 110:40–46

Mann EE, Thompson LC, Shannahan JH et al (2012) Changes in cardiopulmonary function induced by nanoparticles. Wiley Interdiscip Rev Nanomed Nanobiotechnol 4:691–702

Mora-Huertas CE, Fessi H, Elaissari A (2010) Polymer-based nanocapsules for drug delivery. Int J Pharm 385(1–2):113–142

Mukherjee SP, Davoren M, Byrne HJ (2010) In vitro mammalian cytotoxicological study of PAMAM dendrimers—towards quantitative structure activity relationships. Toxicol In Vitro 24:169–177

Pohlmann AR, Fonseca FN, Paese K, Detoni CB, Coradini K, Beck RCR, Guterres SS (2013) Poly (ε-caprolactone) microcapsules and nanocapsules in drug delivery. Expert Opin Drug Deliv 10 (5):623–638. doi:10.1517/17425247.2013.769956

Pramanik D, Campbell NR, Das S et al (2012) A composite polymer nanoparticle overcomes multidrug resistance and ameliorates doxorubicin-associated cardiomyopathy. Oncotarget 3:640–650

Qiu L, Chen Y, Gao M et al (2013) Phagocytic uptake and ROS-mediated cytotoxicity in human hepatic cell line of amphiphilic polyphosphazene nanoparticles. J Biomed Mater Res A 101:285–297

Qureshi ZP, Seoane-Vazquez E, Rodriguez-Monguio R et al (2011) Market withdrawal of new molecular entities approved in the United States from 1980 to 2009. Pharmacoepidemiol Drug Saf 20:772–777

Roberts JC, Bhalgat MK, Zera RT (1996) Preliminary biological evaluation of polyamidoamine (PAMAM) Starburst dendrimers. J Biomed Mater Res 30:53–65

Russel W, Burch R (eds) (1959) The principles of humane experimental technique. Methuen & Co., London

Sadekar S, Ghandehari H (2012) Transepithelial transport and toxicity of PAMAM dendrimers: implications for oral drug delivery. Adv Drug Deliv Rev 64:571–588

Santos NPS, Nascimento SC, Wanderley MS et al (2006) Nanoencapsulation of usnic acid: an attempt to improve antitumour activity and reduce hepatotoxicity. Eur J Pharm Biopharm 64:154–160

Sayes CM, Reed KL, Warheit DB (2007) Assessing toxicity of fine and nanoparticles: comparing in vitro measurements to in vivo pulmonary toxicity profiles. Toxicol Sci 97:163–180

Schnackenberg LK, Sun J, Beger RD (2012) Metabolomics techniques in nanotoxicology studies. Methods Mol Biol 926:141–156

Semete B, Booysen LI, Kalombo L et al (2010) In vivo uptake and acute immune response to orally administered chitosan and PEG coated PLGA nanoparticles. Toxicol Appl Pharmacol 249:158–165

Shell SA, Wappel R, Turner JR et al (2011) Early safety testing for oncology therapies. Drug Discov Dev 14:24–26

Wang K, Fu SZ, Gu YC et al (2009) Synthesis and characterization of biodegradable pH-sensitive hydrogels based on poly(3-caprolactone), methacrylic acid, and poly(ethylene glycol). Polym Degrad Stab 94:730–734

Wesche DL, Schuster BG, Wang WX et al (2000) Mechanism of cardiotoxicity of halofantrine. Clin Pharmacol Ther 67:521–529

Worle-Knirsch JM, Pulskamp K, Krug HF (2006) Oops they did it again! Carbon nanotubes hoax scientists in viability assays. Nano Lett 6:1261–1268

Xiao K, Luo J, Li Y et al (2011) PEG-oligocholic acid telodendrimer micelles for the targeted delivery of doxorubicin to B-cell lymphoma. J Control Release 155:272–281

Yang B, Papoian T (2012) Tyrosine kinase inhibitor (TKI)-induced cardiotoxicity: approaches to narrow the gaps between preclinical safety evaluation and clinical outcome. J Appl Toxicol 32:945–951

Zhang Y, Hu Z, Ye M et al (2007) Effect of poly(ethylene glycol)-block-polylactide nanoparticles on hepatic cells of mouse: low cytotoxicity, but efflux of the nanoparticles by ATP-binding cassette transporters. Eur J Pharm Biopharm 66:268–280

Zheng L, Gou M, Zhou S et al (2011) Antitumor activity of monomethoxy poly(ethylene glycol)-poly (epsilon-caprolactone) micelle-encapsulated doxorubicin against mouse melanoma. Oncol Rep 25:1557–1564

Chapter 15
Cyto-, Geno-, and Ecotoxicity of Copper Nanoparticles

Mahendra Rai, Avinash Ingle, Indarchand Gupta, Swapnil Gaikwad, Aniket Gade, Olga Rubilar, and Nelson Durán

Abstract Nanotechnology has been playing a crucial role in twenty-first century in solving various problems particularly in the field of agriculture, medicine, and electronics. Nanotechnology is a broad and interdisciplinary area of research that has been growing explosively worldwide. Due to the small size, surface tailor ability, improved solubility, and multifunctionality of nanoparticles it may open up new research avenues. Nanoparticles are being viewed as fundamental building blocks of nanotechnology. The synthesis of nanoparticles is an important component of rapidly growing research efforts in nanoscale science and engineering. The synthesis of copper nanoparticles (CuNPs) by physical method involves the mechanical grinding of bulk metals. Subsequently the resulting nanoparticles are to be stabilized by the addition of protecting agents. Whereas in case of chemical methods of synthesis copper salts use to be reduced to copper nanoparticles. Moreover, biological method of synthesis make use of biomolecules for the reduction and stabilization of nanoparticles.

The increasing use of nanoparticles leads to the release and accumulation of these particles in soil, air, and aquatic environment. Therefore, evaluation of nanotoxicity to the ecosystem must be considered. Nanoparticles with their distinct properties require development of methods, which will assess the possible benefits to possible risks and health hazards associated with exposure to nanomaterials as they are used in manufacturing and medicine. Common methods for toxicity evaluation include MTT Assay, Neutral Red Assay, LDH Assay, Comet Assay, and ROS Assay.

M. Rai (✉)
Department of Biotechnology, Sant Gadge Baba Amravati University, Amravati 444 602, Maharashtra, India

Biological Chemistry Laboratory, Institute of Chemistry, Universidade Estadual de Campinas, Campinas, Sao Paulo, Brazil
e-mail: mahendrarai@sgbau.ac.in

N. Durán et al. (eds.), *Nanotoxicology*, Nanomedicine and Nanotoxicology, DOI 10.1007/978-1-4614-8993-1_15,

In the present chapter, we have briefly discussed the methods of synthesis of CuNPs, toxicity evaluation and harmful effects of CuNPs on plants, mice, fishes, and worms.

15.1 Introduction

Copper nanoparticles, due to their unique physical and chemical properties and the low cost of preparation, have been of great interest (Varshney et al. 2012). These are being used for many different industrial applications (Malathi and Balasubramanian 2012). Current usage includes lubricants, polymers, plastics, and metallic coatings and inks (Chen et al. 2006). Copper nanoparticles possess superior mending effects (Liu et al. 2004). One study showed that copper nanoparticles effectively decreased wear and friction and mended worn surfaces when used as oil additive (Tarasov et al. 2002). Copper nanoparticles have also been used as a bimetallic catalyst on activated carbon to effectively reduce elevated levels of nitrate in water (Barrabés et al. 2006). Chen and Hong (2005) reported that the addition of 15 or 30 nm copper nanoparticles greatly increased the ductility of diamond-like carbon (DLC) nanocomposite films. Copper nanoparticles have also been used in electrically conductive polymer composites as fillers (Zhang et al. 2007). Moreover, the multiple uses of copper nanoparticles extend beyond industrial applications.

Copper is already known as an effective antibacterial agent due to its ability to combine with the –SH enzyme group and lead to protein inactivation (Yoon et al. 2007). The authors further reported the use of copper nanoparticles (100 nm) as antibacterial agents against *Escherichia coli* and *Bacillus subtilis*. The latter demonstrated the greatest susceptibility when exposed to the copper nanoparticles. In addition, copper-fluoropolymer (Cu-CFx) nanocomposite film layers have been effectively utilized as a bioactive coating to inhibit the growth of microorganisms including *E. coli*, *Lysteria*, and *Staphylococcus aureus* (Cioffi et al. 2005).

Chatterjee et al. (2012) developed a method for preparation of CuNPs by reduction of $CuCl_2$ in the presence of gelatin as a stabilizer (50–60 nm in size). The bactericidal activity of this CuNPs on Gram-negative *Escherichia coli* was verified by the methods of agar plating, flow cytometry, and phase contrast microscopy. A low minimum inhibitory concentration showed that CuNPs are highly effective against *E. coli*. The same effect was also observed in Gram-positive *Bacillus subtilis* and *Staphylococcus aureus*, with values of inhibition similar to *E. coli*.

Despite the potential benefits of nanotechnology, the knowledge base of human health and environmental effects in the production and use of nanomaterials is insufficient. Human and environment can get exposed to nanoparticles via spillage during shipping and handling of nanobased products (Chen et al. 2006). Nanoparticle toxicity is associated with its smaller size, the efficiency with which it enters the cells, and its increased surface area (Dowling 2004). Like any other novel products, the acceptance of nanotechnology based products will also rely on promising benefits versus the potential risks associated with them (Tsuji et al. 2006). Even though different nanoparticles display similar properties, not all nanoparticles can be treated as equal and must be studied individually (Holsapple et al. 2005). In addition, many nanomaterials present unique properties because of the type of surface coating applied on the material (Thomas and Sayre 2005).

The impact of nanoparticles on health has attracted considerable attention of the public and government worldwide. So far, most of the nanotoxicity research has been focused on different routes of exposures. Oberdorster et al. (2005) has defined nanotoxicology as the "science of engineered nanodevices and nanostructures that deals with their effects on living organisms." In regard to research on the health effects of nanomaterials, the inhalation (via the respiratory tract) exposure route has been extensively studied as compared to the ingestion (via the gastrointestinal tract) or skin absorption routes of exposure (Tsuji et al. 2006). More research is needed to determine if nanoparticles can penetrate the skin (Tsuji et al. 2006), because little information exists as to whether nanoparticles can be absorbed through the stratum corneum (Holsapple et al. 2005).

Both in vivo and in vitro studies have been conducted on the toxicology of metal nanoparticles including CuNPs found in environmental and occupational settings to include effects on the respiratory system and extra-pulmonary organs (Oberdorster et al. 2005). Many factors contribute to the pulmonary toxicity of nanoparticles to include size, dose to target organ or tissue, surface coating or treatment, degree of aggregates formed, surface charge, and shape (Tsuji et al. 2006). However, Colvin (2003) remarked in "the potential environmental impact of engineered nanomaterials," that, although research has been focused on the inhalation exposure route, the more common exposure route may be dermal absorption or oral injection exposure because many nanomaterials are prepared in liquids and agglomerate strongly, making it difficult for them to become respirable nanoparticles.

15.2 Synthesis of CuNPs

Nanotechnology is an enabling technology that deals with nano-sized materials in different field of science. Instead of availability of physical and chemical methods for synthesis of metal nanoparticles, biological synthesis is an important method for development of clean, nontoxic, and environmental-friendly procedure. Nanobiosynthesis by using various biological agents is a rapidly emerging method for synthesis of nanoparticles with well-defined shapes and controlled monodispersity.

15.2.1 Biological Synthesis

Extensive research has been performed by scientists all over the world in the field of biosynthesis of metal nanoparticles using different biological systems and the main focus is always on silver and gold nanoparticles. There are a few reports on the biosynthesis of copper and copper oxide nanoparticles (Thakkar et al. 2010). However, a thorough survey of literature on synthesis of CuNPs provides evidence that CuNPs can be easily synthesized in their oxide form instead of simple CuNPs

(Hasan et al. 2008; Singh et al. 2010) because copper is well known to be susceptible to oxidation. Therefore, synthesis of pure metallic CuNPs in aqueous phase is still an open challenge. However, there are few reports concerning biogenic synthesis of copper nanoparticles (Varshney et al. 2010, 2011; Valodkar et al. 2011; Majumber 2012; Harne et al. 2012; Ramanathan et al. 2013).

The mechanism for the synthesis of nanoparticles using biological systems has not been elucidated, so far. However, it is suggested that different biomolecules are responsible for the synthesis of nanoparticles. The use of fungi is potentially exciting since they secrete large amount of enzymes and are simple to deal within the laboratory (Mandal et al. 2006). But in extracellular biosynthesis of nanoparticles, especially in the case of fungi, several reducing and capping agents secreted are possibly involved which also effects the reducing agents on the shape and size of nanoparticles (Mohanpuria et al. 2008; Mukherjee et al. 2001). Lee et al. (2011) reported the biosynthesis of CuNPs using plant leaf extract *Magnolia* as a reducing agent. After treating aqueous solutions of $CuSO_4 \cdot 5H_2O$ with leaf extract, stable CuNPs were formed, which were ranged in size from 40 to 100 nm.

15.2.2 Chemical Synthesis

Many methods have been proposed for the chemical synthesis of CuNPs. Among these, aqueous reduction method is most commonly employed because of its benefits, like, simple operation, high yield and quality, limited equipment requirement, and simplicity of control. In the year 1997, CuNPs were prepared by the reduction of copper (II) acetate in water and 2-ethoxyethanol using hydrazine under reflux. The synthesized nanoparticles demonstrated different absorption peaks in the spectral range of 572–582 nm. Average size varies from 6.6 to 22.7 nm in ethoxyethanol and from 15.5 to 30.2 nm in water after addition of various amounts of a protective polymer, i.e., poly-(*N*-vinylpyrrolidone) (Huang et al. 1997). Different reducing agents have been tried and used for the synthesis of CuNPs to achieve a high degree of synthesis (i.e., higher yield) and to develop rapid, easy and eco-friendly method. Xiong et al. (2013) obtained water-soluble and stable dispersions of copper nanoparticles in the presence of dopamine as reductant and capping agent and copper chloride of around 2 nm of size.

Zhang et al. (2010) used potassium borohydride as reducing agent for the synthesis of high dispersive CuNPs by chemical reduction method. In addition, they studied the effect of different parameters like reactant ratio, concentration of copper sulfate, reaction temperature, and dispersant on the size of CuNPs and formation rate. The shape and size of nanoparticles were characterized by scanning electron microscopy. The average particles size of CuNPs with spherical shape was reported to be about 100 nm.

Pure metallic CuNPs at a high concentration were synthesized by the reduction of cupric chloride with hydrazine in the aqueous cetyl trimethylammonium bromide (CTAB) solution (Wu and Chen 2004). They used ammonia solution for the adjustment of solution pH up to 10 and hydrazine as a reducing agent in a capped reaction bottles which were found to be crucial for the synthesis of pure CuNPs. The synthesized nanoparticles were characterized by UV–Vis absorption spectrophotometry, transmission electron microscopy (TEM), electron diffraction pattern, energy dispersive X-ray analysis (EDX), X-ray diffraction (XRD), and X-ray photoelectron spectroscopy (XPS). The nanoparticles were found to be with diameter of 5.1 nm (Wu and Chen 2004).

Because of the strong reducing ability of $NaBH_4$, it is widely used as a reductant for this aqueous reduction process. Based on these findings, Liu et al. (2012) prepared CuNPs by reducing Cu^{2+} ions with $NaBH_4$ in alkaline solution. They also studied the effects of $NaBH_4$ concentration and dripping rate on the formation of CuNPs.

15.2.3 Physical Synthesis

For the preparation of CuNPs, several physical synthesis routes have been described, such as precipitation (Zhu et al. 2004), thermal decomposition (Kim et al. 2006), microemulsion (Lisiecki and Pileni 1993), surfactants solution (Wu and Chen 2004), microwave-assisted techniques (Nakamura et al. 2007b), and vacuum vapor deposition (Liu and Bando 2003).

In 2007, Nakamura and collaborators in their two different studies produced Ag core-Cu shell nanoparticles (denoted as Cu/Ag nanoparticles) having 10–40 nm in size by a microwave (MW)-assisted alcohol reduction process. And CuNPs were synthesized via a microwave-assisted alcohol reduction process with or without the surface plasmon absorption using copper (II) octanoate and copper (II) myristate. These nanoparticles were found to be in size range of 5–6 nm (with the surface plasmon absorption) and 2–3 nm (without the surface plasmon absorption) (Nakamura et al. 2007a, b).

In another similar study, Blosi et al. (2011) synthesized colloidal CuNPs using microwave heating by a polyol method that exploits the chelating and reducing power of a polidentate alcohol (diethylenglycol). The synthesis was performed by using eco-friendly additives, for example, ascorbic acid (reducing agent) and polyvinylpyrrolidone (chelating polymer) to improve the reduction kinetics and sols stability. The synthesized nanoparticles were reported to be stable for several months in spite of the high metal concentration.

15.3 Methods of Toxicity Evaluation

Apart from their potential use, these nanomaterials are striking both basic science and technology. The increasing human exposure to them together with their distinct properties requires development and validation of nanobased product along with their predictive toxicity, and hazard capabilities. The increasing human exposure to them together with their distinct properties requires development and validation of nanobased product along with their predictive toxicity, and hazard capabilities.

These methods must assess the possible benefits to possible risks and health hazards associated with exposure to nanomaterials as they are used in manufacturing and medicine. Most common methods of toxicity evaluation are MTT Assay, Neutral Red Assay, LDH Assay, Comet Assay, ROS Assay, Chlorophyll fluorescence, and AlgaToxKit assay.

15.3.1 MTT Assay [3-(4,5-Dimethylthiazol-2-yl)-2,5-Diphenyltetrazolium Bromide, a Yellow Tetrazole]

The MTT tetrazolium salt colorimetric assay measures cytotoxicity and cell proliferation (Mosmann 1983). In this assay, the tetrazolium salt 3-(4,5-dimethylthiazol-2-yl)-2,5-diphenyltetrazolium bromide (MTT) is taken up into the cells. It is then reduced to an insoluble purple formazan by cleavage of the tetrazolium ring by succinate dehydrogenase within the mitochondria. The resulting formazan product accumulates in cells as it cannot pass through the plasma membrane. The MTT assay was tested for its validity in various cell lines (Mosmann 1983).

This method was further explored to embrace its application to the measurement of cell activation. The level of MTT cleavage by viable cells of several origins was found to be directly correlated to the number of viable cells. In this method, the cell are plated into the 96-well microplates and incubated in the presence and absence of the test compound for desired time period. Then, MTT is added and after a short time period the medium is removed from the well. It is then followed by addition of dimethyl sulfoxide (DMSO), which dissolves MTT. The amount of MTT dissolved is determined using spectrophotometer at 560 nm, and the results are presented as percentage of cell survival.

A new modification of this method is established in the form of XTT. It is 2,3-bis (2-methoxy-4-nitro-5-sulfophenyl)-5-[(phenylamino)carbonyl]-2H-tetrazolium hydroxide. XTT is metabolized to a water soluble formazan product and thus eliminates the solubilization step required for MTT assay.

Hernández-Ortiz et al. (2012) proved by MTT assay that as the cell exposure time to nanoparticle increases, cell viability decreases. The 2D cell cultures do not effectively represent the widespread cell–cell, cell–matrix interaction and also different transport circumstances. Hence testing cytotoxicity in 2D cell culture is not an accurate method to evaluate actual toxicity of nanoparticles. To attain more satisfactory and detailed information, Lee et al. (2009) introduced 3D-spheroid-culture-based NP toxicology testing system. They verified toxicity of CdTe and Au nanoparticles in both 2D and 3D spheroid cultures. In which it proved that toxic effects are significantly reduced in the 3D spheroid culture when compared to the 2D culture system.

Advantages:

1. It is relatively simple, rapid, sensitive, and cheaper assay.
2. The use of ELISA reader allows the processing of large number of samples.

3. It could be used to monitor cell activation independently of proliferative activity (Mosmann 1983).

Disadvantages:

1. The results are sensitive to environmental conditions.
2. The assay depends on cell metabolism for the formation of formazan. The background use to be variable due to protein precipitation on adding an organic solvent like DMSO to dissolve the blue formazan product.
3. The resulting product has low solubility (Denizot and Lang 1986).
4. In particular copper nanoparticles interfere the results of MTT assay (Jose et al. 2011).

15.3.2 Neutral Red Assay

The Neutral Red Assay is also used to measure cell viability. It has been used as an indicator of cytotoxicity in cultures of primary hepatocytes (Fautz et al. 1991) and other cell lines (Morgan et al. 1991). The Neutral Red Assay is based on the protocol described by Borenfreund and Puerner (1984). It determines the accumulation of the neutral red dye in the lysosomes of viable, uninjured cells. Neutral red is the marker of cell viability. It has been shown to be selectively retained by the lysosomes of living cells because of the differential pH between the inside of the lysosome and the surrounding cytoplasm. The amount of neutral red taken up by the population of cells is directly proportional to the number of viable cells in the culture (Barstad et al. 1991).

In this method, the cells are seeded in 96-well plates and are treated by the compound for appropriate time period. Then plates are incubated for 2 h with neutral red containing medium. The cells are subsequently washed and the dye is extracted in each well. The absorbance of neutral red extract is to be measured at 540 nm (Repetto et al. 2008).

The test material-induced cytotoxicity is measured over a wide range of concentrations, and the concentration yielding a 50 % reduction in neutral red uptake is used as the measure for comparison between the test materials. The "percentage of untreated control" values versus the test agent concentrations are plotted, and determine the 50 % inhibition of neutral red uptake by the cells.

Advantage:

1. This assay is cheaper and more sensitive than other cytotoxicity tests.

Disadvantages:

1. Prolonged exposure of the cells to the fixative can result in discharge of the dye into the solution.
2. They may get interfered by precipitation of neutral red reaction medium.

3. The test materials that have absorbance maxima near 540 nm may interfere the assay design and results (Johnson et al. 2009).

15.3.3 Lactate Dehydrogenase (LDH) Assay

The LDH leakage assay is based on the measurement of lactate dehydrogenases activity in the extracellular medium. In this assay lactate is converted into pyruvate in the presence of LDH with parallel reduction of nicotinamide adenine dinucleotide (NAD) to NADH. This reaction thus results in a change in absorbance at 340 nm. Thus, LDH released into the media acts as a marker of dead cells. Reliability, speed, and simple evaluation are some of the characteristics of this assay (Decker and Lohmann-Matthes 1988). It is widely accepted method as a marker of cell death, however, it should be noted that this test signifies only the cell membrane integrity (Hillegass et al. 2010).

Advantage:

1. It is among the most sensitive nondestructive assay, mainly used for short-term exposure studies (Bopp and Lettieri 2008).

Disadvantage:

1. The uncertainties in this assay increase at lower compound concentrations (Bopp and Lettieri 2008).

15.3.4 Comet Assay

Comet assay is commonly used to assess the genotoxic potential of chemicals and environmental contaminants. This method is also called as single cell gel assay (SCG) and microgel electrophoresis (MGE). It was firstly established by Ostling and Johanson (1984) as a technique to directly visualize DNA damage in individual cells. In this method, individual cells are encapsulated in a thin layer of low melting point agarose gel on a microscope slide. These cells are then lysed followed by the electrophoresis of DNA. Under the electric charge, intact, unbroken DNA travels modestly due to its large size, but if present, small DNA fragments migrate much further resulting in a comet shape with an extended tail, containing damaged DNA towards the anode. The DNA is detected by staining ethidium bromide or propidium iodide staining. Here, the length and fluorescence intensity of the comet tail is directly proportional to the amount of DNA damage. It is known as a simple, sensitive, and rapid method for measuring single/double strand breaks in DNA in each cell.

Advantages:

1. The comet assay is a sensitive and rapid method for DNA strand break detection in individual cells (Fairbairn et al. 1995).
2. Small amount of sample is required for the assay.
3. The assay requires relatively short time period (Tice et al. 2000).

Disadvantages:

1. The comet assay requires individual cells. Some cell types like blood cells can be directly used for the study. However, for the other tissues, cells need to be individualized (Cotelle and Ferard 1999).
2. The assay requires viable single-cell suspension. However, if the sample contains the necrotic or apoptotic cells, then it will mislead the results (Olive and Banáth 2006).

15.3.5 Reactive Oxygen Species Assay

A key mechanism thought to be responsible for the genotoxic effects exerted by nanomaterials which comprise oxidative stress, referring to a redox imbalance intracellularly (Singh et al. 2009). It occurs usually as a result of increased intracellular reactive oxygen species (ROS) and decreased antioxidants species. ROS are highly reactive molecules that contain the oxygen atom. They are highly reactive due to the presence of unpaired valence shell electrons. They can disturb the homeostasis of the intracellular environment by reacting unfavorably. During any environmental stress, ROS levels can increase dramatically leading to significant damage to cell structures. The resulting situation is known as oxidative stress. ROS are also produced by exogenous sources such as ionizing radiation.

The expansion of results from in vitro experiments to predict the in vivo toxicity is problematic. It is due to the fact that in vitro exposure conditions usually feature much higher concentration and exposure times than found in the cellular environment in vivo (Marquis et al. 2009).

ROS assay involves various types of assays. Most important of them involves the use of dichlorofluorescin diacetate (DCFDA). It is a widely used fluorescence-based probe for in vitro and in vivo measurement of ROS. This method involves the deacetylation of DCFDA by endogenous esterases to dichlorofluorescein (DCFH), which later reacts with ROS to form the DCF, a fluorophore (Wang and Joseph 1999).

15.3.6 Chlorophyll Fluorescence Assay

The absorbed light energy that is not used for photochemistry of Photosystem II (PSII) can be dissipated by chlorophyll a fluorescence of PSII. As the fluorescence can be used to study the electron transport and energy transfer in photosynthesis, it can also be used to study the effects of pollutants and environmental conditions on photosynthesis. The use of chlorophyll fluorescence at 685 nm as an indicator of toxicity of any product is carried out with an algae (*Scenedesmus obliquus*). This bioassay could be used to detect environmental stress (Caux et al. 1992; Willemsen et al. 1995).

Advantages: This is a simple assay and fast method to measured toxicity products (Zemri et al. 2012).

15.3.7 AlgaToxKit-F

Blaise (1991) analyzed many different test protocols with bacteria, protozoa, microalgae, invertebrates, and fish cell lines and on base five criteria (availability in kitformat, portability, maintenance-free bioindicator, performance in microplates, minimaltraining, and equipment requirement); he concluded that the Toxkit tests are the only types of bioassays that fulfil all these five features. In a special kit called the Algaltox kit F contains all the materials, including the test species *Selenastrum capricornutum* and presently as *Pseudokirchneriella subcapitata* immobilized in algal beads, to perform algal growth inhibition assays, by internationally accepted standard methods using absorbance at 670 nm of chlorophyll for its growth inhibition (http://www.biohidrica.cl/assay_algaltoxkit.htm).

Advantage: All costs associated with maintaining algal cultures are eliminated as the algae are provided in the kit as algal beads. These algal beads can be stored for several months and therefore there no investment is required in culturing facilities and equipment.

15.4 Toxicity of CuNPs

15.4.1 Toxic Effects of CuNPs Administrated In Vitro and In Vivo

Due to many unexpected harmful effects of nanomaterials observed globally, now-a-days the nanoparticle toxicity is considered a major concern. The growing use of nanoparticles leads to the release and accumulation of those particles in soil, air, and aquatic environment. Therefore, evaluation of nanotoxicity to the ecosystem is a challenging task. There are many studies currently in process; however, the present knowledge about their ecotoxicity is still inadequate. Consequently, like other nanoparticles, CuNPs also have the toxic effects on different environmental systems. It is necessary to study the toxic effects of CuNPs on different organisms of various habitats. Therefore, in the next section, a briefly discussion about the harmful effects of CuNPs on plants, mice, fishes, and worms have been incorporated.

15.4.2 *Toxicity of CuNPs to Plants*

Copper is a well-known essential micronutrient for virtually all plants. Therefore, it has been widely used for such purpose. After its entry into the environment, it can eventually get oxidized to form copper oxide (Lide 1994). Otherwise, the CuNPs can be formed naturally at the plant root–soil interface (Manceau et al. 2008). In fact, a condition may arise where CuNPs concentration will rise over certain toxic level. Higher plants interact strongly with their surrounding environment and hence they have the risk to get affected by the exposure of nanoparticles. Although, there has been growing amount of research based on the toxicity of nanoparticles to animal and bacteria, very limited study has been performed by using higher plants. The nanotoxicological studies on plants will therefore help us to understand the flow of nanoparticles leading to their accumulation in food chain.

CuNPs are insoluble in water and therefore, there are certain limitations in their use for toxicity testing assays as those assays require different reagents made in aqueous base. In order to study the toxic effects and bioaccumulation of CuNPs on plant seedling, a plant agar test can be performed for homogenous exposure of CuNPs onto the plant *Phaseolus radiatus* (mung bean) and *Triticum aestivum* (wheat) (Lee et al. 2008). The plant agar is soft gel which permits the dispersion and thereby avoiding precipitation of nanoparticles. The study reported the toxicity of CuNPs at 335 and 570 mg/L, respectively, against *P. radiatus* and *T. aestivum*. These effects were only due to the CuNPs and the contribution of copper ions to the toxicity was negligible. The authors concluded that the phytotoxicity and bioaccumulation of CuNPs in these experimental plant species was dose-dependent. CuNPs reduce the length of root emerging from plant seeds (Stampoulis et al. 2009). This study was performed on a terrestrial plant *Cucurbita pepo* (zucchini) grown in hydroponic solution.

15.4.3 *Toxicity of CuNPs to Animals*

15.4.3.1 CuNPs Toxicity to Mice

Exposure of nanoparticles through gastrointestinal tract can lead to various harmful effects (Jani et al. 1994). However, there are very limited reports based on such aspect of copper nanoparticle toxicity. Mice orally exposed to copper nanoparticles showed acute toxicological effects and heavy injuries in kidney, liver, and spleen (Chen et al. 2006). Furthermore, these effects were sex-dependent, where a male mouse has more toxic effects compared to female mice. Similar results were obtained in mice exposed through nasal route by CuNPs. Inhalation exposure of CuNPs (23.5 nm) at very high dose was reported to get translocate to metabolically active body organs like liver and kidney. Inhalation of high dose (40 mg/kg) of CuNPs by mice caused a hydropic degeneration around the central vein, spotty necrosis of hepatocytes, swelling in glomerulus, and severe lesions in olfactory system. It further resulted into the reduced body weight (Liu et al. 2009). This study demonstrates that the high dose of CuNPs could cause harmful effects to the viscera of the exposed mice.

From these studies it is imperative to understand the reasons why CuNPs induce toxic effects to liver and kidney. The answer lies within the activity of these organs. The homeostasis of copper ions is retained in vivo (Jesse and Mary 2004). They are metabolized in the liver where they are transferred to reduced glutathione (GSH). Whenever there is excess of copper in the body, the GSH depletion consequently gives the enhanced toxicity (Freedman et al. 1989; Steinebach and Wolterbeek 1994). Copper ions in excess are removed from the body through kidney (Turnlund et al. 1997). Since the CuNPs have strong ionization potential, increase in their amount leads to excess release of their ions. These copper ions then cause renal inflammation, which results into the reduced glomerular filtration rate. In another study, Meng et al. (2007) correlated the ultrahigh reactivity of nanocopper with the particle size/specific surface area. They showed that as compared to microcopper (17 μm), nanocopper (23.5 nm) interacts rapidly with artificial gastric acid juice. It leads to the formation of ionic copper with ultrahigh reactivity. Additionally, the study also reported the metabolic alkalosis and accumulation of copper in mice kidneys after oral exposure of CuNPs.

15.4.3.2 CuNPs Toxicity to Fish

CuNPs after its dissolution in water, probably release its particles in aqueous environment. Therefore, its dissolution in water is a major concern. As some dissolution of metal particles in water will occur, it is essential to discriminate that whether the toxic effects are due to nanoparticles themselves or if they are due to dissolved metals. Owing to this problem, Griffitt et al. (2007) have designed their experiment by studying the acute toxicity of soluble copper and 80 nm CuNP suspensions on *Danio rerio* (Zebrafish) embryo. The study concluded that the CuNP has acute toxic effect at 1.5 mg/L. The study reported that soluble copper as well as nanocopper induced drastic changes in the gill morphology (see Fig. 15.1). The activity of Na^+/K^+ pump, found in gill region, was prevented by the dissolved copper (Li et al. 1998). Therefore, the gills of zebrafish are prime targets of CuNPs, which result in severe injury to gills. These results were further confirmed by fact that CuNPs significantly induced the hypoxia-inducible factor 1 (HIF-1), heat shock protein 70 (HSP-70), and copper transport regulatory protein (CTR) genes. Exposure of embryonic zebrafish by CuNPs causes the morphological

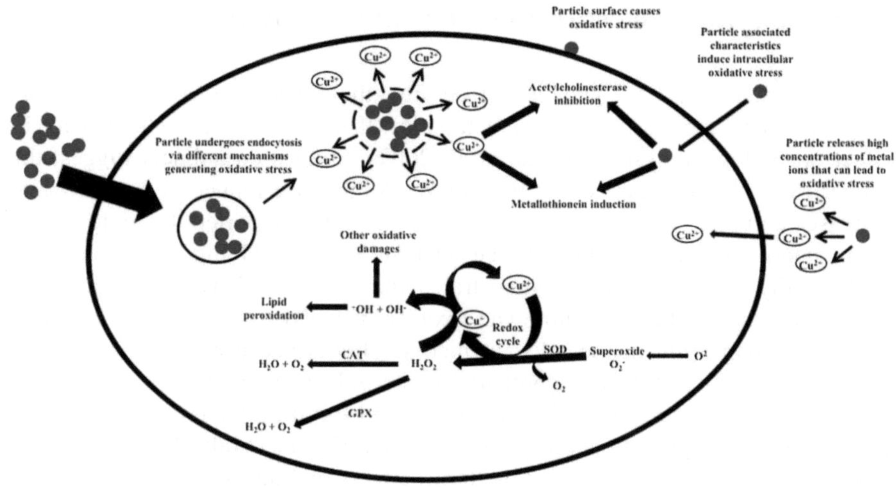

Fig. 15.1 Schematic representation showing the various pathways by which copper nanoparticles can induce oxidative stress on the cell (reprinted with permission from Gomes et al. 2011, copyright (2011) American Chemical Society)

malformation of the larvae (Bai et al. 2010). At some elevated concentrations (0.1 mg/L), the CuNPs even kill the zebrafish embryos in their gastrula stage.

15.4.3.3 CuNPs Toxicity to Worms

There are very few studies that reported toxic effects of CuNPs to the terrestrial environment. Unrine et al. (2010) studied the correlation between harmful effects of CuNPs on the earthworm *Eisenia fetida*. They exposed the worm to CuNPs, in a series of concentrations, and studied its effect on growth, mortality, reproduction, and expression of various genes associated with metal homeostasis, general stress, and oxidative stress. The study reported that the CuNPs up to 65 mg/kg caused no adverse effects on the ecologically relevant endpoints. The study further extends that copper ions and CuNPs were taken up by the earthworms. Therefore, this study suggests that CuNPs may enter the food chains from soil; however, it would be harmful to earthworm only at higher concentrations. In another report, Amorim and Scott-Fordsmand (2012) studied the survival, reproductive output, and avoidance behavior of CuNPs on *Enchytraeus albidus*. The study indicated that CuNPs cause a reduced reproductive output, as well as high avoidance.

15.5 Size and Concentration-Dependent Toxicity

As nanoparticles have the greater surface area, they have distinctive physical and chemical characteristics. These characteristics, therefore, influence their behavior effect on biological systems (Zhao et al. 2007). Studies on animals suggest that copper exerts toxicity depending on its size. As the size reduces, copper toxicity increases abruptly. The study led by Chen et al. (2006) reported the toxic dose of ionic, nano-, and microcopper to be 110, 413, and 500 mg/kg, respectively. Additionally, this study also demonstrated that CuNPs release the toxic cupric ions to the surrounding medium. The toxicity of CuNPs is thought to be due to release of copper ion in the relevant media. This is because nano-sized copper releases more copper ions as compared to its bulk counterpart. This effect was confirmed by higher degree of DNA damage and significant cell death in the cells exposed to nanocopper, in comparison with microCuNPs (Midander et al. 2009).

Prabhu et al. (2010) compared the toxic effects of CuNPs on neuron of dorsal root ganglion (DRG), at concentrations from 10 to 100 μM. The study reported that exposure of CuNPs at 10–20 μM concentrations did not cause any morphological changes in the neuron, whereas the concentration in the range from 40 to 100 μM showed the harmful effect like shrinkage of cell, generation of vacuoles in cell cytoplasm.

Owing to the smaller size, compared to microsized CuNPs (17 μm), nanosized (23.5 nm) CuNPs interact fast with artificial gastric acid juice and thereby release cupric ion, having a very high activity, causing the metabolic alkalosis (Chen et al. 2006). CuNPs also get accumulated in the plant system as consequence of exposure to them. Their bioaccumulation was reported to increase with the increase in exposure concentration (Lee et al. 2008). Similar results were obtained by Kim et al. (2011), where the authors demonstrated that increasing doses of instilled CuNPs produced an increasing concentration of copper, measured in Bronchoalveolar lavage (BAL) fluids of mice lung. All these studies therefore, suggest that the concentration and dose of CuNPs play an important role in determining their level of toxicity. Higher CuNPs concentrations, higher would be their toxic effect. Moreover, smaller size CuNPs exert greater harmful effects to the exposed biological system.

15.6 Mechanism of Copper Nanoparticle Toxicity

As that of any nanomaterial, CuNPs have potential effects on plants, animals, and even microbes. However, very limited studies have been performed till date on deciphering the mechanisms of copper nanoparticle toxicity. Understanding the mechanisms of causing such toxicity is very important in order to develop nanoparticles for a targeted application. This scientific literature is a primary source of information for understanding the mechanisms of copper nanoparticle toxicity.

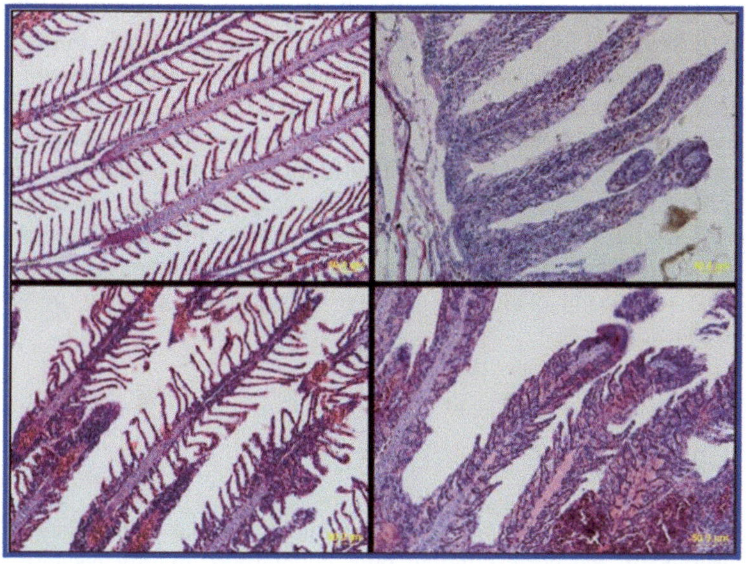

Fig. 15.2 Micrographs showing gill injury induced by 48 h copper exposure. Soluble copper and nanocopper induced dramatic changes in gill morphology. Clockwise from *top left*: control, 0.25 mg/L soluble Cu^{2+}, 1.5 mg/L nanocopper, 0.25 mg/L nanocopper (reprinted with permission from Griffitt et al. 2007)

Exposing cells with CuNPs also increased the level of prostaglandin. Some proinflammatory mediators have impact on the restrictive characteristics of blood–brain barrier. After exposure to CuNPs, cells also respond by elevated production of these proinflammatory mediators like TNF-α and IL-1β (Trickler et al. 2012). Furthermore, like any other metal nanoparticles, CuNPs interact with cell membranes. This interaction can damage plasma membranes (Minocha and Mumper 2012). The copper nanoparticles might interact with the –SH groups of the protein thereby leading to the denaturation of the protein contents of cell membranes. Additionally, it is also reported that the exposure of cells to transition metals results in the disturbance of intracellular redox status or alteration of protein conformation and thereby leading to the inhibition of protein function (Kawata and Suzuki 1983; Li et al. 1994).

Apoptosis includes the series of molecular mechanisms which can lead to death of aged or diseased cell. This process includes various biochemical changes like caspase activation, DNA fragmentation, etc. (Shimizu et al. 2004). The apoptosis can be executed by various pathways and can be induced by any kind of stress (Rastogi et al. 2009). In this context, Sarkar et al. (2011) reported that CuNPs alter the levels of oxidative stress by increasing the production of ROS and reactive nitrogen species (RNS). Several studies have suggested that oxidative stress can induce the lipid peroxidation of the mitochondrial membrane, which can ultimately result into the decoupling of oxidative phosphorylation, interference of electron transport, and a reduction in mitochondrial membrane potential (Freedman et al. 1989; Saris and Skulskii 1991; Mattie and Freedman 2001). Further studies also revealed that the nanocopper exposure induces ROS through various mechanisms as shown in Fig. 15.2 (Gomes et al. 2011). It also affects the mitochondrial membrane potential following the release of cytochrome c from mitochondria to cytosol. The exposure finally results in the induction of apoptosis through the activation of caspase 3 pathway, caspase 8, Fas, and tBid (Sarkar et al. 2011). Earlier to this study, Prabhu et al. (2010) while studying the effect of CuNPs on somatosensory neuron of rat, showed the reduction of mitochondrial activity by interfering with the activity of reductase enzyme presented in mitochondria.

15.7 Conclusion

Nanoparticles exhibit novel properties and functions that differ distinctly from those observed in the corresponding bulk counterpart, mainly because of their small size, large surface area, solubility, shape, and aggregation. There are concerns that the same properties that make nanoparticles so unique could also be responsible for their potential toxicity.

In view of increasing uses of CuNPs, the potential for adverse health effects due to prolonged exposure at various concentration levels in biotic components (plants, mice, fishes, and worms) and abiotic components of the ecosystem (air, water, and soil) have not yet been established. Moreover, the environmental impact of CuNPs is expected to increase considerably in the coming years. The behavior of the CuNPs inside the cells is still a mystery and metabolic pathway interference or immunological responses induced by the CuNPs are yet to be understood completely. Cytotoxic and genotoxic studies will take up this challenge to decipher the molecular events that could regulate bioaccumulation of CuNPs by prying the normal metabolic pathway. The importance of metal nanoparticles to our well-being is beyond argument; however, their potential adverse impacts need to be studied in more detail. Nanotoxicology as a new discipline will make significant contribution to the development of a sustainable and safe nanotechnology.

15.8 Future Perspective

Although considerable progress has been made in recent years in terms of studies based on cytotoxic and genotoxic effects of CuNPs, some key questions are yet to be answered like:

- In what extent the ions released from the nanoparticles are responsible for toxic effects?

- What are the effects of size, shape, and agglomeration of CuNPs on their toxicity?
- What are the accumulation studies and interference with normal metabolic activity?
- The necessity to develop in vitro test for risk assessment, etc.

The expanding uses of CuNPs in commercial products and their increase in consumption, makes obligatory for the understanding of genotoxic and cytotoxic studies of CuNPs. To address this issue, there is a need for concerted efforts from industry, academia, government, and public areas. Their combined efforts for initiating research and public communication at the same time for improving risk perception of the public towards the CuNPs would be helpful. In this context, academia can focus on basic research whereas government and industry can provide monetary and infrastructure support. The findings of such research will be later communicated to the public to make them aware about the cytotoxic and genotoxic effects of CuNPs in the product if any.

Acknowledgments Support from FAPESP, CNPq, Brazilian Network on Nanotoxicology (MCTI/CNPq), INOMAT (MCTI/CNPq), and NanoBioss (MCTI) are acknowledged. IRG thanks Council of Scientific and Industrial Research, New Delhi, India for providing Junior Research Fellowship (CSIR-09/996(001)/2009-EMR-I).

References

Amorim MJ, Scott-Fordsmand JJ (2012) Toxicity of copper nanoparticles and $CuCl_2$ salt to *Enchytraeus albidus* worms: survival, reproduction and avoidance responses. Environ Pollut 164:164–168

Bai W, Tian W, Zhang Z, He X, Ma Y, Liu N, Chai Z (2010) Effects of copper nanoparticles on the development of zebrafish embryos. J Nanosci Nanotechnol 10(12):8670–8676

Barrabés N, Just J, Dafinov A et al (2006) Catalytic reduction of nitrate on Pt-Cu and Pd-Cu on active carbon using continuous reactor: the effect of copper nanoparticles. Appl Catal B Environ 62:77–85

Barstad R, Cortesi J, Janus J (1991) Use of Clonetics' neutral red bioassay to optimize components of serum-free medium for normal human anchorage-dependent cells. In Vitro Cell Dev Biol 27:160–165

Blaise C (1991) Microbiotests in aquatic ecotoxicology: characteristics, utility, and prospects. Environ Toxicol Water Qual 6:145–155

Blosi M, Albonetti S, Dondi M et al (2011) Microwave-assisted polyol synthesis of Cu nanoparticles. J Nanopart Res 13:127–138

Bopp SK, Lettieri T (2008) Comparison of four different colorimetric and fluorometric cytotoxicity assays in a zebrafish liver cell line. BMC Pharmacol 8:8. doi:10.1186/1471-2210-8-8

Borenfreund E, Puerner JA (1984) A simple quantitative procedure using monolayer culture for toxicity assays. J Tissue Cult Methods 9:7–9

Caux PY, Blaise C, Le Blanc P et al (1992) A phytoassay procedure using fluorescence induction. Environ Toxicol Chem 11:549–557

Chatterjee AK, Sarkar RK, Chattopadhyay AP et al (2012) A simple robust method for synthesis of metallic copper nanoparticles of high antibacterial potency against *E. coli*. Nanotechnology 23:085103

Chen CC, Hong FC (2005) Structure and properties of diamond-like carbon nanocomposite films containing copper nanoparticles. Appl Surf Sci 242:261–269

Chen Z, Meng H, Xing G et al (2006) Acute toxicological effects of copper nanoparticles *in vivo*. Toxicol Lett 163:109–120

Cioffi N, Ditaranto N, Torsi L et al (2005) Analytical characterization of bioactive fluoropolymer ultra-thin coatings modified by copper nanoparticles. Anal Bioanal Chem 381:607–616

Colvin VL (2003) The potential environmental impact of engineered nanomaterials. Nat Biotechnol 21:1166–1170

Cotelle S, Ferard JF (1999) Comet assay in genetic ecotoxicology: a review. Environ Mol Mutagen 35:206–221

Decker T, Lohmann-Matthes M (1988) A quick and simple method for the quantitation of lactate dehydrogenase release in measurements of cellular cytotoxicity and tumor necrosis factor (TNF) activity. J Immunol Methods 15:61–69

Denizot F, Lang R (1986) Rapid colorimetric assay for cell growth and survival. modifications to the tetrazolium dye procedure giving improved sensitivity and reliability. J Immunol Methods 89(2):271–277

Dowling AP (2004) Development of nanotechnologies. Mat Today 7:30–35

Fairbairn DW, Olive PL, O'Neill KL (1995) The comet assay: a comprehensive review. Mutat Res 339:37–59

Fautz R, Husein H, Hechenberger C (1991) Application of the neutral red assay (NR Assay) to monolayer cultures of primary hepatocytes: rapid colorimetric viability determination for the unscheduled DNA synthesis test (UDS). Mutat Res 253:173–179

Freedman JH, Ciriola MR, Peisach J (1989) The role of glutathione in copper metabolism and toxicity. J Biol Chem 264:5598–5605

Gomes T, Pinheiro JP, Cancio I, Pereira CG, Cardoso C, Bebianno MJ (2011) Effects of copper nanoparticles exposure in the mussel *Mytilus galloprovincialis*. Environ Sci Technol 45 (21):9356–9362

Griffitt RJ, Weil R, Hyndman KA et al (2007) Exposure to copper nanoparticles causes gill injury and acute lethality in zebrafish (*Danio rerio*). Environ Sci Technol 41:8178–8186

Harne S, Sharma A, Dhaygude M et al (2012) Novel route for rapid biosynthesis of copper nanoparticles using aqueous extract of *Calotropisprocera L.* latex and their cytotoxicity on tumor cells. Colloids Surf B Biointerfaces 95:284–288

Hasan S, Singh S, Parikh RY et al (2008) Bacterial synthesis of copper/copper oxide nanoparticles. J Nanosci Nanotechnol 8:3191–3196

Hernández-Ortiz M, Acosta-Torres LS, Hernández-Padrón G et al (2012) Biocompatibility of crystalline opal nanoparticles. Biomed Eng Online 11:78. doi:10.1186/1475-925X-11-78

Hillegass JM, Shukla A, Lathrop SA et al (2010) Assessing nanotoxicity in cells *in vitro*. Wiley Interdiscip Rev Nanomed Nanobiotechnol 2:219–231

Holsapple MP, Farland WH, Landry TD et al (2005) Research strategies for safety evaluation of nanomaterials, part II: toxicological and safety evaluation of nanomaterials, current challenges and data needs. Toxicol Sci 88:12–17

Huang HH, Yan FQ, Kek YM et al (1997) Synthesis, characterization, and nonlinear optical properties of copper nanoparticles. Langmuir 13:172–175

Jani PU, McCarthy DE, Florence AT (1994) Titanium dioxide (rutile) particle uptake from the rat GI tract and translocation to systemic organs after oral administration. Int J Pharm 105:157–168

Jesse B, Mary RL (2004) Maintaining copper homeostasis: regulation of copper-trafficking proteins in response to copper deficiency or overload. J Nutr Biochem 15:316–322

Johnson MD, Schilz J, Djordjevic MV, Rice JR, Shields PG (2009) Evaluation of *in vitro* assays for assessing the toxicity of cigarette smoke and smokeless tobacco. Cancer Epidemiol Biomarkers Prev 18(12):3263–3304

Jose GP, Santra S, Mandal SK, Sengupta TK (2011) Singlet oxygen mediated DNA degradation by copper nanoparticles: potential towards cytotoxic effect on cancer cells. J Nanobiotechnol 9:9. doi:10.1186/1477-3155-9-9

Kawata M, Suzuki KT (1983) Relation between metal and glutathione concentrations in mouse liver after cadmium, zinc or copper loading. Toxicol Lett 15:131–137

Kim YH, Lee DK, Jo BG et al (2006) Synthesis of oleate capped Cu nanoparticles by thermal decomposition. Colloid Surf A Physiochem Eng Aspects 284–285:364–368

Kim JS, Adamcakova-Dodd A, O'Shaughnessy PT et al (2011) Effects of copper nanoparticle exposure on host defense in a murine pulmonary infection model. Part Fibre Toxicol 8:29. doi:10.1186/1743-8977-8-29

Lee WM, An YJ, Yoon H et al (2008) Toxicity and bioavailability of copper nanoparticles to the terrestrial plants mung bean (*Phaseolus radiatus*) and wheat (*Triticuma estivum*): plant agar test for water-insoluble nanoparticles. Environ Toxicol Chem 27:1915–1921

Lee J, Lilly GD, Doty RC et al (2009) *In vitro* toxicity testing of nanoparticles in 3D cell culture. Small 5:1213–1221

Lee HJ, Lee G, Jang NR et al (2011) Biological synthesis of copper nanoparticles using plant extract. Nanotechnology 1:371–374

Li W, Kagan HM, Chou IN (1994) Alterations in cytoskeletal organization and homeostasis of cellular thiols in cadmium resistant cells. Toxicol Appl Pharmacol 126:114–123

Li J, Quabius ES, Bonga SEW et al (1998) Effects of water-borne copper on branchial chloride cells and Na+/K+-ATPase activities in Mozambiquetilapia (*Oreochromis mossambicus*). Aquat Toxicol 43:1–11

Lide D (1994) CRC handbook of chemistry and physics: a ready-reference book of chemical and physical data. CRC Press, Boca Raton, FL

Lisiecki I, Pileni MP (1993) Synthesis of copper metallic clusters using reverse micelles as microreactors. J Am Chem Soc 115:3887–3896

Liu Z, Bando Y (2003) A novel method for preparing copper nanorods and nanowires. Adv Mater 15:303–305

Liu G, Li X, Qin B et al (2004) Investigation of the mending effect and mechanism of copper nano-particles on a tribologically stressed surface. Tribol Lett 17:961–966

Liu Y, Gao Y, Zhang L et al (2009) Potential health impact on mice after nasal instillation of nano-sized copper particles and their translocation in mice. J Nanosci Nanotechnol 9:6335–6343

Liu Q, Zhou D, Yamamoto Y et al (2012) Preparation of Cu nanoparticles with $NaBH_4$ by aqueous reduction method. Trans Nonferrous Met Soc China 22:117–123

Majumber DR (2012) Bioremediation: copper nanoparticles from electronic-waste. Int J Eng Sci Technol 4:4380–4389

Malathi S, Balasubramanian S (2012) Synthesis of copper nanoparticles and their biomedical applications. Lambet Academic Publishers, Saarbrucken

Manceau A, Nagy K, Marcus M et al (2008) Formation of metallic copper nanoparticles at the soil–root interface. Environ Sci Technol 42:1766–1772

Mandal D, Bolander ME, Mukhopadhyay D et al (2006) The use of microorganisms for the formation of metal nanoparticles and their application. Appl Microbiol Biotechnol 69:485–492

Marquis BJ, Love SA, Braun KL et al (2009) Analytical methods to assess nanoparticle toxicity. Analyst 134:425–439

Mattie MD, Freedman JH (2001) Protective effects of aspirin and vitamin E (alpha-tocopherol) against copper and cadmium-induced toxicity. Biochem Biophys Res Commun 285:921–925

Meng H, Chen Z, Xing G et al (2007) Ultrahigh reactivity provokes nanotoxicity: explanation of oral toxicity of nano-copper particles. Toxicol Lett 175:102–110

Midander K, Cronholm P, Karlsson HL et al (2009) Surface characteristics, copper release, and toxicity of nano- and micrometer-sized copper and copper (II) oxide particles: a cross-disciplinary study. Small 5:389–399

Minocha S, Mumper RJ (2012) Effect of carbon coating on the physico-chemical properties and toxicity of copper and nickel nanoparticles. Small 8(21):3289–3299

Mohanpuria P, Rana NK, Yadav SK (2008) Biosynthesis of nanoparticles: technological concepts and future applications. J Nanopart Res 10:507–517
Morgan CD, Mills KC, Lefkowitz DL et al (1991) An improved colorimetric assay for tumor necrosis factor using WEHI 164 cells cultured on novel microtiter plates. J Immunol Methods 145:259–262
Mosmann T (1983) Rapid colorimetric assay for cellular growth and survival: application to proliferation and cytotoxicity assays. J Immunol Methods 65:55–63
Mukherjee P, Ahmad A, Mandal D et al (2001) Bioreduction of $AuCl_4$ ions by the fungus, *Verticillium sp.* and surface trapping of the gold nanoparticles formed. Angew Chem Int Ed 40:3585–3588
Nakamura T, Tsukahara Y, Sakata T et al (2007a) Preparation of monodispersed Cu nanoparticles by microwave-assisted alcohol reduction. Bull Chem Soc Jpn 80:224–232
Nakamura T, Tsukahara Y, Yamauchi T et al (2007b) Preparation of Ag core–Cu shell nanoparticles by microwave-assisted alcohol reduction process. Chem Lett 36:154–155
Oberdorster G, Oberdorster E, Oberdorster J (2005) An emerging discipline evolving from studies of ultrafine particles. Environ Health Perspect 113:823–839
Olive PL, Banáth JP (2006) The comet assay: a method to measure DNA damage in individual cells. Nat Protoc 1(1):23–29
Ostling O, Johanson KJ (1984) Microelectrophoretic study of radiation-induced DNA damages in individual mammalian cells. Biochem Biophys Res Commun 123:291–298
Prabhu BM, Ali SF, Murdock RC et al (2010) Copper nanoparticles exert size and concentration dependent toxicity on somatosensory neurons of rat. Nanotoxicology 4:150–160
Ramanathan R, Field MR, O'Mullane AP et al (2013) Aqueous phase synthesis of copper nanoparticles: a link between heavy metal resistance and nanoparticle synthesis ability in bacterial systems. Nanoscale 5(6):2300–2306. doi:10.1039/c2nr32887a
Rastogi RP, Richa, Sinha RP (2009) Apoptosis: molecular mechanisms and pathogenicity. EXCLI J 8:155–181
Repetto G, Anadel P, Zurita JL (2008) Neutral red uptake assay for the estimation of cell viability/cytotoxicity. Nat Protoc 3:1125–1131
Saris NE, Skulskii IA (1991) Interaction of copper with mitochondria. Acta Chem Scand 45:1042–1046
Sarkar A, Das J, Manna P et al (2011) Nano-copper induces oxidative stress and apoptosis in kidney via both extrinsic and intrinsic pathways. Toxicology 290:208–217
Shimizu T, Numata T, Okada Y (2004) A role of reactive oxygen species in apoptotic activation of volume-sensitive Cl− channel. Proc Natl Acad Sci U S A 101:6770–6773
Singh N, Manshian B, Jenkins GS et al (2009) NanoGenotoxicology: the DNA damaging potential of engineered nanomaterials. Biomaterials 30:891–3914
Singh AV, Patil R, Anand A, Milani P, Gade WN (2010) Biological synthesis of copper oxide nano particles using *Escherichia coli*. Curr Nanosci 6:365–369
Stampoulis D, Sinha SK, White JC (2009) Assay-dependent phytotoxicity of nanoparticles to plants. Environ Sci Technol 43:9473–9479
Steinebach OM, Wolterbeek HT (1994) Role of cytosolic copper, metallothionein and glutathione in copper toxicity in rat hepatoma tissue culture cells. Toxicology 92:75–90
Tarasov S, Kolubaev A, Belyaev S et al (2002) Study of friction reduction by nanocopper additives to motor oil. Wear 252:63–69
Thakkar KN, Mhatre SS, Parikh RY (2010) Biological synthesis of metallic nanoparticles. Nanomed Nanotechnol Biol Med 6:257–262
Thomas K, Sayre P (2005) Research strategies for safety evaluation of nanomaterials, part I: evaluating the human health implications of exposure to nanoscale materials. Toxicol Sci 87:316–321
Tice RR, Agurell E, Anderson D, Burlinson B, Hartmann A, Kobayashi H, Miyamae Y, Rojas E, Ryu JC, Sasaki YF (2000) Single cell gel/comet assay: guidelines for in vitro and in vivo genetic toxicology testing. Environ Mol Mutagen 35(3):206–221

Trickler WJ, Lantz SM, Schrand AM et al (2012) Effects of copper nanoparticles on rat cerebral microvessel endothelial cells. Nanomedicine (Lond) 7:835–846

Tsuji JS, Maynard AD, Howard PC et al (2006) Research strategies for safety evaluation of nanomaterials, part IV: risk assessment of nanoparticles. Toxicol Sci 89:42–50

Turnlund JR, Scott KC, Peiffer GL et al (1997) Copper status of young men consuming a low-copper diet. Am J Clin Nutr 65:72–78

Unrine JM, Tsyusko OV, Hunyadi SE et al (2010) Effects of particle size on chemical speciation and bioavailability of copper to earthworms (*Eisenia fetida*) exposed to copper nanoparticles. J Environ Qual 39:1942–1953

Valodkar M, Ravirajsinh NJ, Menaka CT et al (2011) Biocompatible synthesis of peptide capped copper nanoparticles and their biological effect on tumor cells. Mater Chem Phys 128:83–89

Varshney R, Bhadauria S, Gaur MS et al (2010) Characterization of copper nanoparticles synthesized by a novel microbiological method. JOM J Miner Met Mater Soc 62:102–104

Varshney R, Bhadauria S, Gaur MS et al (2011) Copper nanoparticles synthesis from electroplating industry effluent. Nano Biomed Eng 3:115–119

Varshney R, Bhadauria S, Gaur MS (2012) A review: biological synthesis of silver and copper nanoparticles. Nano Biomed Eng 4:99–106

Wang H, Joseph JA (1999) Quantifying cellular oxidative stress by dichlorofluorescein assay using microplate reader. Free Radic Biol Med 27(5–6):612–616

Willemsen A, Vaal MA, de Zwart D (1995) Microbiotests as tools for environmental monitoring. Report No. 9, 607,042,005. National Institute of Public Health and Environmental Planning (RIVM), The Netherlands, p 39

Wu SH, Chen DH (2004) Synthesis of high-concentration. Cu nanoparticles in aqueous CTAB solutions. J Colloid Interface Sci 273:165–169

Xiong J, Wu X-D, Xue Q-J (2013) Biomolecule-assisted synthesis of highly stable dispersions of water-soluble copper nanoparticles. J Colloid Interface Sci 390:41–46

Yoon KY, Byeon JH, Park JH et al (2007) Susceptibility constants of *Escherichia coli* and *Bacillus subtilis* to silver and copper nanoparticles. Sci Total Environ 373:572–575

Zemri K, Amar Y, Boutiba Z et al (2012) Use of chlorophyll fluorescence to evaluate the effect of chromium on activity photosystem II at the alga *Scenedesmus obliquus*. Int J Res Rev Appl Sci 12(2):304–314

Zhang XF, Dong XL, Huang H et al (2007) High permittivity from defective carbon-coated Cu nanocapsules. Nanotechnology 18:275701. doi:10.1088/0957-4484/18/27/275701

Zhang QL, Yang ZM, Ding BJ et al (2010) Preparation of copper nanoparticles by chemical reduction method using potassium borohydride. Trans Nonferrous Met Soc China 20: s240–s244

Zhao Y, Meng H, Chen Z et al (2007) Dependence of nanotoxicity on nanoscale characteristics and strategies for reducing and eliminating nanotoxicity. In: Zhao YL, Nalwa HS (eds) Nanotoxicology. American Scientific, Stevenson Ranch, CA

Zhu J, Li K, Chen H et al (2004) Highly dispersed CuO nanoparticles prepared by a novel quick precipitation method. Mater Lett 58:3324–3327

Chapter 16
Assessing the Erythrocyte Toxicity of Nanomaterials: From Current Methods to Biomolecular Surface Chemistry Interactions

Luis A. Visani de Luna, Diego Stéfani T. Martinez, and Oswaldo L. Alves

Abstract The challenge of nanotoxicology implies development of new methods and improvements of current ones to study the interface of nanomaterials and toxicological science. The nanoscale of materials provides new insights regarding their effects to cells, tissues, and organisms. The blood biocompatibility of nanomaterials is in the light of medical applications and hemolysis assay is pointed out as an important tool to bring toxicological information. Cell membrane disruption by nanomaterials is easily detected by free hemoglobin from red blood cells (RBCs), called erythrocytes. Moreover, nanomaterials' physicochemical properties play a key role for understanding erythrocyte toxicity as well as their biomolecular surface chemistry interactions in biological fluids. In this way, this chapter aims to give an overview of nanomaterials toxicity to RBCs, focusing the current methods in nanotoxicology and exploring the advances in bionanointerface.

16.1 Introduction

The nanotechnology has been employed in many industrial activities and emerged as a strategic point in manufactured products. In this way we can consider the development and production of nanomaterials as an emerging activity at concern (Fairbrother and Fairbrother 2009). Then, the toxicological studies of new materials are considered a core study attempting a sustainable production and to understand adverse effects of these materials in nanoscale upon different levels of living world. The diversity of nanomaterial characteristics like the increase in surface area and

D.S.T. Martinez (✉)
Solid State Chemistry Laboratory, Institute of Chemistry, Universidade Estadual de Campinas—UNICAMP, Cidade Universitária "Zeferino Vaz"—Barão Geraldo, Campinas, Sao Paulo 13083-970, Brazil
e-mail: diegostefani.br@gmail.com

N. Durán et al. (eds.), *Nanotoxicology*, Nanomedicine and Nanotoxicology,
DOI 10.1007/978-1-4614-8993-1_16,

chemical reactivity is far more complex than in the *bulk* state (Ashby et al. 2009). From these properties arise new questions of nanomaterials when in contact with biological systems. New research fields were created in order to attend the demands of bio-nanotechnology, such as nanotoxicology and nanomedicine, both involving scientists from interdisciplinary fields like material and biological sciences. The nanotoxicology framework has been dedicated to improve the safety of nanomaterials for human health as well as the environmental quality, providing data for regulation, products labeling, and *know-how* for medical and ecological applications (Oberdorster et al. 2005; Krug and Wick 2011). The nanomedicine framework has been focused in improvements in drug delivery systems (Hu et al. 2011), nanobioconjugates (Garabagiu 2013), imaging contrast via safe fluorescent nanomaterials (Rothen-Rutishauser et al. 2006), biosensors for improved clinical assays (Song and Park 2011), and treatment of diseases (Huh and Kwon 2011).

Blood is a fluid present in animals and delivers substances and oxygen to the cells and transports metabolic waste away from the same cells. The vertebrate's blood is composed of red blood cells (RBCs) called erythrocytes and white blood cells (WBCs) called leukocytes, suspended in liquid called plasma. The plasma components are 55 % blood fluid, whose 92 % is water and proteins, mainly albumin, glucose, ions, hormones, and carbon dioxide (Alberts et al. 2007). The deep knowledge of the interaction of nanomaterials with blood is fundamental for medical application. The erythrocytes are the most abundant cells of blood and are useful as a cytotoxicity model due to the membrane mechanical stability which indicates either the chemical substances or nanomaterials reactivity and their biocompatibility. Gerashchenko et al. (2002) used RBCs for probing the surface properties of silica and concluded that this model provides a basis for a novel system able to examine the surface of other substances of interest. Recently, researches proposed the use of RBCs membrane for coating nanoparticles (NPs) and evade the immune system promoting a longer retention time in bloodstream. This new method would consist of RBCs isolated from the patient, emptied to leave only the cell membranes decorated with patient's proteins for using them as natural NPs carrier and prevent the host's defense mechanism (Zhang et al. 2012b). Taking into account the nanomaterials toxicity, the hemolysis assay is highly recommended as a reliable test for assessing material surface reactivity and biocompatibility (Dobrovolskaia et al. 2008; Zhao et al. 2011a; Kunzmann et al. 2011; Yu et al. 2011). In this sense, the American Society for Testing and Materials (ASTM) has published a standard test method for analysis of hemolytic properties of nanoparticles (ASTM 2008).

16.2 Toxicity of Nanomaterials to Erythrocytes

Nanomaterials' physicochemical properties and relationship with cytotoxicity have been well documented in the literature (Sharifi et al. 2012). Considering the cells as models for toxicity assays, it is generally accepted that NPs toxicity depends on reactivity of the material components, their surface area, and the surface chemistry in contact with the cell, and the NPs' morphology characteristics such as size, shape, surface chemistry, state of aggregation/agglomeration, and the interactions with components of cell culture media must be taken into account as well (Zhao et al. 2011b). For designed nanomaterials used as medical devices the physicochemical characterization and the biocompatibility assessment are crucial steps for insuring the safe devices. When designing nanomaterials the parameters cited above have to be considered in order to prevent toxicity. The size, for example, can affect the biodistribution as well as the uptake by different cells. Quantum dots (QDs) are crystals (~10 nm), shape and size of which affect the quantum properties of their electrons and determine the color of light that the material can emit. These materials can be used to target the histones in the DNA and map proteins and other biological molecules (Valizadeh et al. 2012). The QDs composition is also important, viewing that metal QDs such as cadmium are very toxic. Titanium dioxide NPs have their toxicity based on their crystalline structure and the toxicity of these NPs is related to the reactive oxygen species (ROS) generation promoting inflammation and other adverse effects (Zhang et al. 2012b).

The degradability is another important factor when developing new nanomaterials and with either low or no degradability these NPs can accumulate and cause toxicity. But when NPs are biodegradable the concern is the products generated from these materials which can be toxic. The surface charge of NPs also has a great influence in the cytotoxicity assessment which may cause an oxidative stress leading to cell apoptosis (Zhang et al. 2012a). A summarized data is shown in Table 16.1 when many authors correlated the erythrocyte toxicity (hemolysis) and NPs' characteristics as mentioned above. By analyzing these results reported in the literature, we can observe that the physicochemical properties and dispersibility in biological buffers are crucial determinants for the toxic effects of nanomaterials to RBCs. However, an integrated and systematic understanding of nanomaterial hemolytic structure–activity relationship needs to be addressed (Lu et al. 2009; Zhao et al. 2011a; Zook et al. 2011, 2012).

16.2.1 Silica Case

Silica is one of the most studied elements on Earth and the application of silica NPs in medical practices is extensive. Zhang et al. (2012a) showed that silica NPs with comparable particle size (~16 nm) and hydroxyl coverage have different toxicity profiles. The particle types used in this study were silica NPs synthetized in

Table 16.1 Nanomaterials hemolytic activity assessment reported in literature

Material	Size (nm)	BET ($m^2\ g^{-1}$)	Conc. ($\mu g\ mL^{-1}$)	Hemolysis (%)	References
Amorphous silica	–	–	100	44	Slowing et al. (2009)
Mesoporous silica (Si-OH)	40–80	855	500	60	
Mesoporous silica ($Si\text{-}NH_2$)	40–80	624	500	7.0	Paula et al. (2012)
Mesoporous silica [Si-P (CH_3)O_3H]	40–80	773	500	20	
Silver nanoparticles	30–50	40	700	50	Choi et al. (2011)
Silver microparticles	800–3,000	–	700	12	
Cerium oxide	20–30	24	50	8.0	Lu et al. (2009)
Titanium oxide	5.0	259	50	2.0	
Nickel oxide	10–20	92	50	20	
Alumina	20–30	35	50	2.0	
Cobalt oxide	20–30	24	50	8.0	
Iron oxide	10	–	30	2.8	Creanga et al. (2009)
CdTe quantum dots	2.8	–	120	9.5	Liu and Yu (2010)
CdTe/ZnS quantum dots	3.7	–	120	3.9	
Polystyrene nanoparticles	50	–	500	5.0	Barshtein et al. (2011)
Polystyrene nanoparticles	107	–	500	2.0	
Polystyrene nanoparticles	250	–	500	2.0	
Gold nanoparticles (GNP)	50	–	100	~50	Khullar et al. (2012)
Albumin-GNP	50	–	100	<1.0	
Carbon nanoparticles (CNP)	3–10	–	100	~20	Chandra et al. (2011)
Fluorescein-CNP	3–10	–	100	~8.0	

low-temperature colloidal (e.g., Stöber silica) or high-temperature pyrolysis (e.g., fumed silica) routes. The Stöber silica has low hemolytic activity; on the other hand fumed silica has a robust dose-dependent hemolytic activity. The authors also described that heating fumed silica, the hemolytic activity was reduced, possibly due to the progressive dehydroxylation of its surface. The concentration of hydroxyl groups (silanol groups) and surface defects have intrinsic interactions with cell membranes resulting in different patterns of silica hemolytic effect. Another determinant factor considering the surface reactivity of silica NPs and their toxicity is the ROS production. RBCs are very susceptible to oxidative damage which could lead to hemolysis and fumed silica was considered a potent hydroxyl radicals generator. Regarding the size and the pore ordering of silica NPs, according to Lin and Haynes (2010), mesoporous silica nanoparticles (MS) have lower hemolytic activity than nonporous silica NPs of similar size. The hemolysis effect is size-dependent only when silica NPs have long-range and ordered porous structure. A nonlinear hemolysis curve was observed when working with MS of 25, 42, 93, 155, and 225 nm (Fig. 16.1). The unexpected behavior of MS-25 indicates

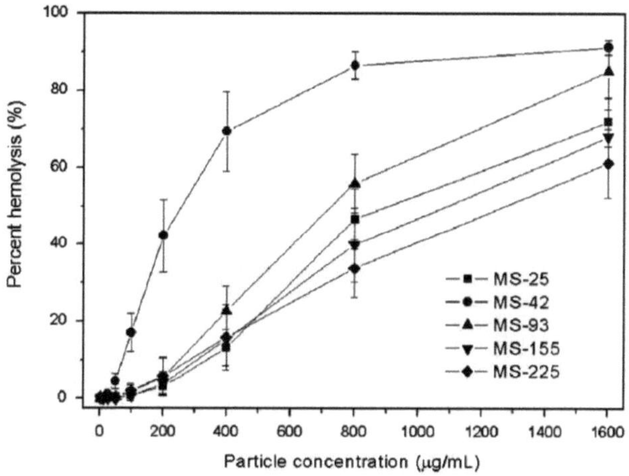

Fig. 16.1 (**a**) Percentage of hemolysis of RBCs in the presence of five sizes of MS NPs at different concentrations ranging from 3.125 to 1,600 μg mL^{-1} for 3 h. Data represent the mean ± SD from at least three independent experiments. Modified from Lin and Haynes (2010) by permission of American Chemical Society

that the size-dependent hemolytic activity stands only for MS which have well-ordered mesoporous structure and the MS-25 probably has lower hemolytic activity due to larger pore size and greater primary pore volume compared to the larger diameter of other MS which reduce the number of cells in contact with silanol groups on MS-25. To further explore the role of mesopore silica and pore stability on RBC toxicity, Lin and Haynes incubated MS silica in PBS (Dulbecco's phosphate-buffered saline, pH = 7.2) solution for 6 days in order to simulate particle modification in body fluids and they observed that this incubation was enough to promote silica NPs' surface changes promoting an increase in particle hemolytic activity. The silica NPs with pore collapsed after PBS aging released silicic acid in solution, and firstly thought as the agent of the hemolytic activity, but after the incubation of dissolved silicic acids with erythrocytes no hemolytic activity was seen. Transmission electron microscopy (TEM) images revealed that incubated NPs had their surface modified (Fig. 16.2). Then, hemolytic activity of MS is strongly correlated to the porous structure, surface groups, and cell-contactable surface area. In this context, our group has recently demonstrated that the hemolytic effect of spherical mesoporous silica nanoparticles is related with particle surface charge. By using specific-surface area in normalized experiments, we observed that particles negatively charged containing silanol or phosphonate groups generated similar hemolytic effect. On the other hand, positively charged particle containing amine groups showed very little toxicity to RCBs (Paula et al. 2012).

16.3 Current Methods and Protocols Used for Assessing the Erythrocyte Toxicity of Nanomaterials

The hemolytic assay provides a fast and practical method to study the effects of nanoparticles on biological membranes since the cell membrane is the first surface which enters in contact with nanomaterials. The first method considering specific

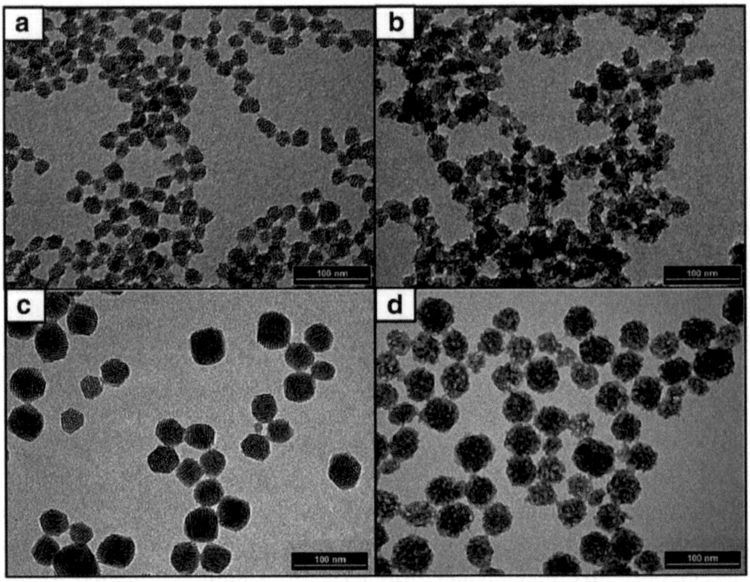

Fig. 16.2 TEM images of surfactant-free (**a**) MS-25 without PBS aging; (**b**) MS-25 after 6-day PBS aging; (**c**) MS-42 without PBS aging; and (**d**) MS-42 after 6-day PBS aging. Reproduced from Lin and Haynes (2010) by permission of American Chemical Society

controls for nanoparticles in the hemolytic assay was published as an ASTM protocol in 2008 (Dobrovolskaia et al. 2008). This method is employed as a preclinical safety testing of nanomaterials and also provides an insight of potential aspects of nanomaterials interference and highlights the importance of previously characterized particles. Limiting aspects of working with hemolytic assay and NPs were highlighted by Choi et al. (2011); among them: the aggregation inside the tube after NPs incubation with DPBS (Dulbecco's phosphate-buffered saline, free Ca^{2+}/Mg^{2+}), the reactivity of NPs with proteins and hemolytic suppression, blood coagulation around particles avoiding RBCs and NPs interaction, and hemoglobin "capture" by NPs after hemolysis reducing the absorbance after centrifuging the samples leading to an underestimated hemolysis effect. For example, Meng et al. (2012) have demonstrated that hemolysis induced by long and short functionalized multiwalled carbon nanotubes was not available to be measured quantitatively in the conventional hemolytic assay due to hemoglobin carbon nanotube surface chemistry interactions. Another toxicological approach has been widely studied to evaluate nanomaterials toxicity, the exerted oxidative stress known to cause cell toxicity, hemolysis, DNA damage, protein denaturation, inflammation, and lipid peroxidation. Extracellular fluids like blood plasma protect against oxidative damage quenching free radicals. Rogers et al. (2008) assessed the capacity of ferric reducing ability of serum (FRAS) assay to screen potential oxidative stress activity of NPs. This method uses ferric ions (Fe^{3+}) and

4,6-tripyridyl-s-triazine (TPTZ) in a buffer solution of acetate mixed with the blood serum dosages and NPs' dosages. The Fe^{3+}-TPTZ complex formed in the solution is reduced by antioxidants from blood serum forming Fe^{2+}-TPTZ, a colored complex measurable at 593 nm. The proposed method used human serum instead of plasma in order to avoid the possible interference of anticoagulants present in the blood sample. This assay is relevant since the oxidative stress generated by nanomaterials may also affect other biological systems.

Predicted toxic effects of nanomaterials are determinant for development of safe NPs; then biomarker studies are required in order to provide valuable information of early adverse effects associated with manufactured nanomaterials in different exposure scenarios. The Hemopexin is a heme-biding plasma glycoprotein produced in the liver against hemoglobin-mediated oxidative damage during intravascular hemolysis. This protein binds free hemoglobin released into the plasma after RBCs lysis. Higashisaka et al. (2012) proved that Hemopexin is a useful biomarker of silica NPs exposure and erythrocyte toxicity using mammalian mice as a model organism. Brazhe et al. (2009) observed the optimized conditions of SERS (surface-enhanced Raman spectroscopy) technique when studying the interaction of NPs with erythrocytes. The phenomenon SERS occurs only when NPs and the molecule studied are close to each other (~15–20 nm) due to plasma resonance and charge transfer effect. This study showed no interference of SERS technique upon living cells when studying NPs–cells interaction. Then, further application of SERS technique can help to elucidate the influence of nanomaterials in living cells.

16.4 Interaction of Biomolecules from Human Blood Plasma with Nanomaterials and the Influence on Erythrocyte Toxicity

The biodistribution aspect of nanomaterials in vivo is a challenge for the design of NPs as a safe drug delivery system (Walkey and Chan 2012). The phenomenon of biomolecules adsorption onto NPs surface consequently gives origin to a particle corona complex (Lundqvist et al. 2008). In biological fluid, proteins can associate with NPs and the characteristics of this association lead to a protein "corona" that largely defines the biological identity of the particle and its interaction with cells (Monopoli et al. 2012). Labarre et al. (2005) examined the in vitro interactions of core–shell poly(isobutylcyanoacrylate)-polysaccharide-synthetized NPs with blood proteins and the potential of these NPs to evade the immunosystem via NP–C3 protein interaction. The C3 protein is present in blood plasma and it is involved in the nonspecific recognition, like opsonization of foreign bodies. The NPs were synthetized via a polymerization mechanism (Redox Radical—RAD and Anionic—An) in the presence of Dextran (Dex), Heparin, and Dextran plus Heparin, resulting in three different spatial arrangements of polysaccharide chains at NP surface (Loops, Trains, and Brush). Considering the hydrodynamic diameters and the shell characteristics of synthetized NPs, the NP-Dex 71 kDa were larger than NP-15 kDa and NPs prepared from Heparin were the smallest ones. A detailed analysis was performed by using a two-dimensional polyacrylamide gel electrophoresis (2D PAGE) technique in order to identify the absorbed serum proteins onto the mentioned NPs surface, previously incubated in citrated serum. The C3 protein was found strongly adsorbed onto NPs surface avoiding the complement activation and possible particle opsonization. Taking into account the protein adsorption onto NPs surface, Dex 71 kDa-RAD was more effective protecting its surface, in the Dex 15 kDa-RAD was found twice the amount of proteins desorbed from its surface, and in Dex 15 kDa-An was found four times more proteins than its Dex counterpart.

The steric repulsion is responsible for avoiding protein interaction with NPs surface and the longest chains were more effective in this repulsion. The fibrinogen was the predominant protein found adsorbed onto NPs Dex 15-RAD/An surfaces, but was hardly detected in Dex 71-RAD surface. Fibrinogen is a protein abundant in blood plasma and is associated with coagulation and platelet aggregation. No correlation was established between the adsorption patterns of the plasma proteins and NPs' physicochemical properties. Moreover, the adsorption phenomena resulted from a combination of steric repulsion, ionic interactions, and the affinity between proteins and the polysaccharides.

The mechanism of plasma protection from NPs-induced hemolysis was investigated by Shi et al. (2012). Silica NPs were incubated with 3 and 10 % human plasma for 1 h and then washed three times and re-suspended in saline buffer (pH 7.4) in order to assure the protein corona formation. The protein content in NPs was determined by a commercial protein assay kit. The hemolytic activity of silica NPs was firstly performed in different temperatures; 37 °C to verify the human body environment and 40 °C to verify the behavior of NPs in fever conditions. The hemolysis activity of silica NPs increased in high temperatures showing that hemolytic activity is largely dependent on temperature. The author also considered the alkalosis conditions of human blood which could decrease the hemolytic activity of silica due to changes in the particle surface in high pH values, particularly silanol groups, because the ionization of these groups (SiO^-) in pH close to 9 provokes electrostatic repulsion between RBCs and NPs. The corona complex in silica NPs inhibited significantly the hemolytic activity. In this case, the covered NP in a protein layer was responsible for suppressing the hemolytic activity confirmed by phenol extraction of protein corona which restored the potential hemolytic activity of silica NPs. The in vitro findings of the "masking effect" promoted by corona show changes in the interactions of NPs and cells and illustrate potential clinical benefits of coated particles due to the enhanced biocompatibility. Paula et al. (2012) also observed the erythrocyte toxicity suppression by protein corona acting as shield, isolating the microchemical environment of the silica nanoparticles regardless of their surface charge (Fig. 16.3).

According to Dobrovolskaia et al. (2009), when in bloodstream, NPs never interact "naked" with tissues and cells. It means that most of the NPs are quickly coated by serum proteins which give origin to either soft or hard corona. When

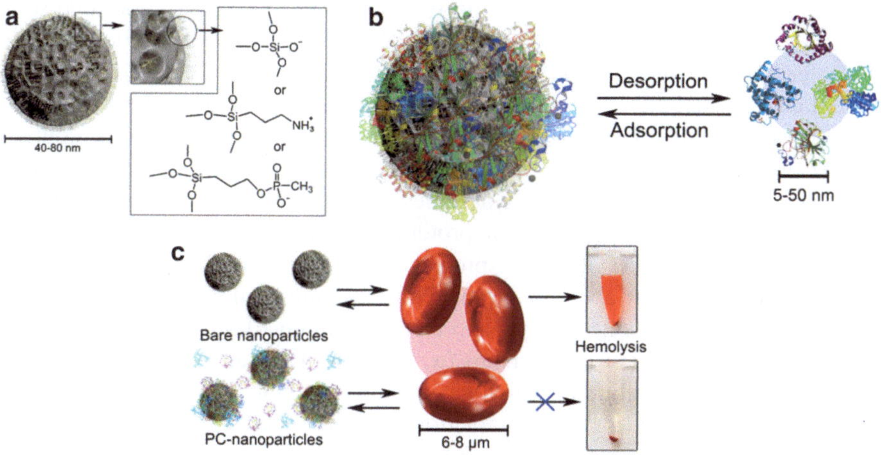

Fig. 16.3 (**a**) Schematic diagram representing a silica nanoparticle functionalized with silanol, propylamine, and propylmethylphosphonate. (**b**) Dynamic interaction of nanoparticles with proteins of human blood plasma and the "corona effect." (**c**) Hemolytic effect induced by bare porous silica nanoparticles or by protein-coated (PC) nanoparticle (objects in diagram are not in scale). Modified from Paula et al. (2012) by JBCS

studying the interaction of gold NPs with human blood Dobrovolskaia and co-workers highlighted the main technique assemblages in order to examine the molecular identities present in plasma protein which can bind the colloidal NPs and possible changes in the agglomeration state and particle size. In this way the main techniques selected for this study are described below.

- Transmission electron microscopy (TEM): to assess the particle dispersion in atomic scale and visually interpret it. However, protein and other biological compounds are invisible without adequate metal staining procedures.
- Atomic force microscopy (AFM): provides information about colloid size and agglomeration of nanoparticles, but it requires dried samples and presents low resolution of flexible compounds such as proteins.
- Dynamic light scattering (DLS): measures the hydrodynamic diameter of nanoparticles and is very sensitive to biological molecules like polymers, proteins, and antibodies.
- Two-dimensional polyacrylamide gel electrophoresis (2D PAGE): isolates proteins attached to nanoparticles.
- Mass spectrometry (MS): used to identify the proteins attached to the nanoparticles surface.

Two sizes of gold NPs were selected (30 and 50 nm) and incubated with plasma; then, the hydrodynamic size was evaluated. After plasma incubation and clean-up step both NPs increased their hydrodynamic size to 76.1 and 100 nm, respectively. The changes in particle size and agglomeration state can affect the uptake from bloodstream and its fate inside the body. The protein-coat stabilized the gold NPs by steric hindrance and no aggregation was observed in PBS solution, otherwise when digested with trypsin enzyme, a black solution was observed in PBS solution, characteristic of aggregated gold NPs. The surface charge of NPs changed after protein coating and NPs became less negative which may modify the interactions with other biomolecules, cells, and tissues. Although human plasma contains ~3,700 proteins (Omenn et al. 2007), the most common protein found to bind gold NPs was fibrinogen, in both particles (30 and 50 nm), but gold colloidal 30 nm NPs bind a greater range of proteins significantly different from 50 nm NPs. Albumin is another abundant protein in blood plasma but was not seen in abundance over NPs surface. The protein conformation is an important factor to be considered in this case when the protein configuration facilitate the bind event; the fibrinogen is composed of three chains (α, β, γ) and it has an elongated configuration in comparison to albumin which has a globular single chain structure. Taking into account the electrostatic interactions of NPs and proteins, pre-plasma-incubated NPs were highly negative charged and the majority of bound proteins found in gold particles after plasma incubation are neutral or positively charged at physiological pH. A dynamic coating model of protein binding resolves the multiple protein charge interactions found in gold NPs surface. A multilayered interaction was proposed by the authors in that a cationic protein binds anionic gold particles at one side or cationic protein initially binds the colloid and covers the charged surface of NPs permitting anionic proteins to bind the cationic coat.

Tenzer et al. (2011) have studied the human blood plasma corona in amorphous silica NPs of different sizes (20, 30, and 100 nm). In this study, by using proteomic analysis and bioinformatics, the authors concluded that (a) the blood plasma corona is highly complex; (b) binding of proteins does not simply correlate with their relative abundance in the plasma; (c) neither protein size nor charge significantly determines the protein composition (fingerprints), indicating that electrostatic effects alone do not constitute the major driving force regulating the protein corona formation; (d) bioinformatics classification revealed an enrichment of lipoproteins, which are involved in coagulation and complement pathway, whereas immunoglobulin and acute phase response proteins displayed lower affinities for the particles; and (e) particle size critically determines quantitatively but not qualitatively the binding of 37 % of all identified proteins.

Finally, despite any empirical evidence, a clear relationship between the synthetized nanomaterials and the structure and composition of the protein corona is not quite understood. The physicochemical parameters of the nanomaterials and the complexity of the bionanointerface, along with variable experimental strategies, make concrete relationships elusive. General empirical qualitative trends have been found, such as, when the high hydrophobicity, charge density, and curvature of the nanomaterial increase, a dense and thicker protein corona is observed; on the other hand, the nanomaterial curvature decreases both the affinity of the corona for the nanomaterial and the corona conformational change (Walkey and Chan 2012).

16.5 Biomolecular Erythrocyte–Nanoparticle Surface Chemistry Interactions

The RBC membrane and its contents are the first elements in contact with nanomaterials; however, this may affect one of the most important properties of RBC, its deformability (ability of erythrocytes to deform and pass through the microcirculation). Zhao et al. (2011c) evaluated the interaction of mesoporous silica nanoparticles with well-ordered cylindrical pore structures (s-MSN, 100–200 nm and *l*-MSN, 300–600 nm) with RBC membranes and suggested a set of conditions enabling a more adequate estimation of the hemocompatibility of nanoparticles: (a) hemolytic potential, (b) propensity to induce RBC membrane deformation or morphological alteration, and (c) tendency to impair RBC deformability. Two main processes involving mesoporous silica and RBC were contrasted; first of all, binding of the silanol-rich surface of NPs and phosphatidylcholine RBC membrane and the bending of the RBC membrane to adapt the rigid surface of silica NP. The model mentioned above initially happens when particle interacts with a flat cell membrane, driven by a local reduction in free energy, the phospholipids in the RBC surface, and close to the site of particle interaction drawn to the particle surface, leading to a membrane wrapping event and eventually encapsulation (Fig. 16.4). This phenomenon is different from the phagocytosis system because it happens without active uptake and energy spends. Taking into account the NPs surface functionality and their effects on RBC, mesoporous silica was labeled with fluorescein isothiocyanate in order to enable the particle tracking by flow cytometry and confocal fluorescence microscopy. The mesoporous silica NPs were then functionalized with aminopropyl, polyethylene glycol (PEG), and carboxyl groups. The functionalized NPs demonstrated lower affinity to RBC than no functionalized particles, independent of the surface charge. The NPs functionalized with amino groups, well known to facilitate the interaction with several biomolecules, unexpectedly reduced the binding of mesoporous silica NPs with RBC. The electrostatic interaction between silanol groups on silica surface and the amino groups diminished the accessibility of silanol groups to the erythrocyte membrane. Regarding the deformability of RBC, a positive correlation between mesoporous silica NPs concentration and increased deformability was observed. The NPs functionalized with aminopropyl and PEG did not show any alteration in the deformability of RBCs.

Erythrocyte membrane contains proteins with ATPase activity that are responsible for keeping an unequal gradient of cations inside the cell and the extracellular media involving cost of energy (ATP). The impairment of Na^+/K^+-ATPase function can lead to a necrotic cell death. Grebowski et al. (2013) studied the influence of fullerenols (fullerene nanoparticles containing hydroxyl groups which increase their hydrophilicity and enable these particles as a useful drug deliver) on the activity of human RBCs' membrane ATPases and the fluidity of the plasma membrane. High hydrophilicity of fullerenol NPs did not bind strongly to the RBCs membrane, differently from fullerene hydrophobic NPs which can easily

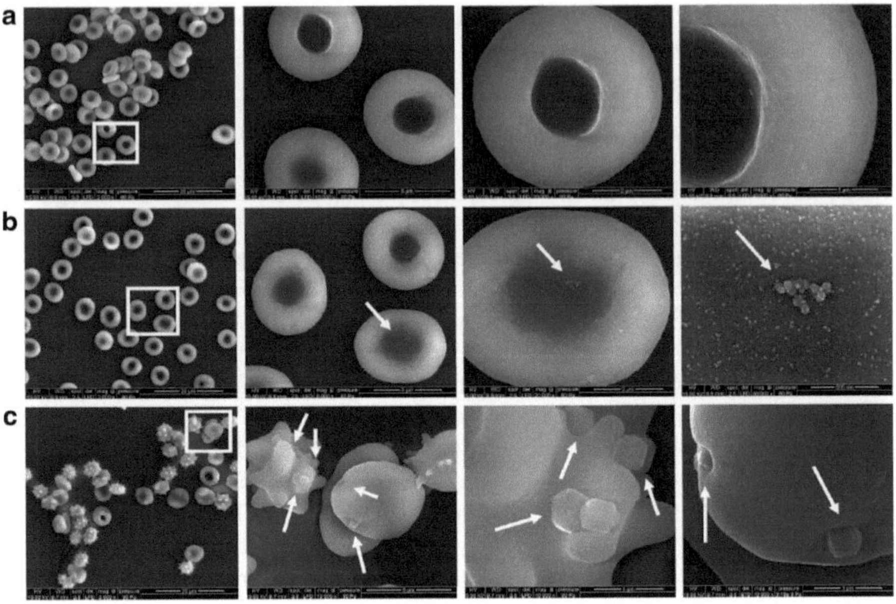

Fig. 16.4 Scanning electron images of RBCs (5 % hematocrit) incubated for 2 h at room temperature with (**a**) PBS as control, (**b**) 100 μg mL^{-1} of s-MSN, and (**c**) 100 μg mL^{-1} of *l*-MSN. Images increase in magnification from *left* to *right* with features highlighted with *white squares* or *arrows*. The nanoparticles attached on the cell surface are distinguished by the particle shape and surface textural difference between the particles and RBCs. Permission from Zhao et al. (2011c) by American Chemical Society

penetrate the lipid bilayer. The incubation of fullerenol NPs with erythrocytes decreased the Na^+/K^+-ATPase activity proportionally to particles concentration. Despite the low toxicity of fullerenol NPs, the interactions between fullerenol and membrane ATPases may cause abnormalities in cell welfare.

16.6 Comments and Future Perspectives

Nanomaterial application demands safety approval regarding its field of interest. If so nanotoxicology emerged as an important area of research covering the toxicity of the nanomaterials. Toxicology as a fundamental science is always looking forward for representative models, in such a way that hemolytic assay became a crucial step to intravenous application of nanomaterials. Silica nanoparticles depicted as potential drug carrier system, and then questions about its toxicity were extensively documented in literature which brought new information, methods, and technologies in attempt to elucidate the toxicity of nanoparticles to RBCs as previously reported in this chapter. Regarding the physicochemical characterization of nanoparticles and its role in the erythrocyte toxicity, an adequate characterization of nanoparticles is a crucial step for posteriorly determining nanotoxicity mechanisms. The topography of nanoparticles is also relevant when studying their effects on RBCs (Zook et al. 2011, 2012).

The interaction of biomolecules with nanomaterials is in growing prominence since cells and tissues possibly will never find naked nanomaterials. A dynamic nanoparticles coating model was proposed by Dobrovolskaia et al. (2009), and asserts about proteins with different charges binding nanomaterials and gives origin to a multilayer coat. Other authors related the corona formation with hemolysis suppression unveiling how cells see nanoparticles after the coating event, but also revealed this as reversible process. The corona can lead to a better dispersion of particles avoiding agglomeration, and modify the nanoparticles surface chemistry and their biological interactions. In this sense, the understanding of protein corona formation is a key point to be addressed, because biomolecular protein corona provides the *biological identity* of nanomaterials in physiological environment (Tenzer et al. 2011; Monopoli et al. 2012).

Finally, erythrocyte toxicity assay demonstrated to be an important tool in order to provide valuable information about nanoparticle toxicity and reactivity, as it is feasible and comparable, and provide many parameters like membrane deformability, membrane protein activity, and at last its lysis.

Acknowledgments The authors thank CNPq, FAPESP, INCT-Imonat, and the Brazilian Nanotoxicology Network—CIGENANOTOX.

References

Alberts B, Johnson A, Lewis J, Raff M, Roberts K, Walter P (eds) (2007) Molecular biology of the cell, 5th edn. Garland Science, New York

Ashby MF, Ferreira PJ, Schodek DL (2009) Nanomaterials, nanotechnologies and design: an introduction for engineers and architect, 1st edn. Butterworth-Heinemann, Oxford, UK

ASTM (2008) Standard test method for analysis of hemolytic properties of nanoparticles E2524-08

Barshtein G, Arbell D, Yedgar S (2011) Hemolytic effect of polymeric nanoparticles: role of albumin. IEEE Trans Nanobioscience 10(4):259–261

Brazhe NA, Abdali S, Brazhe AR et al (2009) New insight into erythrocyte through in vivo surface-enhanced raman spectroscopy. Biophys J 97(12):3206–3214

Chandra S, Das P, Bag S et al (2011) Synthesis, functionalization and bioimaging applications of highly fluorescent carbon nanoparticles. Nanoscale 3:1533–1540

Choi J, Reipa V, Hitchins VM et al (2011) Physicochemical characterization and in vitro hemolysis evaluation of silver nanoparticles. Toxicol Sci 123(1):133–143

Creanga DE, Culea M, Nadejde C et al (2009) Magnetic nanoparticle effects on the red blood cell. J Phys Conf 170(1):1–5

Dobrovolskaia MA, Clogston JD, Neun BW et al (2008) Method for analysis of nanoparticle hemolytic properties in vitro. Nano Lett 8(8):2180–2187

Dobrovolskaia MA, Patri AK, Zheng J et al (2009) Interaction of colloidal gold nanoparticles with human blood: effects on particle size and analysis of plasma protein biding profiles. Nanomedicine 5:106–117

Fairbrother A, Fairbrother JR (2009) Are environmental regulations keeping up with innovation? a case study of the nanotechnology study. Ecotoxicol Environ Saf 72(5):1327–1330

Garabagiu S (2013) Gold nanorods-hemoglobin bio-conjugate: spectroscopy studies. J Mol Struct 1031:216–220

Gerashchenko BI, Gun'ko VM, Gerashchenko II et al (2002) Probing the silica surfaces by red blood cells. Cytometry 49(2):56–61

Grebowski J, Krokosz A, Puchala M (2013) Membrane fluidity and activity of membrane ATPases in human erythrocytes under the influence of polyhydroxylated fullerene. Biochim Biophys Acta 1828(2):241–248

Higashisaka K, Yoshioka Y, Yamashita K et al (2012) Hemopexin as biomarkers for analyzing the biological responses associated with exposure to silica nanoparticles. Nanoscale Res Lett 7 (555):1–9

Hu C-MJ, Zhang L, Aryal S et al (2011) Erythrocyte membrane-camouflaged polymeric nanoparticles as a biomimetic delivery platform. Proc Natl Acad Sci U S A 108 (27):10980–10985

Huh AJ, Kwon YJ (2011) Nanoantibiotics: a new paradigm for treating infectious diseases using nanomaterials in the antibiotic resistant era. J Control Release 156(2):128–145

Khullar P, Singh V, Mahal A et al (2012) Bovine serum albumin bioconjugated gold nanoparticles: synthesis, hemolysis and cytotoxicity toward cancer cell lines. J Phys Chem C 116 (15):8834–8843

Krug HF, Wick P (2011) Nanotoxicology: an interdisciplinary challenge. Angew Chem Int Ed 50 (6):1260–1278

Kunzmann A, Andersson B, Thurnherr T et al (2011) Toxicology of engineered nanomaterials: focus on biocompatibility, biodistribution and biodegradation. Biochim Biophys Acta 1810 (3):361–373

Labarre D, Vauthier C, Chauvierre C et al (2005) Interactions of blood proteins with poly (isobutylcyanoacrylate) nanoparticle decorated with a polysaccharidic brush. Biomaterials 26 (24):5075–5084

Lin YS, Haynes C (2010) Impacts of mesoporous silica nanoparticle size, pore ordering, and pore integrity on hemolytic activity. J Am Chem Soc 132(13):4834–4842

Liu YF, Yu JS (2010) In situ synthesis of high luminescent glutathione-capped CdTe/ZnS quantum dots with biocompatibility. J Colloid Interface Sci 351(1):1–9

Lu SL, Duffin R, Poland C et al (2009) Efficacy of simple short-term in vitro assays for predicting the potential of metal oxide nanoparticles to cause pulmonary inflammation. Environ Health Perspect 117(2):241–247

Lundqvist M, Stigler J, Elia G et al (2008) Nanoparticle size and surface properties determine the protein corona with possible implications for biological impacts. Proc Natl Acad Sci U S A 105 (38):14265–14270

Meng J, Cheng X, Liu J et al (2012) Effects of long and short carboxylated or aminated multiwalled carbon nanotubes on blood coagulation. PLoS One 7(7):e38995

Monopoli MP, Aberg C, Salvati A et al (2012) Biomolecular coronas provide the biological identity of nanosized materials. Nat Nanotechnol 7:779–786

Oberdorster G, Oberdorster E, Oberdorster J (2005) Nanotoxicology: an emerging discipline evolving from studies of ultrafine particles. Environ Health Perspect 113(7):823–839

Omenn GS, Aebersold R, Paik YK (2007) HUPO plasma proteome project. Mol Cell Proteomics 6:2252–2253

Paula AJ, Martinez DST, Araujo Júnior RT et al (2012) Suppression of the hemolytic effect of mesoporous silica nanoparticle after protein corona interaction: independence of the surface microchemical environment. J Braz Chem Soc 23(10):1807–1814

Rogers EJ, Hsieh SF, Organti N et al (2008) A high throughput in vitro analytical approach to screen for oxidative stress potential exerted by nanomaterials using a biologically relevant matrix: human blood serum. Toxicol In Vitro 22(6):1639–1647

Rothen-Rutishauser BM, Schürch S, Haenni B et al (2006) Interaction of fine particles and nanoparticles with red blood cells visualized with advanced microscopic techniques. Environ Sci Technol 40(14):4353–4359

Sharifi S, Behzadi S, Laurent S et al (2012) Toxicity of nanomaterials. Chem Soc Rev 41 (6):2323–2343

Shi J, Hedberg Y, Lundin M et al (2012) Hemolytic properties of synthetic nano- and porous silica particles: the effect of surface properties and the protection by plasma corona. Acta Biomater 8 (9):3478–3490

Slowing II, Wu CW, Vivero-Escoto JL et al (2009) Mesoporous silica nanoparticles for reducing hemolytic activity towards mammalian red blood cells. Small 5(1):57–62

Song HS, Park TH (2011) Integration of biomolecules and nanomaterials: towards highly selective and sensitive sensors. Biotechnol J 6(11):1310–1316

Tenzer S, Docter D, Rosfa S et al (2011) Nanoparticle size is a critical physicochemical determinant of the human blood plasma corona: a comprehensive quantitative proteomic analysis. ACS Nano 5(9):7155–7167

Valizadeh A, Mikaeili H, Samiei M et al (2012) Quantum dots: synthesis, bioapplications, and toxicity. Nanoscale Res Lett 7(1):480. doi:10.1186/1556-276X-7-480

Walkey CD, Chan WC (2012) Understanding and controlling the interaction of nanomaterials with proteins in a physiological environment. Chem Soc Rev 41(7):2780–2799

Yu T, Malugin A, Ghandehari H (2011) The impact of silica nanoparticle design on cellular toxicity and hemolytic activity. ACS Nano 5(7):5717–5728

Zhang H, Dunphy DR, Jiang X et al (2012a) Processing pathway dependence of amorphous silica nanoparticle toxicity: colloidal vs. pyrolytic. J Am Chem Soc 134(38):15790–15804

Zhang QX, Xu X, Bertrand N et al (2012b) Interactions of nanomaterials and biological systems: implications to personalized nanomedicine. Adv Drug Deliv Rev 64(13):1363–1384

Zhao ML, Li DJ, Yuan L et al (2011a) Differences in cytocompatibility and hemocompatibility between carbon nanotubes and nitrogen-doped carbon nanotubes. Carbon 49(9):3125–3133

Zhao F, Zhao Y, Liu Y et al (2011b) Cellular uptake, intracellular trafficking and cytotoxicity of nanomaterials. Small 7(10):1322–1337

Zhao Y, Sun X, Zhang G et al (2011c) Interaction of mesoporous silica nanoparticles with human red blood cells membranes: size and surface effects. ACS Nano 2(5):1366–1375

Zook JM, MacCuspie RI, Locascio LE et al (2011) Stable nanoparticle aggregates/agglomerates of different sizes and the effect of their size on hemolytic cytotoxicity. Nanotoxicology 5 (4):517–530

Zook JM, Halter MD, Cleveland D et al (2012) Disentangling the effects of polymer coatings on silver nanoparticle agglomeration, dissolution, and toxicity to determine mechanisms of nanotoxicity. J Nanopart Res 14(1165):1–9

Chapter 17
Toxicity of Nanomaterials to Microorganisms: Mechanisms, Methods, and New Perspectives

Andreia Fonseca de Faria, Ana Carolina Mazarin de Moraes, and Oswaldo L. Alves

Abstract In the last few years, several nanomaterials with unique physicochemical properties have been developing. Specially, nano-sized materials such as silver and zinc nanoparticles, carbon nanotubes, and graphene oxide have been attracting great attention due to their potential as novel antimicrobial agents. Worldwide, the constant and indiscriminate use of conventional antibiotics has been responsible for the development of several resistant microbial species. In this context, there is a real and increasing demand for new antimicrobial agents. Nanomaterials offer several benefits due to their small size (high aspect volume/area) that provides to nanoparticles great ability to get through physical barriers such as membranes and cellular walls. Henceforth, the aim of this present chapter is to discuss the toxicological aspects of nanomaterials to microorganisms, describing the methods to evaluate their antimicrobial activity and highlighting their implications on the microbial communities of soil and water environments. We also stress the main industrial applications of antimicrobial-engineered nanomaterials.

17.1 Introduction

Nanotechnology comprises the production and application of materials in the nanoscale range (1–100 nm). The rapid growth of nanotechnology was characterized by the development of several nanomaterials with interesting properties for application in medicine, food science, textile, and cosmetic industries (Salata 2004; El-Rafie et al. 2010; Mu and Sprando 2010; Duncan 2011). Comparing to the bulk form, the nanoscale materials have gained great attention due to their

A.F. de Faria (✉)
Solid State Chemistry Laboratory, Institute of Chemistry, Universidade Estadual de Campinas-UNICAMP, Cidade Universitária "Zeferino Vaz"—Barão Geraldo, Caixa Postal 6154, Campinas, Sao Paulo 13083-970, Brazil
e-mail: an.ffaria@gmail.com

N. Durán et al. (eds.), *Nanotoxicology*, Nanomedicine and Nanotoxicology,
DOI 10.1007/978-1-4614-8993-1_17,

unique physicochemical properties such as large specific surface area and higher reactivity (Li et al. 2008; Brar et al. 2010).

Besides the physicochemical properties, nanomaterials also have been displaying several interesting biological properties (Dobrovolskaia and McNeil 2007; Li et al. 2008; Barreto et al. 2011). The antimicrobial property has been extensively explored due to the growing demand for newer and more effective therapies (Blecher et al. 2011). Recently, some nanomaterials have exhibited strong antimicrobial activity toward monocultures of microorganisms (bacteria, fungi, and viruses), including single-walled carbon nanotubes (SWCNTs) (Kang et al. 2007) and graphene oxide (Liu et al. 2011) against *Escherichia coli*; fullerene C_{60} aggregates against *Bacillus subtilis* (Lyon et al. 2006); silver nanoparticles toward *E. coli*, *Vibrio cholerae*, *Pseudomonas aeruginosa* and *Staphylococcus typhus* (Morones et al. 2005); and titanium oxide nanoparticles (TiO_2) against some viruses such as hepatitis B virus (Zan et al. 2007). In general, the antimicrobial activity of nanomaterials is associated to the alteration of permeability and disruption of cell membranes as well as oxidation of cellular components caused by the generation of reactive oxygen species (ROS) (Marambio-Jones and Hoek 2010; Musee et al. 2011).

The antimicrobial nanomaterials have been applied in the development of new functional materials and composites with potential application in textiles industries (Durán et al. 2007), food packing (Emamifar et al. 2010), and water disinfection processes (Mauter and Elimelech 2008). As these nanomaterials and nanocomposites are being incorporated in commercial products, the scientific concerning about their toxicology aspects has grown in the last years (Colvin 2003; Zhao et al. 2008; Meng et al. 2009; Lee et al. 2010; Kunzmann et al. 2011; Klaine et al. 2012). Despite these studies explored the toxicity of nanomaterials to some model microorganisms, they do not reproduce the entire microbial life conditions into the natural environments since the microorganisms are arranged in communities (Kitts 2001; Li et al. 2008). In this context, few studies have been carried out in order to evaluate the impacts of silver nanoparticles (AgNPs) (Sun et al. 2013), carbon nanotubes (Goyal et al. 2010), fullerenes (Tong et al. 2007), and TiO_2 (Nogueira et al. 2012) on soil and waste water-activated sludge microbial communities structure. The understanding about the impact on microbial communities is relevant because manufactured nanomaterials will inevitably reach the environment, thus contaminating water streams, waste water treatment plants, and soil systems (Turco et al. 2011). However, further studies might be performed in order to evaluate the impact of long-term exposition and interactions of nanomaterials with microbial communities in complex environments (Tong et al. 2007; Musee et al. 2011).

17.2 Microorganisms

Microorganisms and their components are measured at micrometer and nanometer scales. The prefix “micro” is equal to 0.000001 m (10^{-6} m) or one part of a million. They can be visualized by several microscopy techniques such as optical, scanning, and transmission electron and as well as by confocal and fluorescence microscopy when the cells are stained by a fluorescent dye. Despite of their cellular structure, the microorganisms can be classified into two main groups: the prokaryotes and eukaryotes. In general, the prokaryotes are structurally simpler than eukaryotes. The DNA (deoxyribonucleic acid) in prokaryotes contains a single and circular chromosome that is not surrounded by a membrane but it is dispersed in the cellular cytoplasm in an irregular shaped structure called of nucleoid (Tortora et al. 2010). Otherwise, eukaryotes have a multiple chromosome enclosed by a membrane, thus forming a separated nucleus that contains the genetic information. In the microbial world, the bacteria and archaea are prokaryotes while filamentous fungi yeast, protozoa, and algae are classified as eukaryotes. Viruses are not included in these systems of classification of living cells (Tortora et al. 2010). They are noncellular particles that replicate but are unable to perform the usual biochemical activities of living cells. Herein, we will emphasize some functional and structural attributes of bacteria, fungi, and viruses.

17.2.1 Bacteria

Bacteria constitute a large domain of prokaryotic microorganisms typically measuring a few micrometers in length. Bacteria are adapted for living in different environments such as soil, water, acid hot springs, ocean deep, volcanic areas, radioactive waste, and eventually in the live bodies of plants and animals (Maier et al. 2009). In these different media, bacteria play an important role in the nutrient recycling process and in the phosphorous, nitrogen, and carbon cycles (Alongi 1994; Maier et al. 2009). The numerous species of bacteria are differentiated by several factors such as morphology (shape), physiological characteristics, nutrient requirements, demand of oxygen, and source of energy (sunlight or chemicals). The extracellular components of bacteria are glycocalyx, cellular wall, flagella, fimbriae, and pili. The glycocalyx is an extracellular polymeric substance (EPS) constituted by polysaccharides and polypeptides that surrounds the cellular wall, providing protection against invasive agents and enabling the microbial adhesion in solid surfaces (Donlan 2002). Flagella, fimbriae, and pilli structures are distributed over the cell surface offering motility and adhesion ability (Pratt and Kolter 1998).

The cellular wall is a complex and semi-rigid structure responsible for maintaining the shape of the cell. The cell wall surrounds the plasmatic membrane and provides structural support and protection from adverse changes in the external environment (Koch 2003). The major function of cell wall is to prevent the cell rupture when the osmotic pressure is greater inside than outside of the cell. The cell wall is composed by peptidoglycan, which is formed by repeated chains of disaccharide cross-linked by polypeptides. According to the structural and chemical composition of cell wall, the bacteria group is subdivided into other two groups: the Gram-negative and the Gram-positive (Martinko and Madigan 2005; Tortora et al. 2010). In Gram-positive bacteria, the cell wall is a thick and rigid structure constituted by many layers of peptidoglycan (Schleifer and Kandler 1972). In contrast, Gram-negative bacteria have a thin cell wall consisting of a few layers of peptidoglycan surrounded by a second lipid membrane containing lipopolysaccharides (LPSs) and lipoproteins. These differences in cell wall structure can interfere in the microbial susceptibility to antibiotics and antibacterial nanomaterials (Kim et al. 2007; Jung et al. 2008; Martínez-Castañón et al. 2008). The cell wall is the first layer that provides protection to the cells. If the cell wall is damaged by chemicals or nanomaterials, the cellular contents may remain intact if the structural integrity of the cytoplasmic membrane is not compromised.

The cellular membrane consists in a layer composed primarily by phospholipids and proteins. The cytoplasmic membrane encloses the cell contents and acts as a selective barrier to nutrients, ions, proteins, and/or other essential molecules. The energy generation occurs through a gradient of concentration across the cell membranes. In other words, the electron transport chain (ETC) is located across the cell membrane. If the ETC structure is damaged, it can affect the ATP synthesis and consequently the energetic metabolism of the cell (Martinko and Madigan 2005). Moreover, if an antibacterial nanomaterial damages the cell membrane, it can cause leakage of the intracellular contents and subsequent cell death (Kamat et al. 2000; Wiesner et al. 2006). Intracellular components are constituted by mesosomes (an extension of the cell membrane that folds into the cytoplasm, increasing its surface area), granules/inclusion (dense crystals and particles that store nutrients), and ribosomes (sites of protein synthesis) (Gitai 2005).

Bacteria reproduce by binary fission, an asexual reproduction where one cell is divided into two identical cells (Angert 2005). The bacterial growth is described by four distinct phases (Martinko and Madigan 2005). The first phase is called "Lag phase" and it is considered an adaptive step. In the Lag phase, the bacterial cells are adapting high concentrations of nutrients and preparing for a second phase of rapid growth. The second phase is known as logarithmic phase or "Log phase." This phase is marked by a rapid exponential growth. During the Log phase, the nutrients are quickly consumed and the metabolic activity is accelerated. After the multiplication phase, the nutrient concentration is depleted and the bacteria enter in the "Stationary phase." In the Stationary phase, the number of dead cells is proportional to the number of new cells and the population stabilizes. If the number of death is greater than the new cells, the bacterial population starts a "Death phase" or logarithmic death phase. This phase can continue until the complete dissemination of the bacterial population. In the laboratory conditions, bacteria are usually grown using a solid or liquid media (Maier et al. 2009). Solid growth media are used to isolate pure cultures of a bacterial strain and liquid growth media are used when is necessary to cultivate the bacteria in large volumes or when great amount of cells is required (Martinko and Madigan 2005; Maier et al. 2009; Tortora et al. 2010).

A wide range of bacteria has been used as model microorganisms in studies of antibacterial activity of nanomaterials. *Escherichia coli* (Sondi and Salopek-Sondi 2004), *Staphylococcus aureus* (Chou et al. 2005), and *Bacillus subtilis* (Yoon et al. 2008) are very common microorganisms utilized to study the microbial toxicity of nanomaterials. However, great attention has been directed to the development of nanomaterials with biocide effect against multi-resistant bacteria (Huh and Kwon 2011).

17.2.2 Fungi

Fungi are a large group of eukaryotes that include yeasts, filamentous fungi (molds), as well as mushrooms. They have a fundamental role on the ecosystem due to their ability to recycle nutrients on soil environment, to increase plant growth through mycorrhizae associations, to degrade pollutants molecules, and to produce beneficial compounds as ethanol, industrial enzymes, and antibiotics (Blackwell et al. 2006; Desprez-Loustau et al. 2007). They are heterotrophic organisms, since they depend on the availability of organic or energy-rich nutrients. In consequence, fungi can be found living in the nature as saprophytes, parasites, or symbionts, thus removing nutrients from dead substrates or beings (animals or plants) (Osiewacz 2002; Maier et al. 2009). The fungi organisms can be divided into two main groups categorized as yeasts (unicellular) and the filamentous fungi (multicellular).

Differently from bacteria, fungi are organisms that contain specialized organelles enclosed within membranes. The major characteristic that separates the fungi from bacteria group is the presence of a nucleus or nuclear envelope, where the DNA (genetic material) is found within (Maier et al. 2009; Tortora et al. 2010). The cytoplasm of fungal cell is more complex in terms of number and types of organelles than the cytoplasm of prokaryotes. Fungi possess ribosomes (synthesis of protein), Golgi complex (packages proteins inside the cell before they are sent to their destination), rough endoplasmic reticulum (surface to anchoring ribosomes during the protein manufacturing), and nucleus (contains DNA); and all of these components are membrane-bound. Mitochondria are important organelles, since most of ATP produced in the cell is synthesized by the inner mitochondrial membrane (Tielens et al. 2002). The cellular wall of filamentous fungi is composed by a thick layer of cellulose or chitin and a thinner outer layer of glycoproteins and polysaccharides (glucans) and other biopolymers (Bartnicki-Garcia 1968; Maier et al. 2009), and there is evidence of a extensive cross-linking among these structures (Adams 2004).

Unlike from bacteria that are unicellular, filamentous fungi are multicellular organisms formed by microscopic structures called hyphae (Harris 2008). The combination of a set of hyphae leads to the formation of a mycelium, which is characterized by an interconnected network of hyphae (Maier et al. 2009). The hyphae are tubular, elongated, and branched microstructures, which contain multiple nuclei and a set of vesicle structures consisting of proteins, lipids, and other organic molecules, and organelles dispersed in the cytoplasmic material (Hawker 1965). In most of filamentous fungi, the hyphae can possess septa, which is a cross-wall structure that almost divides the hypha into two parts. Each part of hyphae is able to grow as a new individual when a fragment breaks off, and this process is categorized as asexual reproduction (Paul 2007; Tortora et al. 2010). Therefore, the part of hyphae responsible for the fungus growth and absorption of nutrients is called vegetative hypha and the portion associated to the reproduction is known as reproductive or aerial hypha, which frequently produces spores. The spores present an important role on life cycle of the fungi, since both sexual and asexual reproduction can occur through spores. In the asexual reproduction, a portion of hyphae or a single spore can germinate and produce a new individual identical to the parent. Otherwise, in the sexual production two identical parts (hypha or spore) may combine by fusing their structures to form an interconnected network that will divide by meiosis to originate a new organism (Tortora et al. 2010). When cultivated in laboratory, filamentous fungi show spongy and cottonous morphological characteristics.

Fungi are generally adapted to live in environments that would be hostile to bacteria (Jay et al. 2005; Maier et al. 2009). As mentioned above, fungi are chemoheterotrophs, since they absorb nutrients and energy from organic compounds present in soil, water, animal, plants, and food. Most of them are strictly aerobic and usually grow better in environments with pH about 4–5, which are too acid to almost group of bacteria (Jay et al. 2005). They are very versatile with respect to the environmental conditions of growth, being able to grow under very low amount of moisture. Moreover, they are resistant to high osmotic pressure and require less nitrogen than bacteria strains. Fungi are capable of metabolizing complex sources of carbon such as lignin (a wood component), because they produce a sort of extracellular enzymes that digest the substrate aiming the release of substances that could be absorbed by the hyphae (Martinez et al. 2004; Maier et al. 2009).

The yeasts are nonfilamentous, unicellular fungi that commonly present oval or spherical morphology. The yeast can be often found growing over fruit leafs, flowers, and sugar liquids as a white coating. Yeasts can reproduce asexually by two different processes: budding and fission. In the budding process, a protuberance is formed, resulting in a bud outer the cell while the genetic material is duplicated. One of the two nucleic acids is transferred to the bud and the daughter cells separate from the parent cell. The budding process can sometimes produce structures in form of chain which are denominated pseudohyphas (Veses and Gow 2009; Tortora et al. 2010). In contrast, in the fission process the nucleus is duplicated and the cell is elongated and divided into two other daughter cells; a similar process previously described to bacteria. When cultivated in agar medium the yeasts produce colonies with morphologically similar to bacteria. The yeasts are facultative anaerobic, able to growing in the presence and absence of oxygen (Rodrigues et al. 2006). This is a valuable attribute, because it allows the yeasts survive in environments with different contents of oxygen. For instance, if yeasts consume oxygen, they metabolize the carbon source by respiration mechanism producing water and carbon dioxide (CO_2); however, in the absence of oxygen they ferment carbohydrates and produce a bio-metabolite (such as ethanol) and CO_2.

17.2.3 Viruses

Viruses are small infectious agents (size of 20–300 nm) that possess a single type of nucleic acid molecules (either DNA or RNA) (Martinko and Madigan 2005). In the microbial world, viruses are considered smaller microorganisms than bacteria, fungi, and protozoa (Maier et al. 2009) and their visualization is only possible by using scanning or transmission electron microscopy (SEM/TEM) (Elechiguerra et al. 2005; Miyako et al. 2008). Generally, the genetic material is covered by a protein coat, which can also be surrounded by an extra envelope of lipids, carbohydrates, and proteins.

Compared to other cells such as bacteria and fungi, viruses present a simpler cellular structure because they have no organelles as ribosomes or mitochondria, and consequently have not biochemical mechanisms (own enzymes, proteins, and ATP generation) necessary for maintaining the cellular metabolism. They are considered obligatory intracellular parasites, since they require living host cells for their growth and replication (Maier et al. 2009). In other words, viruses are unable to grow by themselves, but they can multiply inside the cells by using the over-synthesizing machinery of the host cell (Tortora et al. 2010). Once inside the cell, the replication capacity of a virus is surprising: a single virus can multiply in thousands of new entities in few hours. In this way, from a biochemical point of view viruses are not considered living organisms, because outside a host cell they are inert (Tortora et al. 2010). However, from a clinical point of view viruses are considered alive organisms because they cause infection diseases in healthy cells.

The viruses are able to infect living cells of all domains (Eukarya, Archaea, and Bacteria) and their attachment is determined by the chemical interaction between the surface of virus and specific receptor sites at surface of the host cell. Thus, the virus anchoring is guaranteed by weak bonds between the components of virus and surface of host cells. Succinctly, the virus structure is composed by three essential parts: (1) *acid nucleic*: in virus, the genetic material can be represented by DNA or RNA- but never both; (2) *capsid*: a protein coat that covers and protects the nucleic acid and provides sites for host attachment; (3) in some viruses, the capsid can be enclosed by an extra *envelope*, which consists of a combination of lipids, proteins, and carbohydrates. The viruses are classified on basis of the host cell they infect as well as according to their capsid architecture (helical, polyhedral, enveloped virus, and complex virus) (Carter and Saunders 2007; Tortora et al. 2010). They can infect: (1) plants, causing disease in tomatoes, tobacco, and potatoes crops (Roossinck 2010); (2) human, causing herpes, smallpox, meningitis, and HIV

(Strauss and Strauss 2008); and (3) bacteria, known as bacteriophage or phage (Maier et al. 2009; Armon 2011). These infectious agents are conveyed through aerosols, contaminated food, and water or fomites (Maier et al. 2009). Their growth cycle is described in five different steps: (1) adsorption (attachment of the virus to specific host cells), (2) penetration (move in into host cells by fusion and endocytosis), (3) replication (viruses use the host cell machinery to replicate), (4) maturation and release (the replicated genetic material is assembled in capsids and releases through breakdown of the host cell wall) (Maier et al. 2009).

Most information regarding viruses is provided by the study of bacteriophages (virus that infects bacteria), which can be easily cultivated in laboratory using agar plate or liquid culture medium. Shortly, the viruses and bacteria samples are mixed and each virus is able to infect a bacterial cell, thus multiplying and releasing viral particles that will infect other adjacent bacterial cells. For example, in plaque count assay each infected cell will produce a lysis point or a clear zone, which corresponds to a single viral particle from the original sample. The clear points are counted and the concentration of viral particles is expressed as plaque-forming units (PFU) (Tortora et al. 2010).

17.2.4 Microbial Communities and Biofilms

In natural environments, microorganisms are rarely found as unique and isolated cell. In contrast, microorganisms often construct and live within or associated with microbial communities.

Numerous advances in the microbiology field were performed through the study of pure cultures isolated under laboratorial conditions (Kitts 2001). However, the advances on molecular biology enabled a better understanding about the structure, dynamic, and complexity of natural microbial communities. Unfortunately, most of microorganisms living in microbial communities can not be cultivated under conditions established in laboratory. Therefore, the characterization of 16S rDNA sequence and DNA-based techniques have offered valorous information regarding the diversity of these uncultured microorganisms (Tringe et al. 2005). For instance, one gram of soil is estimated to have up to ten billion microorganisms and the majority of these organisms are uncultivated under standard laboratory conditions (Torsvik et al. 1996; Tringe et al. 2005).

Thus, communities are defined as an integrated set of microbial populations occupying a specific habitat. Microbial communities are described in terms of species diversity and genetic diversity and it is associated to genetic variety of individuals (richness), their relative abundance (evenness), and their distribution among taxa (Liu et al. 2006). There are evidences that microorganisms are able to live in the form of microbial communities in a diverse range of habitats including soil, seawater, freshwater, and wastewater sludge. For example, it has been estimated that 80–90% of the cells in porous soil media are attached on surface and the remainder are free-living (planktonic stage) (Maier et al. 2009). The knowledge regarding the microbial community structure and their diversity is fundamental to understand the environmental impact of chemical contaminations (nanomaterials, oil, hydrocarbons, and pesticides) on the ecosystem (Torsvik et al. 1996; Curtis and Sloan 2004).

Commonly, these microbial communities are most frequently found in association with interface and surfaces forming the so-called biofilms (Costerton et al. 1995; Branda et al. 2005). The microbial adherence is proportioned through the EPS (polymeric extracellular matrix) production, which may contribute significantly to the cell protection from environmental stresses and also allows an improved supply of nutrients (Branda et al. 2005). According to Stoodley et al. (2002), prokaryotic biofilms in a range of ecosystems are composed by multispecies communities, which are arranged in a highly complex structure interconnected through channels that allow the flux of nutrients and oxygen.

Microorganisms growing in communities are genetically and physiologically distinct from their planktonic cell counterparts (Stewart and Franklin 2008). Differently from planktonic cells, microorganisms in biofilms present a wide range of metabolic activity or physiological states. Simultaneously, a group of cells can be expressing a specific metabolic pathway (to adapt to a stress condition) and producing an essential compound (Chmielewski and Frank 2003; Stewart and Franklin 2008). According to Stewart and Franklin (2008), microscale chemical gradients, adaptation to local environmental condition, genetic expression, and genotypic variation caused by processes of selection and mutation are mechanisms that contribute to the genetic and physiologic heterogeneity within biofilms. Moreover, microorganisms growing in form of biofilm or complex communities usually have more resistance to antibacterial agents than individual planktonic cells of the same species (Donlan 2002). In consequence, biofilms may infect a variety of surfaces, increasing the chances of transmission of infectious diseases by offering to microbial community high protection against antibiotics (Branda et al. 2005).

17.3 Current Methods to Accessing the Toxicity of Nanomaterials to Microorganisms

The microbial growth or microbial population can be measured using different methods (Maier et al. 2009; Martinko and Madigan 2005). In general the methodologies are classificated in direct and indirect methods, since direct methods are able to measure the number of cells in a milliliter of a liquid or in a gram of solid material. In other hand, the indirect method measures the total mass of the population, which is directly proportional to cell members. As the microbial population is very large, most of methods apply a mathematic equation for determining the size of the microbial population. In the same way, the antimicrobial activity of nanomaterials can be determined by different techniques that measure the microbial cells viability after a specific time of exposure. Due to the increasing development of new nanobiotic agents, this topic will describe the current methods used in the literature to access the microbial viability. In addition, this topic will also specify the advantages and disadvantages of each methodology.

17.3.1 Turbidity

Turbidity is a practical way for monitoring the bacterial concentration/growth in a liquid suspension. This method has been extensively used to explore the antibacterial activity of nanomaterials such as fullerenes C_{60} (Lyon et al. 2006), zinc oxide nanoparticles (Raghupathi et al. 2011), silver nanoparticles (Raffi et al. 2008), and carbon nanotubes (Ahmed et al. 2012).

The cellular concentration is estimated by using a spectrophotometer where the cell concentration is related to the optical density in a specific wavelength (nm). The typical wavelength varies from 500 to 600 nm but, if necessary, other wavelengths can be used. The suspension changes its color and transparency when the bacterial is growing in a liquid medium. In the spectrophotometer, a beam of light is transmitted through the bacterial suspension to a light-sensitive detector. If the bacterial proliferation increases in the cell culture media, the amount of light that cross the microbial suspension decreases (Tortora et al. 2010; Pelczar et al. 1993). This amount of light will be detected as percentage of transmission (T%) that can be converted to absorbance (or optical density) by the equation $\text{Abs} = 2 - \log(\text{T\%})$ (Atkins and Jones 2010). Using this technique, the bacterial growth can be estimated by measuring the absorbance in different times of exposure. If the density of bacteria is high and it is blocking the passage of light, the sample should be diluted until the absorbance reaches values from 0.2 to 0.8, according to the Lambert Beer's law. The optical density measuring can be performed directly from the sample and compared with other samples.

Therefore, if the optical density value matches with the plate counts of the same bacterial strain, this correlation can be used for further estimations of bacterial population using the turbidity measuring (Tortora et al. 2010). The turbidity is a simple methodology that does not require the use of reagents or complicated processes. However, this technique is not applied for measuring microbial population when the culture medium contains low concentrations of bacteria. Another complication is related to nanomaterials that possess high intensity of color as carbon nanotubes. In this case, the presence of the nanomaterials can contribute with the optical density, leading to an untrusted value of absorbance (Seil and Webster 2012). One suggestion to resolve this interference is preparing a standard curve with the pure nanomaterials and subtract the value from those values of the suspension containing the mixture of microbial cells and nanomaterials (Seil and Webster 2012).

17.3.2 Agar Diffusion Test

The agar diffusion test is a fast and simple assay to evaluate the sensibility of a bacterial strain toward an antimicrobial agent. This method is widely used to investigate the biocide activity of nanomaterials as silver and copper nanoparticles (Ruparelia et al. 2008; Nanda and Saravanan 2009), graphene oxide (Das et al. 2011), nanocrystalline zinc oxide (Applerot et al. 2009), and fullerenes (Lyon et al. 2006); however, it allows only a qualitative result about the sensibility of the microbial strain (Pelczar et al. 1993; Tortora et al. 2010). This method is based on the diffusion of the nanomaterial from high concentrations (filter or disk) to the agar surface. Thus, a disk or a piece of filter paper is immersed into the antimicrobial agent solution and placed over the agar previously inoculated with the bacteria of interest. The standardized bacteria inoculum can be added to the agar plate using a Drygalski-spatel or a sterile swab. The plates are then incubated at a specific temperature for 12 or 24 h. After incubation, if the nanomaterial has activity, clear zones will be observed concentrically around the disk, proving the biocide effect of the sample (Martinko and Madigan 2005; Tortora et al. 2010). In general, inhibition halos are detected when the concentration of the antimicrobial compound is equal or lower than MIC (minimum inhibitory concentration). If the inhibition halo is correlated with MIC, the inhibition zone size will be inversely proportional to MIC values. Moreover, the inhibition zone will be large or small depending on the bacterial susceptibility and efficiency of the antibacterial agent. Other factors as diffusion rate, physicochemical properties of the antimicrobial agent, agar thickness, and composition of the medium culture can also interfere on the halo size. The halo size can be measured by using a simple rule, expressed as millimeters (mm) and compared with other values previously described in the literature. In addition, bacteria strains that present slow growth may show higher susceptibility because the nanomaterial has more time to disseminate through the agar surface (Tortora et al. 2010). Thus, the bacterial samples can be categorized in sensitive, resistant, or intermediate, according to the halo inhibition size (Martinko and Madigan 2005). In some situations, the test of diffusion in agar is not capable is enough to explore the microbial sensibility in the presence of an antibacterial agent. Henceforth, in most of cases it is necessary to apply a quantitative method to determine the susceptibility in terms of the MIC. Overall, this approach can be applied to study the susceptibility of bacteria (Yang et al. 2012), yeasts (Li et al. 2013) and filamentous fungi (Zare-Zardini et al. 2013) to the presence of nanomaterials.

17.3.3 Counting Plate

The most frequent method used to measuring the bacterial populations is the counting plate method (Tortora et al. 2010). This methodology is based on the quantification of the viable cells (capable of reproduction) in a sample. Plate count assumes that each viable cell is able to produce a single colony. However, this principle is not always true, because sometimes the bacteria can grow in form of chain or small agglomerates (Maier et al. 2009; Tortora et al. 2010). Nevertheless in some cases, an isolated colony is not generated from a single cell but from a short group of cells (Seil and Webster 2012). These cellular agglomerates may be disrupted by gentle agitation and low intensity ultrasound treatment of the sample. The number of colonies is an important issue when we use the counting plate method. If the plate contains many colonies, the cells certainly will overlap each other, causing inaccuracies in the count. The Food Drugs and Administration (FDA) stipulates that the counting should be performed in plates containing from 25 to 250 colonies, but some microbiologists prefer plates with 30–300 colonies (Martinko and Madigan 2005; Maier et al. 2009; Tortora et al. 2010). The procedure involves the collect of aliquots (usually 1 mL) from the liquid sample. The aliquot is transferred to tubes containing a buffer or saline solution (0.9% NaCl) and then serially diluted. The dilution process is important because if you plate 1 mL directly from microbial suspension to the agar plate, this will produce an uncountable number of colonies. For example, if a bacterial sample with 10,000 cells per milliliter are plated, that will produce 10,000 colonies on agar surface (Tortora et al. 2010). However, if the same 10,000 colonies/mL are diluted three consecutive times it will present 100 colonies/mL on the agar plate, and it is a countable value (Tortora et al. 2010). After the process of serial dilution, an aliquot (usually 0.1 mL) of each dilution tube is spread over a solidified agar plate using a sterile Drygalski-spatel. The spread plate procedure allows the fixation of the cells on the surface, thus avoiding the contact between the colonies. However, the counting plate can also be conducted by the pour plate procedure. In the pour plate method, the diluted aliquots are introduced into the petri plates and the nutrient medium containing agar (maintained at about 50 °C) is poured over the plate which is gently agitated. Using the pour plate, the colonies will grow into the agar medium and over the agar surface too. Therefore, this technique is not useful when the microorganisms are sensitive to relative high temperatures because the pre-heated agar can cause structural damages to bacteria cells (Martinko and Madigan 2005; Tortora et al. 2010). In both tests, the plates are incubated at appropriate temperature for 18–24 h. As a colony is usually originated from an individual organism, the total number of colonies corresponds to the number of viable cells in the sample. The plates are counted and the results expressed as colony forming units per milliliter (CFU/mL), according to the equation: $\mathrm{CFU/mL} = (N \times \mathrm{DF})/v_{\mathrm{aliq}}$, where N is the number of counted colonies, DF is the dilution factor (that corresponds to dilution 10^{-1}, 10^{-2}, 10^{-3}, etc.), and v_{aliq} is the volume of the aliquot inoculated on the Petri dish. Moreover, this methodology is widely used due to its high sensitivity, enabling the detection of low numbers of cells. However, as the counting plate is subjected to large experimental errors, the performing of three or more replicates of each experimental condition is recommended. Besides, this procedure is laborious and requires the use of a great amount of reagents and laboratorial materials. A significant number of works involving antibacterial activity of nanomaterials shows the application of the counting plate method to quantify the number of viable bacterial cells after exposure to graphene oxide (Liu et al. 2012), silver nanoparticles (Liu et al. 2009), SWCNTs (Liu et al. 2010), and C_{60} fullerenes (Brunet et al. 2009). One disadvantage in use liquid systems to accurate the antibacterial effect of nanomaterials is associated to the limited solubility of the nanomaterials in culture media. Usually, nano-structured materials are insoluble in aqueous system and have a tendency to form agglomerates when introduced in cell culture medium. Therefore nanomaterials may interact with proteins present in the culture medium, thus providing the formation of a protein layer that coats the surface of the nanoparticles. The presence of this protein layer so-called as "corona" contributes positively for the nanoparticle agglomeration process and as well as reduces the direct contact between the nanomaterials and microbial cells (Seil and Webster 2012).

17.3.4 Direct Microscopic Counting

The direct microscopic counting is a conventional technique that allows to estimate easily and quickly the total number of cells in the microbial population without previous treatment (Martinko and Madigan 2005; Tortora et al. 2010). This method is based on filling a cavity in a microscope slide with a microbial suspension. This cavity located in the center of the slide is divided into squares with well-defined dimensions and known volume. In general, a volume of 0.01 mL from microbial suspension is added over the square slide and posteriorly stained to provide the cells visualization on the microscope. Petroff-Hausser cell counter and Neubauer chamber are examples of slides used in the direct microscopic counts. Using the microscope, it is possible to count the number of cells present in each square, and thus knowing the number of cells per unit of volume. The number of cells is counted in several fields and the average value of each field is used to calculate the cellular concentration of the sample. The total number of cells per milliliter on the original sample is calculated multiplying the average counted cells times the inverse of the square volume (Tortora et al. 2010). For example, if the number of counted cells is 18 and the square volume is 1 mm^3 (1 $mm^3 = 0.0001$ mL $= 10^{-4}$), the final cellular concentration is $18 \times (10^4) = 1.8 \times 10^5$ cells per milliliter. If the microbial suspension is diluted before the counting, the value of 1.8×10^5 should be multiplied by the dilution factor. If the sample is diluted 1:100, the value of the dilution factor is 100 and the final bacterial concentration will be 1.8×10^7 cells per milliliter. This direct counting method has the advantage of providing an almost immediate result, not requiring incubation time as described for the counting plate method. However, the disadvantages include no distinction between live and dead cells (Tortora et al. 2010). Moreover, very small cells can be difficult to distinguish leading to an imprecise result; and motile microorganisms are difficult to count by this procedure. In addition to these disadvantages, this method is not suitable for measuring low cellular concentrations (Martinko and Madigan 2005; Tortora et al. 2010).

17.3.5 MTT Assay

This is a colorimetric assay often used to measure the cellular viability via reduction of MTT (3-[4,5-Dimethylthiazol-2-yl]-2,5-diphenyltetrazolium bromide) to purple formazan (Mosmann 1983; Berridge et al. 2005) by the mitochondrial enzyme succinate-deydrogenase (Sylvester 2011). In this assay, it is assumed that MTT is reduced to formazan in the presence of reductase enzymes produced by living cells. The reduction of MTT by living cells produces a blue-purple color formazan product which can be quantified by UV–vis spectroscopy when they are in the dissolved form. The absorbance of the blue solution can be quantified by measuring the optical density at wavelength from 500 to 600 nm and the quantity of viable cells is directly proportional to the amount of formazan. In other words, this methodology measures the metabolic activity (via production of oxireductase enzymes) that can be related to the number of living cells (Sylvester 2011). However, cells with low metabolic activity or resting cells reduce much less MTT when compared to cells in phase of proliferation (Mosmann 1983). The great advantage of this method is that the MTT reduction takes place only in metabolically active cells (Mosmann 1983; Sylvester 2011). Thus, this approach can be extensively used to evaluate the cytotoxic effects of potential antimicrobial agents and toxic nanomaterials against microorganisms. Krishnaraj et al. 2010 used MTT assay to evaluate the cytotoxic effects of biogenic silver nanoparticles against waterborne pathogens (*E. coli* and *Vibrio cholera*). Hernandez-Delgadillo et al. (2012) applied the MTT assay to investigate the antibacterial activity of bismuth nanoparticles and Pakrashi et al. (2011) explored the toxicity of Al_2O_3 nanoparticles toward *Bacillus licheniformis* and cellular viability was determined using the MTT assay. In addition, the MIC values can also be measured using this technique. For this purpose, this test is frequently carried out using a microliter plate (96 well plate) that allows the study of several experimental conditions at the same plate. For example, the maximum concentration of a nanomaterial is added in the first line and then diluted in the subsequent lines to test several concentration of the material at the same plate. The inoculum containing the microorganism of interest is added in each well and the plate is incubated at appropriated conditions of temperature for 24 h. After incubation, the MTT may be added and the plate is incubated again for a couple of hours. As the viable cells are able to reduce MTT to formazan producing a purple color, the line that shows absence of color is considered the MIC (minimum inhibitory concentration). The analysis can be performed visually or quantified by using a microplate reader.

17.3.6 Live/Dead Cell Staining

The microbial cell is considered viable if the membrane is not damaged and it can assume to be dead if the membrane structure is corrupted. The membrane integrity is evaluated by the capacity of the cells to exclude compounds, not allowing the unselective entrance of molecules inside the cell. In contrast, compromised membranes can passively diffuse molecules through the cell wall, thus indicating loss of cellular viability. The live/dead cell staining assay is a fluorescent test used to analyze the bacterial viability, allowing researchers to distinguish living from dead bacterial cells almost instantly (Boulos et al. 1999). This method is able to detect planktonic or bacteria cell living in form of biofilms, even if the population contains different bacterial strains. The assay utilizes a mixture of two fluorescent nucleic acid-binding stains: a green (Syto 9) and a red (propidium iodide—PI). These dyes possess different spectral features and distinct ability to penetrate in bacterial cells (Boulos et al. 1999). The Syto 9 is permeable to all cell membranes and stains the cells as green. Syto 9 is able to bind on both DNA and RNA of intact and damaged cells (living and dead cells), while PI stain penetrates only bacteria with damaged membranes. When these two stains are used together, it is possible to visualize dead and living cells, simultaneously. So, living and dead bacteria will produce green fluorescence, whereas dead or bacteria with damaged membranes will produce fluorescent red. The percentage of viable cells is determined through the ratio of green to red fluorescence intensities. The fluorescent signals can be measured by fluorescence microscopy or confocal fluorescence microscopy. Brady-Estévez et al. (2008) used the live/dead assay (Syto 9/PI) to evaluate the antimicrobial activity of a SWCNTs filter against *E. coli* K12 and Liu et al. (2011) also demonstrated the antibacterial activity of graphene oxide (GO) and GO decorated with silver nanoparticles against *E. coli* cells by using the live/dead test. Similarly, Jung et al. (2008) applied the live/dead assay to investigate the toxicity of silver nanoparticles toward *S. aureus* and *E. coli* strains.

Otherwise, this fluorescent assay has been very useful to investigate the anti-adhesion and anti-biofilm property of nanomaterials. In the last decades, the quantification of the biofilms has been performed using conventional techniques as counting plate (CFU/mL), which are slow and time-consuming (Costerton et al. 1995; Donlan 2002). Nowadays, a promising approach to evaluate the growth of microorganisms on surfaces is the application of Syto 9 and PI stains and further analysis by fluorescence microscopy technique. If more details about the biofilm are required, the confocal fluorescence microscopy provides data about the three-dimensional structure and thickness of the biofilms (Seil and Webster 2012). One of the major advantage of this method is the possibility to analyze the biofilm in situ, without the limitations of the counting plate method, which requires exhaustible steps of sample preparation. In contrast to SEM and TEM, the fluorescence technique is nondestructive and allows the examination of living biological samples, even if the biofilm possesses a large thickness. Park et al. (2011) studied the inactivation of biofilms formed by *Pseudomonas aeruginosa* PA01 using the thermal stress derived from super-paramagnetic iron oxide nanoparticles (SPIONs). To evaluate the bacterial cell viability, the authors stained the cells using Syto 9/PI and the viable and damaged bacterial cells were distinguished through confocal laser-scanning microscopy. Choi et al. (2010) evaluated the effect of silver nanoparticles against biofilm of *E. coli*. At the same bacterial concentrations (3×10^8 CFU/mL), biofilms were about four times more resistant to nanosilver than planktonic cells. The biofilm analysis was carried out by the application of Syto 9/PI dyes and the fluorescence signal was measured using a laser-scanning confocal microscope. Rodrigues and Elimelech (2010) evaluated the impact of SWCNTs on the different stages of biofilm formation by *E. coli* K12. The authors observed that the presence of EPS (exopolymeric substances) in mature biofilms played an important role in mitigating the toxic effects of SWCNTs. The biofilm formation was stained using Syto 9/PI and the image acquisition of the biofilms was performed by confocal microscopy. Additionally, other studies have been described the anti-biofilm activity of functionalized multi-walled carbon nanotubes (Qi et al. 2011), graphene oxide (Carpio et al. 2012), and silver nanoparticles (Dror-Ehre et al. 2010; Fabrega et al. 2011) using the live/dead assay as protocol to quantify the percentage of living cells.

The microbial concentration can also be quantified through the measuring of the metabolic activity. This method assumes a direct correlation between the amount of some products (organic acid or CO_2) with the number of microorganisms in a population. For example, the monitoring of CO_2 production by respirometry was extensively used to evaluate the impact of hazardous hydrocarbons to the soil microbial diversity (Boonchan et al. 2000; Silva et al. 2009; Sayara et al. 2011) and as well as can be applied to study the impact of nanomaterials to microorganism in soil and water environments. The measurement of the dry weight is another way to quantify the microbial growth. This procedure is very popular to determine the biomass of filamentous fungi. The fungus mycelium is removed from the growth medium, filtered to remove the excess of water, and dried in desiccator/oven until constant weight. This method can also be applied to quantify the bacteria growth, but it is a less usual approach (Tortora et al. 2010).

17.3.7 Culture-Independent Methods of Community Analysis

Microorganisms play an important role in the natural ecosystems and most of them are uncultivated in laboratory using traditional microbiological techniques. Generally, these conventional methods are not capable to reproduce the very specific conditions that some microorganisms require to proliferation in natural habitats (Muyzer and Ramsing 1995). Traditionally, the culturing methods are time- and supplement-consuming, because they use a variety of culture media to maximize the recovery of microorganisms from the microbial communities (Hill et al. 2000). However, to overcome these problems, molecular techniques have been used to detect and characterize uncultivated microorganism in natural environments

(Su et al. 2012). Furthermore, the molecular methods allow microbiologists the understanding of natural microbial structure and dynamics (Kitts 2001; Dahllöf 2002). The microbial communities can be subjected to direct analysis (in situ methods), but in most of the cases the nucleic acid is extracted and specific DNA regions are amplified using polymerase chain reaction (PCR). The amplified genes are potentially evaluated using techniques of pattern (fingerprint) analysis, cloning and sequencing, probe hybridization, and microarrays (Dahllöf 2002). Some effective PCR-based methods as DGGE (denaturing gradient gel electrophoresis), T-RFLP (terminal restriction fragment length polymorphism), ARDRA (amplified ribosomal DNA restriction analysis), and a PCR-independent method known as FISH (fluorescent in situ hybridization) have been widely applied to characterize and to access a greater proportion of the microbial community from soil and water systems. Most of PCR-based techniques utilize the sequencing of ribosomal RNA genes (genes that coding for small structural portions of ribosomes), more specifically a subunit 16S rDNA or 16S rRNA to characterize the microbial diversity in natural environments. 16S is a structural component of the 30S subunit of prokaryotic domain ribosomes and it has been referred as 16S rDNA (Woese and Fox 1977). The development of methods based on the 16S rDNA genes sequencing have some advantages including the fact that these genes are present in all known microorganisms; they have conserved sites that can be used as specific target sites for PCR primers; in addition, they contain variable regions that provide specific signature sequences for microorganism identification; they have enough sequence information to be used as a phylogenetic target and the genes are not transferred horizontally among species (Muyzer and Ramsing 1995). Herein, we will introduce some general theoretical principles of DGGE, T-RFLP, and FISH methods.

DGGE analysis was proposed by Muyzer et al. (1993) and this methodology is based on the separation of DNA molecules using a special type of electrophoresis denominated as DGGE. The separation of DNA fragments amplified by PCR occurs in a polyacrylamide gel that contains an increasing denaturing gradient. This method allows the separation of DNA fragments that possess the same length, however with different base-pair sequences (Muyzer et al. 1993; Muyzer and Ramsing 1995; Hill et al. 2000). During the electrophoretic procedure, these DNA fragments conserve their original structure until they reach a denaturing concentration able to breakdown their double-stranded structure (Hill et al. 2000). Consequently, when the double structure is undone the electrophoretic mobility is reduced and the DNA fragment melts, forming a characteristic band on the polyacrylamide gel (Torsvik et al. 1996; Hill et al. 2000). As the melting point is dependent on the nucleotide sequence, the individual bands will match with a specific and particular sequence. Henceforth, DNA fragments with same length but with different base-pair can be separated using DGGE method (Muyzer and Ramsing 1995; Muyzer and Smalla 1998). Summarizing, the resulting electrophoretic gel shows the presence of several bands, which are related to predominant species within the microbial communities. In general terms, the total extracted DNA from soil, water, or activated sludge could be amplified by PCR and resolved using DGGE approach. The DNA fragments separated by DGGE can be further sequenced and the similarity of the nucleotide sequence compared with other sequences deposited on public-domain databases (NCBI GeneBank http://www.ncbi.nlm.nih.gov) (Hill et al. 2000). DGGE analysis of PCR-amplified 16S rDNA has been successfully used to evaluate the environmental impact of nanomaterials to microbial communities of soil (Nogueira et al. 2012) and wastewater sediments (Nyberg et al. 2008).

Terminal Restriction Fragment Length Polymorphisms (T-RFLP) is also a PCR-based tool to study the structure and function of microbial communities. In this technique, the DNA target gene (commonly 16S rDNA) is amplified by using primers functionalized with a fluorescent dye in their terminal parts. These DNA fragments are then digested by restriction enzymes (usually with 4pb recognition size), which are able to cut DNA in several terminal restriction fragments (TRFs). Further, these TRFs are subjected to electrophoresis in polyacrylamide or capillary gel. Usually, the electrophoresis apparatus contains an automated DNA sequencer (equipped with a fluorescent detector) in order to identify only the size of the fluorescently labeled TRFs. The further step is carried out by automated processes that provide the fragment length (in base-pair) based on the TRFs retention when compared with internal markers (fluorescently labeled DNA size standard) (Marsh 1999; Kitts 2001). Succinctly, these softwares integrate the fluorescent signal, thus providing electropherograms which are composed by several peaks with discriminated height and area. Each peak represents a TRF on the electrophoresis gel. The pattern peaks can be compared among samples using statistical methods, and the peak can be identified through comparison with a preexisting database of sequences. According to Marsh (1999), more than 60 terminal fragment sizes can be detected in a T-RFLP analysis of a soil microbial community. This technique is a sensitive and rapid approach to obtain a distinctive fingerprint of a microbial community as well as perform comparative studies across communities (Marsh 1999; Liu et al. 1997). Generally, T-RFLP method possesses higher resolution than DGGE, for example. However, one disadvantage is the selection of the restriction enzymes because even when the enzymes used are optimized, many microbial species may present the same TRFs length (Marsh 1999; Dahllöf 2002). In the nanotoxicology context, T-RFLP analysis has been applied to evaluate the environmental impact of silver nanoparticles on marine bacteria biofilms (Fabrega et al. 2011) and toxicological effect of ZnO and TiO_2 nanoparticles on soil bacteria communities (Ge et al. 2011).

The nucleic acid hybridization offers a simple and effective approach to identify microorganism in complex microbial communities. The hybridization methods can be performed on previously extracted DNA/RNA or using in situ hybridization (FISH), allowing the identification of the microorganism within natural ecosystems (Amann et al. 1995; Ludwig et al. 1997). Unlike the other techniques described above, FISH is a non-PCR-based methodology that has been useful to visualize uncultured microorganisms as well as to study their spatial distribution in the environment (Muyzer and Ramsing 1995; Hill et al. 2000). As the hybridization process is performed directly on the habitat, procedures of DNA extraction and PCR amplification are unnecessary. The detection of microorganisms through FISH is carried out using specific and synthetic oligonucleotides labeled with a fluorescent dye (Moter and Göbel 2000), irradiative element (^{33}P) (Lee et al. 1999), or enzymes that transform colorless substrates in colored precipitates (Pernthaler et al. 2002). For instance, successfully identification of prokaryotes has been achieved using oligonucleotide probes able to bind selectively to the target 16S rRNA of ribosomes. The abundance of the 16S rDNA target within growing cells contributes for more sensitive results (Muyzer and Ramsing 1995). On the other hand, microbial cells with low physiological activity (slow growth or dormant cells) may not be suitable for FISH detection because of the low content of ribosomes (Muyzer and Ramsing 1995; Amann and Fuchs 2008). Besides, problems with fixation and probe permeabilization inside cells may occur, decreasing the FISH accuracy (Gao and Tao 2012; Amann and Fuchs 2008). In order to improve the sensitivity of FISH, Morris et al. (2002) utilized a mixture of four labeled probes aiming to improve the detection signal. The probes were able to improve the detection by targeting different sites of 16S rRNA of alphaproteobacteria living in a oligotrophic marine surface area. In addition, the improvement of FISH sensitivity has been achieved from the combination of peroxidase-labeled probes with catalyzed reported deposition of fluorescently labeled tyramides (FISH-CARD) (Schönhuber et al. 1997; Pernthaler et al. 2002). The microorganism distribution can be revealed by using scanning confocal, flow cytometry, and epifluorescence microscopy (Amann and Fuchs 2008; Hill et al. 2000). In the nanotoxicology field, Mu and Chen (2011) have been applied FISH technique to evaluate the long-term effect of ZnO nanoparticles on the abundance of methanogenic Archaea in anaerobic-activated sludge. Zheng et al. (2011) used FISH to analyze the long-term effects of TiO_2 nanoparticles on the abundance of nitrifying bacteria in activated sludge.

17.4 Nanomaterials: Antimicrobial Properties and Mechanisms of Action

Herein, we described the antimicrobial aspects of fullerenes, carbon nanotubes (single and multi-walled), graphene and graphene oxide, silver nanoparticles, titanium oxide, and zinc oxide nanoparticles. The main studies about antimicrobial activities of nanomaterials are summarized in Table 17.1. Moreover, we highlighted some important parameters involved in the antimicrobial activity of these nano-structured materials such as morphology, size, crystallinity, and concentration as well as their possible mechanisms of action on microbial cells.

Fullerenes: the discovery of fullerene in 1985 opened up a wide range of applications for this third allotropic form of carbon which can vary from nanoelectronics to drug delivery systems (Li et al. 2008). C_{60} is the most abundant fullerene, highly symmetrical and reactive due to its spherical molecule arrangement (Benincasa et al. 2011). On the other hand, C_{60} is insoluble in water but it can

Table 17.1 Summary of selected scientific reports concerning antimicrobial properties of nanomaterials

Nanomaterial	Microorganism tested	Typical size	Main findings	Reference
nC_{60}	*Bacillus subtilis*	Small aggregates (~2 nm) and large aggregates (97–140 nm)	MIC values raging from 0.5 to 0.95 μg/mL; membrane disruption caused by the generation of ROS	Lyon et al. (2006)
C_{60} derivative	*Escherichia coli*	Not specified	ROS inhibited the respiratory chain at a concentration of 5 μM	Nakamura and Mashino (2009)
C_{60} derivative	Herpes simplex virus	Not specified	Antiviral activity at concentration lower than 10 μg/mL	Fedorova et al. (2012)
SWCNTs	*Escherichia coli*	tube diameters of 0.75–1.2 nm	Loss of viability dose independent (1–50 μg/mL); membrane damage by direct contact	Kang et al. (2007)
SWCNTs/ MWCNTs	*Escherichia coli*	SWCNTs diameter of 1.2 nm/ length of 17.8 nm MWCNTs diameter of 17 nm/ length of 91 nm	91% of *E. coli* inactivation after explosion to SWCNTs/ MWCNTs filter	Brady-Estévez et al. (2008)
SWCNTs/ SWCNTs	*Escherichia coli*	SWCNTs diameter of 0.9 nm/ length of 2 nm MWCNTs diameter of 30 nm/ length of 70 nm	SWCNTs 80% of *E. coli* inactivation and MWCNTs 24% of *E. coli* inactivation; membrane damage by direct contact of CNT	Kang et al. (2008)
Graphene oxide	*Escherichia coli*	Nominal size of 0.3 μm	92% of loss of viability at 80 μg/mL and 10.5% at 5 μg/mL; antibacterial activity time-dependent; disruption of cell membranes by direct contact	Liu et al. (2011)
Graphene oxide	*Escherichia coli*	Not specified	25 μg/mL GO-stimulated bacterial growth; no biocide effect	Ruiz et al. (2011)
TiO_2	*Escherichia coli*	12–707 nm	Nanoparticles (smaller than 25 nm) showed 80% of	Simon-Deckers et al. (2009)

(continued)

Table 17.1 (continued)

Nanomaterial	Microorganism tested	Typical size	Main findings	Reference
			cellular inhibition after 24 h at concentration of 500 μg/mL; ROS generation and disturb of membrane integrity	
TiO_2	*Bacillus subtilis* *Escherichia coli*	Average size of 330 nm	99% growth reduction of *B. subtilis* at 2,000 ppm TiO_2 and 72 % growth reduction of *E. coli* at 5,000 ppm; photocatalytic ROS	Adams et al. (2006)
TiO_2	*Escherichia coli*	Mean size of 79 nm	75% of cellular inhibition at 100 ppm; photo catalytic production of ROS	Brunet et al. (2009)
ZnO	*Escherichia coli*	10–50 nm	Bactericidal efficiency of 85% at 1 mM ZnO suspension; MIC value of 80 μg/mL; cell damage and inactivation by ROS generation	Padmavathy and Vijayaraghavan (2008)
ZnO	*Escherichia coli* *Staphylococcus aureus*	20–25 nm	100% of *S. aureus* growth inhibition and 79 % of *E. coli* growth inhibition, both in 24 h	Mirhosseini and Firouzabadi (2012)
ZnO	*Bacillus subtilis* *Escherichia coli*	Average size of 480 nm	98% growth reduction of *B. subtilis* at 50 ppm ZnO and 48 % growth reduction of *E. coli* at 1,000 ppm; photocatalytic ROS production	Adams et al. (2006)
AgNP	*Escherichia coli*	average diameter of 12 nm	100% of cellular inhibition at concentrations of 50–60 μg/mL and 70% of inhibition at 10 μg/mL; degradation of cell membrane by the formation of irregular-shaped	Sondi and Salopek-Sondi (2004)

(continued)

Table 17.1 (continued)

Nanomaterial	Microorganism tested	Typical size	Main findings	Reference
			pits, changing membrane permeability by the release of lipopolysaccharide molecules and proteins	
AgNP	Nitrifying bacteria	9–21 nm	50% of nitrifying cells inhibition at 0.14 mg/L of AgNP; AgNP induced intracellular ROS generation and accumulation	Choi et al. (2010)
AgNP	*Escherichia coli*	Average size of spherical nanoparticles of 39 nm; rod-shaped ranging 133–192 nm in length; truncated-triangular of 50 nm	Shape-dependent interaction: 100% inhibition of bacteria growth by spherical AgNP at 50–100 μg; 100% inhibition of bacteria growth by triangulat AgNP at 1 μg and rod-shaped at 100 μg presented some colonies. Bacterial cell in contact with AgNP inhibits a respiratory enzyme(s), facilitating the generation of ROS and damaging the cell	Pal et al. (2007)
AgNP	*Escherichia coli* *Vibrio cholera* *Pseudomonas aeruginosa* *Staphylococcus typhus*	1–100 nm	75 μg/mL AgNP and general average size of 21 nm inhibited 100% of all bacteria strains; AgNPs can attach to the cell membrane and disturb its function; AgNPs can penetrate thc bacteria and interact with sulfur and phosphorous compounds and AgNPs can release silver ions	Morones et al. (2005)

become soluble by chemical modifications (Li et al. 2008). Fullerenes can also form water-soluble-C_{60} aggregates, which are C_{60} dissolved in a suitable solvent followed by water addition in order to remove the same solvent, giving rise to nanoscale C_{60} aggregates in suspension, denominated nC_{60} (Lyon et al. 2006). These aggregates are the most significant environmental form of C_{60}. The toxicity of C_{60} depends on the fullerene derivative and nC_{60} has shown high antibacterial activity (Lyon et al. 2006). In this way, Lyon et al. (2006) reported the antimicrobial activity of nC_{60} against *Bacillus subtilis*. The authors prepared four different nC_{60} samples: in THF solvent (THF/nC_{60}); C_{60} sonicated in toluene/water (son/C_{60}); C_{60} powder mixed in water (aq/nC_{60}); and C_{60} solubilized in polyvinylpyrrolidone (PVP/C_{60}). The toxicity assays revealed MIC values ranging from 0.5 to 0.95 μg/mL, except for the THF/nC_{60} sample, which achieved the lowest MIC value of 0.09 μg/mL. Moreover, they separated small (average size of 2 nm, except for THF/C_{60}, which was 40 nm) and large (sizes varying from 97 to 140 nm) aggregates through centrifugation in order to investigate the aggregates size/morphology influence on the antibacterial activity and a nonlinear correlation was found. The mechanism responsible for the antimicrobial activity is quite unclear, but Lyon et al. (2006) believe that toxic ROS (reactive oxygen species) may disrupt cell membranes causing oxidation of several organelles. The C_{60} powder itself showed no antibacterial response.

Afterwards, Lyon et al. (2008) studied the THF/nC_{60} mechanism of antibacterial activity toward *Escherichia coli* and the results indicated that the mechanism cannot be attributed to the ROS production. The result is interfered by direct nC_{60} oxidation of dyes, leading to false ROS positive. They found that there is a lack of protein oxidation and the remaining of antimicrobial property in the absence of either light or oxygen. Thus, nC_{60} played a ROS-independent oxidative stress to the cells.

As aforementioned, water-soluble fullerene derivatives can behave different from unmodified C_{60}. Nakamura and Mashino (2009) showed that cationic fullerenes derivative C_{60}-bis (*N,N*-dimethylpyrrolidinium iodide) inhibited the metabolism of *E. coli*. The tests of antibacterial activity were performed measuring the oxygen (O_2) consumption in the presence of glucose. The mechanism of action was suggested to be associated with the inhibition of the respiratory chain caused by the ROS produced by the cationic fullerenes. In addition, the *E. coli* growth was completely inhibited at a fullerene derivative C60-bis concentration of 5 μM. Fullerene triazine derivatives also showed good antibacterial properties toward *E. coli*, *Staphylococcus aureus*, *Bacillus subtilis*, *Bacillus pumilis*, *Pseudomonas aeruginosa,* and *Klebsiella pneumoniae*. The antibacterial screening was conducted by disc diffusion method (Kumar and Menon 2009). *Bacillus subtilis* strains showed the lowest MIC results, ranging from 0.08 to 1.0 μg/mL. The triazine moiety is found to disrupt the cell membrane. Polycationic fullerene derivatives have demonstrated light-activated antimicrobial activity against *Candida albicans* through ROS production upon illumination. This yeast was completely eliminated at 100 μM and showed considerable toxicity in dark conditions (Huang et al. 2010).

Antiviral activity of fullerene derivatives was reported by Fedorova et al. (2012). They showed that water-soluble polycarboxylic C_{60} derivatives inhibited herpes simplex virus into fibroblasts culture cells. All compounds presented anti-virus activity at concentrations lower than 10 μg/mL. Despite the fullerenes and fullerenes derivatives are potential materials for application in nanomedicine, their antimicrobial properties are still under investigation.

Carbon nanotubes: carbon nanotubes (CNTs) have excellent electronic, optical, and mechanical properties. They can be classified according to the number of graphitic-rolled sheets into single-walled (SWCNT) and multi-walled (MWCNT) nanotubes. Therefore, CNTs have been explored in the manufacture of different devices, e. g. gas sensors, biological sensors, atomic force microscopy probes, nanocomposites, and so forth (Souza-Filho and Fagan 2007).

Considering the diameter, SWCNTs have shown to be more toxic toward bacteria than MWCNTs and fullerenes (Musee et al. 2011). In this direction, Kang et al. (2007) reported that highly purified SWCNTs exhibited strong antimicrobial activity. Pristine SWCNTs damaged the outer *E. coli* cell membranes by direct contact, causing intracellular content release and consequent inactivation. The loss of viability of *E. coli* cells was independent of the SWCNTs dose, which varied from 1 to 50 μg/mL. Oppositely, the percentage of the loss of viability was dependent on incubation time, reaching 88% for cells exposed to SWCNTs for 120 min.

In a latter study (Kang et al. 2008), the same team compared the antimicrobial activity of purified SWCNTs and MWCNTs and found that SWCNTs exhibited much stronger biocide activity toward *E. coli*. SWCNT aggregates at a concentration of 5 μg/mL inactivated 80% of *E. coli* cells, whereas MWCNTs inactivated 24%. The bacteria cell membranes are damaged by the direct contact with SWCNT (Fig. 17.2A) (Kang et al. 2008). The major toxicity of SWCNTs may be due to their smaller diameter/length, large surface area, and unusual physicochemical properties which facilitate the penetration of the tubes into the cells.

Brady-Estévez et al. (2008) demonstrated the antiviral properties of a hybrid water-filter of SWCNTs-MWCNTs. The filter is composed by a thin SWCNTs layer (0.05 mg/cm^2) on top of a thicker MWCNTs layer (0.27 mg/cm^2), the latter responsible for the majority of viral removal. The hybrid filter presented high removal of several model viruses (MS2, PRD1, and T4 bacteriophages). The filter was also able to inactivate bacteria strains of *E. coli* and *Staphylococcus epidermidis*, mainly on the top SWCNTs layer.

Carbon nanotubes can undergo chemical functionalization in order to facilitate the interaction of these nanostructures with organic and biological molecules, and with other chemical groups, e.g., drugs and even viruses and bacteria. Therefore, MWCNTs can have their weak antimicrobial activity boosted by chemical modifications. Qi et al. (2011) prepared a MWCNTs composite based on the covalent incorporation of nisin, an antimicrobial peptide, on its surface, using poly(ethylene glycol) (PEG) as linker. The novel composite MWCNTs-nisin has superior antibacterial and anti-biofilm properties compared to pristine MWCNTs, resulting in 1.5–7 times higher antibacterial activity against *E. coli*, *P. aeruginosa*,

B. subtilis, and *S. aureus* than pristine MWCNTs. The MIC of MWCNT-nisin was 10 μg/mL and it is worth to say that the anti-biofilm activity was reached even at lower values of 1.25, 2.5, and 5 μg/mL. The anti-biofilm activity of MWCNT-nisin can be attributed to the decrease of the cellular viability and motility, caused by the interaction between the composite and the bacterial cells.

Non-covalent functionalization can also occur. For instance, Gao et al. (2012) prepared oxidized-MWCNTs decorated with silver nanoparticles (AgNP) by UV light reduction method. The spherical AgNP with average size from 5 to 10 nm were adsorbed onto the MWCNTs surfaces. The antibacterial activity was investigated by plate counting and disc diffusion method. A 100% of inhibition of the Gram-negative *Chlorophenols Arthrobacter* was achieved in the condition of 1.6 mg of MWCNT-AgNP in the bacterial culture medium. Inhibition halos showed that pristine MWCNTs presented lower activity compared to the composite. The antibacterial activity can be caused by a combination of different mechanisms, for AgNP and MWCNTs. Furthermore, functionalized carbon nanotubes can present antifungal activity as reported by Benincasa et al. (2011). The authors prepared SWCNTs and MWCNTs conjugated with Amphotericin B (AMB), yielding water-soluble conjugates. All conjugates presented broad-spectrum antifungical activity, toward strains of *Candida spp.* and *Cryptococcus neoformans*. Interestingly, the conjugates also exhibited high activity against the resistant strains *C. albicans* and *C. famata*. The mechanism of action of the conjugates was suggested to be a nonlytic mechanism, since the materials showed permeabilizing effect on the fungal strains only after a long period of incubation.

Although the antimicrobial activity of carbon nanotubes cannot be attributed to specific factors, parameters as size (diameter/length), microchemical environment on the surface, and aggregation/dispersion state are correlated to their microbial cytotoxicity (Musee et al. 2011).

Graphene: graphene is a single atom-thick sp^2-hybridized carbon layer arranged in a 2D-hexagonal network. Graphene oxide (GO) is a water-soluble chemically modified graphene which contains a wide range of reactive functional groups, e.g., hydroxyl, carboxyl, carbonyl, and epoxy. The elimination of oxygenated groups by a reduction process results in graphene sheets (reduced graphene oxide). These related-graphene nanomaterials show singular electronic, mechanical, and thermal properties. Their field of applications includes nanocomposites, supercapacitors, nanoelectronics, sensors, molecular vehicles, and nanomedicine platforms (Faria et al. 2012).

Currently, there are a few numbers of antimicrobial studies involving graphene-based materials in the literature (Das et al. 2011; Liu et al. 2011; Ruiz et al. 2011). Liu et al. (2011) reported the graphene oxide antibacterial activity against *E. coli* cells. The toxicity assays were carried out by the counting plate method. GO showed strong biocide effect, reaching a loss of *E. coli* viability of 70%. Time-dependent experiments pointed out that the loss of viability of the bacteria cells increased with time, ranging from 49% after 1 h of incubation to 90% after 4 h, for GO dispersion. Furthermore, the bactericidal activity is dose-dependent, e.g., GO dispersion at concentration of 5 μg/mL led to 10% of loss of viability, whereas a concentration of 80 μg/mL resulted in 92% of killing. By SEM images, it was found that GO caused irreversible damages to the cells and the bacteria cells were found wrapped into the GO thin layers. Thus, the antibacterial mechanism can be assigned to membrane oxidative stress by the direct contact of graphene-based materials with bacteria cells, resulting in the release of intracellular material and consequent inactivation of the cells. In a previous work, Hu et al. (2010) demonstrated that the GO nanosheets at a concentration of 85 μg/mL caused 99% loss of *E. coli* viability after 2 h of exposure time. Moreover, the GO dispersion (85 μg/mL) induced the decrease of bacteria metabolic activity to 13%. As aforementioned, SEM images showed clearly that the membrane integrity of *E. coli* cells was severely damaged, leading to release of the cytoplasm content. This effect may be attributed to the oxidative stress and physical disruption of cellular membranes.

Antibacterial activity can be related to the lateral dimension of the GO sheets, as reported by Liu et al. (2012). Larger GO sheets (average size of 0.7 μm^2) in aqueous suspensions (40 μg/mL) indicated a loss of viability of 98% toward *E. coli* cells, whereas the smaller ones (average size of 0.01 μm^2) resulted in 45% of loss. The possible mechanism involves the ability of larger sheets in covering cells more easily, causing the inactivation of the cell by the blockage of active sites on membranes.

Moreover, the antibacterial activity of GO is an issue under debate by far, and further investigation is required due to the lack of consistent results. For instance, Ruiz et al. (2011) conducted a study to elucidate the antimicrobial activity of GO and, oppositely, the results suggested that GO enhanced *E. coli* cellular growth, favoring the formation of dense biofilms. Thus, the GO sheets acted as a surface for bacterial cell attachment and proliferation.

Recently, graphene-based materials nanocomposites have attracted great attention regarding to their antimicrobial activity. For instance, Das et al. (2011) prepared a nanocomposite based on the decoration of graphene oxide sheets with silver nanoparticles (GO-Ag) using $NaBH_4$ as a reducing agent. *Pseudomonas aeruginosa* showed to be more sensitive to the GO-Ag toxicity since its inhibition zones were larger when compared to *E. coli*. Otherwise, GO itself did not show antibacterial activity toward both bacteria strains. The inhibition mechanism is attributed to the contact of silver nanoparticles, Ag^+ ions, and ROS with proteins and enzymes of the respiratory chain.

Ma et al. 2011 also evaluated the antibacterial activity of graphene oxide decorated with silver nanoparticles against *E. coli* cells. The nanocomposite (GO-Ag) showed high antibacterial activity in a slurry reactor, since no more viable *E. coli* cells were detected after 10 min of contact with the nanocomposite. The authors performed TEM images in order to investigate the morphology alterations on *E. coli* cells after direct contact with GO-Ag nanocomposites, as shown in the Fig.17.2D (Ma et al. 2011).

Park et al. (2010) produced a biocompatible paper-like material composed by Polyoxyethylene sorbitan laurate (TWEEN) and reduced graphene oxide (RGO). TWEEN-RGO papers inhibited the growth of *Bacillus cereus* cells. In contrast, RGO-paper without the presence of TWEEN showed a nonspecific bacterial binding. Some et al. (2012) synthesized a novel graphene-based composite through electrostatic interaction between GO and poly (L-lysine) (PLL). In fact, the novel composite exhibited high antibacterial activity toward *E. coli* cells, whereas only GO samples did not present any antimicrobial activity, in accordance to previous works. As exposed above, the biological properties of graphene derivatives still require further investigation. The toxicity toward bacteria strains is under debate hitherto and a possible reason may be attributed to the inclusion of contaminants during the preparation of graphene-based nanomaterials.

Silver nanoparticles: metallic nanoparticles play an important role in the biological and biomedical field, and their applications have been gaining popularity due to their antimicrobial properties (Matthews et al. 2010). Some metals display antibacterial properties in their bulk form, e.g., zinc, silver, and copper, whereas other materials exhibit antibacterial activity only in their nanoscale form. The major variables that influence the antimicrobial activity may be related to the particle size/morphology, zeta potential, and chemistry of surface. Thus, the mechanisms of bactericidal action can vary according to the nanoparticle nature (Seil and Webster 2012).

Silver antimicrobial activity has been known for over centuries and its scaling down to the nanoparticles provides enhanced biocidal properties due to the higher specific surface area when compared with their bulk counterparts (Matthews et al. 2010; De Lima et al. 2012).

Silver nanoparticles with average diameter of 12 nm were prepared through the reduction of Ag^+ ions by ascorbic acid and their antibacterial activity was tested against *E. coli* cells (Sondi and Salopek-Sondi et al. 2004). AgNPs inhibited *E. coli* growth by a dose-dependent relation: AgNPs concentration of 10 μg/mL inhibited bacterial growth by 70%, whereas concentrations of 50–60 μg/mL resulted in 100% of inhibition of bacterial growth. The degradation of the membrane structure may be explained by the formation of irregular-shaped pits on the cell membrane, which are able to modify the membrane permeability through the release of LPS molecules and proteins (Fig.17.2B) (Sondi and Salopek-Sondi et al. 2004). Morones et al. (2005) reported that the antibacterial activity of AgNPs is size-dependent; nanoparticles ranging from 1 to 100 nm interacted preferentially with bacteria. At the concentration of 75 μg/mL and general average size of 21 nm, there was no growth of any strains (*E. coli*, *V. cholera*, *P. aeruginosa,* and *S. typhus*.).

Silver nanoparticles have also displayed antifungal activity toward pathogenic *Candida* spp. (Panacek et al. 2009). The lowest MIC was found against *C. albicans II* at a value of 0.21 mg/L. The highest MIC value of 1.69 mg/L was showed against *C. parapsilosis*, showing be the less sensitive yeast. In addition, silver nanoparticles are able to inhibit some viruses (Elechiguerra et al. 2005). According to the authors, the antiviral activity is provided by the attachment of the AgNPs on the host cells surface and also on the gp120 glycoproteins localized on the virus capsid. Figure 17.1 illustrates the main mechanisms of action of AgNPs on bacterial cells. According to the Fig. 17.1, three different mechanisms can explain the antibacterial activity of the silver nanoparticles: (1) Ag^+ release which can bind to enzymes, proteins, and nucleic acids. The interaction of Ag^+ and proteins is

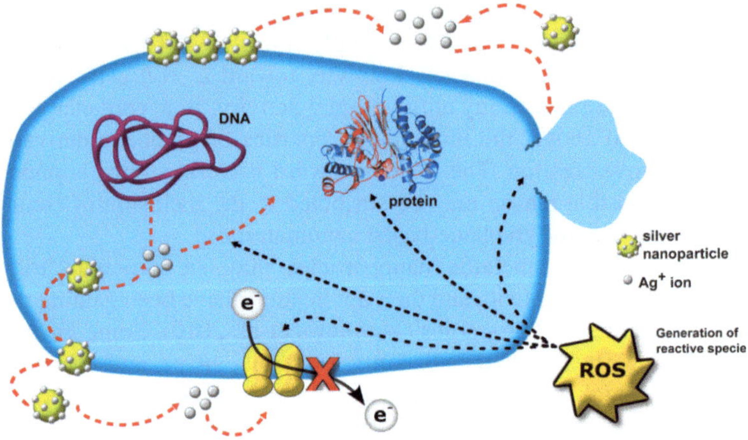

Fig. 17.1 Illustration showing the possible mechanisms of action of silver nanoparticles on bacterial cell. Similar pictures have been published in Marambio-Jones and Hoek (2010) and Li et al. (2008)

facilitated by the chemical affinity existent between thiol groups (SH) of proteins and the silver ions (Durán et al. 2010); (2) generation of ROS (reactive oxygenated species) in the presence of oxidant conditions. ROS can act as an electron acceptor, causing irreversible damages to protein and DNA structures. Outside the cell, ROS may affect cell membrane and membrane proteins (Park et al. 2009; Marambio-Jones and Hoek 2010); (3) Direct interaction of silver nanoparticles with wall and cell membranes. The nanoparticles are also able to accumulate and adhere to the membrane, forming holes and altering its permeability, leading to the release of the cytoplasmic content (Sondi and Salopek-Sondi 2004; Raffi et al. 2008).

Titanium oxide nanoparticles (TiO_2): these nanoparticles are known as photocatalytic material and their bactericidal effects have been investigated for few years. A range of TiO_2 nanoparticles (average sizes varying 12–707 nm) were exposed to *E. coli* strain. The smallest and round-shape particles (lower than 25 nm) showed enhanced antibacterial activity, reaching 80% of loss of cell viability at the concentration of 500 μg/mL after 24 h. TiO_2 increased the ROS generation and disturbed the membrane integrity (Simon-Deckers et al. 2009).

Generally, the antibacterial activity of TiO_2 nanoparticles are associated to ROS production, mainly hydroxyl free radicals and peroxide generated under UV-A irradiation. On the other hand, photocatalytic activated by sunlight would be more attractive for commercial applications (Li et al. 2008). The doping of TiO_2 can lead to this situation. The antibacterial activity of titania silver-doped coatings was screened toward *S. aureus*, *E. coli,* and *B. cereus* strains (Page et al. 2007). The films have superior photocatalytic and antimicrobial properties than titania coatings. Moreover, no activity was found in the dark, which means silver ions are not responsible for antimicrobial action. The coatings presented enhanced photocatalytic inactivation against *S. aureus* and *B. cereus*, achieving values of 99% of inhibition after 6 h of illumination (Page et al. 2007). TiO_2 nanoparticles have also showed activity against

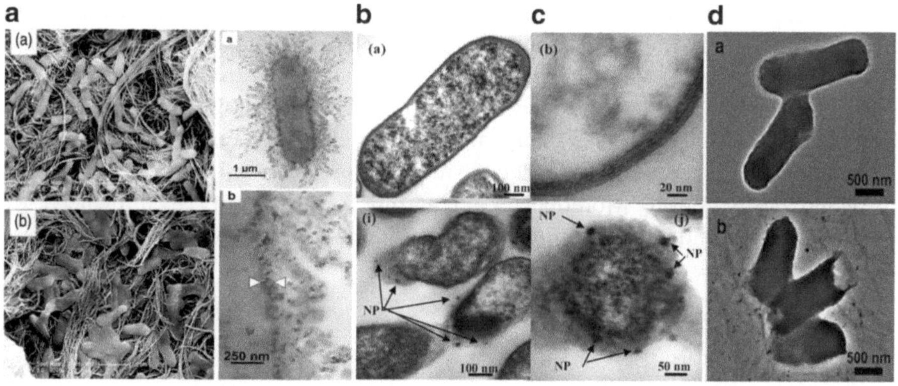

Fig. 17.2 (**A**) Scanning electron microscopy (SEM) of *E. coli* cells after 60 min of exposition to (*a*) multi-walled carbon nanotubes (MWCNTs) and (*b*) single-walled carbon nanotubes (SWCNTs) (Kang et al. 2008); (**B**) (*a*) transmission electron microscopy image (TEM) of *E. coli* cells exposed to 50 µg/mL of silver nanoparticles, (*b*) enlarged cellular membrane showing the attachment of silver nanoparticles on cellular membrane (Sondi and Salopek-Sondi 2004); (**C**) TEM images of *E. coli* thin section: (*a, b*) *E. coli* cells grown in Luria Bertani Broth (LB) and (*i, j*) *E. coli* cells after contact with ZnO nanoparticles (Brayner et al. 2006); (**D**) TEM images of *E. coli* cells: (*a*) control (without exposure to nanomaterial) and (*b*) treated with GO-Ag dispersion for 5 min (Ma et al. 2011)

some viruses; e.g., poliovirus 1, hepatitis B virus, Herpes simplex virus, and MS2 bacteriophage (Li et al. 2008).

Zinc oxide nanoparticles (ZnO): these particles find applications on sunscreens, paints, and food safety due to their elevated UV absorption. Moreover, ZnO nanoparticles present high antimicrobial activity against a range of bacteria strains, in particular *S. aureus* (Li et al. 2008). Mirhosseini and Firouzabadi (2012) investigated ZnO antibacterial activity against *E. coli* and *S. aureus*. As aforementioned, ZnO nanoparticles were more effective toward *S. aureus* strain, showing inhibition zones of 10 and 12 mm at the concentrations of 5 and 10 mM, respectively. Moreover, ZnO completely inhibited *S. aureus* growth after 24 h and suppressed the growth of *E. coli* in 79% at the same time of exposure. ZnO synthesized in polyol medium (di (ethylene glycol)-DEG) also showed damaging effects toward Gram-negative *E. coli* cells, as demonstrated by Brayner et al. (2006). The presence of ZnO nanoparticles at concentrations higher than 1.3×10^{-3} M increased the membrane permeability, allowing the leaking of cell contents and nanoparticles accumulation on the bacterial membrane (Fig. 17.2C) (Brayner et al. 2006).

In a study designed to investigate the particle size effect on the antibacterial activity of ZnO, Padmavathy and Vijayaraghavan (2008) determined that nano-sized ZnO (average size from 10 to 50 nm) presents more effective biocidal activity toward *E. coli* compared to ZnO bulk (2 µm). ZnO nanoparticles with 12 nm at concentration of 1 mM showed biocide efficiency of 85%. Those particles presented MIC value of 80 µg/mL. They concluded that smaller ZnO nanoparticles displayed an improved biocide activity against bacteria cells due to their greater surface area. The ZnO antibacterial activity was found to be related to ROS production, which caused damages on cell membrane and further inactivation (Padmavathy and Vijayaraghavan 2008). In another report, nano-sized ZnO with 12 nm inhibited *S. aureus* and *E. coli* growth in 98% and 100%, respectively. The antibacterial activity of ZnO nanoparticles was also associated with the ROS production, suggesting that the mechanism of action is not only restricted to direct mechanical disruption. Moreover, ZnO antimicrobial activity is influenced by several factors, and the literature has been suggesting that smaller particles are more effective due to their large surface area.

17.5 Antimicrobial Nanomaterials and Industrial Applications

A great variety of nanomaterials such as silver, TiO_2 and ZnO nanoparticles, etc., present industrial applications in detriment of their antimicrobial properties. Among these nanomaterials, the silver nanoparticles are the most used and powerful antimicrobial nanomaterial by far, exhibiting high toxicity against a broad range of microorganisms and low human toxicity (Dallas et al. 2011). AgNPs display more than 250 available commercial products which can vary from antimicrobial surface coatings (appliances, food storage containers, cutting boards, etc.) to textiles and hygiene products (Duncan 2011).

Products such as home water purification systems Aquapurey®, Kinetico®, and QSI-Nano® are able to kill an average of 99% of the microorganisms by using membranes impregnated with silver nanoparticles (Maynard 2007). AgNPs have been used regularly in clinical dressings, creams, gel, and burns treatment, decreasing several infections in chronic wounds (Rai et al. 2009). Furthermore, the AgNPs can be employed as coating agents to produce catheters which prevent the microbial contamination and consequently reduce the transmission of infectious diseases, (Dallas et al. 2011). AgNPs may also be incorporated in other medical devices such as surgical masks, cardiovascular implants, esophageal tubes, bandages, sutures, and other implantable devices (Rai et al. 2009; Duncan 2011).

In the field of textiles, AgNPs have been applied in a large number of commercial biocidal synthetic or natural fibers and yarns. For instance, Fresh®, Silpure®, and Bioactive® products display AgNPs incorporated into polyester fabrics. SmartSilver® are applied to wool, retaining the handling and dyeability properties of the original wool. Such antimicrobial textiles can prevent the development of skin diseases/contamination and avoid the generation of odors. Besides, these materials show a controlled release of the antibacterial agent in order to extend the biocidal activity even after multiple cleaning cycles (Gao and Cranston 2008).

AgNPs incorporated into polymers have been showed potential application as food packaging material. Polymers such as poly(acrylamide), poly(vinyl acetate) (PVA), polyvinylpyrrolidone (PVP), polyethylene (PE), poly(methyl methacrylate)

(PMMA), polyurethane (PU), polyethylene oxide (PEO), alginate, silicon elastomer, cellulose, and chitosan can be functionalized with AgNPs, producing antimicrobial composites able to improve the food shelf-life (Duncan 2011).

Titanium dioxide (TiO_2) has been applied in consumer products due to its promising antimicrobial activity through photocatalysis. It finds applications as air purifiers to remove organic volatiles and kill bacteria, for example the product 3Q™ Multi Stage Air Purifier and Nanobreeze® and water purification systems (eg. Purifics-r) (Li et al. 2008). TiO_2 has also food packaging potential through the incorporation of TiO_2 nanoparticles in polymers such as polypropylene, inhibiting microbial growth onto vegetables surfaces. Regardless their antimicrobial properties, the nano-sized TiO_2 is mainly applied in the degradation of organic contaminants (Duncan 2011).

Carbon nanotubes-based materials find potential application in drinking water treatment/disinfection due to their effective ability to remove pathogens. Microfilters based on SWCNTs and MWCNTs showed high removal and inactivation rate of viruses and bacteria pathogens (Vecitis et al. 2011). Moreover, CNTs may act as scaffolds for antimicrobial AgNPs or TiO_2, enhancing their properties aforementioned (Mauter and Elimelech 2008). Fullerene derivatives, e.g., fullerol and C_{60} aggregates, can also be applied to water/wastewater treatment due to their antimicrobial activity toward viruses and bacteria (Mauter and Elimelech 2008).

Zinc oxide nanoparticles (ZnO) incorporated into polymers also revealed antimicrobial activity, finding applications mainly as food packaging material. ZnO can form antimicrobial composites based on glass, low density polyethylene (LDPE), polypropylene (PP), polyurethane (PU), paper, and chitosan (Espitia et al. 2012).

17.6 Toxicological Aspects of Nanomaterials to Microorganisms in Water and Soil Environments

The development of nanomaterials has been rising in the last few years and a large number of marketed products have gained applications in textiles, water disinfection (Savage and Diallo 2005), antimicrobial coatings (Furno et al. 2004), cosmetics/hygiene (Mu and Sprando 2010), and so forth. The successful use of nanomaterials in industrial processes is in expansion, since U$147 billion was reported for nano-structured products in 2007 and this value can reach U$3.1 trillion up to 2015 (Jusko 2009). On the other hand, concerning about the impact of nanomaterials on human and environment health is also growing rapidly, especially for regulatory agencies (Klaine et al. 2008). Even though nanomaterials have been applied to manufactured products, their toxicological parameters and environmental consequences are still unclear (Turco et al. 2011). The entry of nanomaterials into the environment can occur through the release of compounds originated from the recycling or decomposition of complex nanocomposites. Besides that, nanomaterials incorporated in commercial products such as paints, fabrics, and cosmetics may access the environment by their use and/or laundering procedures (Klaine et al. 2008). In contrast, the direct introduction of pure forms of nanomaterials into the environment is somewhat unlikely, except in the case of industrial spills where raw forms of nanomaterials can contaminate soil and water (Turco et al. 2011). Table 17.2 summarizes scientific reports concerning the impact of nanomaterials on microbial communities in natural environments.

Antimicrobial properties of nanomaterials have been investigated extensively and most of reports have focused on the effect of nanomaterials against pure cultured microorganisms (Sondi and Salopek-Sondi 2004; Morones et al. 2005). However, few studies have been developed in order to understand the impact of nanomaterials on complex microbial communities (Nogueira et al. 2012; Sun et al. 2013). The evaluation of microbial communities is important because they play essential role as facilitators of nutrient recycling, agents of waste decomposition, and degradation of xenobiotic molecules (Kang et al. 2009). In addition, the knowledge about the toxicity of nanomaterials to microbial communities may indicate the real potential impact of nanomaterials on the environment, since the destructuring of these microbial communities may implicate on the equilibrium of the local ecosystem. Moreover, microbial communities can be easily evaluated by using advanced molecular techniques (Terminal Restriction Fragment Length Polymorphism—T-RFLP and Denaturing Gradient Gel Electrophoresis—DGGE) (Muyzer and Ramsing 1995; Kitts 2001).

The impact of nanomaterials on soil microbial communities can be estimated by using basal CO_2 production, enzymatic activity, fatty acid profile, and extraction of total genomic followed by analysis of diversity by DGGE, T-RFLP, and environmental microarrays. Tong et al. (2007) evaluated the impact of fullerene C_{60} and C_{60} aggregates in suspension (nC_{60}) on a soil microbial community. They treated soil microcosms with 1,000 μg of granular C_{60} per gram of soil and nC_{60} at a concentration to achieve 1.0 μg per gram of soil and the environmental toxicity was monitored by soil respiration, enzymatic activities, and diversity analysis using DGGE technique. The results suggested that treated soils with either C_{60} or nC_{60} exhibited a complex and diverse pattern indicating that their addition in aerobic soil microcosms caused little impact on microbial community structure. In the same way, the authors related that the enzyme activity and soil respiration were not adversely affected after introduction of C_{60} and nC_{60} nanomaterials. Nogueira et al. (2012) investigated the effect of polymers, e.g., carboxylmethyl-cellulose (CMC), hydrophobically modified polyethylglycol (HM-PEG), and inorganic nanomaterials ($TiSO_4$ and TiO_2 nanoparticles) on the structural diversity of the soil microbial community through DGGE analysis. The analyses of similarities obtained from DGGE profiles showed that polymers, TiO_2, and gold nanoparticles affected significantly the structural diversity on the microbial community, whereas CdSe/ZnS quantum dots, Fe/Co magnetic fluid, and titanium silicon oxide presented a lower effect. Bradford et al. (2009) studied the impact of silver nanoparticles (AgNPs) on microbial diversity of estuarine sediment samples. The experiment was conducted using AgNPs treatments of 0.0 (control), 25, or 1,000 μg/L. Nevertheless, DGGE profiles revealed insignificant differences in the

Table 17.2 Toxicological aspects of nanomaterials to microbial communities of soil, estuarine, and activated sludge environments

Nanomaterial	Concentration	Environment	Main findings	Reference
C_{60}/*n*C_{60}	1,000 µg of granular C_{60} per gram of soil and 1.0 µg of *n*C_{60} per gram of soil	Soil microbial community	The addition of C_{60} and *n*C_{60} to aerobic microcosms caused little impact. Enzymatic activity and soil respiration was not affected by the introduction of nanomaterials	Tong et al. (2007)
Polymeric nanocomposites	10 g of carboxymehtylcellulose per kg of soil and 10 g of hydrophobically modified polyethylglycol pr kg of soil	Soil microbial community	DGGE profiles showed that the polymeric nanocomposites affected significantly the structural diversity on the microbial community	Nogueira et al. (2012)
TiO_2	5.0 g per kg of soil	Soil microbial community	TiO_2 presented a lower effect on soil	Nogueira et al. (2012)
AgNP	25 µg/L and 1,000 µg/L	Estuarine sediment	DGGE profiles suggested that AgNP presented little or no impact on estuarine sediment bacterial diversity community	Bradford et al. (2009)
AgNP	1.0 mg/L	Activated sludge	Bacteroidetes and Proteobacteria microbial species were sensitive to AgNP treatment	Sun et al. (2013)
SWCNTs	219 mg/L	Activated sludge	The presence of SWCNTs affected the population of sphingomonad	Goyal et al. (2010)

bacterial diversity among the treatments, thus suggesting that AgNPs presented little or no impact on the bacterial diversity community of a estuarine sediment. Kirschling et al. (2010) evaluated the impact of zero valent iron (NZVI) on microbial communities of aquifers contaminated with trichloroethylene (TCE). According to Kirschling et al. (2010), NZVI are promising candidates to reduce TCE contamination in subsurface zones. However, the understanding about the toxicological effects of NZVI on indigenous communities is limited. Henceforth, they used microcosms to determine the effects of NZVI on materials collected from three different aquifers (Alameda Point-CA, Marcelone-MI, and Parris Island-SC). NZVI was added at concentration of 1.5 g/L. DGGE analysis showed significant shifts in eubacteria population only after NZVI exhaustion. Otherwise, quantitative PCR showed an increase on the expression of dsrA gene (sulfite reductases gene) and Archaeal 16S rRNA genes, suggesting that NZVI presence stimulated both sulfate reducer and methanogen communities. The NZVI coated with a biodegradable polyaspartate was also evaluated and the results suggested that the coating presence increased the bacterial population in one order of magnitude when compared to the control.

One of the major difficulties regarding the assessment of nanomaterials toxicity in terrestrial ecosystem is related to their aggregation phenomena. Aggregation processes are dependent on the particle features (size, surface area, and intrinsic magnetic moment) and also can be influenced by the environmental characteristics, including pH, ionic strength, and organic matter contents (Klaine et al. 2008; Yan et al. 2010). Overall, there is a lack of information about the behavior, transport, and fate of nanomaterials through soil profile (Yan et al. 2010).

As aforementioned, nanomaterials are present in commercial products, thus it is inevitable that they will be released to water streams during their life cycle. Consequently, it is clear that nanomaterials might be retained in biological waste water treatment systems (Kiser et al. 2009; Sun et al. 2013). For instance, nanomaterials are able to absorb on the activated sludge which is a mixture of microorganisms, extracellular biopolymers, and organic/inorganic particles. In fact, activated sludge represents the first microbial community to encounter waterborne contaminants.

Sun et al. (2013) evaluated the toxic effect of 1 mg/L of AgNPs on the complex microbial communities present in activated sludge. The toxic effects of AgNPs were analyzed using 16S rRNA gene-based polymerase chain reaction-DGGE. The results demonstrated a reduction on the DGGE gel bands related to certain microbial species (Bacteroidetes and Proteobacteria), concluding that these specific bacterial groups were sensitive to AgNPs treatment. Goyal et al. (2010) studied the impact of SWCNTs on microbial communities of activated sludge batch scale reactors through automated ribosomal intergenic spacer analysis (ARISA). The 16S rRNA gene sequence analysis indicated that the activated sludge was constituted by members of the families Sphingomonadaceae, Cytophagacaceae, and Zoogloea. ARISA results showed that the presence of SWCNTs affected the population of sphingomonad when compared to the experimental control. Oppositely, Nyberg et al. (2008) found no significant effect on the anaerobic

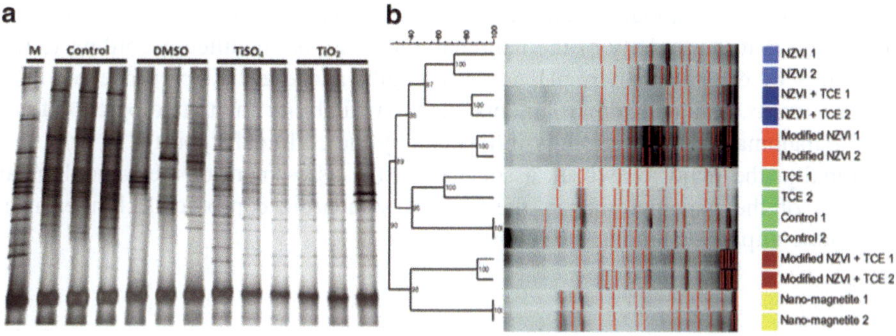

Fig. 17.3 (**a**) DGGE profiles of PCR-amplified 16S rDNA gene fragments of bacterial communities of soil samples treated with water (control), DMSO, $TiSO_4$ nanoparticles, and TiO_2 nanoparticles (Nogueira et al. 2012); (**b**) Eubacterial DGGE profile of Alameda-CA microcosm impacted with NZVI (*light blue*), polyaspartate-NZVI modified (*light red*), modified NZVI plus trichloroethylene (TCE) (*dark red*), NZVI plus TCE (*dark blue*), control (*light green*), TCE (*dark green*), and nanomagnetite (*yellow*), for more information see Kirschling et al. (2010)

communities of Bacteria, Archaea, and Eukarya domains when 50 g of C_{60} was added to 1 kg of wastewater treatment sludge. Figure 17.3 shows the DNA fingerprint of soil samples impacted with $TiSO_4$ and TiO_2 nanoparticles (Nogueira et al. 2012) and with zero valent iron (NZVI) (Kirschling et al. 2010).

One of the key points in the assessment of toxicity of nanomaterials in aquatic ecosystems is that these materials are poorly soluble and susceptible to aggregation phenomenon, thus hindering the quantification of their real concentration in suspension and reducing their bioavailability to the microorganisms (Klaine et al. 2008). Besides, it is important to stress that the physicochemical and the toxicological properties of these nanomaterials will be modified if natural organic matter, polysaccharides, detergents, and other biomacromolecules attach on their surface (Kang et al. 2009).

17.7 Comments and Future Perspectives

Currently, there is a growing interest on the application of nanomaterials as antimicrobial agents. Actually, the use of nanomaterials can contributes to reduce the microbial resistance, usually associated with the frequent use of conventional antibiotics. However, the nanomaterials mechanisms of action are not fully understood and their complete elucidation remains a challenge for future researchers. In addition, only few reports have been focused in study the impact of nanomaterials to the environment and scarce works have been designated to explain the influence of nano-sized materials on crucial ecosystems.

The toxicological effects of nanomaterials to microbial communities have been described, however their long-term effects are still undetermined. However, even though the studies about the environmental impacts are still limited, they already identified that the toxicity of the nanomaterials may be modified according to the presence of organic matter, pH, ionic strength, and eventual presence of other contaminants. In aquatic environments, the bioavailability and toxicological effects of the nanomaterials are strongly influenced by the colloidal stability.

Finally, the major drawback associated to the assessment of the toxicological aspects of the nanomaterials is the lack of standardized protocols, hindering a parallel comparison among the reported data.

References

Adams DJ (2004) Fungal cell wall chitinases and glucanases. Microbiology 150:2029–2035

Adams LK, Lyon DY, Alvarez PJJ (2006) Comparative ecotoxicity of nanoscale TiO_2, SiO_2, and ZnO water suspensions. Water Res 40:3527–3532

Ahmed F, Santos CM, Vergara RAMV et al (2012) Antimicrobial applications of electroactive PVK-SWNT nanocomposites. Environ Sci Technol 46:1804–1810

Alongi DM (1994) The role of bacteria in nutrient recycling in tropical mangrove and other coastal benthic ecosystems. Hydrobiologia 285:19–32

Amann R, Fuchs BM (2008) Single-cell identification in microbial communities by improved fluorescence in situ hybridization techniques. Nat Rev Microbiol 6:339–348

Amann R, Ludwig W, Schleifer KH (1995) Phylogenetic identification and in situ detection of individual microbial cells without cultivation. Microbiol Rev 59:143–169

Angert ER (2005) Alternatives to binary fission in bacteria. Nat Rev Microbiol 3:214–224

Applerot G, Lipovsky A, Dror R et al (2009) Enhanced antibacterial activity of nanocrystalline ZnO due to increased ROS-mediated cell injury. Adv Funct Mater 19:842–852

Armon R (2011) Soil bacteria and bacteriophages. In: Witzany G (ed) Biocommunication in soil microorganisms soil biology. Springer, Berlin, pp 67–112

Atkins PW, Jones L (2010) Chemical principles: the quest for insight, 5th edn. WH Freeman, New York

Barreto JA, O'Malley W, Kubeil M et al (2011) Nanomaterials: applications in cancer imaging and therapy. Adv Mater 23:H18–H40

Bartnicki-Garcia S (1968) Cell wall chemistry, morphogenesis, and taxonomy of fungi. Annu Rev Microbiol 22:87–108

Benincasa M, Pacor S, Wu W et al (2011) Antifungal activity of amphotericin B conjugated to carbon nanotubes. ACS Nano 5:199–208

Berridge MV, Herst PM, Tan AS (2005) Tetrazolium dyes as tools in cell biology: new insights into their cellular reduction. Biotechnol Annu Rev 11:127–152

Blackwell M, Hibbett DS, Taylor JW et al (2006) Research coordination networks: a phylogeny for kingdom Fungi (Deep Hypha). Mycologia 98:829–837

Blecher K, Nasir A, Friedman A (2011) The growing role of nanotechnology in combating infectious disease. Virulence 2:395–401

Boonchan S, Britz ML, Stanley GA (2000) Degradation and mineralization of high-molecular-weight-polycyclic aromatic hydrocarbons by defined fungal-bacterial cocultures. Appl Environ Microbiol 66:1007–1019

Boulos L, Prévost M, Barbeau B et al (1999) LIVE/DEAD BacLight: application of a new rapid staining method for direct enumeration of viable and total bacteria in drinking water. J Microbiol Methods 37:77–86

Bradford A, Handy RD, Readman JW et al (2009) Impact of silver nanoparticle contamination on the genetic diversity of natural bacterial assemblages in estuarine sediments. Environ Sci Technol 43:4530–4536

Brady-Estévez AS, Kang S, Elimelech M (2008) A single-walled-carbon-nanotube filter for removal of viral and bacterial pathogens. Small 4:481–484
Branda S, Vik S, Friedman L et al (2005) Biofilms: the matrix revisited. Trends Microbiol 13:20–26
Brar SK, Verma M, Tyagi RD et al (2010) Engineered nanoparticles in wastewater and wastewater sludge—evidence and impacts. Waste Manag 30:504–520
Brayner R, Ferrari-Iliou R, Brivois N et al (2006) Toxicological impact studies based on Escherichia coli bacteria in ultrafine ZnO nanoparticles colloidal medium. Nano Lett 6:866–870
Brunet L, Lyon DY, Hotze EM et al (2009) Comparative photoactivity and antibacterial properties of C60 fullerenes and titanium dioxide nanoparticles. Environ Sci Technol 43:4355–4360
Carpio IEM, Santos CM, Wei X et al (2012) Toxicity of a polymer-graphene oxide composite against bacterial planktonic cells, biofilms, and mammalian cells. Nanoscale 4:4746–4756
Carter J, Saunders V (2007) Virology: principles and applications. Wiley, West Sussex
Chmielewski RAN, Frank JF (2003) Biofilm formation and control in food processing facilities. Compr Rev Food Sci Food Saf 2:22–32
Choi O, Yu CP, Esteban Fernández G et al (2010) Interactions of nanosilver with Escherichia coli cells in planktonic and biofilm cultures. Water Res 44:6095–6103
Chou WL, Yu DG, Yang MC (2005) The preparation and characterization of silver-loading cellulose acetate hollow fiber membrane for water treatment. Polym Adv Tech 16:600–607
Colvin VL (2003) The potential environmental impact of engineered nanomaterials. Nat Biotechnol 21:1166–1170
Costerton JW, Lewandowski Z, Caldwell DE et al (1995) Microbial biofilms. Annu Rev Microbiol 49:711–745
Curtis TP, Sloan WT (2004) Prokaryotic diversity and its limits: microbial community structure in nature and implications for microbial ecology. Curr Opin Microbiol 7:221–226
Dahllöf I (2002) Molecular community analysis of microbial diversity. Curr Opin Biotechnol 13:213–217
Dallas P, Sharma VK, Zboril R (2011) Silver polymeric nanocomposites as advanced antimicrobial agents: classification, synthetic paths, applications, and perspectives. Adv Colloid Interface Sci 166:119–135
Das MR, Sarma RK, Saikia R et al (2011) Synthesis of silver nanoparticles in an aqueous suspension of graphene oxide sheets and its antimicrobial activity. Colloids Surf B Biointerfaces 83:16–22
De Lima R, Seabra AB, Durán N (2012) Silver nanoparticles: a brief review of cytotoxicity and genotoxicity of chemically and biogenically synthesized nanoparticles. J Appl Toxicol 32:867–879
Desprez-Loustau M-L, Robin C, Buée M et al (2007) The fungal dimension of biological invasions. Trends Ecol Evol 22:472–480
Dobrovolskaia MA, McNeil SE (2007) Immunological properties of engineered nanomaterials. Nat Nanotechnol 2:469–478
Donlan R (2002) Biofilms: microbial life on surfaces. Emerg Infect Dis 8:881–890
Dror-Ehre A, Adin A, Markovich G et al (2010) Control of biofilm formation in water using molecularly capped silver nanoparticles. Water Res 44:2601–2609
Duncan TV (2011) Applications of nanotechnology in food packaging and food safety: barrier materials, antimicrobials and sensors. J Colloid Interface Sci 363:1–24
Durán N, Marcato PD, De Souza GIH et al (2007) Antibacterial effect of silver nanoparticles produced by fungal process on textile fabrics and their effluent treatment. J Biomed Nanotechnol 3:203–208
Durán N, Marcato PD, Conti R et al (2010) Potential use of silver nanoparticles on pathogenic bacteria, their toxicity and possible mechanisms of action. J Braz Chem Soc 21:949–959
Elechiguerra JL, Burt JL, Morones JR et al (2005) Interaction of silver nanoparticles with HIV-1, J Nanobiotechnology 3:1–10

El-Rafie MH, Mohamed AA, Shaheen TI et al (2010) Antimicrobial effect of silver nanoparticles produced by fungal process on cotton fabrics. Carbohydr Polym 80:779–782

Emamifar A, Kadivar M, Shahedi M et al (2010) Evaluation of nanocomposite packaging containing Ag and ZnO on shelf life of fresh orange juice. Innov Food Sci Emerg Technol 11:742–748

Espitia PJP, Soares NFF, Coimbra JSR et al (2012) Zinc oxide nanoparticles: synthesis, antimicrobial activity and food packaging applications. Food Bioprocess Technol 5:1447–1464

Fabrega J, Zhang R, Renshaw JC et al (2011) Impact of silver nanoparticles on natural marine biofilm bacteria. Chemosphere 85:961–966

Faria AF, Martinez DST, Moraes ACM et al (2012) Unveiling the role of oxidation debris on the surface chemistry of graphene through the anchoring of Ag nanoparticles. Chem Mater 24:4080–4087

Fedorova NE, Klimova RR, Tulenev YA et al (2012) Carboxylic fullerene C60 derivatives: efficient microbicides against herpes simplex virus and cytomegalovirus infections in vitro. Mendeleev Commun 22:254–256

Furno F, Morley KS, Wong B et al (2004) Silver nanoparticles and polymeric medical devices: a new approach to prevention of infection? J Antimicrob Chemother 54:1019–1024

Gao Y, Cranston R (2008) Recent advances in antimicrobial treatments of textiles. Text Res J 78:60–72

Gao D, Tao Y (2012) Current molecular biologic techniques for characterizing environmental microbial community. Front Environ Sci Eng 6:82–97

Gao G-H, Lei Y-H, Dong L-H et al (2012) Synthesis of nanocomposites of silver nanoparticles with medical stone and carbon nanotubes for their antibacterial applications. Mater Express 2:85–93

Ge Y, Schimel JP, Holden PA (2011) Evidence for negative effects of TiO_2 and ZnO nanoparticles on soil bacterial communities. Environ Sci Technol 45:1659–1664

Gitai Z (2005) The new bacterial cell biology: moving parts and subcellular architecture. Cell 120:577–586

Goyal D, Zhang XJ, Rooney-Varga JN (2010) Impacts of single-walled carbon nanotubes on microbial community structure in activated sludge. Lett Appl Microbiol 51:428–435

Harris SD (2008) Branching of fungal hyphae: regulation, mechanisms and comparison with other branching systems. Mycologia 100:823–832

Hawker LE (1965) Fine structure of fungi as revealed by electron microscopy. Biol Rev 40:52–91

Hernandez-Delgadillo R, Velasco-Arias D, Diaz D et al (2012) Zerovalent bismuth nanoparticles inhibit *Streptococcus mutans* growth and formation of biofilm. Int J Nanomedicine 7:2109–2113

Hill GT, Mitkowski NA, Aldrich-Wolfeb L et al (2000) Methods for assessing the composition and diversity of soil microbial communities. Appl Soil Ecol 15:25–36

Hu W, Peng C, Luo W et al (2010) Graphene-based antibacterial paper. ACS Nano 4:4317–4323

Huang L, Terakawa M, Zhiyentayev T et al (2010) Innovative cationic fullerenes as broad-spectrum light-activated antimicrobials. Nanomedicine 6:442–452

Huh AJ, Kwon YJ (2011) Nanoantibiotics: a new paradigm for treating infectious diseases using nanomaterials in the antibiotics resistant era. J Control Release 156:128–145

Jay JM, Loessner MJ, Golden DA (2005) Modern food microbiology, 7th edn. Springer, New York

Jung WK, Koo HC, Kim KW et al (2008) Antibacterial activity and mechanism of action of the silver ion in *Staphylococcus aureus* and *Escherichia coli*. Appl Environ Microbiol 74:2171–2178

Jusko J (2009) Information please. Government agencies, concerned about potential health and environmental risks, are stepping up efforts to gather data on nanomaterials. Ind Week 29:1–2

Kamat JP, Devasagayam TP, Priyadarsini KI et al (2000) Reactive oxygen species mediated membrane damage induced by fullerene derivatives and its possible biological implications. Toxicology 155:55–61

Kang S, Pinault M, Pfefferle LD et al (2007) Single-walled carbon nanotubes exhibit strong antimicrobial activity. Langmuir 23:8670–8673

Kang S, Herzberg M, Rodrigues DF et al (2008) Antibacterial effects of carbon nanotubes: size does matter! Langmuir 24:6409–6413

Kang S, Mauter MS, Elimelech M (2009) Microbial cytotoxicity of carbon-based nanomaterials: implications for river water and wastewater effluent. Environ Sci Technol 43:2648–2653

Kim J, Kuk E, Yu K et al (2007) Antimicrobial effects of silver nanoparticles. Nanomed-Nanotechnol 3:95–101

Kirschling TL, Gregory KB, Minkley JREG et al (2010) Impact of nanoscale zero valent iron on geochemistry and microbial populations in trichloroethylene contaminated aquifer materials. Environ Sci Technol 44:3474–3480

Kiser MA, Westerhoff P, Benn T et al (2009) Titanium nanomaterial removal and release from wastewater treatment plants. Environ Sci Technol 43:6757–6763

Kitts CL (2001) Terminal restriction fragment patterns: a tool for comparing microbial communities and assessing community dynamics. Curr Issues Intest Microbiol 2:17–25

Klaine SJ, Alvarez PJJ, Batley GE et al (2008) Nanomaterials in the environment: behavior, fate, bioavailability, and effects. Environ Toxicol Chem 27:1825–1851

Klaine SJ, Koelmans AA, Horne N et al (2012) Paradigms to assess the environmental impact of manufactured nanomaterials. Environ Toxicol Chem 31:3–14

Koch A (2003) Bacterial wall as target for attack: past, present, and the future research. Clin Microbiol Rev 16:673–687

Krishnaraj C, Jagan EG, Rajasekar S et al (2010) Synthesis of silver nanoparticles using Acalypha indica leaf extracts and its antibacterial activity against water borne pathogens. Colloids Surf B Biointerfaces 76:50–56

Kumar A, Menon SK (2009) Fullerene derivatized s-triazine analogues as antimicrobial agents. Eur J Med Chem 44:2178–2183

Kunzmann A, Andersson B, Thurnherr T et al (2011) Toxicology of engineered nanomaterials: focus on biocompatibility, biodistribution and biodegradation. Biochim Biophys Acta 1810:361–373

Lee N, Nielsen PH, Andreasen KH (1999) Combination of fluorescent in situ hybridization and microautoradiography—a new tool for structure-function analyses in microbial ecology. Appl Environ Microbiol 65:1289–1297

Lee J, Mahendra S, Alvarez PJJ (2010) Nanomaterials in the construction industry: a review of their applications and environmental health and safety considerations. ACS Nano 4:3580–3590

Li Q, Mahendra S, Lyon DY et al (2008) Antimicrobial nanomaterials for water disinfection and microbial control: potential applications and implications. Water Res 42:4591–4602

Li C, Wang X, Chen F et al (2013) The antifungal activity of graphene oxide–silver nanocomposites. Biomaterials 34:3882–3890

Liu W-T, Marsh TL, Cheng H et al (1997) Characterization of microbial diversity by determining terminal restriction fragment length polymorphisms of genes encoding 16S rRNA. Appl Environ Microbiol 63:4516–4522

Liu B-R, Jia G-M, Chen J et al (2006) A review of methods for studying microbial diversity in soils. Pedosphere 16:18–24

Liu Y, He L, Mustapha A et al (2009) Antibacterial activities of zinc oxide nanoparticles against *Escherichia coli* O157:H7. J Appl Microbiol 107:1193–1201

Liu S, Ng AK, Xu R et al (2010) Antibacterial action of dispersed single-walled carbon nanotubes on *Escherichia coli* and *Bacillus subtilis* investigated by atomic force microscopy. Nanoscale 2:2744–2750

Liu S, Zeng TH, Hofmann M et al (2011) Antibacterial activity of graphite, graphite oxide, graphene oxide, and reduced graphene oxide: membrane and oxidative stress. ACS Nano 5:6971–6980

Liu S, Hu M, Zeng TH et al (2012) Lateral dimension-dependent antibacterial activity of graphene oxide sheets. Langmuir 28:12364–12372

Ludwig W, Bauer SH, Bauer M et al (1997) Detection and in situ identification of representatives of a widely distributed new bacterial phylum. FEMS Microbiol Lett 153:181–190
Lyon DY, Adams LK, Falkner JC et al (2006) Antibacterial activity of fullerene water suspensions: effects of preparation method and particle size. Environ Sci Technol 40:4360–4366
Lyon DY, Brunet L, Hinkal GW et al (2008) Antibacterial activity of fullerene water suspensions (nC60) is not due to ROS-mediated damage. Nano Lett 8:1539–1543
Ma J, Zhang J, Xiong Z et al (2011) Preparation, characterization and antibacterial properties of silver-modified graphene oxide. J Mater Chem 21:3350–3352
Maier RM, Pepper IL, Gerba CP (2009) Environmental microbiology, 2nd edn. Academic, San Diego
Marambio-Jones C, Hoek EMV (2010) A review of the antibacterial effects of silver nanomaterials and potential implications for human health and the environment. J Nanopart Res 12:1531–1551
Marsh TL (1999) Terminal restriction fragment length polymorphism (T-RFLP): an emerging method for characterizing diversity among homologous populations of amplification products. Curr Opin Microbiol 2:323–327
Martinez D, Larrondo LF, Putnam N et al (2004) Genome sequence of the lignocellulose degrading fungus *Phanerochaete chrysosporium strain RP78*. Nat Biotechnol 22:695–700
Martínez-Castañón GA, Niño-Martínez N, Martínez-Gutierrez F et al (2008) Synthesis and antibacterial activity of silver nanoparticles with different sizes. J Nanopart Res 10:1343–1348
Martinko JM, Madigan MT (2005) Brock biology of microorganisms, 11th edn. Prentice Hall, Englewood Cliffs, NJ
Matthews L, Kanwar RK, Zhou S et al (2010) Applications of nanomedicine in antibacterial medical therapeutics and diagnostics. Open Trop Med J 3:1–9
Mauter MS, Elimelech M (2008) Environmental applications of carbon-based nanomaterials. Environ Sci Technol 42:5843–5859
Maynard AD (2007) Nanotechnologies: overview and issues. In: Simeonova PP, Opopol N, Luster MI (eds) Nanotechnology—toxicological issues and environmental safety and environmental safety. Springer, Dordrecht, The Netherlands
Meng H, Xia T, George S et al (2009) A predictive toxicological paradigm for the safety assessment of nanomaterials. ACS Nano 3:1620–1627
Mirhosseini M, Firouzabadi FB (2012) Antibacterial activity of zinc oxide nanoparticle suspensions on food-borne pathogens. Int J Dairy Technol 65:1–5
Miyako E, Nagata H, Hirano K et al (2008) Photoinduced antiviral carbon nanohorns. Nanotechnology 19:1–6
Morones JR, Elechiguerra JL, Camacho A et al (2005) The bactericidal effect of silver nanoparticles. Nanotechnology 16:2346–2353
Morris RM, Rappé MS, Connon SA et al (2002) SAR11 clade dominates ocean surface bacterioplankton communities. Nature 420:806–810
Mosmann T (1983) Rapid colorimetric assay for cellular growth and survival: application to proliferation and cytotoxicity assays. J Immunol Methods 65:55–63
Moter A, Göbel UB (2000) Fluorescence in situ hybridization (FISH) for direct visualization of microorganisms. J Microbiol Methods 41:85–112
Mu H, Chen Y (2011) Long-term effect of ZnO nanoparticles on waste activated sludge anaerobic digestion. Water Res 45:5612–5620
Mu L, Sprando RL (2010) Application of nanotechnology in cosmetics. Pharm Res 27:1746–1749
Musee N, Thwalaa M, Nota N (2011) The antibacterial effects of engineered nanomaterials: implications for wastewater treatment plants. J Environ Monit 13:1164–1183
Muyzer G, Ramsing NB (1995) Molecular methods to study the organization of microbial communities. Wat Sci Technol 32:1–9

Muyzer G, Smalla K (1998) Application of denaturing gradient gel electrophoresis (DGGE) and temperature gradient gel electrophoresis (TGGE) in microbial ecology. Antonie van Leeuwenhoek 73:127–141

Muyzer G, Waal EC, Uitterlinden AG (1993) Profiling of complex microbial populations by denaturing gradient gel electrophoresis analysis of polymerase chain reaction-amplified genes coding for 16S rRNA. Appl Environ Microbiol 59:695–700

Nakamura S, Mashino T (2009) Biological activities of water-soluble fullerene derivatives. J Phys Conf Ser 159:012003

Nanda A, Saravanan M (2009) Biosynthesis of silver nanoparticles from *Staphylococcus aureus* and its antimicrobial activity against MRSA and MRSE. Nanomed-Nanotechnol 5:452–456

Nogueira V, Lopes I, Rocha-Santos T et al (2012) Impact of organic and inorganic nanomaterials in the soil microbial community structure. Sci Total Environ 424:344–350

Nyberg L, Turco RF, Nies L (2008) Assessing the impact of nanomaterials on anaerobic microbial communities. Environ Sci Technol 42:1938–1943

Osiewacz HD (2002) Genes, mitochondria and aging in filamentous fungi. Ageing Res Rev 3:425–442

Padmavathy N, Vijayaraghavan R (2008) Enhanced bioactivity of ZnO nanoparticles—an antimicrobial study. Sci Technol Adv Mater 9:1–7

Page K, Palgrave RG, Parkin IP et al (2007) Titania and silver–titania composite films on glass–potent antimicrobial coatings. J Mater Chem 17:95–104

Pakrashi S, Dalai S, Sabat D et al (2011) Cytotoxicity of Al_2O_3 nanoparticles at low exposure levels to a freshwater bacterial isolate. Chem Res Toxicol 24:1899–1904

Pal S, Tak YK, Song JM (2007) Does the antibacterial activity of silver nanoparticles depend on the shape of the nanoparticle? A study of the gram-negative bacterium *Escherichia coli*. Appl Environ Microbiol 73:1712–1720

Panacek A, Kolar M, Vecerova R et al (2009) Antifungal activity of silver nanoparticles against *Candida* spp. Biomaterials 30:6333–6340

Park HJ, Kim JY, Kim J et al (2009) Silver-ion-mediated reactive oxygen species generation affecting bactericidal activity. Water Res 43:1027–1034

Park S, Mohanty N, Suk JW et al (2010) Biocompatible, robust free-standing paper composed of a TWEEN/graphene composite. Adv Mater 22:1736–1740

Park H, Park HJ, Kim JA et al (2011) Inactivation of *Pseudomonas aeruginosa* PA01 biofilms by hyperthermia using superparamagnetic nanoparticles. J Microbiol Methods 84:41–45

Paul EA (2007) Soil microbiology, ecology and biochemistry, 3rd edn. Elsevier, Fort Collins, CO

Pelczar MJ, Chan ECS, Krieg NR (1993) Microbiology, 5th edition, McGraw Hill Book Company, NY, USA

Pernthaler A, Pernthaler J, Amann R (2002) Fluorescence in situ hybridization and catalyzed reporter deposition for the identification of marine bacteria. Appl Environ Microbiol 68:3094–3101

Pratt LA, Kolter R (1998) Genetic analysis of *Escherichia coli* biofilm formation: roles of flagella, motility, chemotaxis and type I pili. Mol Microbiol 30:285–293

Qi X, Poernomo G, Wang K et al (2011) Covalent immobilization of nisin on multi-walled carbon nanotubes: superior antimicrobial and anti-biofilm properties. Nanoscale 3:1874–1880

Raffi M, Hussain F, Bhatti TM et al (2008) Antibacterial characterization of silver nanoparticles against *E. coli* ATCC-15224. J Mater Sci Technol 24:192–196

Raghupathi KR, Koodali RT, Manna AC (2011) Size-dependent bacterial growth inhibition and mechanism of antibacterial activity of zinc oxide nanoparticles. Langmuir 27:4020–4028

Rai M, Yadav A, Gade A (2009) Silver nanoparticles as a new generation of antimicrobials. Biotechnol Adv 27:76–83

Rodrigues DF, Elimelech M (2010) Toxic effects of single-walled carbon nanotubes in the development of *E. coli* biofilm. Environ Sci Technol 44:4583–4589

Rodrigues F, Ludovico P, Leão C (2006) Sugar metabolism in yeasts: an overview of aerobic and anaerobic glucose catabolism (chap 6). In: Péter G, Rosa C (eds) Biodiversity and ecophysiology of yeasts, Springer, Berlin, pp 101–121
Roossinck MJ (2010) Lifestyles of plant viruses. Phil Trans R Soc B 365:1899–1905
Ruiz ON, Fernando KAS, Wang B et al (2011) Graphene oxide: a nonspecific enhancer of cellular growth. ACS Nano 5:8100–8107
Ruparelia JP, Chatterjee AK, Duttagupta SP et al (2008) Strain specificity in antimicrobial activity of silver and copper nanoparticles. Acta Biomater 4:707–716
Salata OV (2004) Applications of nanoparticles in biology and medicine. J Nanobiotechnology 2:3
Savage N, Diallo MS (2005) Nanomaterials and water purification: opportunities and challenges. J Nanopart Res 7:331–342
Sayara T, Borràs E, Caminal G et al (2011) Bioremediation of PAHs-contaminated soil through composting: influence of bioaugmentation and biostimulation on contaminant biodegradation. Int Biodeter Biodegr 65:859–865
Schleifer KH, Kandler O (1972) Peptidoglycan types of bacterial cell walls and their taxonomic implications. Bacteriol Rev 36:407–477
Schönhuber W, Fuchs B, Juretschko S et al (1997) Improved sensitivity of whole-cell hybridization by the combination of horseradish peroxidase-labeled oligonucleotides and tyramide signal amplification. Appl Environ Microbiol 63:3268–3273
Seil JT, Webster TJ (2012) Antimicrobial applications of nanotechnology: methods and literature. Int J Nanomedicine 7:2767–2781
Silva IS, Santos EC, Menezes CR et al (2009) Bioremediation of a polyaromatic hydrocarbon contaminated soil by native soil microbiota and bioaugmentation with isolated microbial consortia. Bioresour Technol 100:4669–4675
Simon-Deckers A, Loo S, Mayne-L'hermite M et al (2009) Size composition and shape-dependent toxicological impact of metal oxide nanoparticles and carbon nanotubes toward bacteria. Environ Sci Technol 43:8423–8429
Some S, Ho S-M, Dua P et al (2012) Graphene derivative-poly-L-lysine composites to inhibit bacteria and support human cells. ACS Nano 6:7151–7161
Sondi I, Salopek-Sondi B (2004) Silver nanoparticles as antimicrobial agent: a case study on *E. coli* as a model for Gram-negative bacteria. J Colloid Interface Sci 275:177–182
Souza-Filho AG, Fagan SB (2007) Funcionalização de nanotubos de carbono. Quim Nova 30:1695–1703
Stewart PS, Franklin MJ (2008) Physiological heterogeneity in biofilms. Nat Rev Microbiol 6:199–210
Stoodley P, Sauer K, Davies DG et al (2002) Biofilms as complex differentiated communities. Annu Rev Microbiol 56:187–209
Strauss JH, Strauss EG (2008) Viruses and human disease, 2nd edn. Elsevier Academic Press, Amsterdam
Su C, Lei L, Duan Y et al (2012) Culture-independent methods for studying environmental microorganisms: methods, application, and perspective. Appl Microbiol Biotechnol 93:993–1003
Sun X, Sheng Z, Liu Y (2013) Effects of silver nanoparticles on microbial community structure in activated sludge. Sci Total Environ 443:828–835
Sylvester PW (2011) Optimization of the tetrazolium dye (MTT) colorimetric assay for cellular growth and viability. Methods Mol Biol 716:157–168
Tielens AG, Rotte C, van Hellemond JJ et al (2002) Mitochondria as we don't know them. Trends Biochem Sci 27:564–572
Tong ZH, Bischoff M, Nies L et al (2007) Impact of fullerene (C-60) on a soil microbial community. Environ Sci Technol 41:2985–2991
Torsvik V, Sorheim R, Goksoyr J (1996) Total bacterial diversity in soil and sediment communities—a review. J Ind Microbiol 17:170–178

Tortora GJ, Funke BR, Case CL (2010) Microbiology: An introduction, 10th edn. Pearson Benjamin Cummings, San Francisco

Tringe SG, Mering C, Kobayashi A et al (2005) Comparative metagenomics of microbial communities. Science 308:554–557

Turco RF, Bischoff M, Tong ZH et al (2011) Environmental implications of nanomaterials: are we studying the right thing? Curr Opin Biotechnol 22:527–532

Vecitis CD, Schnoor MH, Rahaman MS et al (2011) Electrochemical multiwalled carbon nanotube filter for viral and bacterial removal and inactivation. Environ Sci Technol 45:3672–3679

Veses V, Gow NAR (2009) Pseudohypha budding patterns of *Candida albicans*. Med Mycol 47:268–275

Wiesner MR, Lowry GV, Alvarez P et al (2006) Assessing the risks of manufactured nanomaterials. Environ Sci Technol 40:4336–4345

Woese CR, Fox GE (1977) Phylogenetic structure of the prokaryotic domain: the primary kingdoms. Proc Natl Acad Sci U S A 74:5088–5090

Yan S, Subramanian S, Tyagi R et al (2010) Emerging contaminants of environmental concern: source, transport, fate, and treatment. Waste Manag 14:2–20

Yang G, Xie J, Hong F et al (2012) Antimicrobial activity of silver nanoparticle impregnated bacterial cellulose membrane: effect of fermentation carbon sources of bacterial cellulose. Carbohydr Polym 87:839–845

Yoon K, Byeon J, Park J et al (2008) Antimicrobial characteristics of silver aerosol nanoparticles against *Bacillus subtilis* bioaerosols. Environ Eng Sci 25:289–293

Zan L, Fa W, Peng TP et al (2007) Photocatalysis effect of nanometer TiO_2 and TiO_2-coated ceramic plate on Hepatitis B virus. J Photochem Photobiol B 86:165–169

Zare-Zardini H, Amiri A, Shanbedi M et al (2013) Studying of antifungal activity of functionalized multiwalled carbon nanotubes by microwave-assisted technique. Surf Interface Anal 45:751–755

Zhao Y, Xing G, Chai Z (2008) Nanotoxicology: are carbon nanotubes safe? Nat Nanotechnol 3:191–192

Zheng X, Chen Y, Wu R (2011) Long-term effects of titanium dioxide nanoparticles on nitrogen and phosphorus removal from wastewater and bacterial community shift in activated sludge. Environ Sci Technol 45:7284–7290

Index

N. Durán et al. (eds.), *Nanotoxicology*, Nanomedicine and Nanotoxicology,
DOI 10.1007/978-1-4614-8993-1, © Springer Science+Business Media New York 2014

MIX
Papier aus verantwortungsvollen Quellen
Paper from responsible sources
FSC® C105338

If you have any concerns about our products,
you can contact us on
ProductSafety@springernature.com

In case Publisher is established outside the EU,
the EU authorized representative is:
Springer Nature Customer Service Center GmbH
Europaplatz 3, 69115 Heidelberg, Germany

Printed by Libri Plureos GmbH
in Hamburg, Germany